T0214570

Lecture Notes in Computer Science 11573

Commenced Publication in 1973
Founding and Former Series Editors:
Gerhard Goos, Juris Hartmanis, and Jan van Leeuwen

More information about this series at http://www.springer.com/series/7409

Margherita Antona · Constantine Stephanidis (Eds.)

Universal Access in Human-Computer Interaction

Multimodality and Assistive Environments

13th International Conference, UAHCI 2019
Held as Part of the 21st HCI International Conference, HCII 2019
Orlando, FL, USA, July 26–31, 2019
Proceedings, Part II

Springer

Editors
Margherita Antona
Foundation for Research
and Technology – Hellas (FORTH)
Heraklion, Crete, Greece

Constantine Stephanidis
University of Crete
and Foundation for Research
and Technology – Hellas (FORTH)
Heraklion, Crete, Greece

ISSN 0302-9743   ISSN 1611-3349 (electronic)
Lecture Notes in Computer Science
ISBN 978-3-030-23562-8   ISBN 978-3-030-23563-5 (eBook)
https://doi.org/10.1007/978-3-030-23563-5

LNCS Sublibrary: SL3 – Information Systems and Applications, incl. Internet/Web, and HCI

This Springer imprint is published by the registered company Springer Nature Switzerland AG
The registered company address is: Gewerbestrasse 11, 6330 Cham, Switzerland

Foreword

The 21st International Conference on Human-Computer Interaction, HCI International 2019, was held in Orlando, FL, USA, during July 26–31, 2019. The event incorporated the 18 thematic areas and affiliated conferences listed on the following page.

A total of 5,029 individuals from academia, research institutes, industry, and governmental agencies from 73 countries submitted contributions, and 1,274 papers and 209 posters were included in the pre-conference proceedings. These contributions address the latest research and development efforts and highlight the human aspects of design and use of computing systems. The contributions thoroughly cover the entire field of human-computer interaction, addressing major advances in knowledge and effective use of computers in a variety of application areas. The volumes constituting the full set of the pre-conference proceedings are listed in the following pages.

This year the HCI International (HCII) conference introduced the new option of "late-breaking work." This applies both for papers and posters and the corresponding volume(s) of the proceedings will be published just after the conference. Full papers will be included in the *HCII 2019 Late-Breaking Work Papers Proceedings* volume of the proceedings to be published in the Springer LNCS series, while poster extended abstracts will be included as short papers in the HCII 2019 *Late-Breaking Work Poster Extended Abstracts* volume to be published in the Springer CCIS series.

I would like to thank the program board chairs and the members of the program boards of all thematic areas and affiliated conferences for their contribution to the highest scientific quality and the overall success of the HCI International 2019 conference.

This conference would not have been possible without the continuous and unwavering support and advice of the founder, Conference General Chair Emeritus and Conference Scientific Advisor Prof. Gavriel Salvendy. For his outstanding efforts, I would like to express my appreciation to the communications chair and editor of *HCI International News,* Dr. Abbas Moallem.

July 2019 Constantine Stephanidis

HCI International 2019 Thematic Areas and Affiliated Conferences

Thematic areas:

- HCI 2019: Human-Computer Interaction
- HIMI 2019: Human Interface and the Management of Information

Affiliated conferences:

- EPCE 2019: 16th International Conference on Engineering Psychology and Cognitive Ergonomics
- UAHCI 2019: 13th International Conference on Universal Access in Human-Computer Interaction
- VAMR 2019: 11th International Conference on Virtual, Augmented and Mixed Reality
- CCD 2019: 11th International Conference on Cross-Cultural Design
- SCSM 2019: 11th International Conference on Social Computing and Social Media
- AC 2019: 13th International Conference on Augmented Cognition
- DHM 2019: 10th International Conference on Digital Human Modeling and Applications in Health, Safety, Ergonomics and Risk Management
- DUXU 2019: 8th International Conference on Design, User Experience, and Usability
- DAPI 2019: 7th International Conference on Distributed, Ambient and Pervasive Interactions
- HCIBGO 2019: 6th International Conference on HCI in Business, Government and Organizations
- LCT 2019: 6th International Conference on Learning and Collaboration Technologies
- ITAP 2019: 5th International Conference on Human Aspects of IT for the Aged Population
- HCI-CPT 2019: First International Conference on HCI for Cybersecurity, Privacy and Trust
- HCI-Games 2019: First International Conference on HCI in Games
- MobiTAS 2019: First International Conference on HCI in Mobility, Transport, and Automotive Systems
- AIS 2019: First International Conference on Adaptive Instructional Systems

Pre-conference Proceedings Volumes Full List

1. LNCS 11566, Human-Computer Interaction: Perspectives on Design (Part I), edited by Masaaki Kurosu
2. LNCS 11567, Human-Computer Interaction: Recognition and Interaction Technologies (Part II), edited by Masaaki Kurosu
3. LNCS 11568, Human-Computer Interaction: Design Practice in Contemporary Societies (Part III), edited by Masaaki Kurosu
4. LNCS 11569, Human Interface and the Management of Information: Visual Information and Knowledge Management (Part I), edited by Sakae Yamamoto and Hirohiko Mori
5. LNCS 11570, Human Interface and the Management of Information: Information in Intelligent Systems (Part II), edited by Sakae Yamamoto and Hirohiko Mori
6. LNAI 11571, Engineering Psychology and Cognitive Ergonomics, edited by Don Harris
7. LNCS 11572, Universal Access in Human-Computer Interaction: Theory, Methods and Tools (Part I), edited by Margherita Antona and Constantine Stephanidis
8. LNCS 11573, Universal Access in Human-Computer Interaction: Multimodality and Assistive Environments (Part II), edited by Margherita Antona and Constantine Stephanidis
9. LNCS 11574, Virtual, Augmented and Mixed Reality: Multimodal Interaction (Part I), edited by Jessie Y. C. Chen and Gino Fragomeni
10. LNCS 11575, Virtual, Augmented and Mixed Reality: Applications and Case Studies (Part II), edited by Jessie Y. C. Chen and Gino Fragomeni
11. LNCS 11576, Cross-Cultural Design: Methods, Tools and User Experience (Part I), edited by P. L. Patrick Rau
12. LNCS 11577, Cross-Cultural Design: Culture and Society (Part II), edited by P. L. Patrick Rau
13. LNCS 11578, Social Computing and Social Media: Design, Human Behavior and Analytics (Part I), edited by Gabriele Meiselwitz
14. LNCS 11579, Social Computing and Social Media: Communication and Social Communities (Part II), edited by Gabriele Meiselwitz
15. LNAI 11580, Augmented Cognition, edited by Dylan D. Schmorrow and Cali M. Fidopiastis
16. LNCS 11581, Digital Human Modeling and Applications in Health, Safety, Ergonomics and Risk Management: Human Body and Motion (Part I), edited by Vincent G. Duffy

34. CCIS 1033, HCI International 2019 - Posters (Part II), edited by Constantine Stephanidis
35. CCIS 1034, HCI International 2019 - Posters (Part III), edited by Constantine Stephanidis

http://2019.hci.international/proceedings

13th International Conference on Universal Access in Human-Computer Interaction (UAHCI 2019)

Program Board Chair(s): **Margherita Antona and Constantine Stephanidis**, *Greece*

- Gisela Susanne Bahr, USA
- Armando Barreto, USA
- João Barroso, Portugal
- Rodrigo Bonacin, Brazil
- Ingo Bosse, Germany
- Anthony Lewis Brooks, Denmark
- Laura Burzagli, Italy
- Pedro J. S. Cardoso, Portugal
- Stefan Carmien, UK
- Carlos Duarte, Portugal
- Pier Luigi Emiliani, Italy
- Vagner Figueredo de Santana, Brazil
- Andrina Granic, Croatia
- Gian Maria Greco, Spain
- Simeon Keates, UK
- Georgios Kouroupetroglou, Greece
- Patrick M. Langdon, UK
- Barbara Leporini, Italy
- I. Scott MacKenzie, Canada
- John Magee, USA
- Alessandro Marcengo, Italy
- Jorge Martín-Gutiérrez, Spain
- Troy McDaniel, USA
- Silvia Mirri, Italy
- Federica Pallavicini, Italy
- Ana Isabel Bruzzi Bezerra Paraguay, Brazil
- Hugo Paredes, Portugal
- Enrico Pontelli, USA
- João M. F. Rodrigues, Portugal
- Frode Eika Sandnes, Norway
- Jaime Sánchez, Chile
- Volker Sorge, UK
- Hiroki Takada, Japan
- Kevin C. Tseng, Taiwan
- Gerhard Weber, Germany
- Gian Wild, Australia
- Ed Youngblood, USA

The full list with the Program Board Chairs and the members of the Program Boards of all thematic areas and affiliated conferences is available online at:

http://www.hci.international/board-members-2019.php

HCI International 2020

The 22nd International Conference on Human-Computer Interaction, HCI International 2020, will be held jointly with the affiliated conferences in Copenhagen, Denmark, at the Bella Center Copenhagen, July 19–24, 2020. It will cover a broad spectrum of themes related to HCI, including theoretical issues, methods, tools, processes, and case studies in HCI design, as well as novel interaction techniques, interfaces, and applications. The proceedings will be published by Springer. More information will be available on the conference website: http://2020.hci.international/.

General Chair
Prof. Constantine Stephanidis
University of Crete and ICS-FORTH
Heraklion, Crete, Greece
E-mail: general_chair@hcii2020.org

http://2020.hci.international/

Contents – Part II

Multimodal Interaction

Assistive Environments

Contents – Part I

Novel Approaches to Accessibility

Universal Access to Learning and Education

Virtual and Augmented Reality in Universal Access

Cognitive and Learning Disabilities

A Collaborative Talking Assistive Technology for People with Autism Spectrum Disorders

Wajih Abdallah[1], Frédéric Vella[1], Nadine Vigouroux[1(✉)],
Adrien Van den Bossche[2], and Thierry Val[2]

[1] IRIT, UMR CNRS 5505, CNRS Université Paul Sabatier, Toulouse, France
{Wajih.Abdallah,Frederic.Vella,
Nadine.Vigouroux}@irit.fr
[2] IRIT, UMR CNRS 5505, Université Jean Jaurès, Toulouse, France
{adrien.vandenbo,thierry.val}@univ-tlse2.fr

Abstract. Autism spectrum disorders (ASD) are characterized by difficulties of socialization, disorders of verbal communication, restricted and stereotyped patterns of behaviors. Firstly, the paper reports tools of the user-centered design (UCD) used as well participants involved in the design of interactive collaborative system for children with ASD. Then, we describe the UCD deployed to design a vocal communication tool (VCT) between an adult with ASD and his family caregivers. The analyses of interviews demonstrate a strong need for a collaborative assistive technology based on voice interaction to avoid the family caregivers repeating the same sentences to the adult with ASD and, to create a friendly atmosphere at home. Observations in a real life environment demonstrate that the VCT is useful and accepted by the adult with ASD and his family. The work is not complete and issues such as designing a spoken dialogue system in the smart home need further works. The study of the type of voice synthesis (human or text-to-speech synthesis) is also an open question.

Keywords: User-centered design · Assistive technology ·
Autism spectrum disorders

1 Introduction

The term "autism" was introduced by the Swiss psychiatrist Bleuler in 1911 to designate people who are schizophrenic folded on themselves, disconnected from reality and excluded from all social life [1].

Autism spectrum disorders are characterized by a triad of impairments [2]: difficulties of socialization, disorders of verbal communication, restricted and stereotyped patterns of behaviors and interests. Firstly, people with ASD suffer from qualitative alterations of social interactions. These disorders are not a lack of interest or willingness on the part of the family to help the person with autistic disorder, but a problem of social skill that prevents him interacting with them. Most people with autism also suffer from communication disorders and these difficulties appear in very different ways. People with autism also exhibit stereotyped and repetitive behaviors. This autism characteristic intervenes differently according to the age or the cognitive abilities of

© Springer Nature Switzerland AG 2019
M. Antona and C. Stephanidis (Eds.): HCII 2019, LNCS 11573, pp. 3–12, 2019.
https://doi.org/10.1007/978-3-030-23563-5_1

people. In this regard, stereotyped body movements or stereo-typical use of objects are more frequently observed.

There are a lot of assistive technologies (AT) devices for children and adults with autism. Putnam et al. [3] have reported on results to elucidate information about software and technology use according domain (education, communication, social skills, therapeutic, entertainment and scheduling) were 31 applications out of 45 are dedicated to the education domain and designed for the personal computer. They also suggested considering sensory integration issues by allowing users to set colors and sounds as design consideration. This study shows that there are few assistive devices to help the family caregivers to help them for activities of daily living. To fill these gaps, we have designed a collaborative voice communication tool (VCT) between an adult with ASD and his or her family by implementing a user-centered design method.

In this paper, we first present the user-centered design tools to design assistive system for people with ASD. Next, the needs and the use context extracted from interviews of family caregivers will be reported as well as the different versions of the prototype. Then lessons from real use observation of VCT will be discussed. Finally, perspectives of this study will be described.

2   The User-Centered Design Implemented for ASD

According to ISO 9241-210 [4], the user-centered design is defined as follows "*Human-centred design is an approach to interactive systems development that aims to make systems usable and useful by focusing on the users, their needs and requirements, and by applying human factors/ergonomics, and usability knowledge and techniques*".

The purpose of the UCD is to respond to better to user needs, is an iterative process as shown in Fig. 1.

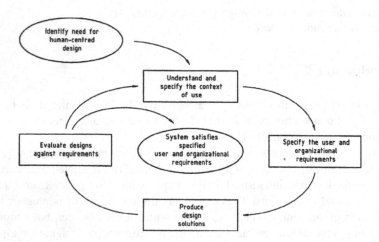

Fig. 1. Diagram of the user-centered design cycle according to ISO 13407 [5].

Philips and Zhao [6] related that almost one third of all users' AT were completely abandoned. This high rate of abandonment ascertained that a large percent of AT devices are not meeting users' needs. Consequently, there is a strong demand for practical, customized and reliable AT. The difficulty to collect and to understand the needs of disabled person is often described in the literature. The ISO 9 241-210 standard [4] and Philips and Zhao [6] recommend an active involvement of persons who will use the AT. Since the persons with ASD cannot express their needs and the context use due to communication and behavior disorder, family and professional caregivers [7] are involved in the UCD. Guffroy et al. [8] suggested involving the family, professional and human environments that are part of his or her ecosystem. One approach is to integrate this ecosystem into the user-centered design method.

The next section reports assistive technologies for person with ASD (See Table 1). Table 2 illustrates UCD tools and actors involved in the design of AT for people with ASD.

Table 1. Examples of assistive technology (AT).

AT	Autism spectrum disorders	Aims of AT
çATED [9]	Learning disabilities Life skill functioning	Digital tablet time management for children with ASD of learning
vSked [10]	Life skill functioning	Interactive and collaborative visual scheduling system on touch screen for autism classrooms
iCAN [11]	Communication skills	Teaching-assistive tablet application for parents and teachers to teach functional communications based on PECS [12]
HANDS [13, 14]	Social interactions Life skill functioning	Mobile cognitive support application for smartphones based on pervasive technology design
ECHOES [15]	Social interactions Learning disabilities	Touch interactive whiteboards allowing multi-modal 3D and socially realistic environment

The increased interest in the potential of technology in the context of autism is motivated by the recognition of autistic people's affinity with computers and more recently tactile devices (See Table 1). The studies presented above have confirmed their utility and efficiency as a means to support and develop social and interactional skills in children with ASD. The five studies have been conducted within a school environment in order to optimize the opportunities for contextual design. These studies were primarily designed to explore usability and acceptability factors of assistive tools designed to help children with ASD with communication, social and life skills development. The ECHOES project has worked on a technology-enhanced learning

Table 2. Examples of tools illustrated by the user-centered method for the AT.

AT	Methods used	Participants in UCD
çATED [9]	Interviews for need expression Observation in the evaluation phase	Interviews with mediator Conception: Engineer… 7 children, a teacher and two auxiliary School Life
vSked [10]	Interviews of therapists and educator: Observations of interactions between teachers and students Design Focus group discussions in the evaluation phase	Educators (n = 10), therapists (n = 3) and autism specialists Designer team plus an autism specialist and a teacher Stakeholders of all types (neuroscientists, special educators, assistive technology specialists and private therapists)
iCAN [11]	Interview questions for need expression Field notes, interviews, and transcripts for the evaluation phase	8 parents and 3 special education instructors 11 children, special education instructors, parents and therapists
HANDS [13, 14]	Teachers questionnaires; Child and parent interviews, first prototype [13] Classroom observations, teacher interviews; teacher questionnaire; semi-structured interviews parents and children in the qualitative interpretivist evaluation, second prototype [14]	9 teachers; 10 children and parents in two schools 15 teachers; 6 parents; 10 children with ASD in four specials schools; 26 children used the second prototype
ECHOES [15]	Series of workshops for idea generation and sensory exploration Observations phases for the design (mok-ups; storyboardings) And reflection phase for the design Observation sessions to collect data in the evaluation phase	Parents, children with ASD and teachers Children who are involved in ECHOES-like activities Practitioners and clinicians providing interpretation of the children's behaviors All stakeholders primarily led by the participatory team (researchers, practitioners, technology experts, plus parents and children with ASD)

environment with the participation of young users with ASD during the participatory design. The HANDS project [13, 14] has adopted the principles of persuasive technology design [16]. cATED [9] and vSked [10] are interactive and collaborative visual scheduling systems which demonstrates that visual schedules have positive impacts on individuals with ASD. Four of the five applications develop tools to support activities of daily living while the iCAN project proposes a tablet-based system that adopts the successful aspects of the traditional PECS (Picture Exchange Communication System) [12] to teach functional communications. All these systems incorporate advantageous features of digitalization, visual and sometimes multimodal (visual and voice) representations. However, no system is dedicated to adults with autistic disorders.

The Table 2 reports the tools used to design assistive technologies. Observations of children with ASD in classroom and interviews of teachers and therapists are the most commonly used tools for understanding the context and expressing the needs. Sometimes parents are involved as in the project HANDS, ECHOES and iCAN. The information gathered from these data made it possible to highlight the functionalities and characteristics of the designed systems. The qualitative evaluation of the first prototype [13] has identified a number of improvements that have been introduced in both the design and implementation for the second version of HANDs prototype [14]. The ECHOES project has implemented a participatory design process that involves young children with autistic spectrum disorders. It is interesting to point that "the children with ASD play the role of informants rather than fully fledged design partners" [15]. In çATED [9], the researchers have integrated the stakeholder mediator who facilitates the communication with the children with ASD and the other participants in the design of the çATED system. The presence of the mediator is very important to present the demands of the user, to choose the solutions and to evaluate the assistive technologies.

These studies also demonstrate the close collaboration between the researchers and the school and medical practitioners in the design process. Another point is to mention the different roles of stakeholders in understanding the demand, the behavior of persons with ASD, participating in the design and evaluation of the designed system. All these systems include enhanced ecological environments with and in schools.

Table 2 shows the importance of taking into account the children's ecosystem with ASD in designing systems that support the development of social and interactional skills. Few systems that relieve caregiver interventions exist for adults with ASD. We propose to describe the approach implemented for the design of a voice communication.

3 Methodology

This part describes the UCD approach implemented for the design of the VCT. Consecutively, interviews, prototyping and observation phases of the UCD are described.

3.1 The End-User

Christophe, 32 years old, is a person with ASD (communication, social participation and behaviors disorders) as defined in [1]. During the week he lives in an adult rehabilitation center for people with ASD. He spends the weekend with his family. His family caregivers wish to have a vocal communication tool for assisting them. Christophe is very soliciting his parents to get permission to do an action. This solicitation is very time-consuming and tiring for them. His family caregivers expressed the need of the VCT.

3.2   Interviews

Firstly, requirements and context use have been defined to specify the utility sought by the end-users of the VCT. Three interviews were conducted with his family (sister, mother and father). The questions were oriented to know his abilities of communication, his social participation skills, his preferences of sounds and colors, his favorite objects and the requirements of the VCT.

For the expressive communication, Christophe speaks in isolated words, for instance "drink", "chocolate", "cake", "pee", etc. He does not have the capabilities to make syntactically correct sentences. The parent's answers are also simple: such as "put" and "yes just go".

His social participation is very limited. Christophe interacts only with people he knows and especially with his mother who loves to touch her. He prefers to play alone even in the living space. He is unable to make decisions alone, he always ask permission to drink, to take an object, … He is always waiting an answer to his requirements from his family environment. He likes to listen to soft music, Christmas songs, and watch TV. He is attracted by the red color that corresponds to the color of his spinning top and loves any shape that rotates such as spinning top and spinner.

The interview's analysis states that Christophe always asks an agreement or a confirmation from this family to do an action. He may repeat the request without break until he or she receives a response from his or her family environment. This social behavior can be generated anxiety, tiredness or cognitive overload to the family.

The analyses of interviews demonstrate a strong need for a collaborative assistive technology based on voice interaction to avoid the family caregivers repeating the same sentences to the adult with ASD and to create a friendly atmosphere at home.

3.3   Design of the VCT Prototype

From the understanding of context use and the needs expressed by the whole family, a prototype has been designed. The first version of VCT (see Fig. 2) consists of:

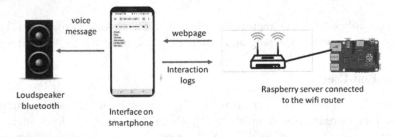

Fig. 2. VCT infrastructure.

- Loudspeakers connected to smartphone which sends the voice message to be broadcast;
- A Webpage as interface;
- A raspberry which hosts the application's web pages and interaction data.

From the need's analysis, only six messages were identified corresponding to the main confirmation messages pronounced by Christophe's family. These speech acts are "put", "yes put", "put!", "of course put", to action such "go to sleep", "yes just go". The Christophe's mother recorded these messages in quiet environment in MP3 format. This design choice is justified because Christophe interacts most often with his mother. Each URL (Uniform Resource Locator) of the web page is linked to one message (see Fig. 3). The url selection is made by a pointing click. This selection can be made by anyone present at Christopher's home. Then the vocal message is sent to the loud-speaker and the interaction log to the raspberry server.

Fig. 3. VCT interface.

4 Experimentation

The aim of this experiment is to: (1) observe the behavior of Christophe when the VCT will answer him instead of his mother; (2) identify new requirements for another version of the VCT prototyping based on these observations.

4.1 Context of Use

The VCT is installed in the mother's office and the loudspeaker is placed on the dining table (see Fig. 4). The mother and the father are doing their daily activities in the kitchen or in the office. Christophe is often in the living room watching television and playing in parallel with his spinning and his hand spinner. He moves around in the whole house and he stays more times in the living room. The wizard of Oz will select the appropriate url to answer to Christophe. The wizard of Oz is also in the living room on the other side of the table (see Fig. 4). He is known to Christophe as a family member to preserve a known family environment. This configuration was chosen to facilitate Christophe's social participation.

The solution based on Wizard of Oz was chosen before implementing VCT based on a spoken dialogue system. These first trials must have before to prove the useful-ness and acceptability of VCT by Christophe. Christophe's parents were instructed not to respond to his confirmation requests to complete the task.

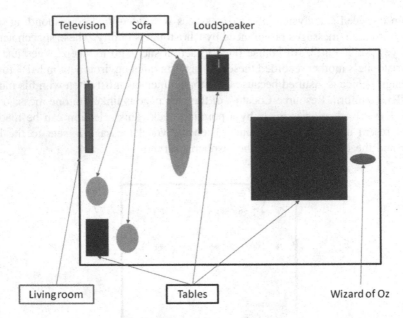

Fig. 4. Observation environment.

4.2   Observation

Observation 1: (2h 30)

The situation caused confusion: in fact, the mother responded spontaneously when she was with her son in the living room. The sound level (high noise level) of the messages was also considered too high by Christophe's parents because it made them startle. The sound level was adjusted by the wizard during the experiment.

However, Christophe obeyed the message content broadcasted by the loudspeaker: twice for the request to put his spinner and twice for the request to put the hand spinner. At this end of this observation, the mother strongly wants messages to be automated broadcasting to relieve her nervous tension due to her son's strong demands.

Observation 2: (3h00)

The experiment conditions are the same than during the previous observation. Christophe's parents haven't replied to Christophe's permission requests. The wizard tells him with the system. Christophe requested more permissions than the previous observation session: "go to the bed", "go to the toilet" and "put the hand spinner" or the "spinning top". During the session, Christophe made 23 requests (78.8% of requests to "put", 17.39% to "go to bed" and 4.34% to "go to the toilet"). He was more and more receptive to the VCT system especially to put something and go to the toilets. For the permission "go the bed", Christophe asked several times his mother who answers him "seat down" whereas the VCT system tells him several times (13 answers during 4 min) "go to sleep". Christophe has to go and see this mother before he accepted the permission given by the system. The parents think Christophe will get accustomed to the system's orders. Experiment 3 could not be carried out because of Christophe's great fatigue.

4.3 Discussion

Both Christophe's parents reported that the VCT is useful and very interesting for them and for Christophe. They also think that Christophe will accept this assistive technology in the future.

Christophe was surprised and worried (observation 1): he tried to identify where the voice message came from (Verbatim of his mother: "No, at first he was looking for where the sound came from. He was looking at me. And then he saw that I wasn't answering. Then he got used to it"). He watched his mother by wincing. Christophe has smiled to the wizard (observation 2) who demonstrates that he has totally under-stood and accepted the VCT. He is also pleased with the relevance of the response. The observations showed that the list of vocal messages is sufficient. Indeed, for the pre-test a loudspeaker, round and red, Christophe's favorite shape and color) was given to him with the aim of accepting it as one of his favorite objects. This loudspeaker was hidden.

These Christophe's behaviors confirm that it is more efficient to integrate loud-speakers in the smart home than to design a spoken dialogue in one of his favorite numeric object. The impact of the presence of the person in the living room, who recorded the voice messages, on Christophe's behavior must be studied. Several possibilities are to be considered: (1) recognition of the person present in the room and broadcasting of recorded sound messages from another family member not present; (2) restitution of messages with voice synthesis from texts with a woman's or man's voice path.

Christophe's parents wish to have a spoken dialogue system that would respond to Christophe's requests, and this in any room of the house.

5 Conclusion

The related works related that there is a great lack in assistive technologies in obtaining consent from adults with ASD to their ecosystem (family and professional caregivers). We conducted a UCD method based on interviews, prototyping phases and some first observations in real-life environments of the assistive technology VCT. VCT is a spoken tool that responds by means of a voice message to authorization requests from a person with social interaction disorders. In the tested version it is a wizard of Oz who selected the appropriate voice message. The empirical observation showed that the VCT is useful and accepted by the person with ASD. The parent interviews also show the usefulness for them, freeing them up for their daily activities.

Clearly, the work is not complete, and issues such as designing a spoken dialogue system in the smart home need further works. In this perspective, we will also have to study the adaptation of the voice of the synthesis system from texts to the people present in the house as well as the sound level according to the ambient noise of the house. It is clear that the study of the use of the VCT tool in rehabilitation center is also a path to explore.

Acknowledgements. We would like to acknowledge Christophe and his family who participated in the UCD approach.

References

1. Spitzer, R.L., Gibbon, M.E., Skodol, A.E., Williams, J.B., First, M.B.: DSM-IV-TR casebook: a learning companion to the diagnostic and statistical manual of mental disorders, 4th edn, text revision. American Psychiatric Publishing, Inc. (2002)
2. American Psychiatric Association, DSM-5 Task Force: Diagnostic and statistical manual of mental disorders, 5th edn. American Psychiatric Publishing, Inc., Arlington (2013)
3. Putnam, C., Chong, L.: Software and technologies designed for people with autism: what do users want? In: Proceedings of the 10th International ACM SIGACCESS Conference on Computers and Accessibility, pp. 3–10. ACM (2008)
4. Bevan, N., Carter, J., Harker, S.: ISO 9241-11 revised: what have we learnt about usability since 1998? In: Kurosu, M. (ed.) HCI 2015. LNCS, vol. 9169, pp. 143–151. Springer, Cham (2015). https://doi.org/10.1007/978-3-319-20901-2_13
5. Jokela, T., Iivari, N., Matero, J., Karukka, M.: The standard of user-centered design and the standard definition of usability: analyzing ISO 13407 against ISO 9241-11. In: Proceedings of the Latin American Conference on Human-Computer Interaction, pp. 53–60. ACM (2003)
6. Phillips, B., Zhao, H.: Predictors of assistive technology abandonment. Assistive Technol. 5 (1), 36–45 (1993)
7. De Leo, G., Leroy, G.: Smartphones to facilitate communication and improve social skills of children with severe autism spectrum disorder: special education teachers as proxies. In: Proceedings of the 7th International Conference on Interaction Design and Children, pp. 45–48. ACM (2008)
8. Guffroy, M., Vigouroux, N., Kolski, C., Vella, F., Teutsch, P.: From human-centered design to disabled user & ecosystem centered design in case of assistive interactive systems. Int. J. Sociotechnol. Knowl. Dev. (IJSKD) 9(4), 28–42 (2017)
9. Mercier, C., Guffroy, M.: Gérer le temps à l'aide d'une application numérique sur tablette pour un public avec autisme. In: Communication présentée à la 7e Conférence sur les environnements informatiques pour l'apprentissage humain, Agadir (Maroc) (2015)
10. Hirano, S.H., Yeganyan, M.T., Marcu, G., Nguyen, D.H., Boyd, L.A., Hayes, G.R.: vSked: evaluation of a system to support classroom activities for children with autism. In: Proceedings of the SIGCHI Conference on Human Factors in Computing Systems, pp. 1633–1642. ACM (2010)
11. Chien, M.E., et al.: iCAN: a tablet-based pedagogical system for improving communication skills of children with autism. Int. J. Hum. Comput. Stud. 73, 79–90 (2015)
12. Bondy, A.S., Frost, L.A.: The picture exchange communication system. Focus Autistic Behav. 9(3), 1–19 (1994)
13. Mintz, J., Branch, C., March, C., Lerman, S.: Key factors mediating the use of a mobile technology tool designed to develop social and life skills in children with autistic spectrum disorders. Comput. Educ. 58(1), 53–62 (2012)
14. Mintz, J.: Additional key factors mediating the use of a mobile technology tool designed to develop social and life skills in children with autism spectrum disorders: evaluation of the 2nd HANDS prototype. Comput. Educ. 63, 17–27 (2013)
15. Porayska-Pomsta, K., Frauenberger, C., Pain, H., Rajendran, G., Smith, T., et al.: Developing technology for autism: an interdisciplinary approach. Pers. Ubiquit. Comput. 16(2), 117–127 (2012)
16. Fogg, B.J.: Persuasive technology: using computers to change what we think and do. Ubiquity 5, 89–120 (2002)

Usability Enhancement and Functional Extension of a Digital Tool for Rapid Assessment of Risk for Autism Spectrum Disorders in Toddlers Based on Pilot Test and Interview Data

Deeksha Adiani[1], Michael Schmidt[1], Joshua Wade[1(✉)],
Amy R. Swanson[2], Amy Weitlauf[2], Zachary Warren[3],
and Nilanjan Sarkar[1]

[1] Adaptive Technology Consulting, LLC, Murfreesboro, TN 37127, USA
josh@innovateatc.com
[2] Treatment and Research Institute for Autism Spectrum Disorders (TRIAD),
Vanderbilt University, Nashville, TN 37212, USA
[3] Pediatric, Psychiatry and Special Education, Vanderbilt University, Nashville,
TN 37212, USA

Abstract. Early accurate identification and treatment of young children with
Autism Spectrum Disorder (ASD) represents a pressing public health and
clinical care challenge. Unfortunately, large numbers of children are still not
screened for ASD, waits for specialized diagnostic assessment can be very long,
and the average age of diagnosis in the US remains between 4 to 5 years of age.
In a step towards meaningfully addressing this issue, we previously developed
Autoscreen: a digital tool for accurate and time-efficient screening, diagnostic
triage, referral, and treatment engagement of young children with ASD concerns
within community pediatric settings. In the current work, we significantly
improve upon and expand Autoscreen based on usability data and interview data
collected in a pilot investigation of pediatric healthcare providers using Autoscreen. The enhanced version of Autoscreen addresses limitations of the previous tool, such as scalability, and introduces important new features based on
rigorous interviews with the target user population. Once validated on a large
sample, Autoscreen could become an impactful tool for early ASD screening
and targeted referral in primary care settings. The comprehensively-enhanced
tool described in the current work will enable the investigative team to achieve
this goal.

Keywords: Autism Spectrum Disorders · Digital screening ·
Usability enhancement · Scalability

© Springer Nature Switzerland AG 2019
M. Antona and C. Stephanidis (Eds.): HCII 2019, LNCS 11573, pp. 13–22, 2019.
https://doi.org/10.1007/978-3-030-23563-5_2

1 Introduction

Autism Spectrum Disorders (ASD) are classified in the Diagnostic and Statistical Manual of Mental Disorders, 5th Edition (DSM-5) as neurodevelopmental disorders marked by (a) deficits in social communication and interaction and (b) repetitive and restrictive patterns of behavior and interest [1]. ASD often negatively affects lifespan outcomes, especially with regards to meaningful social engagement and occupational attainment [2]. Moreover, ASD prevalence has been on the rise for over a decade and, according to the Centers for Disease Control and Prevention, is now at its highest ever rate of 1 in 59 among children in the United States [3]. Therefore, early accurate identification and treatment of young children with ASD represents a pressing public health and clinical care challenge. Given mounting evidence that early, accurate diagnosis of ASD is possible and that very young children who receive intervention can demonstrate substantial gains in functioning, current American Academy of Pediatrics practice guidelines endorse formal ASD screening at 18 and 24 months of age [4, 5]. Unfortunately, large numbers of children are still not screened for ASD, waits for specialized diagnostic assessment can be very long, and the average age of diagnosis in the US remains between 4 to 5 years of age [3].

Continuing our earlier work [6], the current project uses pilot test and interview data to rigorously enhance and extend *Autoscreen*—a digital tool for time-efficient screening, diagnostic triage, referral, and treatment engagement of young children with ASD concerns within community pediatric settings. While technically novel in many respects, Autoscreen is not the first digital application designed to streamline screening of ASD in toddlers in clinical settings. For instance, CHADIS (Child Health and Development Interactive System; chadis.com) [7] is a web-based platform for administering screeners and assessments (e.g., M-CHAT) to a variety of populations, such as children with ASD and adolescent females with eating disorders. Although CHADIS supports some instruments for ASD screening and assessment, only Autoscreen has the specific focus of ASD risk assessment in toddlers based on a very brief (i.e., approximately 15-min) actively-guided interaction with minimal training required for administration. Cognoa [8] is another related application that uses a mobile application to provide a variety of screening and assessment services for individuals with concerns related to ASD and Attention Deficit Hyperactivity Disorder among others (cognoa.com). Cognoa, intended for use by clinicians and parents, uses manually coded video analyses of children's behavior to make risk predictions. However, Cognoa is not designed to facilitate highly rapid administration in constrained clinical settings, as it requires a video review process by a team of remote clinicians.

2 Pilot Investigation of Usability

2.1 Implementation

The technical design of Autoscreen was previously described in [6]. The application was designed using the Unity game engine, which, despite being convenient for rapid deployment across a wide range of platforms (e.g., iOS, Android, and desktop operating systems), is not an ideal tool for designing resource-efficient and scalable mobile applications. However, given these limitations, we successfully implemented a fully-functional prototype which reliably and accurately presented the novel procedures and screener content to pediatric care providers. The prototype application incorporated (a) a guided interface for the interactive administration of the assessment procedures, (b) a built-in behavioral coding interface, (c) automatic report generation detailing assessment results (including both dichotomous and continuous scale risk indices as well as noted areas of concerns and identified strengths), and (d) a video model of Autoscreen's procedures carried out between a provider and a child with ASD concerns (see Fig. 1).

Fig. 1. Screen capture from the video model: the provider and child engage in the turn-taking activity while the tablet running Autoscreen is out of the view of the child.

Providers interacted with Autoscreen remotely using two wireless peripheral devices that connected to the testing tablet via Bluetooth: an earbud and a presentation remote. This approach allowed providers to interact with the application without drawing the child's attention to the presence of the tablet. Providers progressed through test items using the wireless presentation remote which featured buttons for convenient navigation (i.e., forward/backward through test items), timer activation, and audio prompt activation. With the wireless earbud, providers were also given real-time audio instructions by the application, thus allowing providers to obtain testing instructions without distracting the child. The application provided clear, short, and simple instructions to lessen initial training burden, for which a short pre-administration tutorial was embedded into the application. Scoring was completed within the application via prompts which guided the provider to rate the child's responses in easy-to-report Likert categories (often, inconsistent, rare/not observed).

2.2 Usability Testing

In our IRB-approved pilot investigation, n = 32 professionals and paraprofessionals licensed to conduct ASD evaluations (e.g., developmental-behavioral pediatricians, clinical psychologists, speech-language pathologists) were recruited to assess the usability and preliminary validity of Autoscreen. Participants also included n = 32 families, i.e., a parent/caregiver and a child (18–36 months of age) with clinically-verified ASD diagnosis. The study was designed to capture data that would allow the investigative team to identify areas in which to enhance usability and extend functionality to include user-requested features. Data collected from participants included responses to (a) the System Usability Scale (SUS) [9], (b) Acceptability, Likely Effectiveness, Feasibility, and Appropriateness Questionnaire (ALFA-Q) [6], and (c) semi-structured Customer Discovery style interviews [10].

Each session lasted approximately one hour (excluding consent/assent). This hour consisted of 20 min to introduce the provider to the novel tool and its associated screener, 20 min for the core Autoscreen procedures, and a 20 min interview with the provider to complete questionnaires and discuss usability issues.

2.3 Results and Discussion

Providers reported high levels of both usability (mean SUS = 86.93) and acceptability (mean ALFA-Q = 87.50) of Autoscreen. This level SUS is regarded as "excellent" in the literature [11], and the overwhelming majority of providers reported "good" to "excellent" usability or better (see Table 1). The most common issue affecting usability reported by providers concerned minor difficulties involving the wireless earpiece through which audio prompts were given. As a result, we have transitioned to a new earpiece design that securely wraps around the participant's ear.

Table 1. Overview of participant-reported scores on the system usability scale

Adjective	SUS cutoff	% Respondents above cutoff
Worst imaginable	12.5	0%
Awful	20.3	0%
Poor	35.7	0%
OK	50.9	6%
Good	71.4	25%
Excellent	85.5	25%
Best imaginable	90	44%

SUS = System usability scale [9]; See [11] for interpretation of scores

We next sought to identify—through Customer Discovery interviews [10]—updates and features viewed as highly desirable by participants that should be included in the next version of Autoscreen. We discovered three important results. First, we were able to clearly identify a set of meaningful value propositions based on reported clinical care challenges. Interviewees were unanimously positive about the potential of Auto-screen and the vast majority indicated a strong desire to use the tool within their practices. Second, we identified participant-requested features that could provide substantial additional value (e.g., a digital note-taking feature to quickly record behavioral notes on-the-fly). Third, interviewees provided feedback about the broader context into which Autoscreen might ultimately be deployed, as well as how such a system could be optimally integrated within existing systems of care. For example, some interviewees suggested that Autoscreen could function well not just in face-to-face evaluations, but in remote administration scenarios as well (i.e., tele-medicine services for rural families). Additionally, interviewees reported numerous suggestions about aesthetic changes, many of which were incorporated and are discussed in Sect. 3.

3   System Enhancement and Extension

3.1   Frontend

Concurrent with the pilot study in which we collected user feedback about usability and satisfaction with the Autoscreen prototype, we commenced work on the enhanced application. Whereas the prototype was created in the Unity game engine, subsequent development of Autoscreen was performed using Android Studio. At present, Auto-screen can run on both tablets and phones, although the visual content is optimized for tablets 8" or larger. With regards to aesthetic changes, the revised color-palette was selected to achieve optimal readability and legibility [12]; the application has a logo in white and light green, on a backdrop of blue and cyan and now features a consistent visual template (Fig. 2). As requested by providers, the timer screen now features a larger font that is much easier to see at a distance (Fig. 2). After specified intervals, an audio clip is played indicating the time remaining for a particular task. In addition, buttons for timer play/pause and reset are available for use as needed.

Just as in the prototype application, the screening process begins with a form for subject data entry, and metadata such as date and time are automatically captured. Here, the provider enters the subject's anonymized identifier as well as the date of birth using large drop-down menus. The forward and backward arrow buttons permit users to navigate through the Autoscreen application screens, or "Activities" in android terms, and the *onBackPressed()* method has been overridden to switch back to the previous Activity using the physical back button on the tablet. Unlike the prototype application, the revised layout is now consistent across the entire application (see side-by-side comparison in Fig. 2). The home button located in the top left corner of the screen permits the user to pause the current activity in order to restart or go to review tutorial information. This functionality was not available in the Autoscreen prototype. A pro-gress bar is located on the left side of the screen and displays five distinct sections of the application: subject data entry, materials checklist, primary screening activities, post-procedures screener, and risk profile report. The application header indicates at all times the section that the provider is currently in. Each page contains an audio tip button that allows the user to receive specific instruction for that page by clicking the button.

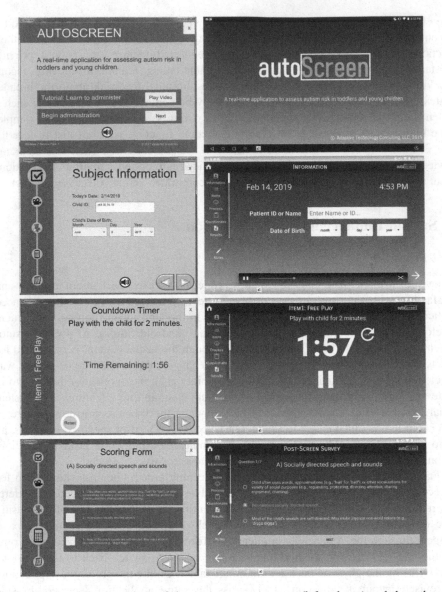

Fig. 2. Side-by-side comparison of the Autoscreen prototype (left column) and the enhanced implementation (right column).

The previously produced video model can now be streamed directly into the application and displayed in a media player. When a video is played, a small media player (Fig. 3) appears and includes functions for play, pause, and stop, and additionally features an interactive progress bar.

A closed caption button was also added to the media player, accommodating users who may prefer to read the text in addition to receiving the audio instructions. Subtitle

Fig. 3. Screen capture showing the newly introduced media player and audio subtitle features of Autoscreen.

text were loaded from *.srt* (SubRip Text) files. Below the media player, there is an additional button labeled "Materials". Tapping on this button produces a pop-up dialog showing pictures of the items required for that particular stage of screening. The dialog box is an unobtrusive feature as it does not interrupt the video currently playing. To minimize the dialog, the user can touch the screen anywhere around the dialog box.

Many providers from the pilot study also requested a feature to facilitate on-the-fly note-taking, which has now been added to the current application. The note-taking widget is a separate layout with a multiline *EditText* box. The user can type notes on any page that they like and at any time, and the save button will store the text in a file associated with the session (see Fig. 4).

3.2   Backend

The current work utilizes Amazon Web Services to host and facilitate a centralized backend model. Enhancements in the current backend model now include the ability to handle requests from frontend clients. The backend is segregated into (a) a REST API endpoint server based on Flask [13], and (b) a predictive model utilizing Numpy (www.numpy.org) for data-processing, and Keras (www.keras.io) for a model-based prediction. The development stack was built entirely in Python 3.6 and included Flask for its straightforward approach to developing functional, full-featured RESTful API endpoints, and Keras for high-level abstraction to design, train, test, and deploy neural networks. Additionally, Keras uses the default TensorFlow backend for its lower-level component and GPU access when available. Numpy handles common data manipulation and is requisite for most data-science applications. Multithreaded applications such as Keras, TensorFlow, and even Python in general are especially problematic for

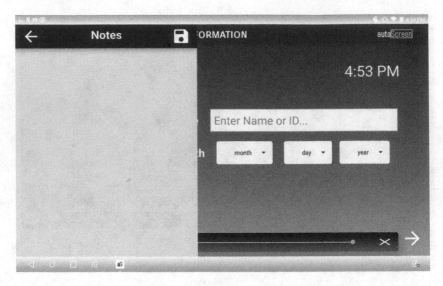

Fig. 4. Newly added feature for on-the-fly note-taking.

asynchronous, event-driven, context-sensitive frameworks such as Flask. As a result, Autoscreen's server splits these into two closely matched, socket-connected daemon applications. Python's *SocketServer* from the standard library provides a high-level abstraction for OS-level system calls to connect each process.

This design has several advantages apart from overcoming some multithreading challenges. First, the system is modular as the API endpoints are not directly tied to any specific predictive model, thus permitting models to be changed or updated over time; Flask interacts with a *SocketServer* and passes parameterized commands for some model to operate on (e.g., to make predictions on batch data). Both platforms must agree on an internal argument or API. In fact, the API-facing interface contains end-points to receive an entire pre-trained model and can be instructed to load a model from an HDF5 file. Future work will implement functionality for sending persistent or "pickled" models. This design will allow the predictive model and API handler to exist on different machines. The API handler could theoretically "round-robin" instructions to predictors on different machines for load balancing via socket connections. Future work will introduce the flexibility to test experimental models and to ensure accountability as models change over time.

Finally, in order to validate the backend architecture's ability to handle client requests, work continued on the development of a predictive model based on a subset of test items used to recognize ASD. A sequential, feed-forward, deep-net makes predictions at a current accuracy of upwards of 95% based on 90/10 testing split for 739 examples, which is based on the exploratory dataset described in our earlier work [6]. The reader should note that this exploratory model does not represent Autoscreen-generated features at this time, and thus is intended as a benchmark for an idealistic upper-bound on the potential real-world performance of Autoscreen. In lieu of a model validated explicitly on Autoscreen data, however, the functional backend architecture represents a major step forward in advancing Autoscreen to the level of scalable deployment. In sum, initial development on hosting servers, installation and configuration of a running environment, and DNS record-keeping have established a robust publicly-accessible interface. Future work will seek to establish authorization mechanisms as well as trust relationships.

4 Conclusions

The current work builds significantly on our previous work, making several new contributions. First, our current frontend was redesigned based on extensive feedback from the target user audience of pediatric care providers who assess risk for ASD. This included both an overall aesthetic update and the addition of many new user-requested features. Second, although our earlier prototype was fully-functional, it lacked the ability to scale given the choice of development environment (i.e., the Unity game engine). Now, our heavy investment in the new backend architecture will facilitate scalable client-server communication for dynamic risk profile generation. While the current data sample collected from tests of Autoscreen is too small to permit real-world deployment at this time, our current system represents a significant step towards achieving that goal. Future research and development will involve continual improvement of both the user interface and the server-side components of Autoscreen, as well as a large-scale study to capture data from a sample sufficiently large for validation of the novel tool.

Acknowledgement. This work was supported by the National Institute of Mental Health of the National Institutes of Health through award number 1R43MH115528. The authors would like to express thanks to the families and pediatric care providers who gave significant time for participation in this study.

References

1. American Psychiatric Association: Diagnostic and statistical manual of mental disorders (DSM-5®). American Psychiatric Publishing (2013). https://doi.org/10.1176/appi.books. 9780890425596
2. Howlin, P., Goode, S., Hutton, J., Rutter, M.: Adult outcome for children with autism. J. Child Psychol. Psychiatry **45**(2), 212–229 (2004). https://doi.org/10.1111/j.1469-7610. 2004.00215.x
3. Baio, J., et al.: Prevalence of autism spectrum disorder among children aged 8 years — autism and developmental disabilities monitoring network, 11 sites, United States, 2014. MMWR Surveill. Summ. **67**(6), 1–23 (2018). https://doi.org/10.15585/mmwr.ss6513a1
4. Dawson, G., et al.: Randomized, controlled trial of an intervention for toddlers with autism: the early start denver model. Pediatrics **125**(1), e17–e23 (2010). https://doi.org/10.1542/ peds.2009-0958
5. Warren, Z., McPheeters, M.L., Sathe, N., Foss-Feig, J.H., Glasser, A., Veenstra-VanderWeele, J.: A systematic review of early intensive intervention for autism spectrum disorders. Pediatrics **127**(5), e1303–e1311 (2011). https://doi.org/10.1542/peds.2011-0426

6. Sarkar, A., Wade, J., Swanson, A., Weitlauf, A., Warren, Z., Sarkar, N.: A data-driven mobile application for efficient, engaging, and accurate screening of ASD in toddlers. In: Antona, M., Stephanidis, C. (eds.) UAHCI 2018. LNCS, vol. 10907, pp. 560–570. Springer, Cham (2018). https://doi.org/10.1007/978-3-319-92049-8_41

7. Sturner, R., Howard, B., Bergmann, P., Stewart, L., Afarian, T.E.: Comparison of autism screening in younger and older toddlers. J. Autism Dev. Disord. 47(10), 3180–3188 (2017). https://doi.org/10.1007/s10803-017-3230-1

8. Kanne, S.M., Carpenter, L.A., Warren, Z.: Screening in toddlers and preschoolers at risk for autism spectrum disorder: evaluating a novel mobile-health screening tool. Autism Res. 11(7), 1038–1049 (2018). https://doi.org/10.1002/aur.1959

9. Brooke, J.: SUS-a quick and dirty usability scale. Usability Eval. Ind. 189(194), 4–7 (1996). https://doi.org/10.1201/9781498710411

10. Blank, S., Dorf, B.: The Startup Owners Manual, vol. 1. K&S Ranch Inc., Pescadero (2012)

11. Bangor, A., Kortum, P.T., Miller, J.T.: An empirical evaluation of the system usability scale. Int. J. Hum.-Comput. Interact. 24(6), 574–594 (2008). https://doi.org/10.1080/10447310802205776

12. Hall, R.H., Hanna, P.: The impact of web page text-background colour combinations on readability, retention, aesthetics and behavioural intention. Behav. Inf. Technol. 23(3), 183–195 (2004). https://doi.org/10.1080/01449290410001669932

13. Ronacher, A.: Welcome—flask (a python microframework) (2010). http://flask.pocoo.org/. Accessed 30 Jan 2019

Understanding How ADHD Affects Visual Information Processing

Yahya Alqahtani$^{(\boxtimes)}$, Michael McGuire$^{(\boxtimes)}$, Joyram Chakraborty$^{(\boxtimes)}$, and Jinjuan Heidi Feng$^{(\boxtimes)}$

Department of Computer and Information Sciences, Towson University, Towson, USA
{yalqah1, mmcguire, jchakraborty, jfeng}@towson.edu

Abstract. Attention Deficit Hyperactivity Disorder (ADHD) is a condition that is characterized by impulsivity, age-inappropriate attention, and hyperactivity. ADHD is one of the most prevalent disorders among children. For a significant number of children whose condition persists into adulthood, ADHD leads to poor social and academic performance. In this paper we present preliminary results of an experiment that investigates how ADHD affects visual information processing under three information presentation methods (textual, graphical, and tabular). The efficiency and accuracy of both the neurotypical group and the group with ADHD were significantly impacted by different information presentation methods. However, the neurotypical group and the group with ADHD showed different patterns in their perceived interaction experience with the three information presentation methods. The result provides insights that might help designers and educators develop or adopt more effective information representation for people with ADHD.

Keywords: ADHD · Information presentation methods · Visual information processing

1 Introduction

One of the most common mental disorders that affect children and adults is Attention-deficit/hyperactivity disorder (ADHD) [1]. People with ADHD experience different symptoms including short attention span, hyperactivity and impulsivity [1]. According to the Centers for Disease Control and Prevention (CDC), the estimated number of children and adolescents who have ADHD is 6.1 million, which counts for 9.4% of children 2 to17 years old [2]. Until today, the real cause of ADHD hasn't been identified and the focus of the treatment is to reduce its symptoms [2]. Treatment for ADHD usually combines psychological and medical interventions. Medical interventions seek to reduce hyperactivity and impulsivity as well as to enhance the ability to focus, learn and work [3]. Although ADHD causes long-term impact on a large number of children and adults, existing research on information technology based solutions to support people with ADHD is rather limited [4].

Analyzing data presented in various forms is crucial for both academic and professional performance as well as everyday life. Different information presentation

© Springer Nature Switzerland AG 2019
M. Antona and C. Stephanidis (Eds.): HCII 2019, LNCS 11573, pp. 23–31, 2019.
https://doi.org/10.1007/978-3-030-23563-5_3

methods can substantially affect task performance. To date, there is no research conducted that investigated the effect of ADHD and information presentation methods on visual interaction when processing data. To fill this gap, we conducted an experiment to better understand the relationship between ADHD and information presentation methods in the context of visual information processing. The specific information presentation methods investigated in the study are textual presentation, graphical presentation, and tabular presentation. The result of the study provides insight on how people with ADHD interact with different information presentation methods as well as their subjective perception regarding the interaction experience.

2 Related Research

2.1 ADHD

ADHD has been established as a brain disorder that affects the development and functioning of both children and adults. The disorder is characterized by inattention because the afflicted individual is unable to sustain attention and is not persistent when performing tasks. It is also presented as hyperactivity because the individual is restless and frequently moves, including excessive mannerisms such as tapping and fidgeting and excessive talking. Impulsivity is yet another characteristic, which is manifested as acting hastily often without prior thought, being intrusive in social situations and interrupting others excessively. [1] ADHD is a prevalent neurobiological condition among children and its behavioral influences are most noticeable in school-age children as it affects between 5% and 8% of these children [5].

In children, ADHD is observed with various behavioral attributes including having a short attention span as these children lose attention after short durations. This problem may lead to poor academic performance, especially in the subjects or activites that require sustenance of attention. It has been reported that children with ADHD underperform in reading and mathematical tasks [5].

2.2 Computer-Based Solutions to Support People with ADHD

Most of the previous work in assistive technologies regarding users with ADHD focused on monitoring or extending the users' attention span or supporting specific daily routines [6, 7]. For example, Beaton et al. developed a reflective mobile application to help young people with ADHD better understand their engagement levels during their daily tasks. The application was used together with an electroencephalographic (EEG) device to collect data about user-specific task engagement. The task engagement data was further analyzed with geographic and temporal data so that the degree of engagement could be interpreted in the context of time and location [6].

Dibia introduced FOQUS, a smartwatch application designed to help people with ADHD increase their attention span and reduce anxiety. The app has functionalities such as Pomodoro time management technique, meditation techniques, positive messages and health tips. 10 participants with ADHD used the application during an evaluation study and eight of those participants reported reduced level of anxiety [7].

Researchers also developed games such as Tarkeezy that makes use of eye tracking technologies to develop engaging behavioral therapy programs. The system captures a user's eye gaze while playing the game and the data is used as an control element in the game's interactive interface [8].

More recently, Asiry et al. designed a system with an adaptable user interface to extend attention span for children with ADHD. In this study, children's attention during reading is tracked through two modalities: their eye movement captured via a webcam and the location of pointing via a mouse. Whenever the system senses that the user is hovering on a pre-defined Area of Interest, one color scheme would be applied to that area in an attempt to keep the user attentive to the content. A user study was conducted involving 21 students with ADHD aged between 10 to 12. It was found that 'highlighting', 'contrast', and 'sharpening' all significant affect the attention span of children with ADHD. However, the 'highlighting' scheme had the highest effect among the three schemes [9].

Analyzing data presented in various forms are crucial for both academic and career development as well as everyday life. It has long been discussed that different information representations such as tables and graphs could affect information processing and decision-making [10]. Given the reported difficulty that people with ADHD experience in reading and mathematical skills [5], insight on how people with ADHD interact with different information representations may help understand the challenges they experience in related tasks. To date, there is no previous research examining this specific topic. Therefore, we conducted a controlled user experiment as the first attempt to start filling this gap.

3 Method

3.1 Experiment Design

In this study we adopted a split-plot design that consists of both between-group and within-group variables. The between-group independent variable is the condition of the participants: the neurotypical group and the group with ADHD. The within-group independent variables are the information presentation methods and the difficulty of the questions. Three information presentation methods were investigated in this study: the textual method, the tabular method, and the graphical method. Participants answered 3 groups of questions with different levels of difficulty based on the visual information presented to them.

Participants' performance was measured through the time it took to complete the task, the quality of the answers provided by the participants, and satisfaction and preference ratings collected through a post-test questionnaire. We also collected the participants' eye movement data through a Tobii X2 60 system to better understand the visual scanning patterns while interacting with different information representations.

3.2 Participants

Twenty-four participants took part in the study. Twelve participants had no cognitive or perceptual impairments and twelve participants were diagnosed with ADHD. The neurotypical participants were recruited through email announcement and flyers. The participants with ADHD were recruited via email announcement through a student support group on campus. In the neurotypical group, three participants were female and nine were male. In the group with ADHD, six were female and six were male. All participants were native English speakers. The age for all participants was between 20 to 24 except for one neurotypical participant who was between the age of 25-30 and one participant with ADHD between the age of 18 to 20. All participants were university students with different majors.

3.3 Scenario and Tasks

During the study, the participants viewed health-related information presented in different methods and answered questions based on the information presented. Health-related information was chosen as the context of the tasks because people constantly browse and search health-related information on the Web. We developed imaginary data for four counties regarding smoking, high cholesterol, diabetes and high blood pressure and their occurrence among four ethnic groups. The amount of information contained in each country was the same. For each country, we developed three questions for participants to answer based on the data presented. The level of difficulty of the questions varied based on the amount of information needed to answer the question. The easy questions required the participant to navigate the information about a specific ethnic group in order to find the answer. For example, in order to answer the following question, the participant only needed to navigate to the data about the male Asian group.

"Among Asian males, which of the 4 health related conditions has the highest percentage of occurrence?"

The medium level questions required the participant to find information about a specific gender. For example, in order to answer the following question, the participant needed to find relevant data about all female ethnic groups.

"Among all female groups, which ethnic group and condition has the highest percentage of occurrence."

The difficult questions required the participant to look at all groups to find the answer. For example, in order to answer the following question, the participant needed to find data about all ethnic groups, both male and female.

"Among all groups, which ethnic group and condition has the smallest difference between male and female?"

The quality of an answer was measured using a numeric rating with three possible values: 0, 0.5, and 1. For the easy questions, the answer only consisted of one specific health condition. An answer with the correct condition would receive a quality rating of 1. An incorrect answer would receive a quality rating of 0. The answer for a medium or difficult question consisted of two components: one specific health condition and one specific ethnic group, each contributing 0.5 point towards the quality rating. An answer containing both the correct health condition and the correct ethnic group would receive a quality rating of 1. An answer containing the correct health condition but the incorrect ethnic group, or vice versa, would receive a quality rating of 0.5. An answer containing incorrect health condition and incorrect ethnic group would receive a rating of 0.

3.4   Information Representation

We developed three sets of webpages that present the same amount of information in three different information presentation methods: textual, graphical, and tabular (Figs. 1, 2 and 3). The three information presentation methods were chosen because they are the most commonly adopted methods to present visual content.

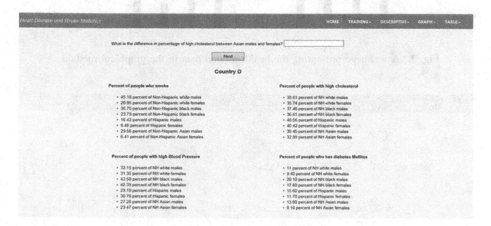

Fig. 1. A webpage presenting the health-related data in the textual method

3.5   Procedure

At the beginning of the study, the participants completed a training session that exposed them to all three information presentation methods. Following the training session, each participant reviewed the health information of three countries presented in three conditions (textual, graphical and tabular) and answered questions based on the information. Under each condition, they viewed the data of a specific country and answered one easy question, one question of medium level difficulty, and one difficult question. Both the order of the information presentation methods and the countries were counterbalanced to control the learning effect. Participants completed a demographic and satisfaction questionnaire at the end of the study. At the end of the experiment, all participants were awarded a $20 gift card for their time and effort.

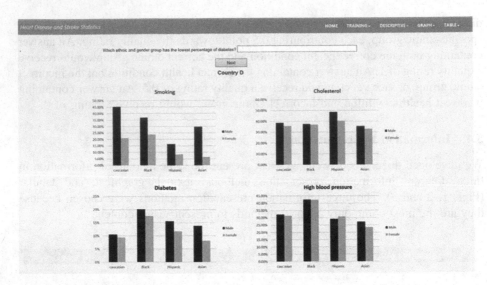

Fig. 2. A webpage presenting the health-related data in the graphical method

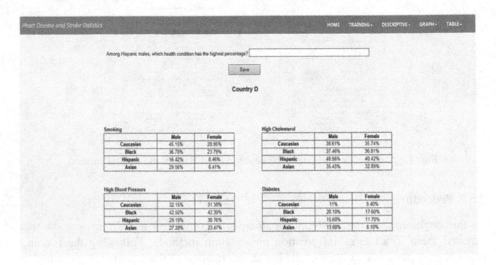

Fig. 3. A webpage presenting the health-related data in the tabular method

4   Results

We conducted a Repeated Measures Analysis of Variance (ANOVA) test with task completion time as the dependent variable and participant condition, information presentation method, and question type as the independent variables. The results suggest that both information presentation method and type of questions have significant effect on task completion time ($F(2, 44) = 42.62$, $p < 0.001$; $F(2, 44) = 7.93$, $p < 0.001$). There is a significant interaction effect between information presentation method, types of questions and task completion time ($F(4, 88) = 10.19$, $p < 0.001$). No significant difference is observed in task completion time between the neurotypical participants and the participants with ADHD ($F(1, 22) = 1.21$, n. s.). As indicated in Fig. 4, participants in both groups spent significantly longer time completing the tasks when the information was presented in text as compared to a graph or a table.

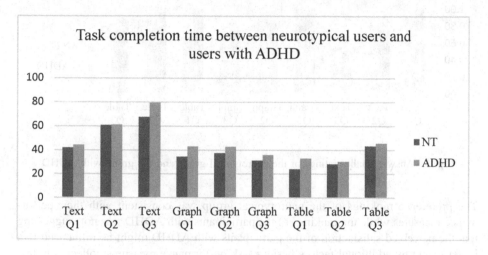

Fig. 4. Task completion time (in seconds) of both participant groups

A Repeated Measures Analysis of Variance (ANOVA) test was also conducted with answer quality ratings as the dependent variable and participant condition, information presentation method, and question type as the independent variables. No significant difference was observed in the quality of answers between the neurotypical participants and the participants with ADHD ($F(1, 22) = 0.59$, n. s.). Both the information presentation method and the type of questions have significant effect on the quality of the answers ($F(2, 44) = 6.23$, $p < 0.005$; $F(2, 44) = 12.29$, $p < 0.001$). There was significant interaction effect between information presentation method, types of questions and quality of the answers ($F(4, 88) = 6.00$, $p < 0.001$). As demonstrated in Fig. 5, neurotypical participants had lower-quality answers when the information was presented in text as compared to table. For participants with ADHD, no significant difference was observed in the answer quality ratings among the three information presentation methods.

At the end of the study, participants ranked their preference for the three information presentation methods. The graphical method was the most preferred among the neurotypical participants (6 out of 12) while the tabular method was the most preferred among the participants with ADHD (8 out 12). Regarding the least preferred method, 9 out of 12 neurotypical participants chose the textual method and 2 chose the graphical method. Interestingly, 7 participants with ADHD chose the graphical method and 4 chose the textual method. A Chi-squared test suggests that the difference in the preference rankings between the 2 groups is significant ($X^2(1) = 4.70$, $p < 0.05$).

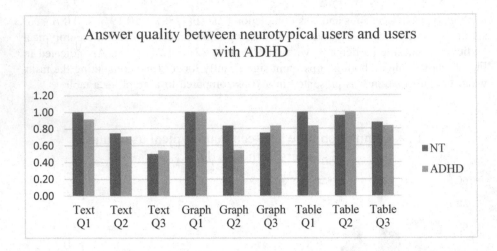

Fig. 5. Answer quality ratings of the neurotypical group and the group with ADHD

The preference rankings of the neurotypical group was consistent with their performance measures while the rankings of the participants with ADHD was not, suggesting that the perceived satisfaction of the participants with ADHD might be influenced in a larger extent by additional factors besides task performance measures collected in this study.

5   Discussion and Conclusion

We examined the impact of three information representations on visual information processing for people with ADHD through a controlled experiment. Compared to the neurotypical participants, the participants with ADHD spent similar amount of time to answer the questions and produced answers of similar quality. For both the neurotypical participants and the participants with ADHD, the textual method required significantly longer time to answer the questions than the tabular method and the graphical method. However, there was a difference in the impact of information representation on the quality of answers between the two groups. The quality ratings of the participants with ADHD were similar among the three information representations while the neurotypical participants had higher ratings in the tabular condition than the textual condition.

There is also significant difference between the two groups of participants in the subjective preference rankings. 9 of the 12 neurotypical participants chose the textual method as the least preferred method, which was consistent with the task performance results. In contrast, only 4 of the participants with ADHD chose the textual method and 7 chose the graphical method as their least preferred method, which was inconsistent with the task performance results. In the future, we will analyze the eye tracking data to examine the visual scanning patterns of the participants. We hope that the eye tracking data will help explain the inconsistency between the performance measures and the perceived satisfaction of the participants with ADHD.

This study provides preliminary understanding regarding how people with ADHD interact with different information presentation methods. However, the results need to be interpreted with caution because of the small sample size. In addition, we didn't collect information about the medication usage of the participants due to IRB restrictions. Some of the participants with ADHD might have taken medication on the day of the study, which could have affected their performance. We are planning a future study involving a larger number of participants with ADHD. We are also revising the IRB application so that medication and other health-related information relevant to task performance could be collected.

References

1. Parekh, R.: What Is ADHD? July 2017. https://www.psychiatry.org/patients-families/adhd/what-is-adhd. Accessed 04 Jan 2019
2. Centers for disease control and prevention: Attention-Deficit/Hyperactivity Disorder (ADHD), 21 September 2018. https://www.cdc.gov/ncbddd/adhd/data.html
3. Rubia, K.: Cognitive neuroscience of attention deficit hyperactivity disorder (ADHD) and its clinical translation. Front. Hum. Neurosci. **12**(March), 1–23 (2018). https://doi.org/10.3389/fnhum.2018.00100
4. Sonne, T., Marshall, P., Obel, C., Thomson, P.H., Grønbæk, K.: An assistive technology design framework for ADHD. In: Proceedings of OzCHI 2016 (2016)
5. Al-Shathri, A., Al-Wabil, A., Al-Ohali, Y.: Eye-controlled games for behavioral therapy of attention deficit disorders. In: Stephanidis, C. (ed.) HCI 2013. CCIS, vol. 373, pp. 574–578. Springer, Heidelberg (2013). https://doi.org/10.1007/978-3-642-39473-7_114
6. Beaton, R., Merkel, R., Prathipati, J., Weckstein, A., McCrickard, S.: Tracking mental engagement: a tool for young people with ADD and ADHD. In: Proceedings of the 16th International ACM SIGACCESS Conference on Computers & Accessibility, pp. 279–280 (2014). https://doi.org/10.1145/2661334.2661399
7. Dibia, V.: Foqus: a smartwatch application for individuals with ADHD and mental health challenges. In: Proceedings of the 18th International ACM SIGACCESS Conference on Computers and Accessibility - ASSETS 2016, pp. 311–312, November 2016. https://doi.org/10.1145/2982142.2982207
8. Dendy, C.A.Z., Zeigler, A.: A Bird's-Eye View of Life with ADD and ADHD. Cherish the Children, Cedar Bluff (2003)
9. Asiry, O., Shen, H., Wyeld, T., Balkhy, S.: Extending attention span for children ADHD using an attentive visual interface. In: 2018 22nd International Conference Information Visualisation (IV), pp. 188–193. IEEE, July 2018
10. Vessey, I.: Cognitive fit: a theory based analysis of the graphs versus tables literature. Decis. Sci. **22**(2), 219–240 (1991)

Attention Assessment: Evaluation of Facial Expressions of Children with Autism Spectrum Disorder

Bilikis Banire[1](✉), Dena Al Thani[1], Mustapha Makki[3],
Marwa Qaraqe[1], Kruthika Anand[2], Olcay Connor[2],
Kamran Khowaja[1], and Bilal Mansoor[3]

[1] Division of Information and Computing Technology,
College of Science and Engineering, Hamad Bin Khalifa University, Doha, Qatar
banire.bilikis.o@gmail.com
[2] Step by Step Center for Special Needs, Doha 47613, Qatar
[3] Mechanical Engineering Program, Texas A&M University at Doha,
Doha, Qatar

Abstract. Technological interventions for teaching children with autism spectrum disorders (ASD) are becoming popular due to their potentials for sustaining the attention of children with rich multimedia and repetitive functionalities. The degree of attentiveness to these technological interventions differs from one child to another due to variability in the spectrum. Therefore, an objective approach, as opposed to the subjective type of attention assessment, becomes essential for automatically monitoring attention in order to design and develop adaptive learning tools, as well as to support caregivers to evaluate learning tools. The analysis of facial expressions recently emerged as an objective method of measuring attention and participation levels of typical learners. However, few studies have examined facial expressions of children with ASD during an attention task. Thus, this study aims to evaluate existing facial expression parameters developed by "affectiva", a commercial engagement level measuring tool. We conducted fifteen experimental scenarios of 5 min each with 4 children with ASD and 4 typically developing children with an average age of 8.8 years, A desktop virtual reality-continuous performance task (VR-CPT) as attention stimuli and a webcam were used to stream real-time facial expressions. All the participants scored above average in the VR-CPT and the performance of the TD group was better than that of ASD. While 3 out of 10 facial expressions were prominent in the two groups, ASD group showed addition facial expression. Our findings showed that facial expression could serve as a biomarker for measuring attention differentiating the groups.

Keywords: Attention · Adaptive learning · ASD · Facial expression · Virtual reality · Affectiva

1 Introduction

Autism Spectrum Disorder (ASD) is a neurodevelopmental disorder characterized by a deficit in social communication and repetitive patterns of behavior [1]. They also exhibit an unusual pattern of attentional behaviors, such as difficulty in sustaining their

© Springer Nature Switzerland AG 2019
M. Antona and C. Stephanidis (Eds.): HCII 2019, LNCS 11573, pp. 32–48, 2019.
https://doi.org/10.1007/978-3-030-23563-5_4

attention [2]. According to reports from Center for Disease Control and Prevention (CDC) of United States, the prevalence of this disorder is relatively high and has been on the increase, moving from 1 in 110 children in the year 2000 to 1 in 68 children in 2014 [3]. The existing method of ASD diagnosis is conducted by a multi-disciplinary team consisting of specialists in a developmental pediatrician, child psychiatrist, and psychologist. Several instruments have been developed for history taking, play-based observation of the child, such as the autism diagnostic interview, the diagnostic interview for social and communication disorders (both semi-structured interviews) and the autism diagnostic observation schedule (a play based interactive assessment) [4]. Other professionals, such as a speech-language pathologist, occupational and behavioral therapists assess the communication, sensory and behavioral difficulties respectively. Attention skills assessments are using observational methods and are often subjective. Therefore, a need for objectively determining the attentional challenges of children with ASD is warranted.

As a result, studies have explored different technologies for educational and behavioral interventions to support attention span through objective measures. For example, Lahiri [5] found out that the individualized viewing patterns, eye movement, and task performance can be used in the design of a virtual reality application for teaching social skills. Using adaptive robotics for teaching social interaction to children with ASD was achieved by using specific head tilting of the children to make the robot understand their needs and respond accordingly [6]. The need for measuring attention for the design of adaptive learning system is not limited to people with attention deficit, but it is also utilized in the typical population as an experimental evaluation. For instance, Szafir [7] showed that adaptive robotic agent using behavioral techniques improved learners recall abilities and this design improved the learning outcomes in both groups. These interventions work differently for the children with ASD as some may require over-stimulating effect and others prefer the opposite [8] due to their high sensory processing demand as compared to the typical population. Therefore, attention assessment has always been a way of evaluating learning intervention and improving the learning experience.

In human-computer interaction, good learning design influence users' positive emotions and supports better learning outcome [9]. The evaluation of good learning outcome is usually measured based on the learners' attention, participation through task accuracy and time taken to finish the task. In this study, we proposed a more objective evaluation of facial expression as a measure of attention during a computer-based attention task in children with ASD and neurotypical peers. Moreover, we discussed how facial movement measure can affect the design of an adaptive learning system for children in the spectrum. This study hypothesized that the existing facial expression for measuring engagement levels by affectiva SDK can apply to typical children and will work differently for children with ASD. This hypothesis tailors our study to 2 research questions:

Research Question 1: What facial expressions are exhibited by children with ASD and neurotypical peers during attention task?

Research Question 2: Can facial expressions during attention task serve as an indicator to differentiate children with ASD and neurotypical peers?

2 Related Work

Attention assessment can provide great insight into how children with ASD learn as well as how and when their attention needs to be supported. Hence, teachers are keen on observing and taking notes of student's attention level and interaction objectively. Recently, objective techniques are being explored to automatically detect the attention of children with ASD during educational activities and ways to support their needs accordingly.

Objective approach is commonly used in adaptive learning (intelligent tutoring system) is becoming popular for children with ASD. In recent studies, different objective techniques are being used in the design of learning application with the ability to detect and reorient attention of children accordingly [5, 6, 10]. This approach of design is not limited to children with ASD but it is also utilized in children with hyperactive attention disorders who have similar attention problem with children with ASD [11] as well as typical children [12].

Several studies have looked at the direct measure of attention through the brain, heart rate and skin conductance using EEG (Electroencephalogram) headset and ECG (Electrocardiography) [13–17]. Primarily, these studies have considered obtrusive technology in typical population which gave better attention assessment, but there are chances of interference with the level of attention measured as the learner could be distracted with the thoughts of a foreign object on their body. A recent study conducted by [18] showed an adaptive learning system could be designed for children with ASD using EEG headset. This system gave a better assessment of attention as it measures their attention directly from the brain, but this approach may prove difficult to implement in children with ASD because of their sensitivity to touch and sensory processing disorder [19].

Other studies have considered touch-free technologies to measure attention such a using camera to measure head tilt [20], eyebrow raise, hand raise count [21] and facial expressions [22, 23]. These studies achieved success with the typical population. Recent technology by affectiva [24] identified ten basic facial expressions for measuring engagement levels. However, these expressions are yet to be explored in children with ASD and how these expressions affect design of adaptive learning system.

3 Method

3.1 Participants

This study used 8 participants, 4 children with ASD (3 boys and 1girl) and 4 neu-rotypical peers (2 boys and two girls) who are within the age range of 7 and 11 years. The ASD participants were recruited through an autism school. The ASD were diag-nosed of moderate ASD and this was verified with the reports with the school man-agement to ensure the students are eligible for our study. Further questions were asked by teachers to verify if the participants are not having any form of visual impairment or physical issues that may hinder the participants from taking the experiment.

3.2 Set-Up

Virtual reality continuous performance task (VR-CPT) was used as an attention task which mimics the conventional computerized version of continuous performance task (CPT) used to assess sustained and selective attention [25]. The first VR-CPT was created by [26] which was implemented with head mounted gear to create an immersive effect. The immersive VR-CPT has been successful in measuring attention in different studies for typical and children with ADHD [27]. Our version of VR-CPT was developed as non-immersive to make the experiment bearable for the population of our participants. Majority of the studies reviewed on virtual reality application for children of ASD used the desktop option to avoid the possibility of "cyber-sickness" and unusual head attachment so as not to influence their outcome of the intervention [28, 29]. The desktop VR-CPT presents the simulation of a conventional classroom which presents a teacher in front of the class, other students seated on the chair and desks, ceiling light, windows, a door and a blackboard where alphabets are displayed for 250 ms. Users are expected to interact with VR-CPT through a keyboard. To avoid further distraction through interaction, we improvised for the conventional keyboard with a simplified keyboard as seen in Fig. 1.

Fig. 1. Experimental set-up

The testing room consists of two monitors 25 in. and 34 in. for the participants and the researcher respectively. We have used a logi-tech webcam which was attached to the top of the 24 in. monitor for the participants. We conducted this experiment in a dimly light-room environment to prevent interference of the ceiling white light or rays of daylight. In addition, this room was isolated and free from external noise or distractions. The screen-based eye tracker and webcam used for objective attention assessment in this study are less likely to interfere with the research outcome due to its unobtrusiveness.

3.3 Task

There were 4 levels of the VR-CPT experiments, where all letters appear for a period of 250 ms on the classroom board. The participants were expected to focus and sustain their attention on boards irrespective of the actions going on in the virtual classroom. The children needed to click the simplified keyboard only when letter X appears. The first level of the VR-CPT had no distraction. The second scenario had minimal distractions with audio and visual such as coughing and students raising hands mainly from the center of the class. The third scenario had a medium level of distractions which were from the center and left-hand side of the classroom while the fourth level has the highest level of distraction from the left, right and center. All the participants took all the 4 experimental tasks except a participant from the ASD group who took 3 levels attention task.

3.4 Procedures

Parents of all participants were given a consent form which was approved by the institution review board committee. The form gave detailed information about the experiments and the rights of the participants throughout our study. After that, an experimental manual was used to describe how the experiment will go and how they are expected to interact with the software and hardware. We ensured all the participants got the same type of instructions and set-up. At the start of the experiment, the researcher welcomes the participant and caregiver to the room and engaged the child and the parents in a discussion to get the participant settled to the room environment. Then, the researcher gave VR-CPT instructions in form visuals and text on a hardcopy to explain to the participants on how to take the test and number of attention tasks they are expected to complete.

We induced the attention of the participants for an attention and non-attention task by presenting them with a blank screen and a VR-CPT on a desktop. All the participants took a demo of the VR-CPT before the main experiment to ensure they understood what they are expected to do. Each of the participants took a 3 to 4 min break after completing the second level and the total time for the attention tasks was 20 min per participants.

3.5   Data Collection

Facial expressions data collection and analysis was carried with a commercial software tool: iMotions embedded with affectiva [30]. The possibility of extending our research to other biometric measures later in future has influenced our choice of iMotions SDK for conducting our research. Affectiva software utilizes Histogram of Oriented Gradient (HOG) and Support Vector Machine Algorithm (SVM) to classify facial expression based on over 10,000 faces encoded manually across the world [31]. This classifier outputs the quantitative values of the facial expression of the participants using a likelihood value of 0–100% where 0 indicates absence of facial expression and 100 indicates the occurrence of an expression.

The sample data of the 15 and 16 experimental sessions from the ASD typical group respectively (31 tests in total) were collected using Logitech webcam and Affectiva SDK. Data from each experiment was sampled at 16 Hz thereby generating 16 samples of data per second, and each test took 300 s. The total sample for all participants across all the attention tasks taken was 148,800 samples (72,200 from ASD and 76,800 from typical groups). Each sample has 21 features of facial expression, 8 basic emotions and 98 raw features on facial landmarks. We investigated and analyzed 10 facial expressions for all the samples from the two groups of participants. The remaining facial expressions and 8 basic emotions were excluded. A custom python script was used to select the desired features and basic statistical analysis for all samples of each participant. The full description of the facial expressions can be found at [24, 30] while the description of the 10 facial expressions we have analyzed are given in Table 1.

Table 1. 10 facial expressions for measuring engagement level by "affectiva"

S/N	Facial expression	Description
1	Brow furrow	When the eyebrows moved closer and lowered together
2	Brow raise	When the eyebrows moved upwards
3	Lip corner depressor	When lip corners dropped downwards
4	Smile	When lip corners and cheek are pulled outwards and upwards
5	Nose wrinkle	When the nose skin is wrinkled and pulled upwards
6	Lip suck	When the lips are pulled inwards to the mouth
7	Lip press	When the lips are pressed together without pushing up the chin boss
8	Mouth open	When the upper and lower lips are apart
9	Chin raise	When the chin boss and the lower lip pushed upwards
10	Lip pucker	When the lips pushed forward

4 Results

We analyzed 15 experimental scenarios conducted with 8 participants each gets a commission score and omission scores which identify the right clicks (i.e., letter X) and wrong clicks (i.e., other letters aside X) respectively. Then we estimated the missed target (i.e., when the participant did not click letter X). These scores were used to evaluate their attention level. These scores were not shown to any of the participants until they completed the whole experiment in other not to influence the study by their mood especially when they have a low score.

4.1 Participants' Characteristics and Data Sample

8 children, 4 with mild ASD and 4 neurotypical peers, were enrolled in the study by taking a total of 31 experimental scenarios of attention tasks, where 7 of the participants took 4 experimental tests each except 1 who took 3, i.e., no-distraction, easy, medium and hard. Each caregiver of the participants presented the diagnosis report to confirm they were eligible for the experiment. All the participants are verbal and passed the basic skills test needed for the attention task which was letter recognition. The inclusion criteria for the ASD participants were diagnosis report and who can differentiate letters while the typical children scored below 15 in a Childhood Autism Spectrum Test (CAST) [32].

The average age and the standard deviation of the ASD group were 8.75 and 1.45 respectively, while those of the typical group were 8.25 and 1.30. The t-test conducted to check for the differences between the two groups was p = 0.68. This result indicated that there was no significant difference between the two groups. The summary of the participants is given in Table 2.

Table 2. Participants' demographics

Participants	Age	Sex	Test (#)	Nationalities	CAST (CPT) score
ASD					
P1	11	M	4	Indonesian	- (28)
P2	8	M	4	Indian	- (37)
P3	7	F	3	Sudanese	- (25)
P4	9	M	4	Lebanese	- (38)
AVG (8.75)					
SD (1.45)					
Typical					
T1	9	F	4	Pakistani	4(40)
T2	7	M	4	Pakistani	5(39)
T3	10	F	4	Yemeni	1(40)
T4	7	M	4	Palestinian	3(35)
AVG (8.25)					
SD (1.30)					

*CAST-(Childhood Autism Spectrum Test), * CPT-Continuous Performance Task

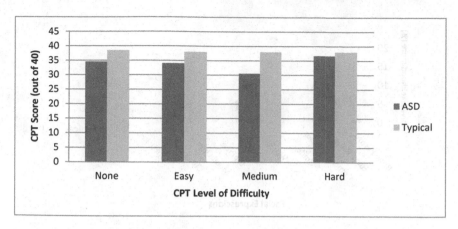

Fig. 2. CPT scores for ASD and typical

4.2 VR-Continuous Performance Task (CPT)

The scores for correct clicks, omitted letters, and incorrect clicks were saved for the analysis of the participants' attention level. We have only considered correct clicks scores in this study and excluded the scores for the omitted letters and incorrect clicks based on the scope of this study. The attention performances for all the participants were all above average as seen in Fig. 2.

In Fig. 2 above, children with ASD performed less than the control group in all the level of the CPT test, and the statistical inference using t-test gave a value of $p = 0.02$ stating there is a significant difference between the two groups when considering CPT (task performance) for measuring the level of attention.

4.3 Facial Expression

This study examined ten facial expressions that are attached to engagement which are: *brow furrow, brow raise, lip corner depressor, smile, nose wrinkle, lip suck, mouth open, chin raise and lip pucker* [affectiva]. This claim was based on the facial expression data for 3.2 million people from 75 countries. The hypothesis of this study is to investigate if the facial expressions attached to engagement apply to children with ASD during an attention task and how these facial expressions can influence the design of adaptive software learning for children in this spectrum. This study captured 4,800 quantitative samples of facial expression with over 40 features per participant during each level of the attention task to answer our research questions.

Research Question 1: *What facial expressions are exhibited by children with ASD and neurotypical peers during attention task?*

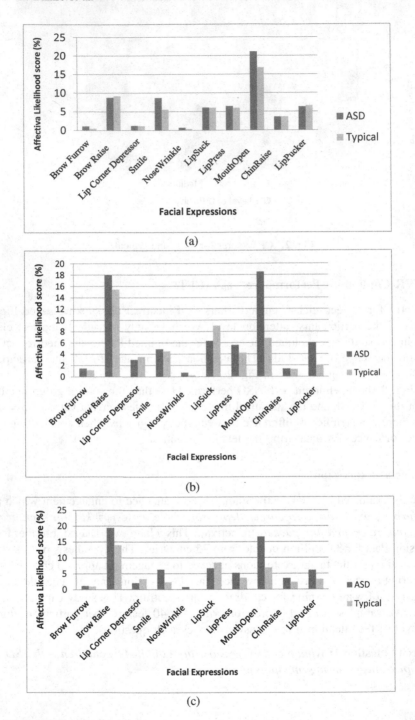

Fig. 3. (a) Facial expressions for both groups during attention task level 1 (none level). (b) Facial expressions for both groups during attention task level 2 (easy level). (c) Facial expressions for both groups during attention task level 3 (medium level). (d) Facial expressions for both groups during attention task level 4 (hard level)

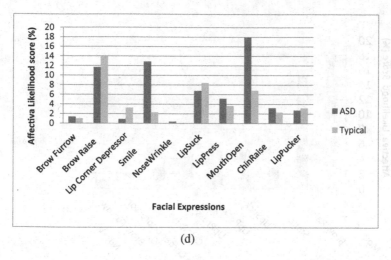

(d)

Fig. 3. (*continued*)

We identified the facial expression of all the participants from the two groups to determine if there were any differences. The result showed that the level of facial expressions differs in the two groups as shown in Fig. 3a–d.

We calculated the average value of all the facial expression likelihood score for all the participants in each group to understand the hierarchical order of the likelihood score. The result showed that brow raise and mouth open were a prominent facial expression in both groups but in opposite orders. The ASD group showed more of a mouth opened than that of the typical group while brow furrow nose wrinkle is less prominent and other varies differently between the groups as shown in Fig. 4a and b.

Research Question 2: *Can facial expressions during attention task serve as an indicator to differentiate children with ASD and neurotypical peers?*

The average value of affectiva likelihood score (between 0 and 100) of over 4,800 samples for each participant was taken to identify the frequencies of the expression. Then we chose the likelihood values that were above the median value for each participant as the prominent facial expressions. Lip press is appeared to be common in all the participants with ASD while it is different in the typical group. Although, lip press is common in half of the typical population the remaining half did not exhibit this action while paying attention. Afterward, we identified the facial expression common to each group and sorted them from largest to smallest as seen in Figs. 5 and 6.

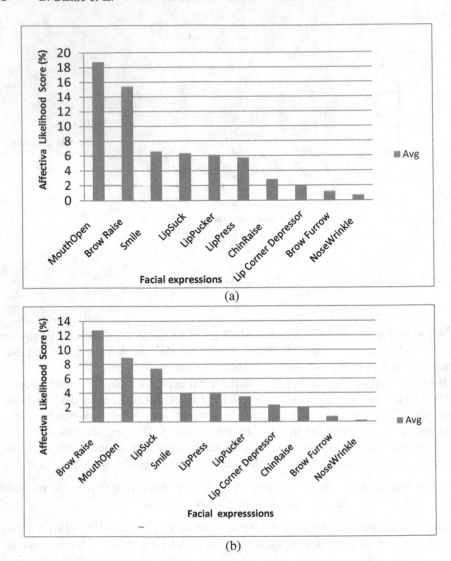

Fig. 4. (a) The hierarchical order of affectiva facial expressions in the ASD group. (b) The hierarchical order of affectiva facial expressions in typical group

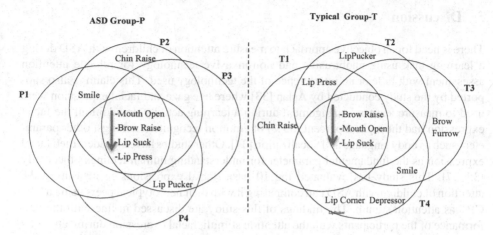

Fig. 5. Common facial expressions exhibited among the ASD and Typical groups for the 31 experiments

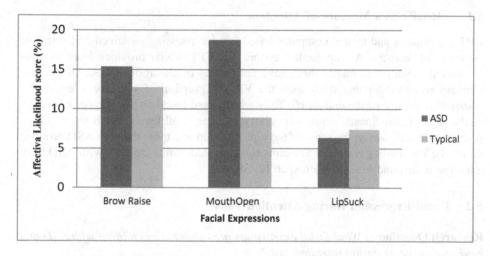

Fig. 6. Comparison of similar facial expression in both groups

In Fig. 6 above, children with ASD exhibited more facial expression with a brow raised and mouth open and less with lip suck than the control group when the "affectiva likelihood score" was averaged for all the entire CPT experimental tests. The statistical inference using t-test gave a value of $p = 0.40$ stating that there was no significant difference between the two groups when considering these prominent facial expressions for measuring the level of attention. However, these findings need to be investigated with a larger sample.

5   Discussion

There is need for an objective approach to measure attention of children with ASD during a learning task using unobtrusive and non-invasive technology for reliable attention assessment with little or no interference of the technology used. This claim is also supported by the study conducted by Aslan [33] where eye gaze and facial expression were used to measure learners' engagement during a learning activity. In addition, the facial expression had the highest percent (55%) of emotion recognition amongst other parameters such as body language and vocalization [34]. Other studies have also identified facial expression as the fundamental parameter for understanding human feelings objectively [35–37]. This study has evaluated the 10 basic facial expressions in measuring the attention of children with ASD and comparing it with their neurotypical peers using a VR-CPT as attention stimuli. The findings of this study are discussed in line with the performance of the participants with the attention stimuli, facial expression during attention task and how these measures differentiate children with ASD from their neurotypical peers. Then we discussed the design implication for an adaptive learning system.

5.1   VR-CPT as a Measure of Attention

CPT is a popular and trusted computer-based test for assessing sustained and selective attention of learners. A step further is the, VR-CPT which provides learners with ecological validity that makes them have the feeling of classroom tasks. All the participants recruited for this study took the VR-CPT version of 4 different levels *(No distraction, easy, medium and hard)*. They were scored based on the number of correct clicks, and all participants' score were above average in all levels which showed they paid attention. However, the score of the typical group was more than the ASD group at all levels. This finding is similar to claims by other studies that children with ASD have attention deficit and low attention span [2, 38, 39].

5.2   Facial Expression During Attention Task

Research Question 1: *What facial expressions are exhibited by children with ASD and neurotypical peers during attention task?*

The findings from this study showed that 4 out of the 10 facial expressions are common in children with ASD which are: mouth open, brow raise, lip suck, and lip press. Similar facial expressions are prominent in the neurotypical group except for lip press. Hence, 3 facial expressions (mouth open, brow raise, lip suck) are common in both groups during an attention task. This result is consistent with the attention measuring states of learners by Asteriadis [40] where mouth open and brow movement were considered for assessing the attention-related states. Another study by Ross [21] found out that eyebrow raise count and hand raise count are useful for measuring attention level. Our findings are in relation with existing studies shows that among the basic regions of the face which are: eyes, nose, cheek and mouth, the important region of the face that correlates attention in children with ASD and neurotypical peers are the *eye region* and *mouth region.*

Research Question 2: Can facial expression during attention task serves as an indicator to differentiate children with ASD and neurotypical peers?

According to the result of our experiments with two sets of participants, 3 facial expressions (mouth open, brow raise, lip suck) are common in both groups during the attention task. However, the ASD group exhibited more of mouth open and brow raise than the neurotypical group but the differences in these 3 facial expressions are not significant to differentiate the groups. This finding also negates our hypothesis as we expected the facial expression of children with ASD to be different from that of neurotypical peers based on the findings of existing studies [41–43]. However, this finding is only based on mild ASD group who require the least support as compared to severe and moderate ASD. There may be differences in the facial expression of moderate and severe ASD with typical children. Nevertheless, lip press was exhibited by all the participants in the ASD group while half of the neurotypical group showed lip press. This finding may give a clearer picture if this study is repeated with more samples from both groups.

5.3 Design Implication

Assessment of attention deficits in children with ASD is often subjective. This pilot study helps bring objectivity to this measure. Though children with ASD have difficulty inferring other people's emotions and expressing the same, there are few facial expressions that are common between children with ASD and typically developing children. Further analysis of these common expressions displays the difference between the two groups which helps pinpoint the difficulties children with ASD face with regards to attention.

The facial expressions exhibited by children with ASD during attention tasks have significant educational and behavioral value, both diagnostically and therapeutically. It brings objectivity to the assessment of attention in children with ASD. Based on this, therapeutic programs (for learning and communication) can be tailor-made for children with ASD keeping in mind the facial expressions. Additionally, facial expressions can be used to design adaptive learning software for children with ASD, which can be very helpful in their academics.

6 Conclusion

This pilot study evaluated how existing facial expression parameters can be used to measure the attention of children with ASD and neurotypical peers. We also looked at how facial expressions can be used in the design of adaptive learning software for children with ASD. In line with our findings, we proposed 4 facial expressions: *mouth open, brow raise, lip suck, and lip press* as parameters for measuring the level of attention in children with ASD during learning. These 4 facial expressions can be embedded in the design of adaptive learning software for children on the spectrum. In the control group (TD), similar facial expression were identified except for lip press which is only associated with ASD group. Although, both groups exhibited almost similar facial expressions during the attention task but differ in one of the expression. Hence, lip press needs to be further investigated as a biomarker for differing ASD and typical children. Considering the small sample size in this study, there is a need for a more detailed evaluation of the effectiveness of facial expressions and the practicality of using the same approach in clinical practice and research. Additionally, the study can be extended to children with Attention Deficit/Hyperactivity Disorder, where attention is the deciding factor.

References

1. American Psychiatric Association: Diagnostic and Statistical Manual of Mental Disorders (DSM-5®). American Psychiatric Publishing (2013)
2. Patten, E., Watson, L.R.: Interventions targeting attention in young children with autism. Am. J. Speech-Lang. Pathol. **20**(1), 60–69 (2011)
3. Centers for Disease Control and Prevention: CDC estimates 1 in 68 children has been identified with autism spectrum disorder, 27 March 2014. https://www.cdc.gov/media/releases/2014/p0327-autism-spectrum-disorder.html
4. Baird, G., Cass, H., Slonims, V.: Diagnosis of autism. BMJ **327**(7413), 488–493 (2003)
5. Lahiri, U., et al.: Design of a virtual reality based adaptive response technology for children with autism. IEEE Trans. Neural Syst. Rehabil. Eng. **21**(1), 55–64 (2013)
6. Esubalew, T., et al.: A step towards developing adaptive robot-mediated intervention architecture (ARIA) for children with autism. IEEE Trans. Neural Syst. Rehabil. Eng. **21**(2), 289–299 (2013)
7. Szafir, D., Mutlu, B.: Pay attention!: designing adaptive agents that monitor and improve user engagement. In: Proceedings of the SIGCHI Conference on Human Factors in Computing Systems. ACM (2012)
8. Wetherby, A.M., Prizant, B.M.: Autism Spectrum Disorders: A Transactional Developmental Perspective, vol. 9. Brookes Publishing, Baltimore (2000)
9. Norman, D.: Emotion & Design: Attractive Things Work Better. Interactions **9**(4), 36–42 (2002)
10. Escobedo, L., et al.: Using augmented reality to help children with autism stay focused. IEEE Pervasive Comput. **13**(1), 38–46 (2014)
11. Sonne, T., Obel, C., Grønbæk, K.: Designing real time assistive technologies: a study of children with ADHD. In: Proceedings of the Annual Meeting of the Australian Special Interest Group for Computer Human Interaction. ACM (2015)
12. Mana, N., Mich, O.: Towards the design of technology for measuring and capturing children's attention on e-learning tasks. In: Proceedings of the 12th International Conference on Interaction Design and Children. ACM (2013)
13. Huang, R.S., Jung, T.P., Makeig, S.: Multi-scale EEG brain dynamics during sustained attention tasks. In: 2007 IEEE International Conference on Acoustics, Speech and Signal Processing - ICASSP 2007 (2007)
14. Ghassemi, F., et al.: Classification of sustained attention level based on morphological features of EEG's independent components, pp. 1–6 (2009)
15. Hamadicharef, B., et al.: Learning EEG-based spectral-spatial patterns for attention level measurement. In: 2009 IEEE International Symposium on Circuits and Systems (2009)
16. Silva, C.S., Principe, J.C., Keil, A.: A novel methodology to quantify dense EEG in cognitive tasks. In: 2017 IEEE International Conference on Acoustics, Speech and Signal Processing (ICASSP) (2017)

17. Belle, A., Hobson, R., Najarian, K.: A physiological signal processing system for optimal engagement and attention detection. In: 2011 IEEE International Conference on Bioinformatics and Biomedicine Workshops (BIBMW) (2011)
18. Zheng, C., et al.: An EEG-based adaptive training system for ASD children. In: UIST 2017 Adjunct - Adjunct Publication of the 30th Annual ACM Symposium on User Interface Software and Technology (2017)
19. Schafer, E.C., et al.: Personal FM systems for children with autism spectrum disorders (ASD) and/or attention-deficit hyperactivity disorder (ADHD): an initial investigation. J. Commun. Disord. 46(1), 30–52 (2013)
20. McCuaig, J., Pearlstein, M., Judd, A.: Detecting learner frustration: towards mainstream use cases. In: Aleven, V., Kay, J., Mostow, J. (eds.) ITS 2010. LNCS, vol. 6095, pp. 21–30. Springer, Heidelberg (2010). https://doi.org/10.1007/978-3-642-13437-1_3
21. Ross, M., et al.: Using support vector machines to classify student attentiveness for the development of personalized learning systems. In: 2013 12th International Conference on Machine Learning and Applications (ICMLA). IEEE (2013)
22. D'Mello, S.K., Graesser, A.: Multimodal semi-automated affect detection from conversational cues, gross body language, and facial features. User Model. User-Adap. Inter. 20(2), 147–187 (2010)
23. Yüce, A., et al.: Action units and their cross-correlations for prediction of cognitive load during driving. IEEE Trans. Affect. Comput. 8(2), 161–175 (2017)
24. Affectiva. Mapping Expressions to Emotions (2018). https://developer.affectiva.com/mapping-expressions-to-emotions/. Accessed 28 Jan 2019
25. Katona, J.: Examination and comparison of the EEG based attention test with CPT and T.O.V.A. In: 2014 IEEE 15th International Symposium on Computational Intelligence and Informatics (CINTI) (2014)
26. Rizzo, A.A., et al.: A virtual reality scenario for all seasons: the virtual classroom. CNS Spectr. 11(1), 35–44 (2009)
27. Díaz-Orueta, U., et al.: AULA virtual reality test as an attention measure: convergent validity with Conners' continuous performance test. Child Neuropsychol. 20(3), 328–342 (2014)
28. Bellani, M., et al.: Virtual reality in autism: state of the art. Epidemiol. Psychiatric Sci. 20(03), 235–238 (2011)
29. Parsons, S., Cobb, S.: State-of-the-art of virtual reality technologies for children on the autism spectrum. Eur. J. Spec. Needs Educ. 26(3), 355–366 (2011)
30. iMotions, iMotions and Affectiva (2012)
31. Senechal, T., McDuff, D., Kaliouby, R.: Facial action unit detection using active learning and an efficient non-linear kernel approximation. In: Proceedings of the IEEE International Conference on Computer Vision Workshops (2015)
32. Williams, J., et al.: The CAST (childhood asperger syndrome test) test accuracy. Autism 9(1), 45–68 (2005)
33. Aslan, S., et al.: Learner engagement measurement and classification in 1: 1 learning. In: 2014 13th International Conference on Machine Learning and Applications (ICMLA). IEEE (2014)
34. Mehrabian, A., Wiener, M.: Decoding of inconsistent communications. J. Pers. Soc. Psychol. 6(1), 109 (1967)
35. Senechal, T., et al.: Facial action recognition combining heterogeneous features via multikernel learning. IEEE Trans. Syst. Man Cybern. Part B (Cybern.) 42(4), 993–1005 (2012)
36. Janssen, J.H., et al.: Machines outperform laypersons in recognizing emotions elicited by autobiographical recollection. Hum.-Comput. Interact. 28(6), 479–517 (2013)

37. Tariq, U., et al.: Recognizing emotions from an ensemble of features. IEEE Trans. Syst. Man Cybern. Part B (Cybern.) **42**(4), 1017–1026 (2012)
38. Higuchi, T., et al.: Spatiotemporal characteristics of gaze of children with autism spectrum disorders while looking at classroom scenes. PLoS ONE **12**(5), e0175912 (2017)
39. Kinnealey, M., et al.: Effect of classroom modification on attention and engagement of students with autism or dyspraxia. Am. J. Occup. Ther. **66**(5), 511–519 (2012)
40. Asteriadis, S., et al.: Estimation of behavioral user state based on eye gaze and head pose—application in an e-learning environment. Multimedia Tools Appl. **41**(3), 469–493 (2009)
41. Bieberich, A.A., Morgan, S.B.: Self-regulation and affective expression during play in children with autism or down syndrome: a short-term longitudinal study. J. Autism Dev. Disord. **34**(4), 439–448 (2004)
42. Czapinski, P., Bryson, S.E.: 9. Reduced facial muscle movements in autism: evidence for dysfunction in the neuromuscular pathway? Brain Cogn. **51**(2), 177–179 (2003)
43. Chu, H.C., Tsai, W.W.J., Liao, M.J., Chen, Y.M.: Facial emotion recognition with transition detection for students with high-functioning autism in adaptive e-learning. Soft Comput. 1–27 (2018)

Improving Usability of a Mobile Application for Children with Autism Spectrum Disorder Using Heuristic Evaluation

Murilo C. Camargo[1](✉), Tathia C. P. Carvalho[2], Rodolfo M. Barros[1], Vanessa T. O. Barros[1], and Matheus Santana[1]

[1] State University of Londrina, Londrina, PR 86057 970, Brazil
murilocrivellaric@gmail.com,
rodolfomdebarros@gmail.com, vanessa@uel.br,
ss.matheus.94@gmail.com
[2] São Paulo State University, Bauru, SP 17033 360, Brazil
tathiacarvalho@gmail.com

Abstract. Autism Spectrum Disorder (ASD) is a complex clinical condition that includes social, behavioral, and communication deficits. As numbers in ASD prevalence rise significantly, the tools for computer-assisted interventions also increase proportionally, which can be confirmed by the growth in the literature body addressing the issue. The development of autism-specific software is far from being straightforward: it often requires a user-centered approach, with a cross-functional team, and a primary focus on usability and accessibility. One of the most popular methods for finding usability problems is the heuristic evaluation, which is performed by having a group of experts testing the User Interface and providing feedback based on predetermined acceptance criteria. Thus, this paper informs on the assessment of a mobile application for autistic individuals using the heuristic evaluation. The software subjected to evaluation – prototyped in a previous study – addresses organization and behavioral patterns in ASD children. Through the heuristic evaluation, improvements could be performed in the application. Also, lessons learned with the evaluation process include recommendations to help the selection of methods and materials, the conduction of the evaluation, and the definition of the follow-up strategy. By describing the method stepwise and sharing lessons learned, the aim is to provide knowledgeable insights for development teams handling autism-specific software.

Keywords: Heuristic evaluation · Usability · Autism Spectrum Disorder · Mobile application

1 Introduction

Autism Spectrum Disorder (ASD) is a cognitive disturbance that refers to a variety of severity degrees and symptoms – including social, behavioral, and communication deficits [1]. The fifth version of the Diagnostic and Statistical Manual of Mental Disorders (DSM-5), provided by the American Psychiatric Association [2], expanded

© Springer Nature Switzerland AG 2019
M. Antona and C. Stephanidis (Eds.): HCII 2019, LNCS 11573, pp. 49–63, 2019.
https://doi.org/10.1007/978-3-030-23563-5_5

the comprehensiveness on the disturbance, in an attempt to add flexibility to the diagnosis criteria and support health care practices. Especially in the last decade, ASD numbers rose significantly. For instance, in the United States, according to the Autism and Developmental Disabilities Monitoring (ADDM) network, about 1 in 68 children aged 8 has been diagnosed with ASD [3].

Studies report that ASD individuals are especially interested in visual media, due to their frequently high visual processing skills [4, 5]. Based on this preference, many studies have been emerging on the development of computer-assisted interventions for ASD individuals, including communication and social interaction support [6, 7], and applying multiple technologies such as Virtual Reality [8] or Kinect [9]. Computer-assisted interventions provide users with a controlled environment, which can be personalized according to their needs, in order to foster independence and motivation. Though the use of computers represents a novel strategy to improve the welfare of people with ASD, the development of accessible technology is complex: it requires a user-centered approach – preferably with a cross-functional team – and a primary focus on usability, accessibility, and the adaptation of internal procedures and policies [10].

In a previous study, we described the first stages in designing a mobile application for ASD children – named DayByDay [11]. Now, considering the importance of usability throughout the entire development process, we intend to go one step further and improve the usability of our application through a heuristic evaluation. The heuristic evaluation is a well-known method for finding usability problems by having a small group of experts testing the interface and describing design issues based on certain rules (the heuristics). The method is sometimes criticized because it focuses on finding problems rather than evaluating the user performance. Of course, it does not replace user testing. However, this approach is particularly interesting in intermediate development stages: it encourages evaluators to describe specific usability problems and connect them with objective assessment criteria, thus guiding the development team towards a high-quality software product [12].

Multiple sets of heuristics are available in the recent literature [12–15]. In general, they derive from the original set of usability heuristics proposed by Jakob Nielsen in 1994. In this study, we use the set of heuristics described by Islam and Bouwman [15]. The authors propose a semiotic framework for interface evaluation consisting of five major areas: syntactic, pragmatic, social, environmental, and semantic aspects. Those major areas are divided in 16 topics of interest. We believe this approach enables evaluators to cover the entire interface in a comprehensive manner, and at the same time provides an easier vocabulary, which avoids interpretation errors. The full description of the usability heuristics is presented within this paper in a section exclusively dedicated to the method.

This work is structured as follows: Sect. 2 provides a brief context on the current status of the mobile application (DayByDay); Sect. 3 describes the evaluation method applied in terms of the usability heuristics set, planning, participants, materials, and procedures; Sect. 4 presents the results of the evaluation process; Sect. 5 presents the improvements performed on the application, and a comparison between the former version of the interface and how it changed after the heuristic evaluation; Sect. 6 discusses the lessons learned with the process; and Sect. 7 provides conclusions and future work directions. By presenting the heuristic evaluation process and discussing lessons learned targeted specifically to the ASD context, this paper can positively assist development teams handling accessibility.

2 Previous Work: DayByDay

The first stage in the development process of our mobile application was described in a previous study [11]. The application – named DayByDay – aims at helping ASD children aged 8 to 12 organize their routine and perform daily activities, such as "have breakfast", "brush the teeth", or "make your bed". Children must select an activity from a list displayed on the screen and follow the steps in order to complete it. For instance, the activity "have breakfast" may include the steps "pick a bowl", "add cereal", "pour some milk", and "eat the meal". By completing activities, children earn points (represented by stars, in this case). After collecting a number of stars, they are rewarded with a real-life token of their choice (a game, a leisure activity, some candy, etc.).

Parents, educators, and therapists are also encouraged to participate. In the "caregiver management system" within the app, parents and educators are asked to tailor content-specific features, such as the activities to be performed, the sub-steps in each activity, and the list of rewards available to the ASD children. In the first version of the application, there are options to upload images, use the built-in camera to take pictures, or record voice messages. In addition, they can monitor the children's progress through push notifications and a control center. One of the main concerns of the earliest version is to provide feedback to the ASD users. This is achieved by two different strategies: one, by earning stars each time they successfully complete an activity (visual feedback); and two, by receiving messages (either written or recorded) and real-life rewards from their caregivers after their performance.

Personalization is also a topic of major importance. Autism deals with a variety of severity degrees, which means there are considerable differences in interests and skills. In this context, parents and educators are entitled to select the most suitable output channels from the multimedia options provided – upload images, take pictures, record voice messages, write memos, organize the scheduled activities, or a combination of those. Besides that, the reward system intends to foster motivation and reinforce desired behavioral patterns when the ASD children successfully complete an activity. This first version of the app was prototyped and presented to parents of ASD children in a focus group session. In addition to that, a questionnaire was also applied – results show a positive acceptance overall, with only a few minor improvement suggestions. A comprehensive explanation on the first version is available in [11].

Figure 1 shows some of the screens designed for the first version. The screen to the left is the list of activities presented to the child. Activities marked with a "check" are completed, whilst the ones with a "padlock" are not available at the moment. In the center, there is an example of the activity "change clothes". In this case, pictures of the pieces of clothes were uploaded to help the child. Finally, the screen to the right is showed after the completion of an activity. It includes the stars earned so far, a message from the parents, and action buttons. The cloudy background was designed to add playfulness and draw the children's attention. Also, the background color would change according to the time of the day: light blue for mornings, sunset orange for afternoons, and a purple sky at night.

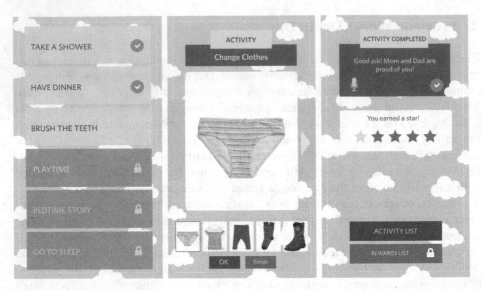

Fig. 1. DayByDay screens, first version. List of activities (left), activity steps (center), and activity completed screen (right).

3 Method

Even though questionnaires and focus groups are important sources of feedback, these approaches have significant limitations. On the one hand, respondents tend to answer favorably to the object of interest (in this case, the interface) when they are confronted face to face. On the other hand, in case interviewees are not able to express their knowledge in words, what they say is not necessarily what they mean [16]. In a heuristic evaluation, evaluators tend to be more comfortable to give their opinions and are more likely to provide "negative" feedback, that is, list usability problems. This method also makes the evaluation process easier, because evaluators are not expected to provide complex explanations, but rather connect the usability problems they find with a standardized list of principles. Furthermore, the first part of the assessment (described in [11]) was conducted only with parents of ASD children. Involving experts and combining multiple evaluation techniques is a good approach to have a multidisciplinary overview from different fields of knowledge [10].

Generally, there are two evaluation methods applying usability heuristics: user testing (testing involves end-users), and usability inspection (also called "expert evaluation", does not involve end-users) [12]. In this study, we conduct a usability inspection because we want to improve usability and prevent errors prior to having end-users testing the interface. The usability inspection involves a small set of evaluators, who assess the interface and judge its compliance with recognized usability principles, namely the heuristics.

3.1 Usability Heuristics Set

Heuristics are general rules that describe common properties of usable interfaces. This study uses the Semiotic Interface sign Design and Evaluation (SIDE) framework, proposed by [15]. SIDE consists of five major areas (syntactic, pragmatic, social, environmental, and semantic aspects), divided in 16 minor topics. The framework properties are summarized below:

Syntactic Evaluation. This area involves the evaluation of primary visual/graphic aspects, namely: interactivity, color, clarity and readability, presentation, context, and consistency.

1. Interactivity: Related to the level of interactivity in the interface and how it connects to the user previous knowledge in digital environments;
2. Color: The colors used in the interface, how light the colors are, and whether there is enough color contrast between elements;
3. Clarity and readability: Elements in the interface must be clear, concise, and easy to understand. Text must be informational and easy to read;
4. Presentation: Related to the overall appearance of the interface and the structure of the elements (layout, arrangement, font size, etc.);
5. Context: Measures the level of accuracy in the interpretation of the web domain and/or the application name, and how much sense users make of it;
6. Consistency: Related to the design patterns used throughout the interface that can help users in the interpretation process.

Pragmatic Evaluation. Also related to visual/graphic interface aspects, this area evaluates how the interface elements influence the user perception of it. Topics of interest include the position, amplification, relations, and coherence.

1. Position: Elements must be placed in common positions, with respect to the user habits in digital environments, in order to favor the user localization;
2. Amplification: This attribute is concerned with elements that can be amplified – such as icons, thumbnails, small images, short text – and whether users understand their "abbreviated" meaning;
3. Relations: Relationships between elements (action buttons, hyperlinks, text/image) must allow a "cause-and-effect" interpretation;
4. Coherence: Digital elements must match real-world conventions, so that users can infer relationships in a logical manner.

Social Evaluation. This area assesses the digital environment in relation to the social context and how well the interface is organized in terms of cultural representations. It includes: cultural marker, matching, organization, and mapping.

1. Cultural marker: Refers to the use of elements (colors, language, iconic representations, etc.) within a specific cultural context;

2. Matching: Proper use of digital metaphors to express reality, conventions, and real-world objects;
3. Organization: Related to how the content is organized in categories within the interface and whether that organization favors an easy and safe navigation;
4. Mapping: In case the interface deals with complex concepts or activities, it must apply a "generalization" (simple-to-complex) approach.

Environmental Evaluation. The environmental part of the evaluation is concerned with the ontologies used to universalize the interface.

1. Ontology: The interface must include universalization traits, so that users are able to interpret the referential meaning of objects.

Semantic Evaluation. Finally, the semantic evaluation includes the interpretation accuracy, which is measured in five levels.

1. Interpretation accuracy: Measures the accuracy of the user interpretation in relation to the interface design. The five possible levels are – accurate, moderate, conflicting, erroneous, or incapable.

3.2 Participants and Materials

The application was prototyped using Marvel [17]. It is an easy-to-use online software that allows static screens to become fairly interactive without codification. Of course, Marvel does not integrate back-end components or server/database solutions. However, it does provide a full front-end experience with all functional and interactive capabilities, thus allowing the evaluator to inspect the application in a smartphone. Working with a mobile prototype enables evaluators to perform a more accurate inspection whereas the end-user will be running the application in mobile devices.

In order to gather feedback from different fields of knowledge, DayByDay evaluation was performed with a cross-functional team, involving design, special education, and computer science experts. Six evaluators (E) formed the evaluation team: (E1) a computer science professional specialized in mobile development, (E2) a project manager dealing with accessible software development, (E3) a user experience designer, (E4) a graphic designer with experience in accessible applications, (E5) a teacher working with accessibility in Basic Education, and (E6) a university professor who researches accessibility for autism.

3.3 Procedures

During the evaluation session, the evaluator is expected to go through the interface several times, inspecting and comparing the elements with a list of recognized usability principles (the heuristics). Prior to the beginning of the evaluation, a hard copy of the sixteen heuristics was provided for each evaluator. Then, the researcher presented directions for the inspection. In this heuristic evaluation, participants are asked to perform activities in two given scenarios.

In the first scenario, evaluators should complete an activity designed for the ASD audience, namely "have breakfast". In order to start, the evaluator should choose the activity. It consists of four steps: "pick a bowl", "add cereal", "pour milk", and "eat the meal". Participants should progress through all the steps in order to complete the activity. In the second scenario, evaluators pretend they are caregivers of ASD children, and are requested to create and upload a new activity to be displayed in the child's application. This task is more complex, since it involves more options and functionalities. First, evaluators must sign in on the application. This action requires filling a form with personal information and choosing between default personalization options. Then, they are asked to create a new activity, which consists of four steps: select a date to insert the activity; write the name of the activity; upload, from the phone gallery, pictures to be used in the step-by-step; and confirm the actions.

In both scenarios, despite the fact that evaluators had specific tasks to complete, they were also encouraged to explore other sections within the interface. They could, for instance, access the "settings" and navigate through the various personalization options available. It was suggested that evaluators performed the tasks in both scenarios at least twice. In the first time, they would have an overview of the application, learn the mechanics and try to complete the task. In the second time, they should inspect every component individually, and provide detailed feedback according to the list of heuristics. However, evaluators could play with the interface as many times as they wished. There was no time limit either. They are allowed to test the prototype freely as long as they still have reasons to do so.

Prior to the beginning of the session, the researcher provided a brief explanation on the application and about the tasks the evaluators were expected to accomplish in each scenario. After that, the evaluation itself proceeded without further assistance during the evaluator's performance. Evaluators inspected the interface individually – that is, one at a time in the evaluation room – and listed the usability problems they found based on the heuristics set. Furthermore, each participant was asked to rate the problems in a scale from 1 to 5, as described below:

- Rate 1: Not a usability problem;
- Rate 2: Cosmetic problem. Should only be fixed if there is extra time available;
- Rate 3: Minor usability problem with low priority to fixing;
- Rate 4: Major usability problem with high priority and important to fix;
- Rate 5: Usability catastrophe that needs urgent fixing.

Feedback collected with each evaluation session was compiled in individual evaluation reports, which were then analyzed by the researchers. Together, they provide meaningful insights on usability problems that may assist the development team in fixing bugs and preventing errors. Results of the evaluation are presented in the next section.

4 Results

By choosing experts from various fields of knowledge, the heuristic evaluation collected high-quality feedback on multiple aspects of the interface, such as graphic elements, implementation issues, non-functional attributes, teaching-learning adaptation, and so on. After the evaluation sessions, results were compiled in evaluation reports and analyzed. Tables 1, 2, 3 and 4 summarize the findings from the heuristic evaluation. In addition to analyzing each evaluator's results individually (E1, E2, E3, E4, E5, E6), we also consider the overall number of problems (simply indicated as E).

Table 1. Number of usability problems divided by evaluator.

Numbers refer to:	E	E1	E2	E3	E4	E5	E6
Child's interface	8	2	0	3	0	3	0
Caregiver system	49	11	7	9	12	5	5
Overall interface	57	13	7	12	12	8	5

Table 1 presents the number of usability problems found by all six evaluators. A total of 57 problems were detected. The caregiver system (used to create and manage activities) had the majority of detected issues (84.2%), whilst the child's interface (used by the ASD children to perform activities) had only 8 problems. E2, E4 and E6 did not find problems in the child's interface. Overall, the problems are fairly distributed among all the evaluators, but are more prominent in the caregiver system, suggesting that this portion of the interface needs priority.

Problems were ranked according to their severity in a scale from 1 to 5. The ranking is demonstrated in Table 2. Most problems were ranked 4 (24.5%) and 1 (22.8%), but ranks 2, 3 and 5 also appeared significantly. Evaluators E1, E3, E4, and E6 indicated most problems were medium to high priority (ranks 3, 4, and 5), in contrast with the feedback received from evaluators E2 and E5, who ranked problems mainly as 1 or 2 (lowest to low priority).

Finally, Tables 3 and 4 present the usability problems ranked and divided in the two parts of the interface. Table 3 summarizes the data related to the child's interface, while Table 4 presents the ranking for the caregiver system. E1 reported to have found only minor problems in the child's interface, but considerably severe problems in the caregiver system (ranks 4 and 5). E2 did not find any problems in the child's interface, and neither did E4 and E6. A few problems were reported in the child's interface by three evaluators (E1, E3, and E5). In the caregiver system, in turn, the number of problems was higher and they were considered more critical (ranks 3, 4, and 5). If we take into account the area of expertise, the majority of problems are related to implementation issues (25 problems – 43.8% – reported by science computer professionals E1 and E4). In second place, reports suggest problems in the educational domain (17 problems, or 29.8%, reported by E3 and E6). Finally, the least amount of problems is from the interface design (15 problems, or 26.3%, reported by E2 and E5).

Table 2. Number of usability problems divided by severity level.

Severity level	E	E1	E2	E3	E4	E5	E6
1 (lowest)	13	2	3	1	2	4	1
2 (low)	10	2	2	1	2	2	1
3 (average)	11	2	0	3	5	1	0
4 (high)	14	4	1	4	2	1	2
5 (highest)	9	3	1	3	1	0	1

Table 3. Usability problems (child's interface).

Severity level	E1	E2	E3	E4	E5	E6
1 (lowest)	1	0	0	0	2	0
2 (low)	0	0	1	0	1	0
3 (average)	0	0	2	0	0	0
4 (high)	1	0	0	0	0	0
5 (highest)	0	0	0	0	0	0

Table 4. Usability problems (caregiver system).

Severity level	E1	E2	E3	E4	E5	E6
1 (lowest)	1	3	1	2	2	1
2 (low)	2	2	0	2	1	1
3 (average)	2	0	1	5	1	0
4 (high)	3	1	4	2	1	2
5 (highest)	3	1	3	1	0	1

In order to be fixed, problems should be prioritized accordingly. Table 5 lists the main problems – divided by severity levels – reported by the evaluators. According to the scale used in this study, rank 1 is the lowest priority level and 5 is the highest. In some cases, the same problem was reported by more than one participant. Furthermore, very similar problems were found in several screens (for instance, a button that does not follow design patterns and appears multiple times throughout the interface). For this reason, similar problems appear only once. In addition, each problem is related to the corresponding principle from the heuristics set.

5   Application Improvements

After the heuristic evaluation, developers analyzed carefully the feedback collected and planned improvements in the application. Unfortunately, some of the expert recommendations could not be performed due to technology and/or human resources limitations. However, the development team did try other solutions when experts' suggestions were not possible to implement. One of the major problems detected

Table 5. Usability problems description.

Severity level	Problem description	Problem refers to:
1 (lowest)	"The interface should include a video tutorial"	Interactivity
	Add options to sign in via Facebook or Google	Interactivity
2 (low)	Some of the icons used in the interface may be confusing	Matching
	"The sign in form is too long and I do not know whether it refers to the caregiver or the child"	Organization
	"The interface could include a gallery with default pictures"	Interactivity
	Font size is too small	Clarity and readability
3 (average)	Review buttons "save" and "share" position	Position
	Some buttons do not follow web conventions	Consistency
	Add filters and organization options to the search tool	Organization
	Replace the button "age" for "date of birth"	Interpretation accuracy
4 (high)	"The colors used in the interface may not be the best to fit every user's preferences"	Color
	"There is too much text information on the interface. Many ASD children cannot read"	Clarity and readability
	Section titles are too long	Clarity and readability
	Include labels to all input fields	Presentation
	Signing in is a long and exhausting task	Mapping
	It is difficult to find the buttons to go to the next step	Organization
5 (highest)	"The rewards system must be revised. It takes too long for the child to get the reward, and they may not be able to associate it with success"	Relations
	Remove "delete" option for images. It may cause an accidental exclusion	Presentation
	"No fill" should not be a color option	Color
	Headings and sub-headings need to be clearer. It is difficult to understand what they imply	Organization

during the heuristic evaluation is the use of text. In fact, autism covers a broad spectrum of symptoms, and ASD individuals are often not able to read. This is why a picture-oriented approach is more effective (Fig. 2).

In the first version of the application, the list of activities was presented mainly with the help of text. In the improved version, however, pictures replaced text, also including icons to help the visualization and the understanding. The interface is as simple as possible, to avoid overexposure to unnecessary information. Pictures were used to replace text throughout the application, keeping written information to a bare minimum. In the activity steps, for instance, images take over almost the entire screen

Fig. 2. DayByDay screens. List of activities, first version (left); list of activities, improved version (center); activity step, improved version (right).

(Fig. 2, picture to the right). This helps ASD children focus on the activity and decreases the need for assistance at the same time.

The colors used in the interface also needed a revision. The first version of the interface used vibrant colors by default, as well as a cloudy background, with hopes to make the interface more attractive. However, experts mentioned that this overload of visual information could potentially be unpleasant to ASD individuals. They might be attracted to vibrant colors, or they could feel anxious or irritating about it. In the second version of the application, the development team tried to overcome this problem by providing a color selection option. In this new functionality, caregivers (or parents, or teachers) choose the colors that best suit the child's preferences (Fig. 3).

In relation to the rewards, experts pointed out that it might not be as effective if they are handed long after the completion of an activity. In our first version, children collected stars as they completed tasks, and only after completing all the activities they would be rewarded. This hinders the association between a successful performance and the pleasure of receiving a reward. In order to solve this problem, the improved version brings an instant rewarding system. Besides a visual feedback displayed on the screen, the child chooses a reward right after the activity. We encourage parents to have at hand some real-life options, but we also provide some digital rewards (mini-games, stickers, etc.). In the new version, rewards are simpler, but more frequent and immediate.

Finally, the long form required to sign in – which, in the first version, included extensive personal information and was split into two screens – was synthesized in only two text input fields in the improved version. This time, users are simply asked to type their e-mail account and choose a password (Fig. 4).

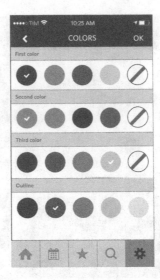

Fig. 3. DayByDay screen, improved version. Color selection.

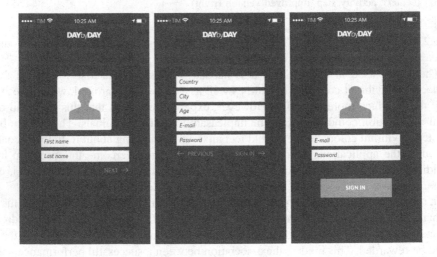

Fig. 4. DayByDay screens. Sign in form, first version (left); sign in form (continued), first version (center); sign in form, improved version (right).

In addition to all the improvements described so far, the development team also performed the adjustments ranked with lower priority levels, including font size, icons design, headings and sub-headings description, layout and elements position arrangements, action buttons patterns, among others. By the end of the development, all of the experts' recommendations had been successfully accomplished, even when the team had to explore alternative solutions due to project limitations.

6 Lessons Learned

Results achieved with the heuristic evaluation show prominent improvements in the User Interface, especially in terms of layout arrangement (elements position, design patterns), visual presentation strategies (use of colors, textual information), and customization options (rewards, color scheme, information management). Although the heuristic evaluation was undeniably helpful in providing high-quality feedback for the development team, dealing with autism added complexity to the process. This is why lessons learned are important: they share the knowledge and experience acquired throughout the process, and might support the development team in future decision-making.

In this study, lessons learned are divided in three main topics: (1) selection of the method and materials, (2) conduction of the evaluation process, and (3) definition of the follow-up strategy. First of all, regarding the selection of the method and materials, it is assumed that developers choose the usability inspection, because this is the object under appreciation in this study. On the one hand, it is necessary to think about the team composition and, on the other, the materials used to perform the evaluation. During this study, it became clear the importance of having a cross-functional team inspecting the interface. Experts from various fields of knowledge bring together their own background and unique points of view that provide high-quality feedback on usability aspects. Thus, developers must consider including experts that make sense to the project scope (in this case, computer science, design, and education).

Furthermore, the materials selected must provide an evaluation scenario that is as close to reality as possible. The technology used in this study (the Marvel prototyping software) allowed experts to test the interface in a mobile device and with most of the functionalities and options available, which increases the evaluation accuracy. However, it does not support cross-platform interaction. This requires a new round of evaluation, since testing the interface in different operational systems could potentially result in different usability problems. Also, the heuristics set must be selected according to the development context and the nature of the software. Many heuristics are available both in literature and across the Internet, so choosing the right one is a matter of research.

In addition to that, clear instructions must be provided for the evaluators. Of course, researchers do not intend to influence the experts' judgment, but they do need to feel confident about the interface and know what is expected from their evaluation. In this sense, a focus group could have been performed prior to the beginning of the evaluation sessions. It allows team members to connect with one another in the first place. Furthermore, researchers could have had the opportunity to provide more detailed explanation on the application (learning objectives, autism specifications, technology limitations, etc.). Development experts, for instance, may have a lot of experience with programming languages and usability issues, but we cannot assume they understand the ASD specific clinical conditions. After the focus group (or another similar method) the evaluation session should proceed individually, so that experts are not influenced by external judgments. Once the evaluation is completed, data is extracted, categorized, and analyzed, which leads to the definition of the follow-up strategy.

The quantitative analysis relating the number of problems to a specific part of the software (as in Tables 1, 2, 3 and 4) allows a systemic view of the usability problems and the identification of deficits in a particular area. For instance, most of the problems were detected in the caregiver system, suggesting that this part of the interface needed more attention. Also, many problems were reported by the computer science and education experts, which suggests that, in terms of design, the software reached a positive acceptance overall. In turn, the qualitative analysis (Table 5) provides a detailed description of each problem individually. In addition, connecting problems with the heuristic principles and ranking them according to a severity scale helps the prioritization process and supports further improvements. After all, the development team is responsible for deciding what is possible based on the financial, technological, and human resources available.

7 Conclusions and Future Work

In order to detect and perform improvements in the interface using heuristic evaluation, this study reported on the evaluation (planning, conduction, and results) of DayByDay, an ongoing mobile application for children with Autism Spectrum Disorder (ASD). With the participation of a cross-functional team of experts, it is concluded that this method, when applied consistently and according to the development context peculiarities, may provide accurate data and insights on usability issues. Feedback received from the experts helped the development team improve the former version of the software. Furthermore, lessons learned during the process are important to raise awareness on the results achieved and future challenges to overcome.

The development of accessible software is complex and requires an iterative process, so that the development team can accomplish the best possible results prior to releasing the software. Therefore, results in this paper are not the final milestone, but rather represent the second phase of an ongoing development process. This means that future upgrades and improvements needs to be performed in the application, which will then be subjected to testing procedures involving end-users. Above all, the sharing of experiences in autism – a complex topic still emerging in literature – helps tailor best practices within the community of developers and shape the way we see autism accessibility in the future.

References

1. Wojciechowski, A., Al-Musawi, R.: Assistive technology application for enhancing social and language skills of young children with autism. Multimedia Tools Appl. **76**, 5419–5439 (2017)
2. American Psychiatric Association: Diagnostic and Statistical Manual of Mental Disorders, DSM-5. Artmed, 5th edn (2013)
3. Wingate, M., Kirby, R.S., Pettygrove, S., Schulz, E., Gosh, T.: Prevalence of autism spectrum disorder among children aged 8 years – autism and developmental disabilities monitoring network, 11 sites, United States, 2010. Morbidity and Mortality Weekly Report, Center for Disease Control and Prevention, pp. 1–21 (2014)

4. Kamaruzaman, M.F., Ranic, N.M., Norb, H.M., Azaharia, M.H.H.: Developing user interface design application for children with autism. Soc. Behav. Sci. **217**(1), 887–894 (2016)
5. Mazurek, M.O., Shattuck, P.T., Wagner, M., Cooper, B.P.: Prevalence and correlates of screen-based media use among young youths with autism spectrum disorders. J. Autism Dev. Disord. **42**(1), 1757–1767 (2012)
6. Alvarado, C., Munoz, R., Villarroel, R., Acuña, O., Barcelos, T.S., Becerra, C.: ValpoDijo: developing a software that supports the teaching of chilean idioms to children with autism spectrum disorders. In: 12th Latin American Conference on Learning Technologies (LACLO) (2017)
7. Zhang, L., Fu, Q., Swanson, A., Weitlauf, A., Warren, Z., Sarkar, N.: Design and evaluation of a collaborative virtual environment (CoMove) for autism spectrum disorder intervention. Trans. Access. Comput. **11**(2), 11 (2018)
8. Winoto, P., Xu, C.N., Zhu, A.A.: "Look to remove": a virtual reality application on word learning for Chinese children with autism. In: Antona, M., Stephanidis, C. (eds.) UAHCI 2016. LNCS, vol. 9739, pp. 257–264. Springer, Cham (2016). https://doi.org/10.1007/978-3-319-40238-3_25
9. Cai, Y., Chia, N.K.H., Thalmann, D., Kee, N.K.N., Zheng, J., Thalmann, N.M.: Design and development of a virtual dolphinarium for children with autism. Trans. Neural Syst. Rehabil. Eng. **21**(2), 208–217 (2013)
10. Panchanathan, S., McDaniel, T., Balasubramanian, V.N.: An interdisciplinary approach to the design, development and deployment of person-centered accessible technologies. In: 2013 International Conference on Recent Trends in Information Technology (ICRTIT), pp. 750–757. IEEE (2013)
11. de Oliveira Barros, V.T., de Almeida Zerbetto, C.A., Meserlian, K.T., Barros, R., Crivellari Camargo, M., Cristina Passos de Carvalho, T.: DayByDay: interactive and customizable use of mobile technology in the cognitive development process of Children with autistic spectrum disorder. In: Stephanidis, C., Antona, M. (eds.) UAHCI 2014. LNCS, vol. 8514, pp. 443–453. Springer, Cham (2014). https://doi.org/10.1007/978-3-319-07440-5_41
12. Pribeanu, C.: A revised set of usability heuristics for the evaluation of interactive systems. Informatica Economică **21**(3), 31–38 (2017)
13. Quiñones, D., Rusu, C., Roncagliolo, S., Rusu, V., Collazos, C.A.: Developing usability heuristics: a formal or informal process? IEEE Latin Am. Trans. **14**(7), 3400–3409 (2016)
14. Sim, G., Read, J.C.: Using computer-assisted assessment heuristics for usability evaluations. Br. J. Edu. Technol. **47**(4), 694–709 (2016)
15. Islam, M.N., Bouwman, H.: Towards user-intuitive web interface sign design and evaluation: a semiotic framework. Int. J. Hum. Comput. Stud. **86**(1), 121–137 (2016)
16. Ardito, C., Buono, P., Caivano, D., Costabile, M.F., Lanzilotti, R.: Investigating and promoting UX practice in industry: an experimental study. Int. J. Hum. Comput. Stud. **72**(6), 542–551 (2014)
17. Marvel Homepage, Free Mobile & Web Prototyping. https://marvelapp.com. Accessed 15 Jan 2019

Learning About Autism Using VR

Vanessa Camilleri, Alexiei Dingli, and Foaad Haddod[(✉)]

University of Malta, Msida, Malta
{vanessa.camilleri, foaad.haddod.15}@um.edu.mt

Abstract. This paper describes a project that was carried out at the University of Malta, merging the digital arts and information technologies. The project, 'Living Autism', uses virtual reality (VR) technologies to describe daily classroom events as seen from the eyes of a child diagnosed with autism. The immersive experience is proposed as part of the professional development program for teachers and learning support assistants in the primary classrooms, to aid in the development of empathic skills with the autism disorder. The VR experience for mobile technologies has also been designed in line with user experience (UX) guidelines, to help the user assimilate and associate the projected experiences into newly formed memories of an unfamiliar living experience. Living Autism, is framed within a 4-min audio-visual interactive project, and has been piloted across a number of schools in Malta with 300 participants. The qualitative results collected gave an indication that the project had a positive impact on the participants with 85% of them reporting that they felt they became more aware of the autistic children's needs in the primary classroom.

Keywords: Virtual reality · Continuous professional development ·
Autism spectrum disorder · Primary education

1 Introduction

Virtual Reality (VR) is the projection of a computer-generated virtual environment that is rendered immersive by its unique affordances of vision, sound, and tracking of user movement. In a VR system, the user would experience an illusion of presence that would aid the user to perceive herself situated in the virtual environment and surrounded with the virtual objects. The sense of presence in addition helps a user experience the events happening in the virtual environment as though these are really happening [21]. The place illusion mechanism in VR allows the users to "be there" [6], and the plausibility illusion mechanism [21] allows players to react to the events happening in the generated environment. The way (VR) makes it possible to induce illusions in which users report and behave as if they have entered into altered situations and identities is through its sensory-coupled stimuli that match the brain's expectations of what should be happening according to the context in which the virtual projections are placed. The effect on the brain can be robust enough for participants to respond realistically to the situations and scenarios in which they are placed, leading to a possible change in behavior in the same way as when the users would have been exposed to the same scenario had this occurred in the physical world.

M. Antona and C. Stephanidis (Eds.): HCII 2019, LNCS 11573, pp. 64–76, 2019.
https://doi.org/10.1007/978-3-030-23563-5_6

The quality and intensity of the VR experience when compared with other simulation technologies employed in education renders this immersive technology more than suitable for teachers' CPD [3]. A number of companies have developed virtual environments in an attempt to replicate and simulate the classroom environment and what a teacher experiences in reality [14]. Yet a combination of animation-like surroundings and cartoon-like avatar characters does not fully equate to the intense experience an educator goes through when facing a real class. The challenge exists in how to provide a more authentic experience that would simulate a realistic classroom environment that would then aid teachers and education professionals to associate their own experiences of the class, to the virtually generated environment.

To overcome the challenge above, the use of real video combined by the natural movement of one's head to inspect the entire classroom and the students around the user embodies an authentic experience that comes as close to the factual thing as possibly can. It provides a visceral sensation that no video on a flat screen can inflict on one's emotional experience. Loewus [14] reports empirical results from a research project managed by a neuro-cognition scientist, Richard Lamb, who analyzed data collected from numerous VR users through heart rate monitors, sphygmomanometers to measure the blood pressure, electrodermal readers to determine the galvanized skin response, and neuroimaging apparatus to picture brain activity. The results were compared to respective data collected when person experience real life situations, and what emerged was that the body in both cases responded in a similar way. His conclusions showed no difference whatsoever between a teacher delivering within a real class and delivering a session through the VR-enabled experience.

Living autism, is a VR experience that uses different film techniques, to transpose the reality a child who is diagnosed with the autism disorder into the virtual classroom with the final aim of then transporting the adult who takes on the experience into a world that he/she may not necessarily be familiar with. Teachers and learning support assistants living through this experience have the possibility to create new associations of what a child in a primary classroom and who is on the autism spectrum might experience. The final scope is that of creating a deeper awareness of actions or stimuli that may have a negative consequence on a child with autism in a way that is not easily explained using any other resource or modality.

By exploring user perceptions, the project described in this paper aims to address three main research questions:

RQ1: How does UX design in VR affect immersion?
RQ2: How is the level immersion related to the manifestation of user behavior?
RQ3: How does immersion relate to empathic behavior in users?

The rest of the paper is structured as follows: Sect. 2 is a review of the literature around VR for continuous professional development (CPD) and user experience (UX) principles for immersive environments. Section 3 gives a deeper insight into the methodology and design of the project 'Living Autism' whilst Sect. 4 presents some of the results obtained during the pilot project. A discussion of the results and an overview of the directions for future work follow in Sects. 5 and 6 of the paper.

2 Literature Review

2.1 Professional Development

One of the attributes of a professional person is a life-long pledge to maintain the highest levels of associated knowledge and skills through education. Continuous Professional Development (CPD) is one systematic way of providing such education, and is essential for any professional who seeks to excel and remain abreast of contemporary information, tools, techniques or methodologies within the field of that same profession. As the field of expertise evolves and rapidly moves ahead, the need and necessity for a professional to follow CPD activities does not cease, in an effort to assist the same professional in diligently perform current duties and enhance career progression opportunities. CPD can take different forms and shapes, be formal or informal, as long as it is relevant to the field of work and focusses on career needs and objectives.

Bosschieter [6] points out that CPD can be a combination of various approaches, ideas and techniques that assist in learning and growing as a professional. She argues that the main focus should be on the end product, namely the benefits yielded in real practical terms. Every professional should be responsible and consciously recognize the need to maintain the highest level of standards in one's vocation to voluntarily seek and feel the need of continuous career development. The use of technologies and digital applications assist and facilitate the CPD process as hand-held devices, apps, and multimedia methodologies allow professionals to access, experience and appreciate such educational materials. Research indicates that the potential of new media plays a crucial part in CPD [7].

The teaching profession is no exception and the need for seasoned educators to pursue their training to ensure they are on their game is always on. Morrison McGill [15] claims that teachers are required to meet the needs of an ever-evolving audience, and thereby they are compelled to regularly practice CPD. However, a variety of factors around educators are continuously changing as Day and Leach [8] list administrative issues like legislation, regulations and teaching conditions, as well as academic matters like curricula, teaching approaches and aids, apart from other broader external factors like environmental, socio-economical, and cultural. The authors identify the development of the system, the educator and the learner as three closely-related purposes why CPD for teachers are instrumental. Yet, if we had to focus specifically on the development of the teacher in isolation then ulterior purposes can emerge, from classroom practical needs to the changing role of the teacher, from evolving school-wide needs to novel government educational policies, apart from personal and professional needs as each educator progresses in life. Jones [10] picks on the first of these purposes as she investigates the challenges teachers encounter when coming in contact with digital and technological tools. She reports that less than 3% of further education institutions in UK had the majority of their academics competent in the use of learning technologies, while 70% of the teacher population confessed that they have a lack of confidence in the use of digital applications. Use of technology should not be done just for the sake of being different, novel or trendy, but educators need to appreciate and comprehend that a particular technology is relevant within the specific contextual circumstances that fit in with the planned lesson and the expected academic objectives and outcomes. Walker et al. [22] conclude in their TEL survey report that teachers will only employ a particular technology only if they envisage it to enhance their current practice as a beneficial academic aid to the educator and student alike.

2.2 Virtual Reality for CPD

To overcome such obstacles and in an attempt to encourage teachers to experiment and experience the benefits of employing different technologies training agencies and CPD providers have reverted to Virtual Reality (VR).

The primary subject of virtual reality is simulating the vision. Every headset aims to perfect an approach to creating an immersive 3D environment. Two screens (one for each eye) are placed in front of the eyes thus, eliminating any interaction with the real world. Two autofocus lenses are generally placed between the screen and the eyes and these adjust based on individual eye movement and positioning. The visuals on the screen are rendered either by using a mobile phone or HDMI cable connected to a PC. To create a truly immersive virtual reality there are certain prerequisites including frame rate and refresh rates, field of view of the headset used. With an inconsistent frame rate and field of view, the user may experience latency with a large time gap between an action and the response from the screen. Apart from the image there are other elements that go into creating an immersive VR experience, making users completely engrossed in the virtual environment and include the impact of sound and eye and head tracking [16].

Researchers at the University at Buffalo in the state of New York have been employing VR to introduce new technologies to educators in an attempt to create a middle-ground safe space whereby educators can freely practice, explore and experiment with what it is like to employ different technologies with non-real kids [1]. The up-and-close experience, that designers compare to a "flight simulator for teachers", employs images of real students recorded inside a classroom with the help of 360-degree high definition cameras, creating a three-dimensional seamless and immersive environment. Teachers are able to enjoy the full capabilities that VR has to offer in other domains like gaming, medicine, automobile industry, tourism and modelling, whereby users can experience incredible events that would either be impossible in real life. Some examples of popular VR experiences include walking through historical sites and buildings, gender switching and embodiment, Formula 1 car driving, free falling from the sky, diving around the Titanic and many more. The maximization of the senses is exploited through a series of emotional phases that facilitate learning, memory, and emotion while maintaining experimental control [11].

VR technology is still evolving to further enhance the user experience by closing the gap between a real-life occurrences and VR-generated ones. Crosswater Digital Media [6] are adding functionality to their 360/VR experience by enabling VR users to move and interact with individual students as they comprehensively collect video data from within a real classroom. Movements include stepping in any direction, kneeling and verbally addressing persons within the video itself. This will eventually render the VR experience even more realistic and genuinely faithful to classroom settings. Teachers can experience their anxieties, challenges and novel situations through a controlled VR recreation session during their CPD in order to master and muscle memorize their response actions when they come across and experience that same stressful pedagogical event [18].

2.3 UX in VR

User Experience (UX) is the concept that brings together various elements of a system or application such as functionality, usability and appeal to enrich the user's practice whilst increasing the effectiveness and efficiency of the task completion [20]. Having the added value of seamlessly integrating the technology into the daily life practices of the user is also an important aspect of UX [10]. Although each user might experience a new technological practice in different ways, the use of VR for professional development by exploiting its immersive affordances is an example of how emotions can be made use of to stimulate corresponding actions and behavior [2]. Although there may be instances where it is not very clear how the UX is affected by VR technologies, studies indicate that each user interprets the VR experience according to personal background and history [17]. Studies by Shin conclude that rather than the 3D technology per se, what affects the user experience is the narrative surrounding the context of the immersive environment. Therefore the content that is developed needs to be not only engaging, but also targeted specifically towards an audience who can fully associate to the context and the 3D scenario. One of the most important aspects of UX is that of achieving the right flow for the user to engage more deeply with the VR narrative in such a way as to establish the perception of 'being there' inside the story. This may in fact depend not just on personality traits but also on the background and past experiences of the user [21].

Bachen et al. [2] theorise that one of the methods in achieving the right flow in 3D environments is that of helping users identify with the avatar or a non-player character (NPC). Although some personalities may be more disposed towards achieving presence, immersion and character identification than others, by designing the narrative in a way that a user can associate to the context and can readily recognize scenarios within familiar environments can increase the sense of presence and flow [12]. For CPD where simulation of a realistic environment is key, it is important to consider elements based on interaction, and behavioral activities [18].

3 Study Design

3.1 Participants

Participants for this experience were recruited on a voluntary basis, from a number of random Maltese primary schools. Researchers visited the schools, and enlisted teachers and learning support assistants who voluntarily enrolled to take on the experience as part of their CPD sessions. A total of 63 participants were recruited by convenience sampling from teachers' continuous professional development sessions, parental meetings and two educator conferences held across the country. 26.2% of the participants were male, whilst 73.8% of the participants were female. 42.2% of the participants were in the 25–34 age range, whilst 29.7% were in the 35–44 age range. Each participant was asked to go through the VR experience using a head-mounted device for mobile. In addition the participants were asked to report their thoughts, perceptions and ideas in a pre- and post-experience survey, giving a measure of (a) their self-presence in the environment and (b) their perceptions about the issues of autism and migration relating to classroom environment. This was also followed by a more in-depth qualitative analysis of the self-reported perceptions through observation, and the level of self-reported empathy with learners who have needs relating to the autism spectrum disorder.

3.2 Measures

The scope of the pre- and post- experience survey was to give a self-reported measure of personality traits relating to empathy. The surveys were adapted from the empathy measures developed by the Research Collaboration Lab responsible for a number of education and professional development projects with teachers [8].

The surveys developed focus on empathy as a measure of the efforts that any given person would make to understand others, as well as the ability of a person to communicate her understanding of somebody's personal situation. For both pre- and post-VR survey, we used 11 empathy items measured using a 5-point Likert Scale from 1 (strongly disagree) to 5 (strongly agree) (Table 1).

Table 1. Empathy quotient items

Item no.	Item description
T1	I see others' point of view
T2	When I don't understand someone's point of view I ask questions to learn more
T3	When I disagree with someone it's hard for me to see their perspective
T4	I consider people's circumstances when talking with me
T5	I try to imagine how I would feel in someone else's situation
T6	When someone is upset I try to remember a situation when I felt the same way
T7	When I am reading a book or watching a movie I think about how I would react if I were one of the characters
T8	Sometimes I wonder how it would feel like if I were a student in a classroom
T9	When I see one of my students upset I try to talk to them
T10	I easily feel sad when people around me are sad
T11	I get upset when I see someone being treated disrespectfully

In addition to the surveys, researchers used visual observation codes to record and register users' behavior and actions during their VR experience. The codes were used to identify user behavior during the session. These included changes in Facial Expressions (F), Sighing (S), Heavy Breathing (B), Crying (C), Visible Emotion (E), Impassivity (I), and Smiling (Sm). Although the observations only provide a superficial measure of the impact of the 3D immersive experience and are not directly related to the levels of empathy, these give an indication of how the experience affected the users through the manifestation of their behaviour.

3.3   Narrative, Film and Audio

UX was achieved through the use of different sensory modalities such as visuals and audio as well as reality matching, surroundness in the field of view and spatial audio, and a strong plot of the storyline with a dynamic unfolding sequence of events. This experience was brought to the user through a first-person narrative as a child on the autism spectrum at a primary school. The application used real film to provide a first-person account of what happens in a typical day at school as the user is surrounded by a cacophony of sounds and visual stimuli.

The narrative evolves over 6 scenarios typical of a day at school. These include class sessions, lunch breaks in the school yard, and transition across corridors (see Fig. 1). All filming was carried out using a 360 Samsung Gear camera, mounted on a tripod and fitted with a binaural microphone. Scenes were filmed on a low mounted camera to convey to the users an experience from the perspective of a child. Actors were briefed and scenes were staged inside real classroom settings. To aid visual and perception cues and decrease adverse VR effects such as motion sickness, stable visual cues were included in the filming whilst movement was restricted to a linear motion at a constant speed.

Fig. 1. The VR classroom scene

An iterative process has been implemented to progress from the planning to the development phase. The transition to VR development involved work that focused on the user experience, with a degree of constant calibration to minimise motion sickness. Head tracking technology was implemented using the Unity 3D platform and the experience used visual and auditory sensory cues. The VR experience was designed for the Samsung Gear VR Headset. This affected the user interface resolution so particular care and attention needed to be given to the UI elements and how these would appear on the screen. Any use of non-diegetic elements that might distract from the overall narrative of the experience were avoided to impart a more improved realistic experience. Sounds were used to help the user direct her gaze to a specific direction but a voice over was also used to articulate user thoughts and help her associate what was happening on-screen to reactions and behavior to strengthen presence and immersion.

4 Results

The results from the pre- and post- VR experience give an indication of the self-reported measures of user empathy levels.

The table of results (see Table 2) shows the findings after comparing the pre- and post- values following the VR experience. The findings showing a mean above 4, indicate a tendency towards strong agreement with the empathy items. Item 3 shows a lower mean as it is presented in the negative form. The standard deviation values indicate that the values are clustered more towards the mean. The p- values indicate that the only items which register a significant statistical difference between the pre- and the post-VR empathy questionnaires are items T5, 6, 7, 8 and 10.

Table 2. Paired samples T-test for the pre- and post- experience survey

T-test	Pre		Post		Paired samples test			
	Mean	Std. dev.	Mean	Std. dev.	95% confidence interval	Std. dev.	t	Sig. (2-tailed)
T1	4.440	0.562	4.460	0.618	−.2, .168	0.729	−0.173	0.863
T2	4.333	0.539	4.365	0.679	−.23387, .17038	0.803	−0.314	0.755
T3	2.540	0.930	2.810	1.216	−.55666, .01697	1.139	−1.881	0.065
T4	4.267	0.578	4.200	0.819	−.12279, .25612	0.733	0.704	0.484
T5	4.318	0.563	4.587	0.558	−.45194, −.08775	0.723	−2.962	0.004
T6	4.191	0.692	4.492	0.564	−.49154, −.11164	0.754	−3.174	0.002
T7	4.079	0.885	4.365	0.848	−.55112, −.02031	1.054	−2.152	0.035
T8	4.127	0.852	4.635	0.517	−.70958, −.30629	0.801	−5.035	0.000
79	4.254	0.740	4.349	0.765	−.34181, .15133	0.979	−0.772	0.443
T10	4.095	0.756	4.413	0.710	−.56449, −.07043	0.981	−2.569	0.013
Til	4.677	0.696	4.532	0.503	−.05447, .34479	0.786	1.454	0.151

N=63, df=62, p<.05

All the items listed as having had a significant statistical change refer to user perceptions of others in different situations especially those which may be in difficult or unhappy predicaments. Although the survey does not give an empirical measure of the changes in levels of empathy, the results give an indication that the VR experience has in no significant way affected the users' personality but it has impacted the way they think about other human beings in difficult circumstances.

In addition to the questionnaire, users' reactions were also observed during their VR experience and the observations were coded as per the measures shown in Sect. 3.2 above.

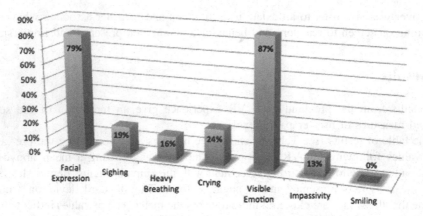

Fig. 2. Observable user behavior during VR experience

The results from Fig. 2 show that 87% of the users going through the 'Living Autism' VR experience showed some visible emotion whilst 79% showed changes in facial expressions. Visible emotion included emotional sounds and signs of some distress and agitation. Changes in facial expressions include observable changes below the head mounted mobile device and therefore concentrate around the mouth area. In addition, a number of users manifested more than one simultaneous behavior.

Following the VR experience participants were also asked to describe their emotional state, and their perception of stimuli effects on autism. All the participants described that the experience made them feel as though they were in another dimension. Even though in the real environment they were used to taking up the central role of the teacher, within the VR environment featuring a familiar class and school setting, they felt they were seeing it and experiencing it from the completely different perspective of a child with autism. The users that showed the most visible emotion reported that they felt the chaos of the class as overwhelming, and they needed to escape to a 'quiet room' without so many stimuli coming at them. Users also reported that now they could really understand what the children who are autistic go through every day as they face the various stimuli, including sounds, and olfactory ones, that are found in the classroom and that their level of understanding was considerably different from that which they had covered in theoretical lessons. All the participants reported that they could associate only too well with what happens in a classroom.

5   Discussion

Overall the results indicate that the VR experience had a degree of impact on the users' perceptions of autism. The qualitative feedback received shows that users could associate and identify with the scenarios presented and the classroom narratives. This association supported self-reported presence and immersion into the VR environment, a concept which tallies with research studies about presence and consciousness in VR [16]. The research questions posed at the start of the project, take into account the UX design of VR experience and how this links together immersion, presence and manifestations of user behavior.

RQ 1 deals with UX design and how this can lead to immersion. Users have reported that they have felt as though they were in another dimension, losing track of time, or place, but having a sense of 'being there' in the virtual world. The mediated sensory experiences have been used to allow users to map their body schema into the contextual framework of the scenarios being presented. The realistic filming, and the spatial audio helped connect pre-existing memory associations to form new ones, thus the illusion of their embodiment in the VR environment extends to the perception of the real world.

Research presented by Banakou et al. [4] had already given similar results that demonstrate that altered body self-representation from an adult to a child, can pose a significant influence on the users' perception of the world and this in turn increases the sense of flow and immersion into the virtual world. UX design principles that have been applied to the VR setting include

- the experience of user comfort: although the experience itself aims to create awareness of adverse effects of audio-visual stimuli on a child with autism, filming and audio were designed and planned to minimize discomfort such as motion sickness, or cyber-tearing;
- the interface design for the start of the mobile application is simple and takes into account usability issues.
- sound: the spatial sound helps users navigate since the interface does not allow for the inclusion of controllers or directions. Sounds direct the users where to look at and also act as a vocal interpretation of thoughts going on inside the child's head.

RQ 2 refers to the manifestation of user behavior and how the level of immersion relates to externalized actions. A relatively high number (87%) of participants displayed visible emotion such as distress and agitation throughout the experience. This also tallies with previous research studies by Kroes et al. [12]. Through the VR experience of 'Living Autism' users were more aware of their own actions and behavior, and their self-reported realization of their own manifestations led them to reflect about the level of immersion into the VR environment.

RQ3 enquires about the relationship between immersion and the empathy. The pre- and post-experience surveys give a measure of the self-reported empathy arising from the experience. The results reported a significant change in the participants' ability to try and imagine how they would feel in another person's situation, or their ability to associate a situation to another person's plight. One of the items on the empathy scale that registered the greatest significant statistical difference for the participants included the ability to understand what it feels like to be a child in a classroom. The results show that following the VR experience the users could report a greater understanding of what it feels like to be a child in a classroom setting. This is interesting more so because all the participants had a direct relationship with the classroom whether these were educators/teachers, or learning support assistants. The inference from this is that there can be instances where teachers and LSAs still do not fully identify with the children as individuals with differing needs, despite the fact that their work environment is the classroom. There may be a number of reasons for this, but one of them could be that the pressures related to the quantity of work that needs to be covered within the academic year, reduces their focus to the academic progress rather than on any other matter. The relation of immersion to empathy also tallies with previous work in the field such as that by [19]. All studies reviewed do not mention changes to the user's inherent personality traits and this is also reflected in the pre- and post-experience survey results presented in this paper.

6   Conclusion

This project stemmed from an identified need about more awareness of how class stimuli can have an adverse effect on a child diagnosed with autism. CPD courses for teachers attempt to bring more awareness about this subject. However, research indicates that when CPD is presented through immersion and simulation, there is deeper engagement which can lead to effective changes in behavior. This paper gave an overview of the results emergent from a VR experience project about 'Living Autism'. The experience designed around UX principles for the design of VR, features daily classroom events for a child diagnosed with autism and presented to an intended audience of teachers and learning support assistants through a first-person narrative. The storyline includes 4 case scenarios that take the users through various learning and socialization instances.

The project enquires into the relationship between UX, immersion and empathy levels in participants taking on the VR experience. Findings from self-reported measures on empathy, as well as qualitative feedback from observation and participant post-experience reflections, indicated that the experience facilitated increased empathy levels with children diagnosed with autism. As a result of this, participants reported more awareness of actions and behavior that might result in a negative reaction of a child with autism.

The project itself had a number of limitations that need to be addressed for future developments. These included the relatively generalised data about autism and how this was integrated in the classroom scenarios. Autism is a complex disorder that manifests itself across a wide spectrum of behavior for different individuals. Individual behavior could not be captured and presented to the users for a variety of reasons including limitations of time and resources. The sample size of the population was another limitation of the project. However, there are sufficient grounds to there is potential for further development. Although this project focused on the context of a classroom, the findings indicate that this can be scaled to the workplace or any other community space, for users to use as part of their CPD. The findings do not give any indication of the long-term effect of the VR experience on the users, or how this will affect their class behavior in the longer term. Future work on this project, may include a longitudinal follow up study to try and understand whether the self-reported changes in empathy would still be significant after a period of time.

Another future direction in this field is the use of neuroscience to understand more about the impact of an immersive experience on the parts of the brain specifically linked to empathy. Results that can empirically link immersion to increased empathic levels would bring great promise to the field of human behavior in VR.

References

1. Anzalone, C.: UB's virtual reality expertise creates simulated classroom environment for aspiring teachers, 28 June 2017. UBNow - Campus News. https://www.buffalo.edu/ubnow/stories/2017/06/vr-teacher-training.html. Accessed 14 Oct 2017
2. Bachen, C.M., Hernández-Ramos, P.F., Raphael, C., Waldron, A.: How do presence, flow, and character identification affect players' empathy and interest in learning from a serious computer game? Comput. Hum. Behav. **64**, 77–87 (2016)
3. Bambury, S.: CPD in VR (2017). VirtualiTeach. https://www.virtualiteach.com/cpd-in-vr. Accessed 15 Oct 2017
4. Banakou, D., Groten, R., Slater, M.: Illusory ownership of a virtual child body causes overestimation of object sizes and implicit attitude changes. Proc. Natl. Acad. Sci. **110**(31), 12846–12851 (2013)
5. Bohil, C., Owen, C., Jeong, E., Alicea, B., Biocca, F.: Virtual Reality and Presence, 21st Century Communication: A Reference Handbook. Sage Publications, Thousand Oaks (2009)
6. Bosschieter, P.: Continuing professional development (CPD) and the potential of new media. Indexer **34**(3), 71–74 (2016)
7. Crosswater: VR/360 (2017). Crosswater Digital Media. https://crosswater.net/. Accessed 15 Oct 2017
8. Day, C., Leach, R.: The continuing professional development of teachers: issues of coherence, cohesion and effectiveness. In: Townsend, T. (ed.) International Handbook of School Effectiveness and Improvement, pp. 707–726. Springer, Dordrecht (2007). https://doi.org/10.1007/978-1-4020-5747-2_38
9. Gaumer, E., Soukup, J., Noonan, P., McGurn, L.: Empathy Questionnaire (2016). Research Collaboration. http://www.researchcollaboration.org/uploads/EmpathyQuestionnaireInfo.pdf
10. Jones, B.: To raise teaching standards we must first improve the use of technology in the classroom, 1 February 2017. The Telegraph Education section. http://www.telegraph.co.uk/education/2017/02/01/raise-teaching-standards-must-first-improve-use-technology-classroom/. Accessed 14 Oct 2017
11. Kim, G.J.: Human-Computer Interaction: Fundamentals and Practice. Auerbach Publications, Boca Raton (2015)
12. Kroes, M., Dunsmoor, J., Mackey, W., McClay, M., Phelps, E.: Context conditioning in humans using commercially available immersive virtual reality. Sci. Rep. **7**(1), 8640 (2017)
13. Kuliga, S.F., Thrash, T., Dalton, R.C., Hölscher, C.: Virtual reality as an empirical research tool—exploring user experience in a real building and a corresponding virtual model. Comput. Environ. Urban Syst. **54**, 363–375 (2015)
14. Loewus, L.: How virtual reality is helping train new teachers. Educ. Week **37**(3), 1–2 (2017)
15. Morrison McGill, R.: Professional development for teachers: how can we take it to the next level? 29 January 2013. The Guardian - Professional Development - Teacher's Blog. https://www.theguardian.com/teacher-network/teacher-blog/2013/jan/29/professional-development-teacher-training-needs. Accessed 13 Oct 2017
16. Rautaray, S.S., Agrawal, A.: Vision based hand gesture recognition for human computer interaction: a survey. Artif. Intell. Rev. **43**(1), 1–54 (2015)
17. Sanchez-Vives, M.V., Slater, M.: From presence to consciousness through virtual reality. Nat. Rev. Neurosci. **6**(4), 332–339 (2005)
18. Shin, D.: The role of affordance in the experience of virtual reality learning: technological and affective affordances in virtual reality. Telematics Inform. **34**(8), 1826–1836 (2017)

19. Shin, D.: Empathy and embodied experience in virtual environment: to what extent can virtual reality stimulate empathy and embodied experience? Comput. Hum. Behav. **78**, 64–73 (2018)
20. Shneiderman, B., Plaisant, C., Cohen, M., Jacobs, S., Elmqvist, N., Diakopoulos, N.: Designing the User Interface: Strategies for Effective Human-Computer Interaction. Pearson, London (2016)
21. Slater, M.: Place illusion and plausibility can lead to realistic behaviour in immersive virtual environments. Philos. Trans. R. Soc. London B: Biol. Sci. **364**(1535), 3549–3557 (2009)
22. Teng, C.: Customization, immersion satisfaction, and online game loyalty. Comput. Hum. Behav. **26**(6), 1547–1554 (2010)
23. Walker, R., Voce, J., Nicholls, J., Swift, E., Ahmed, J., Horrigan, S., Vincent, P.: 2014 Survey of Technology Enhanced Learning for higher education in the UK. UCISA, Oxford (2014)

Breaking Down the "Wall of Text" - Software Tool to Address Complex Assignments for Students with Attention Disorders

Breanna Desrochers, Ella Tuson, Syed Asad R. Rizvi, and John Magee[✉]

Department of Math and Computer Science,
Clark University, Worcester, MA 01610, USA
{bdesrochers,btuson,asrizvi,jmagee}@clarku.edu

Abstract. One undergraduate student's strategy to deal with long assignment instructions is to black out all of the information that they deem to be unimportant in the text, allowing them to focus just on the "important" information. While this technique may work well on paper, it does not naturally transition into a digital format. The student in the example above also identifies as having an attention disorder. In this paper we introduce a Microsoft Word add-in that enables the user to black out selected text using a new menu. Participants used the new Microsoft Word add-in to mark up a sample assignment. They were then asked in a post questionnaire to provide feedback on their experience utilizing the tool. Separately, we also conducted a survey in which we asked undergraduate students about their current strategies to understand long assignment instructions and why those strategies work for them. We then discuss their responses and compare it to the results of the previously mentioned case study.

Keywords: Accessibility · Attention disorders · Text understanding · Education tools

1 Introduction

Attention disorders such as ADHD can make university-level assignments with long textual narratives difficult to understand and follow. We were inspired by a student who used a technique of marking up assignments with a black marker in order to simplify the text and leave information directly relevant to the task visible. Informal discussions with other students indicated that such 'wall of text' assignments might present a challenge to a variety of students.

This paper is divided into two main parts. In the first part (Sect. 2), we present and evaluate a software tool designed to mimic the blacking-out technique of a marker on paper. In the second part (Sect. 3), we conduct and analyze a broader survey to understand the challenges posed by such assignments and the current techniques students use to understand them.

M. Antona and C. Stephanidis (Eds.): HCII 2019, LNCS 11573, pp. 77–86, 2019.
https://doi.org/10.1007/978-3-030-23563-5_7

2   Software Tool Study

An individual undergraduate student has a tried and true method to work on complex college assignments, especially when they are long, wordy, and resemble a 'wall of text', rather than step-by-step instructions. A single page from an assignment identified by this student is shown in Fig. 1. When starting these assignments, their first step is to mark up the complicated project instructions using a marker and paper (Fig. 2). As instructors are increasingly assigning work digitally, the student either needs to print and then mark up all assignments, or find a simple way to perform a similar task digitally. Another advantage of a digital approach, when compared to the analog paper and marker, is the ability to "undo" any unwanted black outs, which is also a problem the student has struggled with.

The particular student referenced in the example above also identified as having an attention disorder. From past research, reading comprehension is a key issue for people with attention-deficit/hyperactivity disorder (ADHD) [1–3].

In addition, there are a number of studies that focus on reading comprehension in digital environments. Some provide feedback on ways that highlighting text affects people's comprehension [4,5]. Other studies provide data related to the text format and explore how a person's reading comprehension changes based upon whether the text is plain or contains emphasis, images, links, etc [5,6].

Our example student's method focuses on blacking out unimportant instructions. This allows them to remove unnecessary parts of the assignment so that they are left with only the information that they find useful. In order to bring the concept of blacking out text on paper to the computer, we created a Microsoft Word add-in tool, the sole purpose of which is to enable users to easily black out any text they choose in the document. We then recruited participants to offer feedback related to how our tool fares against the dreaded 'wall of text'.

2.1   Methods

Tool. In order to mimic blacking out text on paper, we created a Microsoft Word add-in (Fig. 3). The add-in provides users with a new menu on the right of the document that enables them to easily black out selected text. It also provides the option to remove the black out from selected text, similar to an "undo" function.

Participants. The subjects ($N = 5$) ranged from 19 to 23 years old ($M = 21$). Each of the participants was given an opportunity to self identify as having an attention disorder ($N = 2$).

Procedure. Participants were first directed to complete a preliminary questionnaire. They were then directed to a computer science long term programming assignment that included the new Microsoft Word add-in menu bar. Participants were instructed on how to use the new tool and were asked to act as if this was

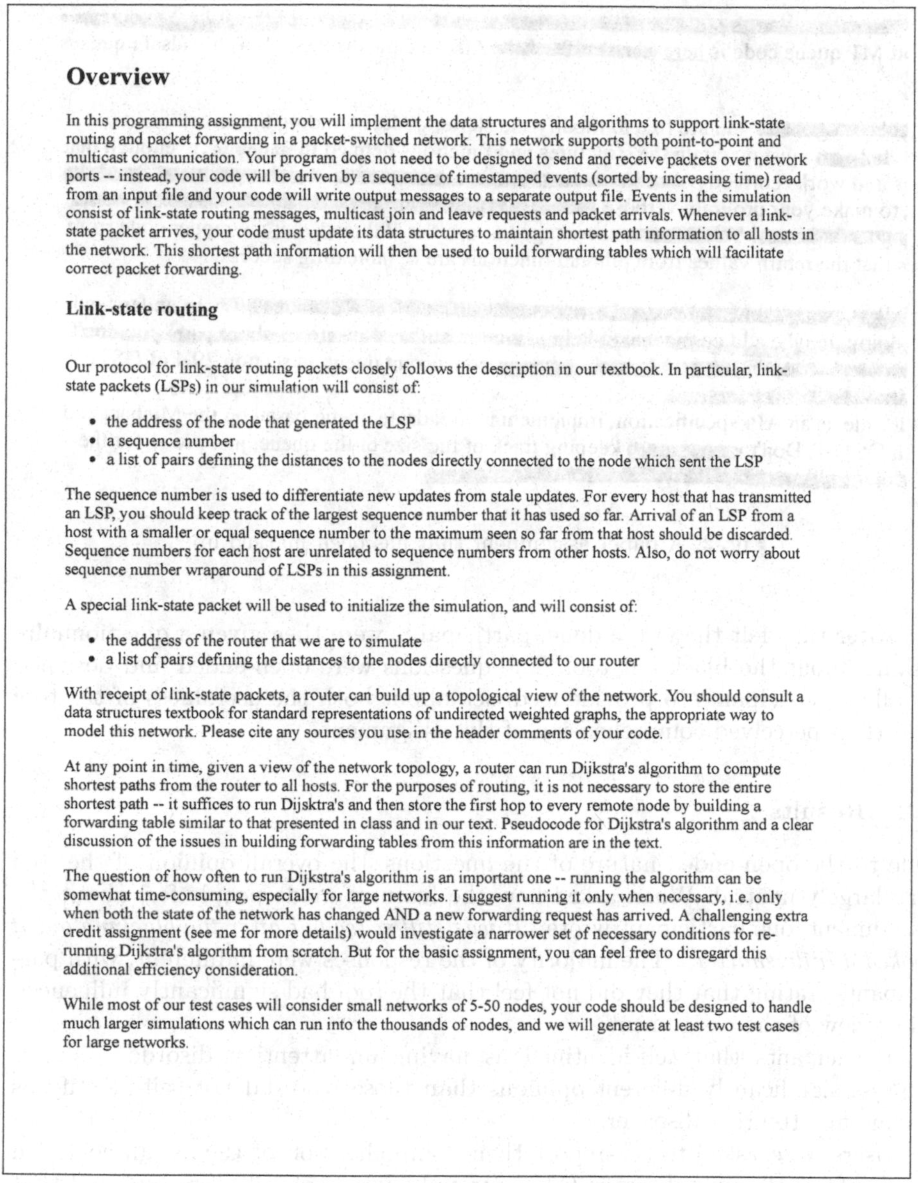

Overview

In this programming assignment, you will implement the data structures and algorithms to support link-state routing and packet forwarding in a packet-switched network. This network supports both point-to-point and multicast communication. Your program does not need to be designed to send and receive packets over network ports -- instead, your code will be driven by a sequence of timestamped events (sorted by increasing time) read from an input file and your code will write output messages to a separate output file. Events in the simulation consist of link-state routing messages, multicast join and leave requests and packet arrivals. Whenever a link-state packet arrives, your code must update its data structures to maintain shortest path information to all hosts in the network. This shortest path information will then be used to build forwarding tables which will facilitate correct packet forwarding.

Link-state routing

Our protocol for link-state routing packets closely follows the description in our textbook. In particular, link-state packets (LSPs) in our simulation will consist of:

- the address of the node that generated the LSP
- a sequence number
- a list of pairs defining the distances to the nodes directly connected to the node which sent the LSP

The sequence number is used to differentiate new updates from stale updates. For every host that has transmitted an LSP, you should keep track of the largest sequence number that it has used so far. Arrival of an LSP from a host with a smaller or equal sequence number to the maximum seen so far from that host should be discarded. Sequence numbers for each host are unrelated to sequence numbers from other hosts. Also, do not worry about sequence number wraparound of LSPs in this assignment.

A special link-state packet will be used to initialize the simulation, and will consist of:

- the address of the router that we are to simulate
- a list of pairs defining the distances to the nodes directly connected to our router

With receipt of link-state packets, a router can build up a topological view of the network. You should consult a data structures textbook for standard representations of undirected weighted graphs, the appropriate way to model this network. Please cite any sources you use in the header comments of your code.

At any point in time, given a view of the network topology, a router can run Dijkstra's algorithm to compute shortest paths from the router to all hosts. For the purposes of routing, it is not necessary to store the entire shortest path -- it suffices to run Dijsktra's and then store the first hop to every remote node by building a forwarding table similar to that presented in class and in our text. Pseudocode for Dijkstra's algorithm and a clear discussion of the issues in building forwarding tables from this information are in the text.

The question of how often to run Dijkstra's algorithm is an important one -- running the algorithm can be somewhat time-consuming, especially for large networks. I suggest running it only when necessary, i.e. only when both the state of the network has changed AND a new forwarding request has arrived. A challenging extra credit assignment (see me for more details) would investigate a narrower set of necessary conditions for re-running Dijkstra's algorithm from scratch. But for the basic assignment, you can feel free to disregard this additional efficiency consideration.

While most of our test cases will consider small networks of 5-50 nodes, your code should be designed to handle much larger simulations which can run into the thousands of nodes, and we will generate at least two test cases for large networks.

Fig. 1. One of three pages from a Computer Networks university-level programming assignment, rendered as a PDF. This assignment was identified by the participant as an example of a challenging "wall of text".

a long term assignment from one of their professors. Each participant was given as much time as they required to read over the assignment and to mark it up as though they would be referring back to it later.

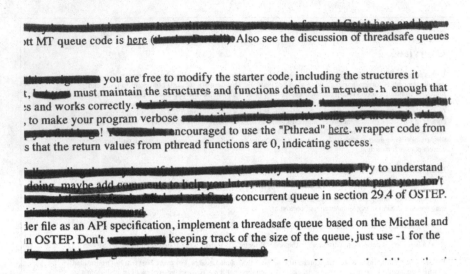

Fig. 2. A paper assignment that has been marked up.

After they felt they were done, participants were then given a questionnaire asking about the black out tool. The questions were open-ended and designed to allow participants to provide feedback about both the usefulness of the tool and their perceived comprehension of the assignment.

2.2   Results

Due to the open-ended nature of the questions, the overall opinion of the tool was largely neutral. When asked how the features made people feel about the assignment, one participant wrote *"It was still scary but after the blacking out it looked a little shorter"*. The majority of the responses were indifferent, with participants stating that they did not feel that the tool had significantly influenced their view of the assignment.

Participants that self-identified as having an attention disorder did not express significantly different opinions than those who did not self-identify as having an attention disorder.

Users were asked to self-report their comprehension of the assignment. On a scale from strongly disagree (1) to strongly agree (5), all users reported that they understood the assignment well ($M = 3.8$).

When asked if they would use the tool again, only one participant stated they would not use the black out tool again. The majority of the participants stated that they might use it in addition to other mark-up approaches, such as highlighting and annotating.

A number of participants expressed interest in the idea of eliminating unwanted text, but most were not comfortable with the idea of blacking out text completely. This appears to be mostly due to the fear of not being able to

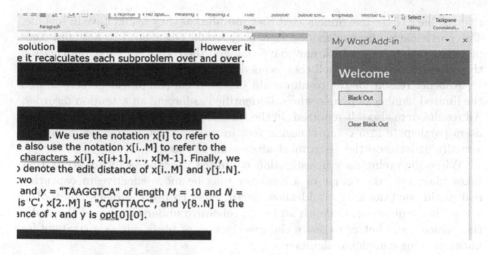

Fig. 3. A Microsoft Word document with the addition of an add-in menu on the right. The new menu has two options that allow users to select text and choose to "Black Out" or "Clear Black Out".

find something later if it was discovered to be useful and one couldn't remember where it had been now that it was blacked out. In fact, one user wrote:

> *Consider alternative styles of de-emphasis. In some cases it might be good to find a middle ground of making text de-emphasized without removing the ability to read it entirely.*

The participant reacted positively towards the idea of de-emphasis, but was hesitant about the permanence of the black out approach. Other users, however, were more interested in the complete removal of text from the assignment, and indicated that this eliminated distractions caused by the presence of unnecessary information. When asked to provide feedback, one user wrote:

> *I think the blacking out feature can be helpful to understand large amounts of text which can be overwhelming for some people.*

While participants provided a mix of both positive and negative feedback, users overall seemed to be largely indifferent towards the digital black out tool.

2.3 Discussion

On the whole, participants reported an overall neutral feeling related to the use of a digital black out tool for assignments. However, based upon the feedback received, it seems clear the concept requires further investigation with a greater scope.

One reason for the indifference of the participants may be because the experiment was not run over a long period of time. As a result, the users never actually referred back to the assignment that they modified as they would have in real life. The participants also understood that they would never need to actually complete the assignment, which may have changed their view on what information they needed, and of how well they needed to comprehend the instructions.

Another reason for the results could be the small sample size received, and the limited number of people who self-identified as having an attention disorder. All results were also self-reported. In the future, it may be beneficial to also have users participate in a comprehension test in order to determine how well they actually understood the assignment after using the tool.

While the preliminary investigation is limited in scope, a future study that takes place over the course of a semester may be more effective in capturing real world conditions. It would allow participants to use the black out tool in a way that requires an in-depth and comprehensive understanding of the material, which could better measure the effectiveness of black out as a strategy for understanding complex assignments.

3   Survey

There are a multitude of strategies that can be used to tackle and deal with long, 'wall of text' assignments, but many of these require a physical copy of the text being read. To validate and enhance the results that were obtained from the software tool study above, we conducted a survey of undergraduate students to assess what strategies they use most frequently. In addition to providing useful statistics in analyzing the applicability of the developed software to our desired demographic, this survey also provided a basis for further work in which strategies can be merged with our software to improve its effectiveness in achieving its goal.

3.1   Participants

We recruited undergraduate college students to take the reading comprehension aid survey. A number of responses were incomplete and did not answer a majority of the questions, therefore we removed these from the dataset. The participants ranged from nineteen to twenty-one years old ($M = 20.53$). They are sophomores ($N = 1$), juniors ($N = 8$), and senors ($N = 8$). The participants studied a wide array of fields but the majority studied computer science ($N = 6$). The majority of participants ($N = 16$) did not self disclose as having ADHD.

3.2   Distribution

Using Facebook and emails, we distributed the survey to an undergraduate population. The survey was hosted on Qualtrics and anonymous data was collected. Participants were asked questions about their current attitudes towards long, 'wall of text', assignments or instructions and what types of strategies they use to deal with them. The questions were a mixture of multiple choice and short answer questions in order to gather this information.

3.3   Results

Participants show a diverse attitude towards 'wall of text' assignments. They were asked to acknowledge if they have struggled with wall of text assignments in the past, using a scale of "Strongly disagree" (1.00) to "Strongly agree" (7.00). The results ranged from 1.00 to 7.00 with a mean of 3.29 and a standard deviation of 1.67. The mode of the results was at somewhat agree (5.00), which indicates that users somewhat agree that they struggle with 'wall of text' assignments. See Fig. 4.

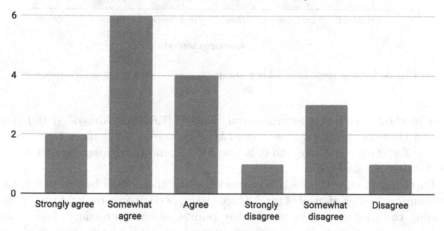

Responses to "I have struggled with understanding long (or 'Wall of text') assignment instructions in the past"

Fig. 4. Survey responses asking about struggles with large assignments.

Participants reported a number of strategies including highlighting text, making comments, rereading certain sections, taking notes, and underlining text. The most used strategy was underlining text ($N = 10$). However, a few individuals ($N = 3$) reported as having no current strategy to deal with long assignment instructions. See Fig. 5.

The participants were then asked to elaborate with their own words their strategies for handling 'wall of text' assignments. The majority of answers ($N = 9$) emphasis the point of the strategies is to identify key information, meaning what they as the student view as the most significant parts of the assignment. Many of them do this by bringing more attention to the parts they deem valuable such as: "I highlight the most important information" and "I often take notes on the side about the most important pieces of information about the assignment and underline information to look at later".

They were then asked about how they implement these tools, digitally or physically with pen and paper. Participants were asked to rank their usage of

Popularity of Annotation Methods

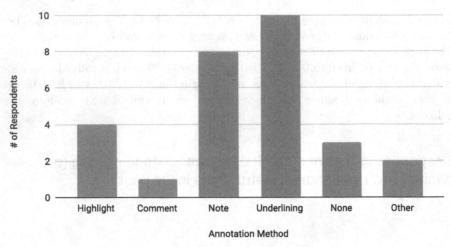

Fig. 5. Survey responses asking about struggles with large assignments.

these methods on a five point scale from "Never" (1.00) to "Always" (5.00). This showed that they used physical tools ($M = 3.56, \sigma = 1.12$) more than digital tools ($M = 2.63, \sigma = 1.05$), and it is notable that no participant reported as to never using physical tools.

Participants were then asked to describe why these tools helped to improve their understanding of 'wall of text'. Once again, participants brought up the idea of using the strategies to emphasize key points, as one user said: "[The strategies] help with picking out important information". One participant said that "[The strategies] break them down into macro information involving steps, while removing the non-critical components on the micro scale like detailed instructions". This answer emphasizes the ease of reading and the removal of information viewed unimportant by the student. Another participant said "It helps to break it into parts that are easier to read, instead of just one huge chunk of text to go through". Overall their reasons fell into four categories: makes the assignment easier to read/understand, emphasizes/extracts important parts, removes non critical information, and breaks down the 'wall of text'.

When asked if they felt their current approach was sufficient for understanding 'wall of text' assignments, the majority ($N = 11$) of respondents said that they did feel their approach was sufficient.

3.4   Discussion

Overall it is clear that 'wall of text' assignments are not something that every student has figured out. Participants' wide range of answers to the question asking if they have struggled with 'wall of text' assignments, shows that there is a large spectrum of competence levels with these assignments.

The attitude paired with the wide array of reported strategies for dealing with the long assignment instructions indicate that there is not a clear and widely used solution at this time. The users that reported not having a current strategy also are not equipped to deal with long assignment instructions, which could then lead to poor reflections of their assignments. Most of the participants report that the strategies they use are there to bring key information to their attention. They use strategies such as highlighting or underlining text in order to make these parts easier to access later. Other participants also use note taking in a similar way in order to pull out the key details and store them in a separate place, thereby making the rest of the 'wall of text' assignment instructions unnecessary. This implies that there is text that the student does not need and could benefit from removing.

Overall, participants reported that they used physical tools, such as pen and paper, more than digital tools in order to handle 'wall of text' assignments. The tools may depend on how the assignment is distributed, whether it is a physical sheet of paper or if they directed to a website for assignment instruction. The method of distribution may affect strategies students feel comfortable using. If there is only one copy students may not feel comfortable permanently marking the instructions which may lead to more strategies such as note taking.

When users were asked how their strategies helped improve their reading, participants' answers mentioned four key themes about their strategies: making the assignment easier to read/understand, emphasizing/extracting important parts, removing non critical information, and breaking down the 'wall of text'. These methods all indicate that there is essential instructions and non essential instructions. Removing instructions in order to emphasize critical information may enhance understanding of assignment instructions.

4   Conclusion

Connecting our software tool study on blacking out text with the survey results, it is evident that all these strategies share similar themes when it comes to comprehending 'wall of text' assignments. The goal is to emphasize key information and remove the focus from non-critical text. The results demonstrate that even though feelings towards this new tool were generally neutral, most participants were open to using this feature for marking up future assignments. Further development of this software can be aimed at providing a way to de-emphasize text without making it unreadable altogether, or a way to hide text temporarily so there is simply less space taken up by the document, making it easier to get through without interrupting the flow of reading.

In terms of the survey that was conducted, the next step would be to introduce those without a current strategy, or those that said their strategies were not effective, to the concept of blacking out text. The software that we introduced in the beginning of the paper could be a good method to introduce the strategy of blacking out text to students who are given digital assignments. Having the students use our software, as well as this technique in general, for the course of a semester would allow the study to take place over a longer period instead of an hour or so and allow students to engage with assignments that are actively assigned to them in a more natural school setting.

Acknowledgments. NSF support for this project is acknowledged and greatly appreciated (#IIS-1551590). The undergraduate authors gratefully acknowledge funding provided by the Clark University LEEP Fellowship program. Research involving human subjects has been reviewed and approved by the Clark University IRB (#2018-030A and #2014-063).

References

1. Pagirsky, M.S., et al.: Do the kinds of achievement errors made by students diagnosed with ADHD vary as a function of their reading ability? J. Psychoeduc. Assess. **35**(1), 124–137 (2017)
2. Miller, A.C., Keenan, J.M., Betjemann, R.S., Willcutt, E., Pennington, B.F., Olson, R.K.: Reading comprehension in children with adhd: cognitive underpinnings of the centrality deficit. J. Abnorm. Child Psychol. **41**(3), 473–483 (2013)
3. Lewandowski, L., Hendricks, K., Gordon, M.: Test-taking performance of high school students with ADHD. J. Attention Disord. **19**(1), 27–34 (2015)
4. Dodson, S., Freund, L., Kopak, R.: Do highlights affect comprehension?: lessons from a user study. In: Proceedings of the 2017 Conference on Conference Human Information Interaction and Retrieval, CHIIR 2017, pp. 381–384. ACM, New York (2017)
5. O'Brien, H.L., Freund, L., Kopak, R.: Investigating the role of user engagement in digital reading environments. In: Proceedings of the 2016 ACM on Conference on Human Information Interaction and Retrieval, CHIIR 2016, pp. 71–80. ACM, New York (2016)
6. Zheng, X., Cheng, W., Fan, Z., Chen, G.: The effect of signals in hypertext reading by tablet computers. In: 2015 IEEE 15th International Conference on Advanced Learning Technologies, Beijing, China, pp. 118–119. IEEE (2015)

Feel Autism VR – Adding Tactile Feedback to a VR Experience

Foaad Haddod[(✉)], Alexiei Dingli, and Luca Bondin

University of Malta, Msida MSD 2080, Malta
{foaad.haddod,alexiei.dingli,
luca.bondin.13}@um.edu.mt

Abstract. Feel Autism VR is a new virtual reality (VR) system which builds upon an Autism VR system which was developed in the past year. The aim of the original system was to design a novel VR environment to boost user's awareness about autism and in so doing, increase the level of empathy towards autistic children. We sought to create a VR environment which provides total immersion to its users. A touching-without-feeling technique was used to send an ultrasound signal to the user's body when the VR experience displays a touching scenario. Sound, vision and virtual touching elements could increase the awareness of presence, of the autistic children. A novel approach has been implemented which will allow users to feel tactile feedback physically without being touched. In so doing, we can recreate the annoyance felt by the autistic child throughout the day. Ultrasound waves will be generated via an ultrasound speaker which send the waves through the air and create pressure to simulate a real-life event. This concept will be inserted in the narrative of the original VR Autism project in order to mimic physical touching between an autistic child and his classmates, teachers, etc. The results were very promising whereby 50% of the users declared that after the experience, they are in a better position to understand children with autism.

Keywords: Autism · Virtual Reality · Tactile · Ultrasonic · Haptogram

1 Introduction

Autism Europe defines autism as a "lifelong disability… People on the autism spectrum experience persistent difficulties with social communication and social interaction, and might display restricted and repetitive patterns of behaviours, activities or interests." Autism Spectrum Disorders (ASDs) are a group of pervasive developmental disorders characterised by core deficits in three domains, i.e. "social interaction, communication and repetitive or stereotypical behaviour" [1], occurring more often in boys then in girls (4:1). While the level of impairment of suffering from autism may differ among different cases of individuals, the bearing on the individuals themselves and the people around them can be considered as being "life-changing" [2]. Parents, for example, describe how children on the autism spectrum may sometimes be left out of social events in schools simply because the environment is not adapted to their needs. This lack of adaptation is the result of a combination of numerous factors. One of these

© Springer Nature Switzerland AG 2019
M. Antona and C. Stephanidis (Eds.): HCII 2019, LNCS 11573, pp. 87–97, 2019.
https://doi.org/10.1007/978-3-030-23563-5_8

challenges is the lack of knowledge of how to deal and how to react when interacting with individuals suffering from ASDs. Teachers face enormous challenges when attempting to integrate children on the autism spectrum in the daily classroom experience. The biggest of these challenges lies in the fact that it is very difficult for teachers to fully comprehend the situations, the triggers and the stress levels that an autistic child may be subject to when interacting in class.

Most studies of ASDs in relation to technology focus on the way different mediums can help the student learn and communicate. They create technology-driven media to aid the student directly by changing the way learning is presented. However little work has been done to explore how such media can use narrative to support empathy. Following on our previous study 'Autism VR' [17], we propose the use of Virtual Reality (VR) coupled with Haptic feedback technology as a new approach that could recreate and repurpose a narrative to immerse adult teachers into the world of a child on the autism spectrum to a degree higher than what was already achieved. VR may be a simulation of the real world or the creation of a completely new world based on computer graphics to provide an experience that allows the user to understand concepts [3]. The realism of the virtual environment allows the user to acquire important skills, increasing the probability to transfer them into real life behaviour [4–6]. Building on 'Autism VR', this study explores various avenues in creating increased immersion in the virtual environment by exploring innovative communication mediums. While virtual reality implementations mostly focus on the visual and audio sensory elicited emotions, the addition of the sense of touch as a continuation of our previous study, should give an increased level of immersion and a provide a more realistic performance to anyone using the system.

The project's scope is that of building upon the use of Virtual Reality as a tool with which to understand better the case of autism in children. More specifically, the research aims at bringing together VR and Haptic Feedback. As the project unfolds and carers experience this VR immersion into autism, we want to ask questions such as Will the VR have a visible impact on the carers' empathy towards children with autism? Also, we want to note how the combination of these two technologies improves one's ability to empathise with children when compared to the results achieved when the standard approach to implementing VR was adopted in earlier research. The next section provides an overview of the research that has already been carried out in the area, highlighting work that has already been done in 'Autism VR'.

2 Literature Review

2.1 Autism and Virtual Reality

Oliver Sacks (1995) describes autism as "an impenetrable world, one which makes it difficult to unveil. The reason for this difficulty lies in the need to truly understand and explain what the relationship is with an autistic child, who, to a certain extent is a strange being who lives on a different plane of existence. In most cases an autistic child is a person whom we cannot connect with" [7]. While it is only recently that Virtual Reality technology applications have delved in autism therapy [13], the use of other technology as a means of helping individuals on the ASD scale has for long been an area of interest for many researchers. Interventions outside the area of Virtual Reality have quantified social skills and cognition over time using a variety of techniques and measures leading to mixed success rates [9]. The use of computers and computer technology for children with autism has been shown by research to have positive and beneficial effects [10]. Due to the increase in diagnosed cases of ASD, software and hardware dedicated to persons with autism have been developed for several decades. These solutions reinforce ASD sufferers' strong points and work on their weaknesses, helping them to increase their vocabulary and communication [12].

A number of studies have been done about the potential benefits of VR in supporting the learning process, particularly in relation to social situations, in children with autism [8]. Among the studies that have been carried out to teach children how to behave in social situation and how to understand social conventions are "Virtual Café" [14] and "Virtual Supermarket" [15]. These studies reported that the user improved in social skills and their performance also increased as the child was able to transfer the learn skills from the VE to real life situations. There have been discussions that these virtual environments can aid the educators, especially with children with special needs. This is because the VE has content which can be controlled. For example by allowing wheelchair users ".. to see how the world looks like from a standing perspective.. [and] to take part in activities or visit places that are inaccessible to real life" [16]. It is on these lines that "Autism VR" [17] drew its inspiration. The experiments carried out in "Autism VR", which brought VR into a classroom environment, prove that Virtual Reality can in fact be used as an effective teaching tool for carers of individuals on the ASD spectrum. Among the participants taking part in the study 50% of the participants reported that they were now able to better empathise with the children.

2.2 Participative Narrative

What differs Virtual Reality form conventional media outlets is the fact that users in a 3D virtual environment play a very important role in the building of the story and their own overall experience since this depends on where they move and look, and their reactions within the world itself [18]. This form of narration is therefore a participative one. Audience participation within virtual reality is an idea recently referred to as 'presence' [19]. Presence has been defined "as the experience of one's physical environment; it refers not to one's surroundings as they exist in the physical world, but to the perception of those surroundings as mediated by both automatic and controlled mental processes [20]". Perception is often discussed in relation to another crucial aspect in the context of effective VR systems [21]. Immersion in the context of virtual reality refers to the amount of sensory input that the virtual reality system creates. An increase in immersion often leads to an increase in the presence felt by the user [22, 23] and while immersion levels in the traditional approaches to VR systems usually involves the use of the visual and audio sensory mechanisms, this research looks at the possibility of incorporating the use of haptic feedback and haptic feedback devices.

2.3   Haptic Technology

Haptic Technology is defined as the technology of virtually touching and feeling objects and forces. It is a new emerging technology from the area of Virtual Reality that allows computer users to use their sense of touch to feel three-dimensional virtual objects using haptic devices. Touch is a very powerful sense and the sensation of touch is the brain's most effective learning mechanism that has for long been neglected in the field of VR. Essentially what this technology gives in terms of benefits, is the ability for a person to feel a simulation of a solid object as if it is really in front of him. Haptic feedback groups together three modalities of feedback, force feedback, tactile feedback, and proprioceptive feedback [24]. More recent research tends to couple force feedback and proprioceptive feedback under the category of kinaesthetic feedback.

2.4   Tactile Feedback

The second type of Haptic feedback is tactile feedback. While kinesthetic feedback is feedback received through the arousal of muscles, joints or tendons, tactile feedback is feedback received on the surface of the skin. Skint issue has a number of different receptors embedded in the skin and right underneath it and these allow the brain to feel things such as vibration, changes in pressure, and touch, for example. The nature of this type of feedback means that in order to trigger such feelings, one does not necessarily require hardware to be worn on the user's person as is the case with kinesthetic feedback. [34] list three types of conventional strategies for tactile feedback. The first strategy is similar to the strategy used for kinesthetic feedback, i.e. attaching devices on user's fingers or palms. This strategy has been explored in research through implementations such as CyberTouch [27], GhostGlove [28], and SaLT [29]. In implementing these approaches researchers made use of numerous pieces of equipment such as vibrotactile simulators, motor driven belts and pin-array units. However, the result of this strategy would mean that the system degrades tactile feelings due to the contact between the skin and the device occurring even when there is no need to provide tactile sensation. A second strategy for tactile feedback is also presented. An implementation of this strategy is seen in [30]. Sato et al. implement a strategy whereby the system controls the positions of tactile devices so that they only make contact with the skin when tactile feedback is required. To achieve this, Sato et al. make use of an exoskeleton master hand that generates encounter-type force feedback. The last is strategy aims at providing tactile feedback without any direct contact to a user's person. In [31], Suzuki and Kobayashi use air-jet to realize noncontact force feedback. However, when it comes to producing detailed texture sensation rather than kinetic feedback, Drif points out [32] that there are at least two major drawbacks. First, air-jet cannot produce localized force due to diffusion. Second, it also suffers from limited bandwidth. In addition, even if multiple air-jet nozzles are used, the variation of the spatial distribution of the pressure is quite limited.

2.5   Tactile Feedback Using Ultrasound

Of the three tactile feedback strategies perhaps the one which promises to take Virtual Reality to the next level of immersion and user interaction is the strategy which requires no actual contact to the user's person. Having already mentioned the use of air-jet for this purpose, other researchers have looked at eliciting tactile feedback through the use of radiation pressure of airborne ultrasound. The use of airborne ultrasound for a tactile display was pioneered by Iwamoto et al. [33]. The potential of using ultrasound for tactile feedback lies in the fact that such feedback can be applied to bare hands in free space with a high spatial and temporal resolution [33]. Whereas other strategies restrict the user in one form or another, in this scenario the user is able to move freely without any hindrance while receiving high fidelity tactile feedback with 3D visual objects.

The principle by which this method is applied in a real-world environment is based on the phenomenon of acoustic radiation pressure. In their study, Iwamoto et al. fabricated a device that comprised of 91 airborne ultrasound transducers set up in a hexagonal arrangement and could produce sufficient vibrations of up to 1 kHz. When the airborne ultrasound is applied on the surface of the skin, almost all the incident ultrasound is reflected back meaning that almost all of the incident acoustic energy is reflected on the surface of the skin. This in turn results in the tactile feedback felt by the end user.

Previous literature in the areas that are to be tackled by this study have shown potential in the approach being proposed. Having reviewed this literature, the next sections introduce the methodology behind the proposed VR implementation using tactile feedback.

3   Methodology

The scope of this project is to implement a novel tactile feeling within a VR experience of the daily life of an autistic child at a school. The suggested approach aims mainly to enhance the interaction between the designed VR environment and potential users. As a result, users are more immersed in the VR experience and can feel empathy towards these disordered children. This section contains three subsections; Haptic interaction, gesturing interaction, and interaction feedback technique to measure the different reactions of users.

3.1   Haptic Interaction

During the first phase of the VR autism application, we encountered a number of challenges when we needed to simulate some human interactions and feelings. Autism is a unique disorder and autistic children behave differently upon the same situations. One of the main interesting findings was that they share tactile frightening issues. During the first stage of the project, we managed to simulate some emotions namely fear, panic, and confusion within 3D sound and make the sound of the thoughts as it comes from the head. Even though the feedback was positive, there is still room for development to improve the experience in general. Simulating the autistic child's thoughts using 3D sound techniques will be more realistic if we manage to make users feel as if they were poked. By synchronizing both interactions, users can definitely better understand the Autistic Children hurt feelings when they are touched. Figure 1 illustrates the integration between the implemented elements that used to develop the VR experience.

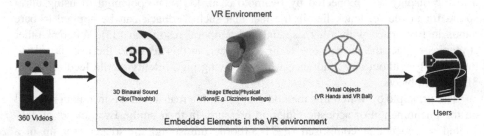

Fig. 1. The VR environment main elements (Source: Self).

3.2   Haptic Interaction

In VR autism experience, users interact with a virtual designed object. The suggested scene includes two virtual hands that simulate the action when the autistic child will be flipping a small ball. In addition, the virtual object was added to the VR environment to help users feel more immersion. While the flipping scene plays, the virtual hands start acting as if the child is actually touching a physical object. In the meantime, the user will be feeling the visualised object in addition to feeling the movement of the actual virtual simulated object. Figure 2 shows a diagram that explains the two actions; the virtual and the physical actions of the user during the VR experience. Haptic interaction will provide an artificial touch sensation that simulates natural human touch. We intend to use more virtual interaction tools to improve users' experience through increased immersion levels.

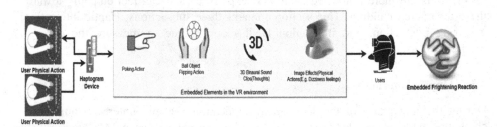

Fig. 2. Virtual and physical interaction in the VR feel autism experience (Source: Self).

3.3 Interaction Feedback

VR technology has shown great potential for simulating how autistic children's private life looks like. The first version of the application was presented in different conferences and global events and showed a great impact on users including parents who have autistic children. During the feedback collection stage, there were some difficulties with measuring direct facial reactions due to the size of the used head-mounted display hardware. Using haptic devices can easily make the feedback measuring process easier. The interaction with the virtual object allows designers observing the gesturing interaction and detect any design flows. Using different sensors, to measure the hand movement, handshaking, and hand sweating, we can collect additional insights about user interaction. Figure 3 illustrates the main used elements to mimic and measure the reactions of autistic children when they are poked.

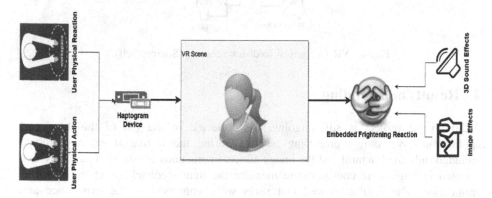

Fig. 3. Tactile virtual and physical scenario (Source: Self).

In addition, using a haptic device to create some artificial touch sensations will help developers when measuring user's physical interactions. Autistic children are extremely sensitive to situations when they are being touched. Such a situation can easily scare them, and for the purpose of this study, it can be simulated by integrating the hepatic interaction, visual effects, and 3D sounds in the head. In addition, autistic children live in repetitive life. One of the observations made when carrying out the previous study was that they like playing with the same object. This could be a toy, a tool, a remote control…etc. This proposed approach is investigating the use of the Haptogram Ultrasound device to simulate playing with a ball scenario. While in the virtual environment there will be virtual hands and a virtual ball object, the user will be able to interact with the same virtual object in reality. The capability of using Ultrasound waves to simulate the tactile experience will add more immersion feeling to the designed VR environment. In addition, users' reactions will be observed and measured during the experience. The Haptogram device will measure the movements of the hands. This will be used as a key factor for measuring the users' reactions when the autistic child responds to the poking action. Figure 4 shows the used feedback measuring technique of the physical actions.

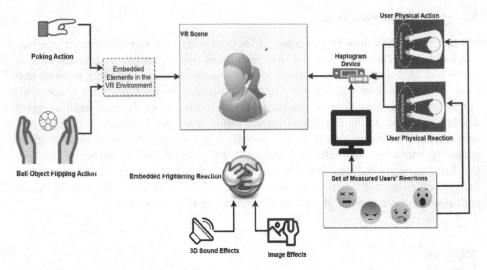

Fig. 4. VR feel autism feedback scenario (Source: Self).

4   Results and Finding

Although the project is still ongoing, the collected results out of the developed application have shown promising results. During the testing stage, a prototype included only the binaural and the image effects interactions elements. Thus, only five selected feelings were considered to measure the final feedback about the designed application. The results showed that users were immersed in the experience and strongly felt: excited, sad, anxious, or surprised. Figure 5 shows the results of users' measured reactions during the experience. Participants stated that it was a very useful application and that it was a high-quality VR experience. At the time of writing, initial

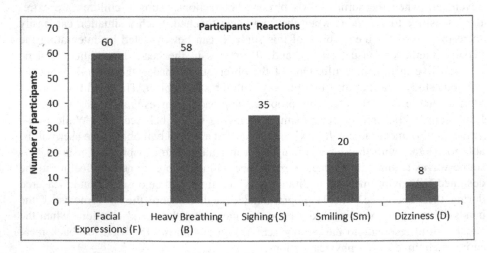

Fig. 5.   Measured feedback physical reactions (Source: Self)

observations have indicated that the used haptic tool is still expensive, and it still has some limited functions. In addition, it requires some extra integration steps to design suitable VR objects. The next section highlights a future plan for the project to overcome the pricing and the scalability of the suggested haptic tool.

5 Future Work

The second phase of the VR Autism project has enabled a new tactile interaction technique to enhance users experience and immersion. Based on the results out of the two designed versions, it is clear that there is a great opportunity to improve the developed VR experience. These improvements should ensure that the experience is capable of simulating more real-life situations and challenges that autistic children might go through when they are at school. The next stage of the project aims to design a Cave VR experience for the different situations of a normal daily life for an autistic child when at school. In addition, work has started on a collaboration endeavour with another group of researches to design a suitable hardware system. This ultrasonic system will be used mainly to simulate real-life tactile actions. The suggested novel approach will investigate the possibility of designing cheaper hardware to be installed in a VR Cave experience and implement real tactile actions. Thus, this will pave the way into adding tactile feedback capabilities in any designed VR application when tactile actions are required.

6 Conclusion

This study presents a new approach into the creation of novel VR systems targeted with the scope of helping autistic children in mind. Previous studies have shown that what we are setting out to achieve can in fact be obtained, and results have been encouraging. Through this study we are aiming to go a step further and introduce haptic technology as a means to further help these individuals in their daily lives. 'Autism VR' provided a first step into what can be achieved through the use of virtual reality as a tool to help teachers adjust their methods to better cope with autistic children. With 'Feel Autism VR' we aim to make an unprecedented leap into the creation of such tools.

References

1. American Psychiatric Association WD, Diagnostic and Statistical Manual of Mental Disorder. American Psychiatric Publishing (2000)
2. Newschaffer, C.J., et al.: The epidemiology of autism spectrum disorders. Annu. Rev. Public Health **28**, 235–258 (2007)
3. Chittaro, L., Ranon, R.: Adaptive 3D Web Sites. University of Udine (2007)
4. Strickland, D.: Virtual reality for the treatment of autism. Stud. Health Technol. Inf. **44**, 81–86 (1997)

5. McComas, J., Pivik, P., Laflamme, M.: Current uses of virtual reality for children with disabilities. Stud. Health Technol. Inf. **58**, 161–169 (1998)
6. Wang, M., Denise, R.: Virtual reality in pediatric neurorehabilitation: attention deficit hyperactivity disorder, autism and cerebral palsy. Neuroepidemiology **36**(1), 2–18 (2010)
7. Hobson, R.P.: Autism and the Development of Mind. Psychology Press (1993)
8. Goodwin, M.S.: Enhancing and accelerating the pace of autism research and treatment: the promise of developing innovative technology. Focus Autism Other Dev. Disabil. **23**(2), 125–128 (2008)
9. Kandalaft, M.R., Didehbani, N., Krawczyk, D.C., Allen, T.T., Chapman, S.B.: Virtual reality social cognition training for young adults with high-functioning autism. J. Autism Dev. Disord. **43**(1), 34–44 (2013)
10. Millen, L., Edlin-White, R., Sarah, C.: The development of educational collaborative virtual environments for children with autism. In: Proceedings of the 5th Cambridge Workshop on Universal Access and Assistive Technology, Cambridge, vol. 1. p. 7 (2010)
11. Ramachandiran, C.R., Jomhari, N., Thiyagaraja, S., Mahmud, M.M.: Virtual reality based behavioural learning for autistic children. Electron. J. e-Learn. **13**(5), 357–365 (2015)
12. Aresti-Bartolome, N., Garcia-Zapirain, B.: Technologies as support tools for persons with autistic spectrum disorder: a systematic review. Int. J. Environ. Res. Public Health **11**(8), 7767–7802 (2014)
13. Wang, X., et al.: Eye contact conditioning in autistic children using virtual reality technology. In: Cipresso, P., Matic, A., Lopez, G. (eds.) MindCare 2014. LNICST, vol. 100, pp. 79–89. Springer, Cham (2014). https://doi.org/10.1007/978-3-319-11564-1_9
14. Mitchell, P., Parsons, S., Leonard, A.: Using virtual environments for teaching social understanding to 6 adolescents with autistic spectrum disorders. J. Autism Dev. Disord. **37**(3), 589–600 (2007)
15. Herrera, G., Alcantud, F., Jordan, R., Blanquer, A., Labajo, G., De Pablo, C.: Development of symbolic play through the use of virtual reality tools in children with autistic spectrum disorders two case studies. Autism **12**(2), 143–157 (2008)
16. Cromby, J.J., Standen, P.J., Brown, D.J.: The potentials of virtual environments in the education and training of people with learning disabilities. J. Intellect. Disabil. Res. **40**(6), 489–501 (1996)
17. Martino, S.D., Haddod, F., Briffa, V., Camilleri, V., Dingli, A., Montebello, M.: Living Autism: An Immersive Learning Experience (2016)
18. Louchart, S., Aylett, R.: Towards a narrative theory of virtual reality. The Centre for Virtual Environments, University of Salford, Manchester (2003)
19. Sanchez-Vives, M.V., Slater, M.: From presence to consciousness through virtual reality. Nat. Rev. Neurosci. **6**(4), 332 (2005)
20. Gibson, J.J.: The Ecological Approach to Visual Perception. Houghton Mifflin, Boston (1979)
21. Gupta, A., Scott, K., Dukewich, M.: Innovative technology using virtual reality in the treatment of pain: does it reduce pain via distraction, or is there more to it? Pain Med. **19**(1), 151–159 (2018)
22. Slater, M., Wilbur, S.: A framework for immersive virtual environments (FIVE): speculations on the role of presence in virtual environments. Presence Teleoper. Virtual Environ. **6**, 603–616 (1997)
23. Chou, R., Gordon, D.B., de Leon-Casasola, O.A., et al.: Management of postoperative pain: a clinical practice guideline from the American Pain Society, the American Society of Regional Anesthesia and Pain Medicine, and the American Society of Anesthesiologists' Committee on Regional Anesthesia, Executive Committee, and Administrative Council. J. Pain **17**, 131–157 (2016)

24. Burdea, G.C.: Force and touch feedback for virtual reality (1996)
25. Jones, L.A.: Kinesthetic sensing. In: Human and Machine Haptics (2000)
26. Rietzler, M., Geiselhart, F., Frommel, J., Rukzio, E.: Conveying the perception of kinesthetic feedback in virtual reality using state-of-the-art hardware. In: Proceedings of the 2018 CHI Conference on Human Factors in Computing Systems, p. 460. ACM, April 2018
27. CyberTouch   (2010).   http://www.est-kl.com/products/data-gloves/cyberglove-systems/cybertouch.html
28. Minamizawa, K., Kamuro, S., Fukamachi, S., Kawakami, N., Tachi, S.: GhostGlove: haptic existence of the virtual world. In: ACM SIGGRAPH 2008 New Tech Demos, p. 18. ACM, August 2008)
29. Kim, S.C., Kim, C.H., Yang, T.H., Yang, G.H., Kang, S.C., Kwon, D.S.: SaLT: small and lightweight tactile display using ultrasonic actuators. In: The 17th IEEE International Symposium on Robot and Human Interactive Communication, RO-MAN 2008, pp. 430–435. IEEE, August 2008
30. Sato, K., Minamizawa, K., Kawakami, N., Tachi, S.: Haptic telexistence. In: ACM SIGGRAPH 2007 Emerging Technologies, p. 10. ACM, August 2007
31. Suzuki, Y., Kobayashi, M.: Air jet driven force feedback in virtual reality. IEEE Comput. Graph. Appl. 25(1), 44–47 (2005)
32. Drif, A., Citérin, J., Kheddar, A.: A multilevel haptic display design. In: Proceedings of the 2004 IEEE/RSJ International Conference on Intelligent Robots and Systems (IROS 2004), vol. 4, pp. 3595–3600. IEEE (2004)
33. Iwamoto, T., Tatezono, M., Shinoda, H.: Non-contact method for producing tactile sensation using airborne ultrasound. In: Ferre, M. (ed.) EuroHaptics 2008. LNCS, vol. 5024, pp. 504–513. Springer, Heidelberg (2008). https://doi.org/10.1007/978-3-540-69057-3_64
34. Hoshi, T., Takahashi, M., Iwamoto, T., Shinoda, H.: Noncontact tactile display based on radiation pressure of airborne ultrasound. IEEE Trans. Haptics 3(3), 155–165 (2010)

Caregivers' Influence on Smartphone Usage of People with Cognitive Disabilities: An Explorative Case Study in Germany

Vanessa N. Heitplatz[1]([✉]) [ID], Christian Bühler[1] [ID],
and Matthias R. Hastall[2] [ID]

[1] Department of Rehabilitation Technology, TU Dortmund University,
Dortmund, Germany
{vanessa.heitplatz, christian.buehler}@tu-dortmund.de
[2] Department of Qualitative Research Methods and Strategic Communication
for Health, Inclusion and Participation, TU Dortmund University,
Dortmund, Germany
matthias.hastall@tu-dortmund.de

Abstract. Intuitive handling, mobile internet access, and a large number of applications make smartphones extremely popular devices. Smartphones promise particularly high potentials for various marginalized groups. This explorative case study examines formal caregivers' attitudes towards smartphone usage and internet access by people with cognitive disabilities. Due to the close relationship to their clients, it is assumed that caregivers support or prevent smartphone usage of people with cognitive disabilities depending on their attitudes and experiences. The aim of this study is to examine which particular factors influence caregiver's attitudes towards smartphone usage. Twenty-four semi-structured interviews with formal caregivers were conducted between January and December 2018 in Germany. This paper discusses the main findings on the background of psychological and technological theories of technology acceptance and personal-growth, including self-determination-theory.

Keywords: Smartphone-usage · People with cognitive disabilities · Caregiver's influence

1 Introduction

Smartphones are extremely popular devices in Germany. In the last five years, smartphone usage among people aged fourteen and more, increased by 34% [1]. Germany is considered as one of the leading four countries regarding to smartphone penetration [2]. Smartphone usage increased rapidly not only in Germany, but also worldwide [2–5]. Internet and smartphone usage are strongly connected [1]. The most common smartphone activities can be grouped into four categories: Communication, entertainment, information research, and facilitation of daily activities [2]. This study focuses on online-based activities of smartphone usage in those four categories.

Not all people have equal opportunities to use smartphones and access the internet, a phenomenon often labeled as "digital divide" [6–8]. These divides depend on income,

© Springer Nature Switzerland AG 2019
M. Antona and C. Stephanidis (Eds.): HCII 2019, LNCS 11573, pp. 98–115, 2019.
https://doi.org/10.1007/978-3-030-23563-5_9

education, age, gender, media literacy or disabilities, among others [6–8]. The Convention on the Rights of Persons with Disabilities (UNCRPD), which was ratified by 177 states including Germany, emphasizes the importance of internet access and participation in a digital society for people with disabilities [9]. Especially people with cognitive disabilities are affected by insufficient internet access and reduced smartphone usage. The resulting disadvantages regarding social participation are manifold and encompass information acquisition, internet communication, dating, and many other aspects of daily living [10–13]. Living situations of people with cognitive disabilities are characterized by strong bonds between these individuals and their caregivers, but also a certain imbalance in power [13]. Therefore, we assume that caregivers' attitudes are affecting smartphone usage of people with cognitive disabilities. This study aims to find out about caregivers' attitudes towards smartphone usage of people with cognitive disabilities.

1.1 Smartphone-Usage and Acceptance in Society

Smartphones are important tools to enhance participation and quality of life; they are easy to use, offer various opportunities of personalization to individual needs [10, 11].

Besides these advantages, phrases like "phubbing" [14] or "nomophobia" [15] signal negative impacts of excessive smartphone usage. Addiction to smartphone usage is a common problem among adults worldwide: "It manifests itself in the excessive usage of their phones, while engaged in other activities such as studying, driving, social gatherings and even sleeping" [16]. Some recent studies also examine links between smartphone usage and negative emotional states such as stress or depression [16, 17].

The great appeal of smartphones becomes visible by people camping hours before official store openings, in order to be among the first people purchasing new models [18, 19]. However, what motivates people to do so? What makes smartphones so popular? One the one hand, some answers can be found in the smartphone characteristics described above [10, 11]. On the other hand, empirically well-tested psychological models such as the Theory of Reasoned Action (TRA) [20] and the Theory of Planned Behavior (TPB) [21], as well as technology acceptance models such as the Technology Acceptance Model 3 (TAM 3) [22] and the Cognitive Affective Normative Model (CAN Model) [23], suggest further aspects related to their popularity. Figure 2 summarizes relevant technology acceptance factors from these models. Yellow main factors are derived from TAM3, while the factor emotions (orange box) is adapted from the CAN Model. "Perceived behavioral control" originates from the TPB. Five surrounding factors are also depicted, which affect the main factors. Main factors and surrounding factors influence individuals' attitudes towards technologies, which affect intentions for technology adoption and use behavior. This integrative view on technology adoption serves as heuristic for the subsequent discussions and analyses of caregivers' attitudes towards smartphone usage.

Based on the TAM3 [22], **perceived usefulness** and **perceived ease of use** are two important main factors that influence adoption and usage behavior [22]. In view of the large amount of features that are operated used through are and touch screen interface, these factors seem particularly relevant for smartphones. The CAN-Model proposes positive and negative **emotions** as important factors influencing usage behavior [23].

People feel positive emotions when being part of social networks or in when communicating via messenger apps to the extent that these activities satisfy their needs for affiliation. Personal assessment of own competences and **perceived behavioral control** are included as main factors in TRA [20] and TRB [21]. Both impact the way people interact with technologies. Positive self-evaluation, based on positive experiences and confirmation of competences, foster usage intentions [20, 21]. As depicted in Fig. 1, other factors such as perceived job relevance, social influence, impacts on images and facilitating conditions [20–23], in turn, affect these main factors. To give an example, Venkatesh and Bala [22] show the importance of experiences and opportunities for testing out new technologies as relevant factors affecting positive or negative emotions. Similarly, social influences, facilitating conditions or perceived relevance can have a positive or negative impact on the main factors, which in turn effects the attitude. On the basis of these factors, people form an opinion which could lead to actual use or rejection of the technology. Hastall, Dockweiler and Mühlhaus [24] describe user acceptance as a dynamic process consisting of distinct phases, which could start with "not being aware of an innovation" (stage 1) and end with "sustained use" (stage 6) or "stopped use" of a technology (stage 7). Stage models like this emphasize the dynamic character of the technology adoption process and are valuable for distinguishing individuals based on their current stage and for developing stage-dependent interventions.

1.2   Caregivers' Influence on Smartphone Usage of People with Disabilities

In Germany, living situations of people with cognitive disabilities are characterized by different grades of control through caregivers. Most people with cognitive disabilities live in residential homes. These settings are characterized by high levels of caregiver control, which result in restricted self-determination and independence of people with cognitive disabilities [25, 10]. A smaller percentage of people with cognitive disabilities are living in so-called outpatient living settings. People with cognitive disabilities in these settings enjoy a larger degree of independence, as they live largely self-determined in their own apartments. Caregivers provide hourly support and assist people with cognitive disabilities in many aspects of daily living [25, 26].

Haage and Bosse [26] observed an association between living in residential homes and digital media access and usage: "Living in care homes [...] does not mean that the individuals there are given any particular help accessing digital media" [26]. In their representative survey of 610 persons with disabilities, 147 persons with cognitive disabilities were asked about their media usage and living situation. Sixty percent of persons with learning disabilities were living in residential homes. Compared to the other groups of persons with disabilities, this group showed the smallest percentage of smartphone access (34%) [26]. A similar result was reported by Zaynel [27], who found out that caregiver's attitudes, social influences and living situations are the most important factors for internet usage of young people with Down Syndrome. Living situations that are characterized by a greater level of control are intended to provide intensive support and a safe living condition particularly for people with more severe

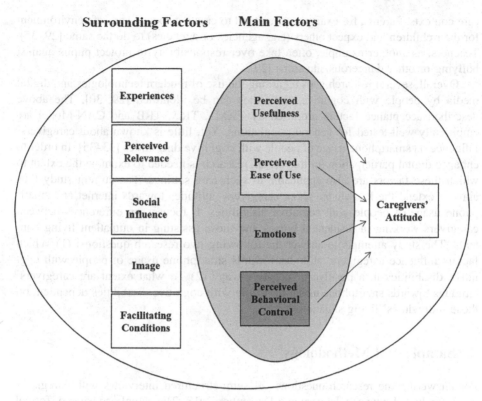

Fig. 1. Factors influencing technology adoption [own figure, based on 20, 21, 22, 23]

forms of cognitive impairment. Hence, caregivers are in a conflicting situation between providing the needed support without restricting to much the autonomy of vulnerable individuals [29–35].

Ideally, the bond between caregivers and people with cognitive disabilities is very close and strong [33]. Often, this implies reduced levels of self-determination and independence for people with cognitive disabilities. This also applies to the usage of new media and technologies like smartphone devices and the internet [29, 30]. People with cognitive disabilities try to "escape the control of the surrounding world. Without having to ask permission, they are all by themselves, capable of deciding which sites they want to visit and with whom they want to communicate. [...] they want to be like everyone else" [30]. Löfgren-Marterson [30] found out that people with cognitive disabilities are using the internet to socialize with others "beyond the control of staff and family members". They feel free to organize, plan and decide by themselves how to arrange meetings. Overall, the more institutionalized facilities are, the stronger they are characterized by heteronomy [29]. Molin [29] emphasizes the importance of caregivers' support for introducing digital media to people with cognitive disabilities: "[P]eople with cognitive disabilities need help in understanding the nature of new apps and pages". Provided support is closely linked to caregivers' levels of skills and familiarity in using new technologies [32, 33]. Similar findings were observed in other care contexts. Parents, for example, often tend to create a special and safe environment for their children and expect others (e.g. agencies or agencies) to do the same [29, 35]. Teachers, as another example, often take over responsibility to protect pupils against bullying or other dangerous situations [29].

Overall, several research gaps regarding the use of modern technologies and digital media by people with cognitive disabilities can be identified [29, 30]. The above described acceptance factors proposed by TAM3, TRA, TRB and CAN-Model are empirically well-tested for general populations. Yet, little is known about caregivers' influence on smartphone usage of people with cognitive disabilities [33–35]. In order to enhance digital participation for this group, research is needed to examine the extent to which these factors are also applicable to such care settings. The current study thus aims to extend the knowledge about caregivers' attitudes towards internet and smartphone usage of people with cognitive disabilities. It focuses on differences between caregivers working in residential homes and those assisting in outpatient living contexts. The study attempts to answer the following two research questions: (1) Which factors influence caregivers' attitudes towards smartphone usage of people with cognitive disabilities in a positive or negative way? (2) To what extent are caregivers' attitudes towards smartphone usage of people with cognitive disabilities dependent on those individuals' living situations?

2   Sample and Methodology

For answering the research questions, 24 semi-structured interviews with caregivers were conducted between January and December 2018. The sample consists of formal caregivers who work in institutions for people with cognitive disabilities in the state Northrine-Westphalia in Germany.

As described in Sect. 1.2, gown-up people with cognitive disabilities in Germany usually live either in residential institutions or in outpatient living contexts. Mirroring this distinction, eight interviewed caregivers (six males, two females) were working in residential homes which are characterized by a high degree of caregiver control. Five caregivers (four females, one male) were working in outpatient living situations, which are characterized by a smaller degree of caregiver control. Additionally, twelve experts (six females, six males) with management responsibilities working in different areas of social welfare were interviewed. All of them work with the target group of this study, but are not directly involved in their daily care. The purpose for conducting these additional interviews was to gain deeper insights into the complexity of caregiver influences from different perspectives.

All caregivers were interviewed using the same interview guide. Participation in this research was voluntary. Caregivers were informed about the interview recording and aims of the study. The interview was structured in four parts. First, caregivers were asked about their function and role in their particular institution. Second, questions about general attitudes towards digitalization and own usage of digital technologies were asked. Third, the main part of the interview guide included questions about the digital infrastructure of their institutions and their attitudes towards the use of digital technologies of their clients. Fourth, respondents were invited to give their personal outlook to the future of digital media use in care settings. The recorded interviews were fully transcribed following the transcription guidelines by Dresing and Pehl [37]. The same evaluation process has been applied to all interviews: During an open coding process of transcripts [38–40], a total number of 35 codes was identified. Following guidelines for axial coding [40], codes were analyzed, rejected or joined together. As a result, ten main codes and 43 subcodes (see Table 1) were identified. Sub codes are divided into two levels. Table 1 shows all identified main categories as well as exemplarily differentiations into sub codes for the categories "internet usage of people with cognitive disabilities", "disability characteristics", and "digital infrastructure". The current analysis is restricted to these three main categories, which emerged as most important categories. Interrelationships between different codes were determined following principles of selective coding [39, 40].

Table 1. Identified results of content analysis (own table)

Main topic	Subtopic level 1	Subtopic level 2
Internet usage of people with cognitive disabilities	- Opportunities - Risks - Suspected problems - Occurred problems	Opportunities: • Autonomy • Communication • Relationships Risks: • Data protection • Liability
Disability characteristics	- Cognitive abilities - Living situation - Income situation - Reading ability - Legal guardians	
Digital infrastructure	- Wireless LAN - Barriers - Implementation - Missing usage opportunities	Barriers: • Data protection • Liability • Access • Costs

3 Results

This discussion focuses on the three main categories "disability characteristics," "digital infrastructure" and "experiences". It is immediately evident that opportunities, support and experiences with smartphones and internet usage differ within the target group of people with cognitive disabilities. These differences can be connected to the degree of (a) institutionalization and (b) caregiver's attitudes. The result discussion below is therefore separated for these two aspects. All quotations from interviewed caregivers in this section were translated from German into English.

3.1   Characteristics of Individuals with a Disability

Caregivers mentioned characteristics of their clients such as income, living situation and cognitive abilities as important factors for their attitudes towards clients' smartphone and internet usage.

Income Situation

Many caregivers indicated barriers originating from the low income of people with cognitive disabilities. Although smartphones are described as highly popular among people with cognitive disabilities, caregivers' attitudes towards the purchase of smartphone devices are rather negative, because clients often do not possess enough money to afford them or their usage. One caregiver stated: "[S]martphones are expensive. Clients do have a maximum of 112 € pocket money per month. Maybe some of them are earning some extra money and get 150 € additional". Furthermore, not only a smartphone is needed, but also infrastructure like wireless LAN, which is often not available (see Sect. 3.2) and could lead to additional cost and efforts for institutions. These results are independent from grade of institutionalization. Low income situations are perceived as general problem that affects basically all people with cognitive disabilities.

Living Situation

Most of the people with cognitive disabilities are spending their whole life in residential institutions with high levels of caregiver control. The interviewed caregivers in those setting stated that people with cognitive disabilities are not particularly interested in using the internet. Yet, this is mainly a function of the age group of individuals living in those settings. As seen in the general population [1], internet usage decreases with higher age. As one caregiver notes: "We have 50 years as average age of residents in our institution. When they were young, they had no contact to digitalization. Therefore, they are not interested in these topics, like every other person over a certain age". Furthermore, it was mentioned that people with cognitive disabilities are satisfied with their offline activities and therefore have no desire to expand their activities into the online world. For this reason, caregivers' support in accessing new media and technologies in those settings was limited. Other tasks such as care assistance or hygiene measures were focused here. Moreover, the available digital infrastructure in many residential living situations did not provide opportunities – even for caregivers – for accessing the internet (Sect. 3.2).

In outpatient living situations, in contrast, caregivers did not see major differences to the general population regarding internet usage and smartphone ownership. People with cognitive disabilities in those settings were reported to use their smartphones mainly for communication, social media consumption, and other tasks of daily living (see Sect. 3.3 for details). All interviewed caregivers stated that all of their clients own a smartphone. The caregivers were even equipped with smartphones to assist their daily work, and reported intensive communication with their clients via messenger tools such as WhatsApp, which included information about upcoming visits or brief discussions.

Cognitive Abilities

Caregivers working in residential institutions argued that persons – due to their extent of cognitive impairments and missing reading abilities – do not sufficiently understand "how things in the online world work". Cognitive abilities to understand and to read texts were often mentioned as important preconditions for using smartphones and the internet. The handling of prepaid cards for mobile internet access was a frequently stated issue. People with cognitive disabilities have problems to understand how pre-paid card works and what needs to be done if no credit is left. Privacy issues when using social network applications like Facebook, Instagram, Snapchat or WhatsApp, and generally low media competencies, are frequently mentioned barriers. While most people in less institutionalized living situations owned a smartphone and used the internet, the opposite is true for people in more institutionalized living situations. Phones without an internet connection are more widespread in the latter context, but still not available for all individuals. This confirms the assumption of a digital disability divide within the target group depending on their living situation. The interviews of carers working in less institutionalized settings suggest that people with cognitive disabilities are capable of using smartphones and accessing the internet. Smartphones thus generally appear as suitable devices for easy access and usage, and some functions can compensate, to some extent, cognitive deficits (e.g., voice input or read-aloud functions). Thus, cognitive abilities might not be the critical limiting factor, as caregivers' attitudes and influences seem to be an even more important factor.

3.2   Digital Infrastructure

Results indicate that four constellations should be distinguished regarding the role of internet access in the care environments: (1) Limited access for employees, but no access for residents, (2) full access for employees, but no access for residents, (3) access for residents under caregivers' control, and (4) self-determined internet usage.

Low Technical Infrastructure
The first scenario is characterized by a low-level technical infrastructure, with no internet access for people with disabilities but limited access for employees. One caregiver reported that a whole team has to share one computer with slow internet access for documentation. Older institutions were often built far away from city centers: "We only have good internet access if the weather is good," one caregiver stated. In this scenario, both clients and caregiver have problems accessing the internet. Caregivers document their activities primarily via paper-and-pencil, and have limited access to digital devices, desktop computers and internet. As a result, people with cognitive disabilities have nearly no opportunities for accessing the internet. Smartphones and digital devices are virtually non-existing for these individuals. Yet, caregivers' attitudes are comparatively open-minded. Many of the interviewed caregivers did not see substantial risks of smartphone usage by their clients. Instead, they considered potential benefits of smartphone usage for their clients as rather high (see also Sect. 3.3).

No Internet Access for People with Cognitive Disabilities
In this second scenario, only employees have internet. The institutions provide a reasonable digital infrastructure, although not for their clients. Employees, in contrast, have internet access, an email address, and a desktop computer to assist their work. Due to regulations regarding data protection, liability, and fears of problems in these areas,

they do not create opportunities for internet usage or provide internet access for their clients. Interestingly, caregivers show more reluctant attitudes towards their clients' internet use, compared to the first scenario. Fears of data protection and problems due to little media literacy of people with cognitive disabilities give rise to more defensive view towards clients' internet and smartphone usage.

Internet Usage Under Caregivers' Control
Institutions in this third scenario try to offer opportunities to access the internet for their residents. Different approaches were employed; in all cases, however, some forms of internet usage monitoring by employees were established. One institution provided one computer with internet access for all clients. The computer is located close to the employees' offices, so that they can visually overlook the internet usage of their clients: "The computer is can be used by all clients. It is aligned in a way we can monitor it out of our offices". In another institution, residents were allowed to use the employees' computers to access the internet. Some organizations attempted to increase clients' media literacy skills by providing workshops for their residents. Almost all caregiver mentioned the shortage of employees and time concerns. In contrast to the last scenarios, people with cognitive disabilities were deemed able to use existing devices such as tablets, desktop computers, and laptops. Opportunities of usage were restricted due to caregivers' time concerns and attitudes. Likewise, attempts to increase clients' media literacy or to introduce them to new media technologies were limited.

Self-determined Usage
This fourth and final scenario was only found in less institutionalized living situations in which people with cognitive disabilities were living mainly self-determined. In those settings, caregiver provided the support to access digital media and the internet. Specifically, caregivers helped their clients to purchase a smartphone or to deal with internet providers. Furthermore, carers often act as contact persons for internet-related and smartphone-related problems. This includes purchases of phones, phone repairs, purchases of prepaid cards for internet access, or acting as peacemaker for conflicts resulting from WhatsApp or Facebook use. In consequence, people with cognitive disabilities in these settings were able to be more self-determined and autonomous in their decisions, because the level of institutionalization was relatively low. People with cognitive disabilities owned smartphones and were using them almost self-determined. Most had stationary internet access and were able to use their own wireless internet connection. Caregivers appeared more open-minded towards clients' smartphone use, and perceived digital media as a great opportunity for people with cognitive disabilities to participate in society. Furthermore, communication with clients became easier (see Sect. 3.1). Nevertheless, caregivers were also aware of risks (see Sect. 3.3) and saw the challenge of increasing their clients' media literacy.

Taken together, these results corroborate the assumption of a digital disability divide between individuals depending on their living settings. In less institutionalized conditions, caregivers accepted clients' internet and smartphone usage, as well as dealing with arising problems, as tasks of their daily work profile. Perceived behavioral control seems to be an important factor in this context. Helping clients with smartphone or internet problems requires technology skills, media competencies and self-confidence: "We are helping, if we are able to. Often they are better informed and have more skills than we have," one caregiver stated.

3.3 Caregivers' Experiences with Clients' Internet Use

Caregivers' attitudes towards clients' smartphone use and internet use likewise depended on previous experiences. While experiences in institutionalized living settings were rare, caregivers appeared general open-minded regarding internet usage of their clients. Yet, caregivers with more experience expressed more negative emotions, mainly due to previous problems and effects on their daily work. Especially in out-patient living situations, caregivers' work routine were strongly influenced by their clients' smartphone usage. They often had to settle disputes resulting from their clients' WhatsApp communication. Incorrect behavior in social media, such as posting bad comments or disclosure of personal data, results to intensive employment to this issues.

Stated opportunities and risks also depend on the grade of control. Caregivers in both settings see disability characteristics such as reading competences and cognitive abilities as most important factor for influencing caregiver's attitude towards this topic.

Risks
Caregivers assumed risks fall in four categories: Data protection issues, liability issues, financial risks for clients who make contracts without having sufficient money, and arising costs for institutions due to missing media literacy of their clients. Caregivers reported clients' unwanted disclosure of personal data on the internet or on social media, as well as risks of cyber-mobbing. In most cases, caregivers see themselves as responsible to solve arising conflicts. Most caregivers assumed a responsibility to protect their clients against financial risks and debts. High initial costs of smartphone purchases and high monthly rates are among the most frequently mentioned risks. Liability issues were also mentioned as barrier for clients' internet usage: What happens if people are unable to pay their rates? Who is responsible for possible expenses, or if people download illegal data or surf on pornography websites? The unclear responsibilities prevented caregivers in enabling smartphone and internet usage for people with cognitive disabilities.

Opportunities
Overall, most caregivers agreed that smartphones offer great opportunities for people with cognitive disabilities. They noted that individuals who own a smartphone became prouder and showed more self-esteem. Smartphones are a status symbol for their clients. Yet, they also can get angry and feel not being taken seriously if parents or legal guardians donate them "special phones" like phones with extra-large keys, for example. Another benefit lies in communication features. Mainly caregivers in less institutionalized living conditions reported that voice messaging provides a great communication opportunity for persons who are not able to read. Self-determination can be enhanced as people are independently able to decide who they want to contact or communicate with. For people who are scared to leave the institutions, this offers opportunities to socialize and get in contact with others.

Occurring Problems

Caregivers who are conversant with smartphone usage of their clients were also asked about actual occurring problems in addition to anticipated problems. Fortunately, no illegal downloads or issues of making contracts without having sufficient money were reported. Instead, caregivers stated that these fears have been so far unjustified. Only one case was mentioned in which a contract was not paid by a client.

More problems arise in contexts of social media usage. Caregivers reported harassments of employees by people with cognitive disabilities: "We had a client, for example, who fell in love with a caregiver. This client made a lot of pictures and posted them on Facebook without asking the affected person for permission". Problems with Facebook, WhatsApp or other social media sites were also reported: "This girl was registered on different pages to find her love on the internet. Without knowing the persons, she established contact and met them, without telling someone".

Overall, positive and negative experiences lead to positive or negative emotions regarding clients' internet and smartphone use. Caregiver in less institutionalized care settings mentioned a great impact of problems that began in the online world (e.g., mobbing, conflicts via WhatsApp), which then moved to the offline world and affected caregivers' daily tasks: "Sometimes I have to solve problems for about two hours before I am able to do my work. Conflicts starting in Facebook or WhatsApp cause conflicts among our residents".

Besides the discussed three central factors of caregivers' technology acceptance, further topics were identified in this study. Table 2 displays all ten main dimensions that emerged as relevant in the interviews. Further research is needed to better understand how these factors interact with each other, and to what extent they can be generalized to other care or usage settings.

4   Discussion

The three discussed categories "disability characteristics," "digital infrastructure" and "experiences" constitute important factors for forming caregivers' attitudes towards smartphone usage of people with cognitive disabilities. While technology acceptance factors are empirically well tested for the general population, this exploratory study examined to which extent they can be applied for understanding caregivers' attitudes towards the smartphone use of people with cognitive disabilities.

The review of existing models showed that **perceived usefulness** is an important influence factor. Results of this study indicate that only a few people with cognitive disabilities in residential institutions use smartphones. Interviewed caregivers of these caregivers reported little **experience** with this topic. Yet, they appeared generally open-minded towards clients' smartphone use, and perceive digital participation as important factor for the future. Caregivers in outpatient living situations, in contrast, had more experiences with clients' internet and smartphone use, and showed a more negative attitude towards it.

Additionally, clients' **living situation** and especially the amount of caregiver control played a major role for influencing caregivers' attitude. As Haage and Bosse [28] stated, usage of digital media depends on living situations. This current study confirms these

Table 2. Overview about identified topics

	Main topic	Subtopic level 1	Subtopic level 2
Target group: People with cognitive disabilities	Internet usage of people with cognitive disabilities	Opportunities Risks Suspected problems Occurred problems	Opportunities: <ul><li>Autonomy</li><li>Communication</li><li>Relationships</li></ul> Risks: <ul><li>Data protection</li><li>Liability</li></ul>
	Smartphone usage	Usability Rules Usage	Usage: <ul><li>Facebook</li><li>WhatsApp</li><li>Online shopping</li><li>Communication</li><li>Sexual interests</li></ul>
	Disability characteristics	Cognitive abilities Living situation Income Reading ability Legal guardians	
	Interest	Extrinsic motivation Intrinsic motivation No interest	
Structural level: Framework conditions	Institutionalization	Control Protection Laws	
	Institution	Outpatient living situation Resident institution Staffing conditions	
	Digital infrastructure	Wireless LAN Barriers Implementation Missing usage opportunities	Barriers: <ul><li>Data protection</li><li>Liability</li><li>Access</li></ul>
	Employees	Engagement Role profile Attitude Media literacy Personal limits Fear Work routine	

(*Continued*)

Table 2. (*Continued*)

Individual level: Own attitudes, experiences and competences	Attitude experts	Attitude Own usage Own media literacy Expert status	
	Wishes for the future	Wishes for the future	Wishes for the future regarding: • Own institution • Digital participation of their clients • Accessibility of digital media

results for people with cognitive disabilities in residential institutions. We examined not only the available digital infrastructure in these institutions, but also focused on caregivers' attitudes towards smartphone usage of their client. Results show that especially the context of institution and caregivers' experiences with smartphone usage in care contexts affect caregivers' attitude regarding technology use of people with cognitive disabilities. Hence, a digital divide among people with cognitive disabilities can be assumed, which largely depends on the living situation. Whereas people in residential institutions have little opportunities for smartphone use, caregivers in outpatient living situations report frequent smartphone usage of their clients. **Facilitating conditions** derived from TAM 3, such as the form of organization or equipment with digital devices (e.g. tablets, laptops, desktop computer) therefore affect the level of digital inclusion.

Another important finding of this study is the relevance of **perceived behavioral control**. As mentioned in Sect. 3.1, caregivers see themselves as contact person for every kind of questions regarding smartphones and internet use. On the one hand, those questions, activities and problems influence caregivers' work routine. On the other hand, it requires competences, technological skills and positive **experiences**, which can strengthen caregivers' self-confidence. Yet, some caregivers might fear a **lack of competences**. This means that caregivers perceive themselves as unable to control or assist activities of people with cognitive disabilities in the internet (see Sect. 3.2). Here, a divergence can be seen between loss of control, strong feelings of responsibility and enhancing self-determination and autonomy. Therefore, strong **feelings of responsibilities**, **protections** and **loss of control** are strong factors that determine caregivers' attitudes towards smartphone usage.

The results of this qualitative exploratory study indicate that neither caregivers nor people with cognitive disabilities are fully satisfied with the current status quo. Contrary to the right of self-determination, autonomy and digital participation, which are included as important goals in the UNCRPD [9], reduced possibilities for self-grow and the fulfilling of the basic human need for participation are still reality in many care settings [9, 31–34]. Nonetheless, institutions for people with cognitive disabilities have a protective function for their clients. It is a narrow ridge between protective functions and giving their clients opportunities in self-determination and personal growth.

This current study reveals gaps between needs, wishes, and rights of people with cognitive disabilities (based on literature results) and feelings of high responsibilities of caregivers that result in attempts "to protect them by restricted internet use [29]. Results of this explorative study suggest that people with cognitive disabilities in residential institutions do not get sufficient opportunities to try out new information and communication technologies. Getting back to the dynamic acceptance process proposed by Hastall et al. [24], people in those living situations are often located between stage one "not being aware" and stage two "forming an opinion about it". Caregivers are influenced by technological (e.g., technology characteristics, perceived usefulness), individual (own experiences, emotions, perceived behavioral control, motivation) and structural factors (digital infrastructure, feelings of responsibilities, institutionalization). Because of strong bonds between caregivers and people with cognitive disabilities, caregivers transfer their own attitudes, experiences and fears to their clients. Problematic are also situations in which clients have developed intentions for smartphone use, but are prevented from using digital technologies solely due to caregivers unjustified negative attitudes.

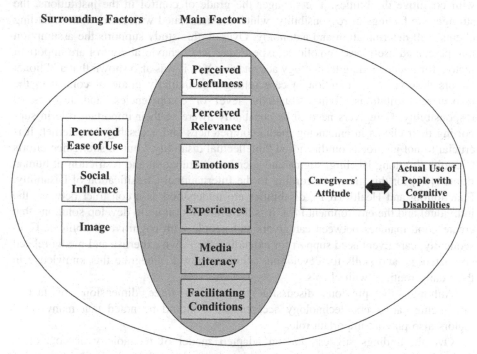

Fig. 2. Revised heuristic of factors influencing the smartphone adoption of people with a cognitive disability (own figure)

Findings also suggest that low perceived usefulness of smartphone usage by people with cognitive disabilities in residential institutions is associated with low experiences by caregivers regarding this topic. Caregivers in this setting show a general open-minded attitude towards smartphone usage by people with cognitive disabilities. In contrast, high perceived usefulness of smartphone usage by people with cognitive disabilities, combined with experiences of caregivers in outpatient living situations, are associated with more critical views of this topic. Depending on the factors described in Fig. 2, caregivers are willing or reluctant to assist people with cognitive disabilities in developing media literacy skills.

Overall, this explorative research illustrates how technology acceptance processes influence smartphone usage of people with cognitive disabilities in two usual settings of living situations in Germany, Caregivers' attitudes differ depending on living situations of their clients. To improve digital participation of people with cognitive disabilities it is therefore important to have influences of living situations and caregivers' attitudes in mind.

5 Conclusion

Smartphones and internet usage are very common in western societies, but still not for many people with cognitive disabilities. Caregivers bear responsibility towards people with cognitive disabilities. The stronger the grade of control in the institutions, the stronger are feelings of responsibility, which are associated with restrictions regarding clients' self-determination and autonomy. Overall, this study supports the assumption that perceived usefulness, emotions, experiences and behavioral control are important factors for understanding technology acceptance. Yet, it was also shown that additional factors deserve more attention in care settings, particularly grade of control in the respective institutions, living situations, level of competencies, and feelings of responsibility. Caregivers need to be constantly aware of their important role in supporting their clients in enhancing media competences and accessing the internet. It is crucial to not just focus on diagnosed "intellectual disability," but to strengthen efforts to avoid "life-long labeling, stigma and social discrimination and restriction of human rights" [41] for this group. According to the International Classification of Disability, Functioning and Health (ICF), disabilities are understood as a construct between the individual and the environment [42]. It is therefore important to develop solutions that create good matches between caregivers and people with cognitive disabilities. Consequently, caregivers need support for extending their own expertise and media-related competences, and skills for developing solutions how to integrate this knowledge in their daily routines with clients.

Although the previous discussion focused on three dimensions of factors influencing caregivers' technology acceptance, it should be noted that many other factors also play an important role.

Overall, findings suggest that an adapted model of technology acceptance is desirable to better explain the complex role of caregivers for individuals with a disability's technology usage in setting with high caregiver control. The current study is a first step towards a better understanding of this phenomenon, and hopefully inspires further research projects that examine ways to reduce digital divides for vulnerable groups in high-control care settings.

References

1. Initiative D21 e.V.: D21 Digital Index 2018/2019: Jährliches Lagebild zur Digitalen Gesellschaft. Stoba Druck, Lampertswalde (2018)
2. Worldatlas. Countries by smartphone penetration. https://www.worldatlas.com/articles/countries-by-smartphone-penetration.html. Accessed 17 Jan 2019
3. Statista. Number of smartphone users worldwide. https://www.statista.com/statistics/330695/number-of-smartphone-users-worldwide/. Accessed 17 Jan 2019
4. Rainie, L., Perrin, A.: 10 facts about smartphones as the iPhone turns 10. http://www.pewresearch.org/fact-tank/2017/06/28/10-facts-about-smartphones/. Accessed 17 Jan 2019
5. Poushter, J.: Smartphone ownership and internet usage continues to climb in emerging economies. http://www.pewglobal.org/2016/02/22/smartphone-ownership-and-internet-usage-continues-to-climb-in-emerging-economies/. Accessed 24 Jan 2019
6. Zwiefka, N.: Informelle Bildung und soziale Ungleichheit im Internet. Fischer, München (2007)
7. Zilien, N., Haufs-Brusberg, N.: Wissenskluft und Digital Divide. Nomos, Baden-Baden (2014)
8. Norris, P.: Digital Divide? Civic engagement, information poverty and the internet worldwide. https://sites.hks.harvard.edu/fs/pnorris/Acrobat/psa2000dig.pdf. Accessed 18 Jan 2019
9. United Nations. https://www.un.org/development/desa/disabilities/convention-on-the-rights-of-persons-with-disabilities.html. Accessed 24 Jan 2019
10. Shpigelman, C.N.: Leveraging social capital of individuals with intellectual disabilities through participation on Facebook. J. Appl. Res. Intellect. Disabil. **31**(1), 79–91 (2018)
11. Sachdeva, N., Tuikka, A.-M., Kimppa, K.K., Suomi, R.: Digital disability divide in information society, a framework based on structured literature review. J. Inf. Commun. Ethics Soc. **13**(3), 283–298 (2015)
12. Dobransky, K., Hargittai, E.: The disability divide in internet access and use. Inf. Commun. Soc. **9**(3), 313–334 (2006)
13. Barkhuus, L., Plichar, V.E.: Empowerment through seamfulness: smart phones in everyday life. Pers. Ubiquit. Comput. **15**(6), 629–639 (2011)
14. Klein, V.: Gemeinsam einsam: Phänomen Phubbing. Untersuchungen zur unangebrachten Smartphone-Nutzung im privaten und öffentlichen Miteinander. Information - Wissenschaft & Praxis **65**(6), 335–340 (2014)
15. Spitzer, M.: Smartphones, Angst und Stress. Nervenheilkunde **34**(8), 591–600 (2015)
16. Alhassan, A.A., Alqadhib, E.M., Taha, N.W., Alahmari, R.A., Salam, M., Almutairi, A.F.: The relationship between addiction to smartphone usage and depression among adults: a cross sectional study. BMC Psychiatry **18**(1), 1–8 (2018)
17. Selvaganapathy, K., Rajappan, R., Dee, T.H.: The effect of smartphone addiction on craniovertebral angle and depression status among university students. Int. J. Integr. Med. Sci. **4**(5), 537–542 (2017)
18. Futurezone. https://www.futurezone.de/digital-life/article215366507/Lasst-euch-Essen-liefern-waehrend-ihr-vor-dem-Apple-Store-auf-das-iPhone-wartet.html. Accessed 18 Jan 2019
19. Müller, S.: https://www.welt.de/wams_print/article988323/Amerikaner-stehen-stundenlang-Schlange-fuer-ein-iPhone.html. Accessed 19 Jan 2019
20. Rossman, C.: Theory of Reasoned Action – Theory of Planned Behavior. Nomos, Baden-Baden (2011)

21. Ajzen, I.: The theory of planned behavior. Organ. Behav. Hum. Decis. Process. **50**(2), 179–211 (1991)
22. Vekatesh, V., Bala, H.: Technoloy acceptance model 3 and a research agenda on interventions. Decis. Sci. **2**(1), 273–315 (2008)
23. Pelegrin-Brondo, J., Reinares-Lara, E., Olarte-Pascual, C.: Assessing the acceptance of technological implants (the cyborg): evidences and challenges. Comput. Hum. Behav. **70**(1), 104–112 (2017)
24. Hastall, M.R., Dockweiler, C., Mühlhaus, J.: Achieving end user acceptance: building blocks for an evidence-based user-centered framework for health technology development and assessment. In: Antona, M., Stephanidis, C. (eds.) UAHCI 2017. LNCS, vol. 10279, pp. 13–25. Springer, Cham (2017). https://doi.org/10.1007/978-3-319-58700-4_2
25. Thimm, A., Rodekohr, B., Dieckmann, F., Haßler, T.: Wohnsituation Erwachsener mit geistiger Behinderung in Westfalen-Lippe und Umzüge im Alter, Ketteler, Bönen (2018). https://doi.org/10.1007/978-3-319-58700-4_2
26. Hasebrink, U., Bosse, I.K., Haage, A.: Mediennutzung von Menschen mit Beeinträchtigungen. Medienbezogene Handlungen, Barrieren und Erwartungen einer heterogenen Zielgruppe. Media Perspektiven **3**, 145–156 (2017)
27. Zaynel, N.: Internetnutzung von Jugendlichen und jungen Erwachsenen mit Down-Syndrom. Springer, Wiesbaden (2017). https://doi.org/10.1007/978-3-658-18405-6_5
28. Haage, A., Bosse, I.K.: Media use of persons with disabilities. In: Antona, M., Stephanidis, C. (eds.) UAHCI 2017. LNCS, vol. 10279, pp. 419–435. Springer, Cham (2017). https://doi.org/10.1007/978-3-319-58700-4_34
29. Molin, M., Sorbring, E., Löfgren-Martenson, L.: Teachers' and parents' views on the Internet and social media usage by pupils with intellectual disabilities. J. Intellect. Disabil. **19**(1), 22–33 (2015)
30. Löfgren-Marterson, L.: Love in cyberspace: swedish young people with intellectual disabilities and the internet. Scand. J. Disabil. Res. **10**(2), 125–138 (2008)
31. Hoppestad, B.S.: Current perspective regarding adults with intellectual and developmental disabilities accessing computer technology. Disabil. Rehabil. Assistive Technol. **8**(3), 190–194 (2013)
32. Chadwick, D.D., Quinn, S., Fullwood, C.: Perceptions of the risks and benefits of internet access and use by people with intellectual disabilities. Br. J. Learn. Disabil. **45**(1), 21–31 (2017)
33. Seale, J.: The role of supporters in facilitating the use of technologies by adolescents and adults with learning disabilities: a place for positive risktaking? Eur. J. Special Needs Educ. **29**(2), 220–236 (2014)
34. McConkey, R., Smyth, M.: Parental perceptions of risks with older teenagers who have severe learning difficulties contrasted with the young people's views and experiences. Child. Soc. **17**, 18–31 (2003)
35. Bühler, C., Dirks, S., Nietzio, A.: Easy access to social media: introducing the mediata-app. In: Miesenberger, K., Bühler, C., Penaz, P. (eds.) ICCHP 2016. LNCS, vol. 9759, pp. 227–233. Springer, Cham (2016). https://doi.org/10.1007/978-3-319-41267-2_31
36. Dirks, S., Bühler, C.: Assistive technologies for people with cognitive impairments – which factors influence technology acceptance? In: Antona, M., Stephanidis, C. (eds.) UAHCI 2018. LNCS, vol. 10907, pp. 503–516. Springer, Cham (2018). https://doi.org/10.1007/978-3-319-92049-8_36
37. Dresing, T., Pehl, T.: Praxisbuch Interview & Transkription. Regelsysteme und Anleitungen für qualitative ForscherInnen, 4 Auflage. Eigenverlag, Marburg (2012)
38. Strauss, A.L.: Qualitative Analysis for Social Scientists. Cambridge University Press, New York (1987)

39. Glaser, B.G., Strauss, A.L.: The Discovery of Groundet Theory. Adline, Chicago (1967)
40. Zaynel, N.: Prozessorientierte Auswertung von qualitativen Interviews mit Atlas.ti und der Grounded Theory. In: Scheu, A.M. (ed.) Auswertung qualitativer Daten, pp. 59–68. Springer, Wiesbaden (2018). https://doi.org/10.1007/978-3-658-18405-6_5
41. Chadwick, D., Wesson, C., Fullwood, C.: Internet access by people with intellectual disabilities: inequalities and opportunities. Future Internet 5, 376–397 (2013)
42. ICF. https://www.dimdi.de/dynamic/de/klassifikationen/icf/. Accessed 2 Jan 2019

The PTC and Boston Children's Hospital Collaborative AR Experience for Children with Autism Spectrum Disorder

David Juhlin[1], Chris Morris[1], Peter Schmaltz[1], Howard Shane[2],
Ralf Schlosser[2,3(✉)], Amanda O'Brien[2], Christina Yu[2],
Drew Mancini[2], Anna Allen[4], and Jennifer Abramson[2]

[1] PTC Inc, 121 Seaport Blvd, Boston, MA 02210, USA
cmorris@ptc.com
[2] Boston Children's Hospital, 9 Hope Ave, Waltham, MA 02453, USA
R.Schlosser@northeastern.edu
[3] Northeastern University, 360 Huntington Ave, Boston, MA 02115, USA
[4] Puddingstone Place LLC, 24 Washington St, Suite 110, Wellesley,
MA 02481, USA

Abstract. Minimally verbal children with Autism Spectrum Disorder often face challenges in the areas of language, communication, and organization. Augmented Reality (AR) may provide a valuable technique to enhance language learning as well as navigating tasks and activities. The "AR experience," developed by PTC Inc with Boston Children's Hospital, was a collaborative effort to provide a working tool for use in studies regarding the effectiveness of AR as a teaching tool for minimally verbal children with Autism Spectrum Disorder. The purpose of this paper is to describe (a) the development of the application using the "Design Thinking Process," (b) describe the features of the resulting AR experience, and (c) present initial evaluation results.

Keywords: Augmented Reality · Autism Spectrum Disorder ·
Language and communication

1 Background

1.1 Children with ASD

Children with ASD tend to have difficulties in the areas of language and communication, and it is therefore only appropriate that these characteristics are critical elements for diagnosing ASD [1]. Delays in language and communication of children with ASD occur in the areas of speech articulation, word use, syntax, comprehension, and pragmatics [2]. Depending on the source consulted, between 20% and about 60% of the population of individuals with ASD are estimated to fail to develop functional speech skills [3, 4].

The expressive language difficulties of children with ASD have been met with considerable intervention efforts [5–9]. Receptive language issues, on the other hand, have received relatively less attention despite considerable difficulties [10]. This includes the understanding prepositions. Its becomes evident when observing how frequently this skill

© Springer Nature Switzerland AG 2019
M. Antona and C. Stephanidis (Eds.): HCII 2019, LNCS 11573, pp. 116–122, 2019.
https://doi.org/10.1007/978-3-030-23563-5_10

is needed on a daily basis [11]. For example, a teacher might (a) refer to the location of an object (e.g., your coat is *next to* the green one), (b) state where an object should be placed (e.g., put your lunch box *on* the shelf), and (c) instruct the child where to position herself (e.g., stand *behind* Oliver). Some children with ASD also have difficulties with transitioning between activities and following step-by-step procedures toward task completion. Hence, they may benefit from visual schedules and task instructions [12]. In general, it has been recognized that children with ASD often prefer and greatly benefit from instruction in the visual modality [13].

1.2 Augmented Reality

Advances in technology have resulted in AR as perhaps the latest and most exciting new approach to instruction, including the supplementing of spoken instruction to children with ASD [14]. Unlike virtual reality, AR is directly related to reality and mixes aspects of reality (e.g., objects and containers on the table top that need to be put in relation to one another based on a prepositional phrase; e.g., "Put the ball in the cup") with computer-generated information (a visual illustration of the action that can be called up by holding the iPad over the object and container). As a result, some argue that AR does "not require as much capacity for abstraction as VR and people with autism who do not have abstraction capacity could benefit from their use" [15]. AR application have been developed and studied in children with ASD in several ways [16–19], but none have focused on receptive language skills which is targeted here.

2 Development Process

2.1 Definition of Strategy and Discovery

The AR experience developed by PTC with Boston Children's Hospital (BCH) was a collaborative effort to provide a working tool for use in studies regarding the effectiveness of Augmented Reality as a learning aid for children with ASD.

The development process used was the Design Thinking Process. This process began with the definition of the strategy as stated previously. This was followed by a discovery research workstream in which a variety of stakeholders were interviewed including the Boston Children's Hospital clinicians from the Autism Language Program, 3 parents of children with ASD, 3 special education teachers and 1 speech-language pathologist. These interviews were supplemented with online research into the effectiveness of techniques for teaching and learning with autistic children, such as the use of video training and other interactive methods. In conclusion of the discovery, the key insights gained were that:

- There is a wide range of ASD that is not in a linear scale, instead there are many dimensions
- Age is defined as chronological and development age. This program would focus on development ages of 3–5 years old
- Six common problems are encountered for children with ASD:
  - Learning to do basic things in their daily life
  - Transferring a skill to a new context
  - Attention

- Learning language & expressing themselves
- Social interactions
- Changes in routine/knowing what to expect
- Additional insights:
  - Multi-sensory learning and limitations
  - Need for motivation
  - Move quickly from one thing to another
  - May need longer time to process
  - Take things literally (piece of cake)
  - Parents are the heroes

2.2 Design Workshop

Design thinking principles were next used to conduct a Define Workshop to identify the best set of user stories that would provide the best outcome. These stories were then reviewed and validated with several groups including BCH clinicians, parents of children with ASD and technology specialist at PTC. Upon determination of the specific user story, a script was built to illustrate the process through which the clinician and child would interact with physical objects in combination with the digital representation in the AR experience.

Key design elements:

- Must be usable by parents in a non-clinical setting
- Parents have limited time
- Limited transferable content from one individual to another
- Treatment plan is not always executed correctly
- Parents do not know what will work for their child

As a result, three high level concepts were developed (see Fig. 1).

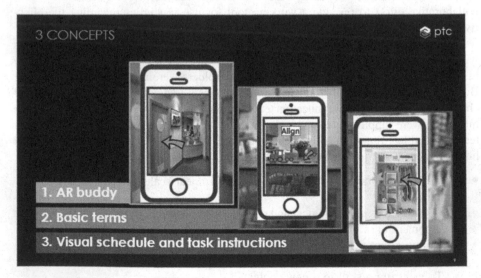

Fig. 1. The AR Buddy, a Visual schedule and tasks instructions, and a Basic terms app.

For each of these concepts a visual story board was produced describing both the interaction as well as the actions within the app to map that action (see Fig. 2).

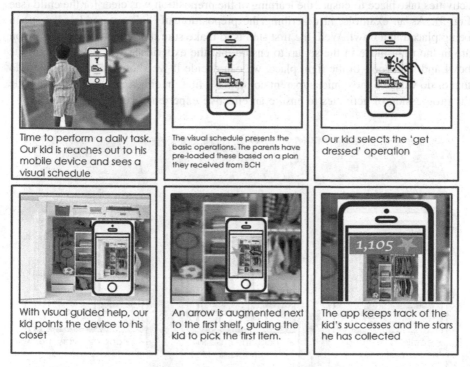

Time to perform a daily task. Our kid is reaches out to his mobile device and sees a visual schedule

The visual schedule presents the basic operations. The parents have pre-loaded these based on a plan they received from BCH

Our kid selects the 'get dressed' operation

With visual guided help, our kid points the device to his closet

An arrow is augmented next to the first shelf, guiding the kid to pick the first item.

The app keeps track of the kid's successes and the stars he has collected

Fig. 2. Visual story boards for each of the three concepts.

Each of these concepts was then validated with the clinical team at the BCH based on three key identified measures of success:

1. The solution holds the child's attention
2. The child can perform the task with assistance of AR
3. The child can eventually perform the task without the aid of AR

This early validation process saved several iterations and thus development costs. The basic terms app was chosen as the right AR experience to develop for phase 1.

2.3 Visualization Process

Continuing with Design Thinking principles a visualization process was initiated based on the decision to model key prepositions. Both physical and digital mockups were used in the evaluation and creation of the user sequences and for the modeling of the User Interface.

Key design considerations were required to ensure a successful AR experience, such as the ability to do comparisons of objects between the digital experience and the physical object. The closer these are in both scale and likeness, the more effective that experience is in teaching the concept by eliminating distractions resulting for differences in objects. Additionally, it was critical that a simple progression of tasks and activities take place to ensure the learning of the preposition was clear for the child (see Fig. 3). As an example, in describing the preposition of "in", we modeled a spoon being placed in a bowl. Well, the first step is to make sure that the bowl and the spoon are in the right place in the set up to ensure that the experience will work. To get the bowl and the spoon in the right place, we can decide if we show the child how to do that or do we have the clinician/parent do that task first. In either case, we had to decide the progression of activities to ensure an effective experience.

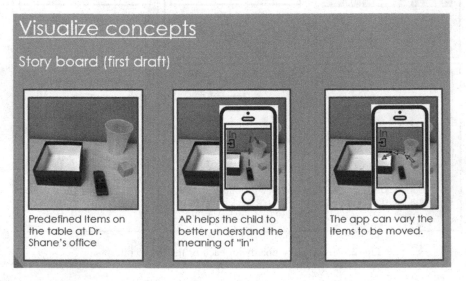

Fig. 3. First draft of a story board to illustrate the directive of placing an object in a container.

2.4   Build Process

The build process began in earnest upon completion of the UX design based on the preliminary visualization story boards created. In each step of the build and design process, the staff at the BCH were providing feedback on the approach and the experience. This was done thru email exchanges, and live meetings in which demonstrations were provided.

The application was built leveraging technology from PTC. The primary product used was Vuforia Studio, which is built on top of the Vuforia Engine (the most common AR development platform). The primary reason for the use of Vuforia studio is the ease of implementation and rapid prototyping. Studio allows for importing of 3D models and placing these models into a 3D space without having to deal with the underlying code.

In addition, Vuforia Studio supports ThingMark technology which is marker-based tracking. The ThingMark technology has an advantage over QR codes due to allowing the association of multiple URLs to a single ThingMark. This allows one single ThingMark to store multiple versions as the application was being developed. The animation of the movements of the objects within the application was built using Java script since it integrates well with Vuforia Studio. Vuforia Studio is hosted in the cloud, but the experiences can also be downloaded to the application and run in an offline mode. To create some of the 3D models, Creo 5.0 was used.

Upon successful completion of Phase 1, a Phase 2 project has begun using the same Design Thinking principles to expand the scope of the experience to use more prepositions and the introduction of several verbs with a child's toy farm set to make the experience a bit more relatable for the children. Initial data evaluating the use of the application by children with ASD are still being collected.

References

1. American Psychiatric Association: Diagnostic and statistical manual of mental disorders, 4th edn. Washington
2. Tager-Flusberg, H., Paul, R., Lord, C.: Language and communication in autism. In: Volkmar, F.R., Paul, R., Klin, A., Cohen, D. (eds.) Handbook of Autism and Pervasive Developmental Disorders, Diagnosis, Development, Neurobiology, and Behavior, vol. 1, 3rd edn. pp. 335–364. Wiley, Hoboken (2005)
3. Peeters, T., Gillberg, C.: Autism: Medical and Educational Aspects. Whurr Publishers, London (1999)
4. Prizant, B.M.: Language, social and communication development in autism. J. Autism Dev. Disord. **26**, 173–178 (1996)
5. Bellini, S., Akullian, J.: A meta-analysis of video modeling and video self-modeling interventions for children and adolescents with autism spectrum disorders. Except. Child. **73**, 264–287 (2007)
6. Mirenda, P., Iacono, T.: Autism Spectrum Disorders and AAC. Paul H. Brookes, Baltimore (2009)
7. Rogers, S.: Evidence-based interventions for language development in young children with autism. In: Charman, T., Stone, W. (eds.) Social & Communication Development in Autism Spectrum Disorders: Early Intervention, Diagnosis, & Intervention, pp. 143–179. Guilford Press, New York (2006)
8. Tager-Flusberg, H., Rogers, S., et al.: Defining spoken language benchmarks and selecting measures of expressive language development for young children with autism spectrum disorders. J. Speech, Lang. Hear. Res. **52**, 643–652 (2009)
9. Schlosser, R.W., Koul, R.: Speech output technologies in interventions for individuals with autism spectrum disorders: a scoping review. Augmentative Altern. Commun. **31**, 285–309 (2015)
10. Sevcik, R.A.: Comprehension: an overlooked component in augmented language development. Disabil. Rehabil. **28**, 159–167 (2006)
11. Egel, A.L., Shafer, M.S., Neef, N.A.: Receptive acquisition and generalization of prepositional responding in autistic children: a comparison of two procedures. Anal. Interv. Dev. Disabil. **3**, 285–298 (1984)
12. Shane, H.C., Laubscher, E., Schlosser, R.W., Flynn, S., Sorce, J.F., Abramson, J.: Applying technology to visually support language and communication in individuals with ASD. J. Autism Dev. Disord. **42**, 1228–1235 (2012)

13. Shane, H., et al.: Enhancing communication for individuals with autism: a guide to the visual immersion system. Paul H. Brookes Publishing Company, Baltimore (2014)

14. Walker, Z., McMahon, D.D., Rosenblatt, K., Arner, T.: Beyond pokemon: augmented reality is a universal design learning tool. SAGE Open 7(4), 1–8 (2017)

15. Herrera, G., Jordan, R., Gimeno, J.: Exploring the advantages of augmented reality for intervention in ASD. In: Proceedings of the World Autism Congress South Africa (2006)

16. Bai, Z., Blackwell, A., Coulouris, G.: Through the looking glass: pretend play for children with autism. In: International Symposium on Mixed and Augmented Reality 2013 Science and Technology Proceedings, pp. 1–4. IEEE, Adelaide (2013)

17. Casas, X., Herrera, G., Coma, I., Fernandez, M.A.: Kinect-based augmented reality system for individuals with autism spectrum disorders. In: Proceedings of GRAPP/IVAPP, pp. 240–246. SciTePress (2012)

18. Escobedo, L., Tentori, M., Quintana, E., Favela, J., Garcia-Rosas, D.: Using augmented reality to help students with autism to stay focused. IEEE Pervasive Comput. 13(1), 38–46 (2014)

19. Hayes, G.R., Hirano, S., Marcu, G., Monibi, M., Nguyen, D.H., Yeganyan, M.: Interactive visual supports for children with Autism. Pers. Ubiquit. Comput. 14(7), 663–680 (2010)

Design of an Intelligent and Immersive System to Facilitate the Social Interaction Between Caregivers and Young Children with Autism

Guangtao Nie[1]([⊠]), Akshith Ullal[1], Amy R. Swanson[2],
Amy S. Weitauf[2], Zachary E. Warren[2], and Nilanjan Sarkar[1,3]

[1] EECS, Vanderbilt University, Nashville, TN 37212, USA
Guangtao.nie@vanderbilt.edu
[2] TRIAD, Vanderbilt University Medical Center, Nashville, TN 37212, USA
[3] Mechanical Engineering, Vanderbilt University, Nashville, TN 37212, USA

Abstract. Children with autism spectrum disorder (ASD) have core deficits in social interaction skills. Intelligent technological systems have been developed to help children with ASD develop their social interaction skills, like response to name (RTN), response to joint attention (RJA), initiation of joint attention (IJA) and imitation skills. Most existing systems entail human-computer interaction (HCI) or human-robot interaction (HRI), in which participants interact with the systems to elicit certain social behaviors or practice certain social skills. However, because the robot/computer being the only therapeutic factor in HRI/HCI systems, this may result in the isolation effect. Therefore, in this work, an intelligent and immersive computer system is proposed for caregivers and their young children with ASD to interact with each other and help develop social skills (RTN and IJA). In this computer assisted HHI setting, caregivers deliver social cues to participants (young children with ASD) and give a decision-making signal to the system. The system also provides different non-social cues, to help caregivers to elicit and reinforce the social behaviors of participants. By including a caregiver in the loop, we hope to ameliorate the isolation effect by creating a more real-world HHI scenario. In this paper, we will show the feasibility of the proposed system and validate its potential effectiveness by both subjective measurements and objective measurements.

Keywords: Autism spectrum disorder (ASD) ·
Computer-assisted human-human interaction · Response to name (RTN) ·
Initiation of joint attention (IJA)

1 Background

1.1 Autism Spectrum Disorder

Autism Spectrum Disorders (ASD) are a group of developmental disabilities characterized by impairments in social interaction and communication [1]. According to estimates from CDC's Autism and Developmental Disabilities Monitoring (ADDM) Network, 1 in 59 children is believed to be identified with ASD [2]. ASD occurs in all racial, ethnic and socioeconomic groups [2]. Intelligent technological systems have

© Springer Nature Switzerland AG 2019
M. Antona and C. Stephanidis (Eds.): HCII 2019, LNCS 11573, pp. 123–132, 2019.
https://doi.org/10.1007/978-3-030-23563-5_11

been developed to help children with ASD develop their social interaction skills, like response to name (RTN) [3], response to joint attention(RJA) [4–6], initiation of joint attention(IJA) and imitation skills [7, 8]. Early intervention for young children with ASD may differ from treatment for older children due to the developmental differences in their social relationships, cognitive and communicative processes and learning characteristics [9]. The proposed system in this paper, which is designed for very young children with ASD, focuses primarily on setting up a natural learning environment for child initiative acts, development of nonverbal intentional communicative acts and reciprocal play with social partners [9]. The two tasks in this system, response to name (RTN) and initiation of joint attention (IJA), were designed based on the developmental considerations and utilize child gaze as the fundamental measurement to influence the process of interaction.

1.2   Computer-Assisted Human-Human Interaction

Over the past several decades, computer-assisted human-human interaction has been developed to facilitate cooperative work and job efficiency. The early and influential survey of computer-supported cooperative work was conducted by R. Johansen [10], who defined and illustrated multiple approaches and applications of computer-supported cooperative work, including software such as the online meeting, screen sharing, project management and calendar management software for groups. Here we specify the HCI scheme we adopted in the proposed system as computer-assisted human-human interaction, which theoretically belongs to an example of the groupware idea of Johansen in [10].

Most existing intelligent systems for children with ASD entail HCI or HRI, in which participants interact with the systems to elicit certain social behaviors or develop certain social skills. However, due to the robot/computer being the only therapeutic factor in HRI/HCI systems, the isolation effect has been reported, where after gaining some social skills within the HRI/HCI systems, one may not be able to transfer the skills back into real world HHI [11, 12]. Therefore, the computer-assisted HHI scheme was adopted here, which incorporated caregivers in the interaction loop to help ameliorate the isolation effect.

2   System Description

2.1   System Architecture and Environment

The proposed computer assisted HHI is depicted in Fig. 1, and the system environment is shown in Fig. 2, and the system architecture is shown in Fig. 3. This system was designed based upon our existing work in [3] which utilized a closed-loop interaction protocol between participants and the system. Introducing caregivers into the system has several advantages. Caregivers can provide real social cues, such as calling a child's name. In the previous work, the system provided the social cue by playing prerecorded audio.

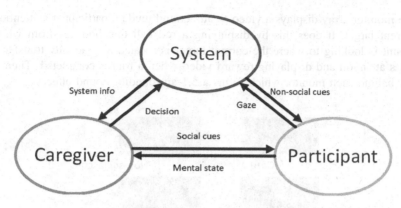

Fig. 1. Computer-assisted HHI

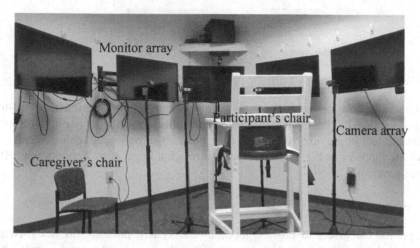

Fig. 2. System environment

We designed our system to reward children for responding to their actual caregivers, so that the elicited and reinforced social behaviors could be more easily transferred back to the real world. Our system also has caregivers use a tablet app to influence the system process based on real-time observation of participant behavior, which makes the system more adaptive and individualized. In our previous work, the system monitored nothing but the gaze of participant to do a closed-loop interaction in a fixed protocol. Also, participant mood and engagement (e.g., gestures, facial expressions) can now be considered based on caregiver input. All of these changes expand upon our previous work to create a more individualized, adaptable, and generalizable system for potential intervention.

From Fig. 2, one can see that the child sits in the center of a camera array in the shape of semi-circle with a radius of 90 cm. The caregiver sits in the small chair under the left most monitor.

The monitor array displays a video to attract and guide a participant's attention to the current target. It does this by displaying a red ball that bounces from where a participant is looking to where the current target is located to gradually transfer participant's attention and displaying reward video when a trial is completed. There is a speaker behind each monitor which forms a 5.1 surrounding sound effect.

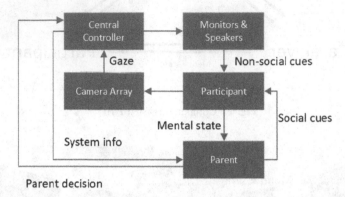

Fig. 3. System architecture

The camera array covers 180° in yaw in front of the participant to track the real-time head pose. This is used as input to the central controller to specify the starting position of the guiding ball or provide feedback to caregivers about participant engagement.

All the modules in Fig. 3 are connected through a Local Area Network with IPV4 Internet protocol.

2.2 Tablet App Design

The app for caregivers to provide input to the system process was designed and implemented using Unity with C#. As we wanted the caregiver to spend as much time as possible on observing participant's mental state and performance, the app was designed to be neutral and simple (e.g., showing only one large button to provide input; showing a question with no more than three response options available). The app had two primary functions:

1. Basic interaction process control function: Call name, Pause video, Reward. These buttons are available at different times, and only one at a time, based on the interaction stage. Details about their availabilities are described in Subsect. 2.4.
2. System prompted question for decision making of non-social assistance (e.g., audio of the video and bouncing ball) and performance feedback (e.g., engagement level and task difficulty level).

2.3 Input and Output of System, Caregiver and Participant

Within our new system, the caregiver observes the participant's mental (such as emotional distress, engagement, and attention) and calls the child's name (the social cue). The system monitors the real time gaze of participant and provides several non-social cues, including:

1. Pictures, audio and video
2. Moving objects (e.g., bouncing ball) across monitors (starting from where participant is looking at to the target monitor)

The information and options that the system provides for caregiver:

1. Task and trial information
2. Calling name/ Pausing video
3. Triggering moving objects

The system input from caregiver:

1. Feedback about participant's mental state (e.g., engagement level, frustration level, etc.)
2. Decision to influence the system process

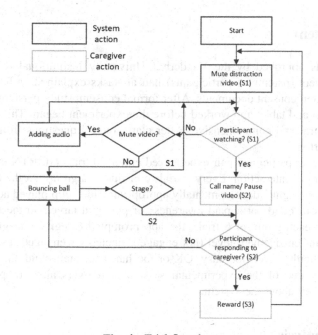

Fig. 4. Trial flowchart

2.4 Task Setup

The detailed information about the interaction scheme implemented in this system is plotted as a flowchart in Fig. 4. Two tasks are defined for this system: response to name (RTN) and initiation of joint attention (IJA). The general procedure of two tasks are described below:

RTN: 1. Play a video clip (e.g., distraction video) on a monitor to distract participant's attention away from caregiver to distraction video.

2. When caregivers press a button to confirm that participants are watching the distraction video, the app prompts the caregiver to call the participant's name.

3. If the child does look, caregivers press a button on the app to confirm that participants respond to them by looking at them. The system then plays a reward video on the monitor just over the caregiver's head and starts the next trial.

IJA: 1. Play a video clip (e.g., distraction video) on a monitor to distract participant's attention away from caregiver.

2. When caregivers press a button to confirm that participants are watching the distraction video, the system will hide the distraction video.

3. When participant responds to caregiver by looking at him/her, the system resumes playing the hidden video on the monitor over caregiver's head.

If a participant doesn't look at the distraction video or caregiver at a certain stage within 7 s, tablet app will prompt the caregiver to trigger non-social cues such as audio or bouncing ball.

3 Experiment

This study was approved by the Vanderbilt University Institutional Review Board (IRB). Caregivers (parents of participants) had all tasks explained verbally and then completed written consent documents. After formal consent, study personnel explained how the system and tablet app worked before the experiment began. This introduction script was neutral and comprehensive. After clarifying all questions from caregivers, the session started.

Each caregiver-participant pair experienced either 20 trials (10 RTN trials+ 10 IJA trials) or 20 min of interaction. Every 5 trials constituted a group. At the beginning of each group, an engaging, developmentally appropriate video clip played across the five target monitors to build participant awareness of potential targets in the system environment. After each group of 5 trials, the app prompted caregivers to give feedback about participant engagement level (not engaged, unclear or engaged) as well as perceived task difficulty level (too easy, OK or too hard) for their children.

At the very end of the experimental session, caregivers also completed a user experience survey about the system.

4 Data Analysis

Six participants (3 TD, 3 ASD) were recruited to validate the feasibility of the system. Their average age was 25.8 months (SD = 8.2). The ratio of male: female was 2:4. In this section, both subjective and objective measurements will be reported to validate the potential intervention effectiveness of the proposed system.

4.1 Objective Measurement

Response time was defined here as the time elapsed from when a caregiver called the child's name/paused the video to the time of caregiver's pressing the button and confirming that participants responded (turned and looked).

The averaged response time of RTN and IJA trials is shown in Figs. 5 and 6, respectively. Note that a shorter response time indicates a better performance.

Based on the Figs. 5 and 6, several preliminary findings are provided here:

1. For both RTN and IJA trials, TD generally performed better than ASD and the performance of both groups fluctuates with the session time going on.
2. From the pre-(Trial #1) and post-(Trial #10) comparison perspective: TD group's performance of RTN and IJA both decreased a bit. ASD group's performance of RTN decreased a lot while that of IJA increased a lot.

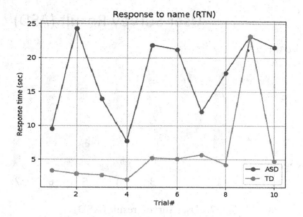

Fig. 5. Response time of RTN trials

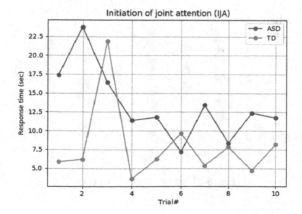

Fig. 6. Response time of IJA trials

4.2   Subjective Measurement

The caregivers' subjective feedback about the system design and user experience are shown in Figs. 7 and 8.

Here are the survey questions:

1. In general, what do you think of the design of the tablet app?
2. How much do you think the system could help you teach your child?
3. Do you like the RTN task design?
4. Do you like the IJA task design?
5. How much do you think your child liked the system?
6. In general, what do you think of the system?
7. Do you think your child's RTN skill improved through this session?
8. Do you think your child's IJA skill improved through this session?

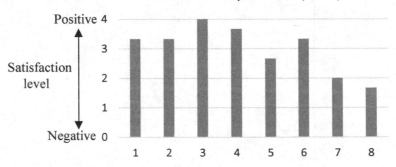

Fig. 7. User survey result (ASD)

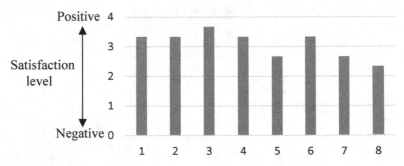

Fig. 8. User survey result (TD)

From the two user surveys above, we can see that caregivers were relatively satisfied with the tablet app design (3.33/4) and overall system design (3.33/4) and caregivers liked both task designs (RTN-3.83/4, IJA-3.5/4).

From the pre- and post- comparison perspective, a correlation test was conducted between the objective performance measurement of skill improvement and subjective performance measurement of skill improvement:

1. The correlation between RTN performance variation between Trial #1 and Trial #10 and user survey question 7: $r = -0.292$, $p = 0.574$
2. The correlation between IJA performance variation between Trial #1 and Trial #10 and user survey question 8: $r = 0.495$, $p = 0.318$.

These two correlation tests indicate that the caregiver's impression of RTN improvement is consistent with objective measurement while caregiver's impression of IJA improvement is inconsistent with objective measurement.

5   Conclusions

5.1   Achievements

We designed and implemented an intelligent and immersive system for caregiver-participant pairs to practice social interaction skills, specifically RTN and IJA. We incorporated caregivers into the system potentially to ameliorate the isolation effect which occurred in other HRI/HCI. We also provided uniform assistance for caregivers to trigger the social event and transfer participants' attention. Standardizing options for caregivers' behaviors within the system enabled us to compare different time and participant groups.

Subjective reports from caregivers showed positive results regarding system and task design. Children with TD performed better than children with ASD across both tasks, as predicted by the literature, which suggests that our system captured real differences in social responsiveness that distinguish diagnostic groups. Additionally, on IJA tasks, children with ASD showed increased performance by the end (Trial #10), compared with baseline (Trial #1), which is a promising result regarding the potential effectiveness of this system.

5.2   Limitations

The weaknesses of this paper are as follows:

First, more participants need to be recruited in the future to have a comprehensive test to validate the effectiveness of the proposed system.

Second, the intervention effect of RTN of ASD were not as promising as IJA of ASD so far, based on the objective analysis in Figs. 5 and 6. If with more caregiver-participant pairs recruited, we still cannot see a promising RTN skill improvement of ASD, we may need to modify our system or task design.

Third, the task designs lacked variation. The lack of variation could lead to participant losing social interest towards the system in a long-term interaction.

References

1. Lord, C., Cook, E.H., Leventhal, B.L., Amaral, D.G.: Autism spectrum disorders. Autism Sci. Ment. Heal. **28**(2), 217 (2013)
2. Baio, J., et al.: Prevalence of autism spectrum disorder among children aged 8 years—Autism and developmental disabilities monitoring network, 11 sites, United States, 2014. MMWR Surveill. Summ. **67**(6), 1 (2018)
3. Zheng, Z., et al.: Design of an autonomous social orienting training system (ASOTS) for young children with autism. IEEE Trans. Neural Syst. Rehabil. Eng. **25**(6), 668–678 (2017)
4. Zheng, Z., Zhao, H., Swanson, A.R., Weitlauf, A.S., Warren, Z.E., Sarkar, N.: Design, development, and evaluation of a noninvasive autonomous robot-mediated joint attention intervention system for young children with ASD. IEEE Trans. Hum.-Mach. Syst. **48**(2), 125–135 (2018)
5. Robins, B., Dickerson, P., Stribling, P., Dautenhahn, K.: Robot-mediated joint attention in children with autism: a case study in robot-human interaction. Interact. Stud. **5**(2), 161–198 (2004)
6. Bekele, E., Lahiri, U., Davidson, J., Warren, Z., Sarkar, N.: Development of a novel robot-mediated adaptive response system for joint attention task for children with autism. In: RO-MAN, 2011, pp. 276–281. IEEE (2011)
7. Greczek, J., Kaszubski, E., Atrash, A., Matarić, M.: Graded cueing feedback in robot-mediated imitation practice for children with autism spectrum disorders. In: The 23rd IEEE International Symposium on Robot and Human Interactive Communication, 2014 RO-MAN, pp. 561–566 (2014)
8. Zheng, Z., Das, S., Young, E.M., Swanson, A., Warren, Z., Sarkar, N.: Autonomous robot-mediated imitation learning for children with autism. In: 2014 IEEE International Conference on Robotics and Automation (ICRA), pp. 2707–2712 (2014)
9. Zwaigenbaum, L., et al.: Clinical assessment and management of toddlers with suspected autism spectrum disorder: insights from studies of high-risk infants. Pediatrics **123**(5), 1383–1391 (2009)
10. Johansen, R.: Groupware: Computer Support for Business Teams. The Free Press (1988)
11. Feil-Seifer, D., Matarić, M.J.: Socially assistive robotics. IEEE Robot. Autom. Mag. **18**(1), 24–31 (2011)
12. Sharkey, N., Sharkey, A.: The crying shame of robot nannies: an ethical appraisal. Interact. Stud. **11**(2), 161–190 (2010)

Taking Neuropsychological Test to the Next Level: Commercial Virtual Reality Video Games for the Assessment of Executive Functions

Federica Pallavicini[✉], Alessandro Pepe, and Maria Eleonora Minissi

Department of Human Sciences for Education "Riccardo Massa",
University of Milan-Bicocca, Piazza dell'Ateneo Nuovo 1, 20126 Milan, Italy
federica.pallavicini@gmail.com

Abstract. Virtual reality and video games are increasingly considered as potentially effective tools for the assessment of several cognitive abilities, including executive functions. However, thus far, only non-commercial contents have been tested and virtual reality contents and video games have been investigated separately. Within this context, this study aimed to explore the effectiveness in the assessment of executive functions using a new type of interactive content - commercial virtual reality games - which combines the advantages of virtual reality with that of commercial video games. Thirty-eight participants completed the Trial Making Test as traditional commonly used assessments of executive functions and then played the virtual reality game Audioshield using an HTC Vive systems. Scores on the Trial Making Test (i.e., time to complete part A and B) were compared to scores obtained on Audio-shield (i.e., number of orbs hit by the players and technical score). The results showed that: (a) performance on the Trial Making Test correlated significantly with performance on the virtual reality video game; (b) scores on Audioshield can be used as a reliable estimator of the results of Trial Making Test.

Keywords: Virtual reality · Virtual reality video games · Video games · Cognitive assessment · Executive functions

1 Introduction

1.1 Current Issues in the Assessment of Executive Functions

The term executive functions refers to a wide range of high-level cognitive processes in the prefrontal lobe [1] associated with reactive inhibition and the regulation of goal achievement behavior [2]. Inhibitory control (i.e., resisting one's initial impulse or a strong pull to do one thing and instead act more wisely) [3], working memory (i.e., holding information in mind while performing one or more mental operations) [4], and cognitive flexibility (i.e., the ability to flexibly adjust to changing demands or priorities, to look at the same thing in different ways or from different perspectives, as required for set shifting or task switching) [5] have been identified as the three main components of executive functions [6]. Executive functions are essential for many everyday life

© Springer Nature Switzerland AG 2019
M. Antona and C. Stephanidis (Eds.): HCII 2019, LNCS 11573, pp. 133–149, 2019.
https://doi.org/10.1007/978-3-030-23563-5_12

activities across a wide range of contexts, such as the ability to work, function independently at home, or maintain social relationships [3, 7]. Furthermore, these cognitive skills have been found to be predictive of health, wealth, academic success, and quality of life throughout life [8].

Given the relevance of executive functions in daily life and the great complexity of cognitive impairments associated with them, an accurate and efficient assessment is of particular relevance [9]. The most widely used tools for the executive functions assessment are paper-and-pencil, and computer-based tests such as the Trial Making Test (TMT) [10], the Stroop test [11], and the Wisconsin Card Sorting Test (WCST) [12, 13]; however, these tests have several important limitations, including:

- Low accessibility: Traditional neuropsychological instruments have been developed to be used in clinical settings; hence, they are costly and require the presence of a health professional (i.e., neurologist or neuropsychologist). Accordingly, these measures are difficult to access and are often applied only retrospectively, that is, after the cognitive impairment has been detected. This delay has a serious negative influence on the effectiveness of the intervention and the treatment options [14]. In addition, most of these tests have been designed and tested on a population with medium educational levels; therefore, they are not suitable for individuals with less education [15];
- Low ecological validity: Most neuropsychological tests are unable to measure complex real-life activities [16] and fail to predict the way in which individuals manage these activities in their daily living [17]. For example, performance on traditional tests of executive functions, such as the WCST and the Stroop test, does not reflect activities of daily living and do not predict functional performance across a range of real-world situations [18]. In addition, although commonly used for clinical and research purposes, traditional neuropsychological tasks are thought to lack the stresses and distractions of real-life settings [19]
- Low quality of assessment data: Neuropsychological tests are often viewed as repetitive, frustrating, and boring, leading to individuals' disengagement, which may negatively affect the quality of assessment data [20].

In response to the limits of conventional assessments of executive functions, neuropsychologists are increasingly emphasizing the need for a new generation of "function-led" tests that would be able to asses real-world functioning and tap into a number of executive domains [21, 22]. A number of ecological tasks have already been developed for the evaluation of executive functions, such as the Multiple Errand Test (MET) [23], or the Behavioral Assessment of the Dysexecutive Syndrome (BADS) [24]. Even if this new type of tasks provided more accurate estimates of the individual functional performance in activities of daily living compared to traditional neuropsychological tests [18] they would have several limitations that restrict their adoption. Among the main one, ecological tasks are time-consuming and require the presence of a specialist [25]. In addition, data collected in the real-world environment are frequently confounded by numerous factors, which cannot be captured by these measures [26].

1.2   Information and Communication Technologies for the Neuropsychological Evaluation of Executive Skills

With the aim of overcoming the existing problem in neuropsychological assessment of executive functions, Information and Communication Technologies (ICT) are increasingly employed, particularly virtual reality and video games, which are described in detail in the following paragraphs.

Virtual Reality. By definition, virtual reality allows users to navigate and interact with a three-dimensional computer-generated environment in real-time [27]. This technology has been successfully applied in the assessment of cognitive skills, including executive functions, for decades [28, 29], when it was still much less advanced from the point of view of both hardware and software. Several studies have reported that virtual reality was effective in offering a more realistic, ecologically valid alternative to traditional neuropsychological tasks assessing executive functions [22, 30], providing a balance between naturalistic observations and the need to control the key variables [31]. Virtual environments, in fact, thanks to the possibility to provide various complex conditions [32] and to record and analyze individual performance [33], combine the rigor and the control of the laboratory measures with simulations that reflect real-life situations [34].

Several ad hoc virtual contents have been developed for the assessment of executive functions. In particular, different virtual shopping environments have been tested over the last decade to evaluate executive functions deficits in people with frontal head injuries or stroke, such as the Virtual Multiple Errand Test (VMET) [35, 36], the Adapted Four-Item Shopping Task [37], and the Virtual Supermarket (VSM) [38]. Another virtual reality tool that has been developed to test EF is the Jansari Assessment of Executive Functions (JEF) [39], which requires individuals to complete different tasks in an office, the Virtual Library Task (VLT) [40], that required individuals to complete tasks as if they were running a library, and RehabCity [41], which assessed executive functions by asking the users to accomplish goals that require problem-solving. Furthermore, a recent study reported the efficacy of three immersive virtual reality tasks (i.e., a seating arrangement task, an item location task, and a virtual parking simulator) in assessing executive functions in a healthy population of younger and older adults [42].

Video Games. Every day, about 2.6 billion people worldwide, 61% male and 39% female, with a mean age of 34 years old, play video games on their personal computers, mobile phones, or consoles [43]. One of the main reason for their success and diffusion is related to the fact that computer games are designed to be motivating and challenging [44], providing easy access to "fun" and to a sense of engagement and self-efficacy [45]. This ability to motivate the user [46], as well as to elicit positive emotions [47] and a short-term increase in subjective well-being [48], has in the last years begun to be exploited for purposes beyond entertainment, including the mental health panorama [49, 50]. In the neuropsychological field, in particular, video games appear relevant also because of their nature of requiring complex cognitive skills, including planning and strategizing, which place executive-functioning demands on players, such as selective attention and inhibition [51].

As early as in 1987, it was observed, for the first time, that famous commercial video games such as *Donkey Kong* (Nintendo) and *Pac-Man* (Atari) can have a positive effect on cognitive skills, improving the reaction times of older adults [52]. A few years later, in 1989, *Space Fortress*, the first of a series of non-commercial computer games designed by cognitive psychologists as a training and research tool [53] - often defined in literature as "serious games" as they use gaming features as the primary medium for serious purposes [54] - became so successful that it was added to the training program of the Israeli Air Force. From that moment on, efficacy has been demonstrated not only for non-commercial video games developed ad hoc, but also for commercial video games, in the training of several cognitive abilities [55], especially perceptual attentional skills [51], mental spatial rotation abilities [56], and executive functions, such as task-switching [57] and working memory [58].

Regarding the use of video games in the assessment of executive functions, especially in these last years, different serious games and game-based versions of neuropsychological tests have been tested. For example, the time to complete an ad hoc low-cost computer game, the *Stroop Stepping Test* (SST), developed using dance pads adapted from exercise-based video games, has been reported to correlate significantly with traditional measure of executive functions (i.e., TMT, Stroop task) among older people [59]. Another serious game has been developed to assess the cognitive status of the elderly through a touch-based device, showing a strong relationship between the video game performance and inhibitory ability as measured in the Stroop task [60]. In a subsequent study, the same authors tested the effectiveness of a serious game developed with a low-cost game engine to assess executive functions skills, reporting strong correlations between classic test (i.e., WCST, Stroop task, N-Back task) and the game performance [61]. Furthermore, a computational model based on game data obtained by playing an ad hoc video game has been showed to be effective in the estimation of TMT performance [62]. Finally, two recent studies have shown a correlation between performance on serious games developed for the assessment of executive functions (i.e., *EXPANSE*, *Kitchen and Cooking*) and scores on TMT [63, 64].

1.3 Aim of the Study

As described in the above sections, several previous studies have provided evidence on the efficacy of both virtual reality content as well as video games in the assessment of executive functions [22]. In particular, they have demonstrated the effectiveness of using virtual reality contents or video games created specifically for the measurement of these cognitive functions in the neuropsychological assessment. However, studies conducted thus far have some limitations, specifically:

- They have focused only on the testing of non-commercial contents: These types of solutions are often expensive and time-consuming in their development [65], as well as focused on a single specific population, which makes it difficult to replicate the study and to standardize the data.
- They have investigated separately virtual reality contents and video games: Thus far, only non-gamified virtual contents [35], or non-immersive video games (i.e., played on a desktop) [62], have been tested.

A new type of interactive content, in particular, virtual reality video games, could address such limitations. Immersive video games, which entered the market in 2016 thanks to the commercial diffusion of products, such as HTC Vive (Valve) and PlayStation VR (Sony Corp.), could be particularly interesting not only for entertainment purposes, but also in many other fields, including psychology and neuropsychology [66, 67]. Thanks to unique features of immersion (i.e., the perception of being physically present in a non-physical world) [68] and interactivity (i.e., the ability of the digital content to respond to the users' actions and be modified by them), virtual reality video games lead in fact to incomparable opportunities, significantly different from those provided by non-immersive gaming, such as traditional console or PC games [67, 69] Interestingly, a previous study [70] has reported the efficacy of a virtual reality video game, in particular, an exercise-based dance game (i.e., *DANCE*), for the training of executive functions in older people. However, this study tested only a non-commercial virtual reality game, and no indications regarding the possible effectiveness of this type of content even for the assessment of executive functions were given.

Within the context described above, this study aimed to compare the effectiveness of a commercial virtual reality video game (i.e., a dance game) in the assessment of executive functions of healthy young adults with the effectiveness of traditional neuropsychological measures. Furthermore, the specific objective of this experiment was to implement a structural equation model to estimate the association of the traditional neuropsychological measures (as a latent variable) with the virtual reality game performance (see the conceptual model in Fig. 1). The main hypotheses explored by the research were:

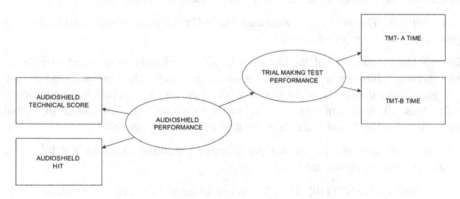

Fig. 1. Conceptual model of relationship among variables.

Hypothesis 1. Performance on the traditional neuropsychological test will correlate significantly with the performance on the virtual reality video game.

Hypothesis 2. The direct path between scores on the virtual reality video game and scores on the traditional neuropsychological test (i.e., time to complete the tasks) will be negative.

2 Materials and Methods

2.1 Participants

38 participants – 13 females (34.2%) and 25 males (65.8%) - were recruited among students and personnel of the University of Milano-Bicocca and of other universities in Milan. Participants' average age was 25.8 ± 4.14 years (min–max = 18–35), average level of education 16 ± 3.05 years (min–max = 11–21) No credits (ECTS) or economic rewards were provided during the research. To be included in the study, individuals had to meet the following criteria: (1) age between 18 and 35 years old; (2) no' major medical disorders (heart disease or high blood pressure, neurological disorders, epilepsy); (3) no left-handed; (4) no presence of pharmacotherapy (psychoactive drugs, anti-hypertensive, anti-depressants); (5) no significant visual impairment (all with normal or corrected-to-normal visual acuity); and (6) not having previous experience with the *Audioshield* game. Before participating, all participants were provided with written information about the study and were required to give written consent in order to be included. The Ethical Committee of the University of Milano-Bicocca approved the study. The research was conducted in accordance with the American Psychological Association ethical principles and code of conduct.

2.2 Measures

Psychological Assessment. The following self-report questionnaires were administered to the participants at the start of the experimental session.

Demographics. The participants were asked to indicate their gender (female/male), age (years old), and years of education.

Gaming Habits and Virtual Reality Knowledge. Individuals were asked to indicate their gaming habits (mean hours spent gaming per week), their previous experience with games included in the study (yes/no), and their previous experience with virtual reality (yes/no); furthermore, they were asked to assess their knowledge of virtual reality on a 7-point Likert scale (from "not at all" to "very much").

Cognitive Assessment. To test the participant's executive functions the following traditional neuropsychological tests was used:

Trial Making Test (TMT) [10]. The TMT is one of the most widely used instruments to measure executive functioning in neuropsychological assessment [71, 72]. In this study, we adopted the Italian version included in the ENB-2 battery [73]. The test consists of two parts (A and B) that clinically assess psychomotor speed (Part A) and mental flexibility (Part B) [74]. The individual performance on TMT is given in the form of a score on TMT-A and on TMT-B, which represents the time (in seconds) taken to complete the task.

2.3   Virtual Reality Video Game

In this study, we adopted *Audioshield*, a virtual reality dance game created by Dylan Fitterer for the HTC Vive (Valve) and launched in April 2016. We selected this particular game among other commercial virtual reality games for two main reasons. First, this genre of video games (i.e., dance game), which incorporates both cognitive engagement and physical activity, may have particular benefits for executive functions [75, 76]. Second, this title has some characteristics that could be closely related to executive functions, such as the inhibition of responses and working memory.

Audioshield, in particular, is a dance game in which orbs come flying towards the player, who needs to follow the beat of the music to successfully hit them. The player uses the HTC Vive's handheld motion-sensing controls to operate two shields, blue and red. Red balls must be deflected with the red shield controlled by the right hand while blue balls must be deflected with the left hand. Purple orbs require a combination of both arms. During the game, the color and the direction of the orbs that the player has to hit changes continuously, for example, the blue balls can come from the right (or red from the left) and require the user to respond correctly very quickly.

As detailed later in the procedure, after a brief explanation of the video game by the experimenter, the participants were asked to practice for about 2 min using the song "Engage" (difficulty: normal; shield: gladiator; environment: horizon). Subsequently, the individuals were asked to complete the song "I drop gems" (difficult: normal; shield: gladiator; environment: horizon) while their performance was evaluated. It took about 5 min to complete the gameplay. The dependent variables used to evaluate the video game performance were: (a) the technical score (i.e., a measure of how many orbs players hit, with 10.00 being a perfect technical score with no misses and 0.00 being all missed), (b) the number of balls the player missed, and (c) the numbers of orbs the player hit.

Regarding the hardware, the virtual reality setting adopted in this study included the following units:

- *HTC Vive system (Valve)*, a consumer-grade virtual reality system designed for the use in video games, consists of an HMD, two controllers, and two infrared laser emitter units. The headset covers a nominal field of view of about 110 (approximately 90 per eye) through two 1080 × 1200 pixel displays that are updated at 90 Hz. Games played with the Vive allow physical movement within a play area that is limited to 4 × 4 m.
- *A portable computer* (MSI GT73 VR, Intel® Core™ i7 processor, GeForce® GTX 1070 8 GB, 17.3" Full HD 1920 × 1080 pixel displays).

2.4   Procedure

After individuals gave written informed consent to participate, they completed the self-report questionnaire assessing demographics, gaming habits, and virtual reality knowledge. Subsequently, the participants underwent the neuropsychological evaluation. They were required to complete the paper-and-pencil version of the TMT (Part A and Part B).

Once this phase was completed, the Vive was connected to the PC through a 5 m cable with an HMDI connection, a USB 2 connection, and power. Participants were asked to wear the HMD and were given the HTC Vive controllers. They were then provided with a training period of about 2 min on *Audioshield* using the song "Engage" to familiarize themselves with the video game tasks and controllers. In the next step, individuals completed the song "I drop gems" while their performance was evaluated. The audio level was set to 45 for all participants.

Subjectively, tracking appeared stable when using this configuration, and the video game was playable with no visible tracker artifacts. All measurements were taken in an 8 × 5 m room with a 3.2 m high ceiling lighted by fluorescent lighting, with no reflective surfaces and no exposure to natural lighting. In the center of this room, a 4 × 4 m grid (i.e., the play area) was drawn on the floor using string and chalk, with grid lines drawn 1 m apart. The entire procedure took about 40 min to complete.

2.5   Analytic Strategy and Data Modelling

This research adopted the structural equation modeling (SEM) approach to analyze the cumulative network of relationships among the variables under study. SEM is a widely used quantitative technique for testing models by allowing a simultaneous estimation of a system of regression equations involving multiple endogenous variables as well as their measurement errors [77]. The conceptual model (Fig. 1) was designed by considering game performance (as measured by two observed variables: *Audioshield* technical score and number of hits) and executive functioning [as measured by two observed variables: psychomotor speed (TMT - Part A) and mental flexibility (TMT - Part B)] as an endogenous latent variable.

The Maximum Likelihood method [78] was adopted to determine the parameters for the SEM analysis. The practical and statistical significance of the model was evaluated using the following goodness-of-fit indices: Root Mean Square Error of Approximation (RMSEA, RMSEA < 0.05) [79]; Standardized Root Mean Square Residual (SRMR, SRMR < 0.05) [80]; Normed fit Index (NFI, NFI > 0.95) [81]; Tucker-Lewis Index (TLI, TLI > 0.95) [81]; and Comparative Fit Index (CFI, CFI > 0.95) [81]. All models were tested using Amos software. In keeping with the current literature (e.g., [82]), we estimated confidence limits using both Monte Carlo simulation and bootstrapping methods with a set of random samples (k = 500) and 95% confidence intervals for unstandardized effects.

3   Results

Main descriptive statistics and zero-order correlations of considered variables are summarized in Table 1.

Table 1. Main descriptive statistics and zero-order correlations for TMT scores and *Audioshield* performance (N = 38). Note: * p < .05, ** p < .01 *** p < .001

Variable	1	2	3	4
1. TMT - task A	–			
2. TMt - task B	.390*	–		
3. Audioshield - Score	−.477***	−.417**	–	
4. Audioshield - Hit	−0.241	−.375**	.582***	–
Mean	12.58	59.87	9.71	555.4
Standard Deviation	7.26	20.26	0.211	449.5
Skewness	0.279	1.25	0.135	1.96
Min - Max	18–35	29–117	9–10	97–2,338

The results of the correlational analysis revealed that, in general, game performance was negatively associated with TMT scores (i.e., time to complete the TMT-A and TMT-B). In particular, statistically significant negative correlations were found between *Audioshield* scores and both psychomotor speed (TMT-part A, r = −.477, p < .001) and mental flexibility (TMT-part B, r = −.417, p < .001). Interestingly, the number of *Audioshield* total hits correlated negatively with mental flexibility (TMT-part B, r = −.375, p < .001) whereas it did not correlate with psychomotor speed (TMT-part A, r = −.241, p = n.s.). A more detailed picture of the set of relationships among variables under study can be obtained by estimating the structural equation model (See Fig. 2).

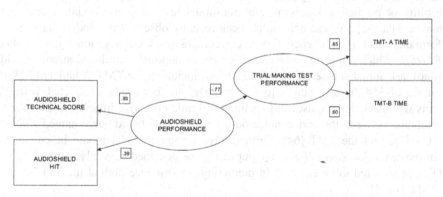

Fig. 2. Results of the structural model estimation. Standardized direct effects were reported, bootstrap estimation (k = 1,500) was conducted along with 95% confidence intervals

The results of the structural equation model revealed an excellent general fit of the conceptual structure with the observed data. In fact, all goodness of fit indexes suggested that the data fit the measurement model well: $\chi^2(1) = 1.41$, p = .235, NC = 1.41, RMSEA = .050, pclose = .257, NFI = .968, NNFI = .988, and CFI = .988, supporting the robust effects among the variables under study both conceptually and statistically. In particular, performance on *Audioshield* was negatively associated with the time to complete the TMT part A and B (β = −.77, p < .01, 95% CI [−1.82; −.099]), with higher scores on the virtual reality game indicating less time to complete the tasks. The results support the H2, suggesting that scores on *Audioshield* can be used to reliably estimate the results of TMT.

4 Discussion

Starting from the first hypothesis of the study, the proposed significant correlation between performance on the virtual reality game and scores on the TMT was supported. A traditional paper-and-pencil task was used to assess various cognitive processes, including psychomotor speed and executive functions (i.e., task-switching, flexibility) [74, 83], especially the part B (TMT-B) [71, 84]. In particular, both the time to complete the TMT-A and the TMT-B correlated significantly with the performance on the virtual reality game (i.e., number of orbs hit by the players and technical score on *Audioshield*), suggesting that the virtual reality game may measure the same aspect of cognition as the traditional paper-and-pencil TMT.

Regarding the second hypothesis, as expected, the structural equation model revealed that performance on the virtual reality game can be used to estimate scores on the TMT. In particular, the results showed that performance on *Audioshield* was negatively associated with the time to complete the TMT part A and B, with higher scores on the virtual reality game indicating less the time to complete the tasks.

Our findings are consistent with previous results showing a correlation between the performance on some video games and the scores on the TMT [63, 64] as well as the possibility of predicting scores on this traditional test using individuals' game performance data [62]. For example, it has been recently observed that individuals' game performance data (i.e., latency times, correct answers) on a serious game called *EXPANSE*, which aims to assesses executive functions, correlated strongly with standard task aimed at assessing these functions, including the TMT-A and TMT-B, in a sample of 354 healthy adults [63]. Similarly, another study conducted with 12 patients with Alzheimer's disease, focusing on 9 mild cognitive impairments, reported a correlation between the performance on a serious game board-game named *Kitchen and Cooking* and the TMT [64]. Furthermore, a computational model based on the performance on *Scavenger Haunt*, a computer game designed heavenly inspired by the TMT, was reported to be effective in predicting performance both at the TMT-A and the TMT-B [62].

However, it is important to underline that our study found important differences compared to the research carried out in the past. First, in this study, we tested a commercial video game instead of a serious game [63, 64] or a game-based version of the TMT [62]. In addition, we tested an immersive video game in this experiment while other studies used non-immersive video games (i.e., played on desktops) [62–64]. Compared to previous literature, in fact, the current research is the first to provide preliminary evidence on the effectiveness of immersive commercial video games, a type of video game that combines the advantages of commercial games with those offered by virtual reality [67].

Although the results of the present study could be interesting for their possible applications, this research has some important limitations that could affect the generalizability of the findings. First, it is important to underline that the obtained results refer to a specific virtual reality video game, in this case, *Audioshield*, a dance game launched in April 2016. In the future, it would be interesting to explore this particular genre of games more in-depth, as these games are deemed interesting with respect to the functioning of executive functions [70]. For example, it would be interesting to test *Beat Saber* (Beat Games), since it has very similar characteristics as the game tested in this study, and it was the most successful virtual reality game title of 2018 on the Steam platform (Vive) [85]. In addition, future studies should investigate the effectiveness of other genres of video games played in virtual reality, for example, puzzle and casual games appear to be particularly interesting, since they are genres with potentially very relevant features in relation to the functioning of executive functions [86, 87].

Second, the results emerging from this study refer to the specific virtual reality system in use [i.e., HTC Vive (Steam)]. Future research should investigate the adoption of other commercial virtual reality systems with different characteristics in terms of immersion and interactions with the game, including other off-the-shelf virtual reality systems, such as Oculus Go (Oculus) or Playstation VR (Sony corp.), for instance. These products are very appealing, since unlike the system tested in this study, they could be used very easily even by non-expert operators and are more budget-friendly.

Third, in this study, a specific test was used to measure executive functions (i.e., the TMT). It would be interesting to adopt also other traditional tests, such as the WCST or the Stroop test, that assess different aspects of executive functions. Finally, another limitation of this study is related to the small sample size and the specific sample included in the study comprising young adults who played often (more than 12 h per week) and had a low knowledge of virtual reality (68% of the sample did not try it before the experiment).

5   Conclusion

Our findings show the feasibility of using a commercial virtual reality game to assess executive functions and cognitive abilities, as measured by the TMT. In particular, user performance on *Audioshield* correlated significantly with time to complete the TMT-A as well as the TMT-B. Furthermore, the most important finding of this exploratory study was that the performance on the immersive video game predicted the TMT scores (i.e., time to complete the TMT-A and TMT-B).

Future studies should investigate the possibility of using commercial virtual reality games to assess executive functions as well as of other cognitive skills, since they represent a very promising tool for the neuropsychological assessment. Using off-the-shelves video games rather than video games created ad hoc could have several advantages for the assessment of executive functions as well as other cognitive processes, in particular:

- Low-cost and ready-to-use: Commercial video games are low-cost and their production time is short, unlike serious games and game-based version of the neuropsychological test.
- Advanced graphic quality and gameplay mechanics: Commercial video games are more costly and a greater number of people is involved in its development compared to ad hoc computer games, for example, *Red Dead Redemption 2* and *GTA V* (Rockstar Games) costed about 944 million and 256 million dollars, respectively, involving each one a team of at least 1.000-person [88, 89] to ensure an astonishing quality of the game, which has a significant effect on the overall gaming experience. This would not be achievable with smaller budgets and staff.
- Possibility to collect the data on very large populations: Commercial video games are played in different cultural and social contexts, which makes it easy to replicate studies in different contexts and to collect the data from large samples.

In addition, virtual reality games have also several advantages compared to video games played on desktops, including:

- Require high physical and cognitive involvement: Immersive video games require a high involvement of the player at both the motor and cognitive level, with a greater engagement compared to non-immersive games [67, 69].
- Elicit more intense positive emotions in the players: Video games played in virtual reality elicit a higher self-reported sense of happiness in users in comparison to non-immersive games [67];
- Allow to gather multiple types of data of the user: Video games played in virtual reality allow the collection of a wide variety of data compared to non-immersive games (e.g., scores, time to complete the task, etc.), such as those about the player's movement within the virtual environment.

Acknowledgments. This work was supported by the "Virtual Video Games" project, which is a part of the research plan of the first author, a post-doctoral research fellow at the University of Milano-Bicocca. The authors would like to thank MSI Italy for supporting the study and providing PCs and hardware for data acquisition. In addition, we thank our Ph.D. Ambra Ferrari and post-graduate students Erica Ilari and Gabriele Barone for helping with data acquisition. Last, we would like to thank all the participants for their willingness to participate and enthusiasm.

Author contributions.
FP proposed the study, supervised the scientific asset, and wrote the first draft of the paper. AP analyzed the data. MEM carried out experiments. All three authors were involved in the drafting, revising, and completing the manuscript.

References

1. Müller, U., Kerns, K.: The development of executive function. In: Handbook of Child Psychology and Developmental Science, pp. 1–53. Wiley, Hoboken (2015)
2. Carlson, S.M.: developmentally sensitive measures of executive function in preschool children. Dev. Neuropsychol. **28**(2), 595–616 (2005)

3. Diamond, A.: Executive functions. Annu. Rev. Psychol. **64**(1), 135–168 (2013)
4. Baddeley, A.D., Hitch, G.J.: Developments in the concept of working memory. Neuropsychology **8**(4), 485–493 (1994)
5. Kiesel, A., et al.: Control and interference in task switching—a review. Psychol. Bull. **136** (5), 849–874 (2010)
6. Miyake, A., Friedman, N.P.: The nature and organization of individual differences in executive functions: four general conclusions. Curr. Dir. Psychol. Sci. **21**(1), 8–14 (2012)
7. Bailey, C.E.: Cognitive accuracy and intelligent executive function in the brain and in business. Ann. N. Y. Acad. Sci. **1118**(1), 122–141 (2007)
8. Moffitt, T.E., et al.: A gradient of childhood self-control predicts health, wealth, and public safety. Proc. Natl. Acad. Sci. U.S.A. **108**(7), 2693–2698 (2011)
9. Martínez-Pernía, D., et al.: Using game authoring platforms to develop screen-based simulated functional assessments in persons with executive dysfunction following traumatic brain injury. J. Biomed. Inform. **74**, 71–84 (2017)
10. Reitan, R.M.: Validity of the trail making test as an indicator of organic brain damage. Percept. Mot. Skills **8**(3), 271–276 (1958)
11. Stroop, J.R.: Studies of interference in serial verbal reactions. J. Exp. Psychol. **18**(6), 643–662 (1935)
12. Grant, D.A., Berg, E.: A behavioral analysis of degree of reinforcement and ease of shifting to new responses in a Weigl-type card-sorting problem. J. Exp. Psychol. **38**(4), 404–411 (1948)
13. Berg, E.A.: A simple objective technique for measuring flexibility in thinking. J. Gen. Psychol. **39**(1), 15–22 (1948)
14. Holtzman, D.M., Morris, J.C., Goate, A.M.: Alzheimer's disease: the challenge of the second century. Sci. Trans. Med. **3**(77), 77sr1 (2011)
15. Ardila, A., Ostrosky-Solis, F., Rosselli, M., Gómez, C.: Age-related cognitive decline during normal aging: the complex effect of education. Arch. Clin. Neuropsychol. **15**(6), 495–513 (2000)
16. Chan, R., Shum, D., Toulopoulou, T., Chen, E.: Assessment of executive functions: review of instruments and identification of critical issues. Arch. Clin. Neuropsychol. **23**(2), 201–216 (2008)
17. Bottari, C., Dassa, C., Rainville, C., Dutil, É.: The criterion-related validity of the IADL profile with measures of executive functions, indices of trauma severity and sociodemographic characteristics. Brain Inj. **23**(4), 322–335 (2009)
18. Burgess, P.W., et al.: The case for the development and use of 'ecologically valid' measures of executive function in experimental and clinical neuropsychology. J. Int. Neuropsychol. Soc. **12**(02), 194–209 (2006)
19. Elkind, J.S., Rubin, E., Rosenthal, S., Skoff, B., Prather, P.: A simulated reality scenario compared with the computerized wisconsin card sorting test: an analysis of preliminary results. CyberPsychol. Behav. **4**(4), 489–496 (2001)
20. DeRight, J., Jorgensen, R.S.: I just want my research credit: frequency of suboptimal effort in a non-clinical healthy undergraduate sample. Clin. Neuropsychol. **29**(1), 101–117 (2015)
21. Jurado, M.B., Rosselli, M.: The elusive nature of executive functions: a review of our current understanding. Neuropsychol. Rev. **17**(3), 213–233 (2007)
22. Parsons, T.D., Carlew, A.R., Magtoto, J., Stonecipher, K.: The potential of function-led virtual environments for ecologically valid measures of executive function in experimental and clinical neuropsychology. Neuropsychol. Rehabil. **27**(5), 777–807 (2017)
23. Alderman, N., Burgess, P.W., Knight, C., Henman, C.: Ecological validity of a simplified version of the multiple errands shopping test. J. Int. Neuropsychol. Soc. **9**(1), 31–44 (2003)

24. Wilson, B.A., Alderman, N., Burgess, P.W., Emslie, H., Evans, J.J.: BADS : behavioural assessment of the dysexecutive syndrome. Harcourt Assessment (1997)
25. Chevignard, M., et al.: An ecological approach to planning dysfunction: script execution. Cortex **36**(5), 649–669 (2000)
26. Valladares-Rodríguez, S., Pérez-Rodríguez, R., Anido-Rifón, L., Fernández-Iglesias, M.: Trends on the application of serious games to neuropsychological evaluation: a scoping review. J. Biomed. Inform. **64**, 296–319 (2016)
27. Burdea, G., Coiffet, P.: Virtual reality technology. Presence Teleoperators Virtual Environ. **12**(6), 663–664 (2003)
28. Schultheis, M.T., Himelstein, J., Rizzo, A.A.: Virtual reality and neuropsychology: upgrading the current tools. J. Head Trauma Rehabil. **17**(5), 378–394 (2002)
29. Rizzo, A.A., Buckwalter, J.G., Neumann, U., Kesselman, C., Thiebaux, M.: Basic issues in the application of virtual reality for the assessment and rehabilitation of cognitive impairments and functional disabilities. CyberPsychol. Behav. **1**(1), 59–78 (1998)
30. Parsons, T.D., Silva, T.M., Pair, J., Rizzo, A.A.: Virtual environment for assessment of neurocognitive functioning: virtual reality cognitive performance assessment test. Stud. Health Technol. Inform. **132**, 351–356 (2008)
31. Korečko, S., et al.: Assessment and training of visuospatial cognitive functions in virtual reality: proposal and perspectivě perspectivě. In: 9th IEEE International Conference on Cognitive Infocommunications (CogInfoCom 2018), pp. 49–43 (2018)
32. Parsey, C.M., Schmitter-Edgecombe, M.: Applications of technology in neuropsychological assessment. Clin. Neuropsychol. **27**(8), 1328–1361 (2013)
33. Weiss, P.L., Kizony, R., Feintuch, U., Rand, D., Katz, N.: Virtual reality applications in neurorehabilitation. In: Selzer, M., Clarke, S., Cohen, L.G., Kwakkel, G., Miller, R. (eds.) Textbook of Neural Repair and Rehabilitation, pp. 198–218. Cambridge University Press, Cambridge (2014)
34. Bohil, C.J., Alicea, B., Biocca, F.A.: Virtual reality in neuroscience research and therapy. Nat. Rev. Neurosci. **12**(12), 752–762 (2011)
35. Raspelli, S., et al.: Validating the neuro VR-based virtual version of the multiple errands test: preliminary results. Presence Teleoperators Virtual Environ. **21**(1), 31–42 (2012)
36. Cipresso, P., et al.: Virtual multiple errands test (VMET): a virtual reality-based tool to detect early executive functions deficit in Parkinson's disease. Front. Behav. Neurosci. **8**, 405 (2014)
37. Nir-Hadad, S.Y., Weiss, P.L., Waizman, A., Schwartz, N., Kizony, R.: A virtual shopping task for the assessment of executive functions: validity for people with stroke. Neuropsychol. Rehabil. **27**(5), 808–833 (2017)
38. Zygouris, S., et al.: Can a virtual reality cognitive training application fulfill a dual role? Using the virtual supermarket cognitive training application as a screening tool for mild cognitive impairment. J. Alzheimer's Dis. **44**(4), 1333–1347 (2015)
39. Jansari, A.S., Froggatt, D., Edginton, T., Dawkins, L.: Investigating the impact of nicotine on executive functions using a novel virtual reality assessment. Addiction **108**(5), 977–984 (2013)
40. Renison, B., Ponsford, J., Testa, R., Richardson, B., Brownfield, K.: The ecological and construct validity of a newly developed measure of executive function: the virtual library task. J. Int. Neuropsychol. Soc. **18**(03), 440–450 (2012)
41. Vourvopoulos, A., Faria, A.L., Ponnam, K., Bermudez i Badia, S.: RehabCity. In: Proceedings of the 11th Conference on Advances in Computer Entertainment Technology - ACE 2014, pp. 1–8 (2014)

42. Davison, S.M.C., Deeprose, C., Terbeck, S.: A comparison of immersive virtual reality with traditional neuropsychological measures in the assessment of executive functions. Acta Neuropsychiatry **30**(02), 79–89 (2018)
43. Entertainment Software Assotiation: Essential facts about the computer and video game industry (2018). http://www.theesa.com/about-esa/essential-facts-computer-video-game-industry/
44. Lorenz, R.C., Gleich, T., Gallinat, J., Kühn, S.: Video game training and the reward system. Front. Hum. Neurosci. **9**, 40 (2015)
45. Green, C.S., Bavelier, D.: Learning, attentional control, and action video games. Curr. Biol. **22**(6), R197–R206 (2012)
46. Gee, J.P., Paul, J.: What video games have to teach us about learning and literacy. Comput. Entertain. **1**(1), 20 (2003)
47. Russoniello, C.V., O'Brien, K., Parks, J.M.: The effectiveness of casual video games in improving mood and decreasing stress. J. Cyber Ther. Rehabil. **2**(1), 53–66 (2009)
48. Przybylski, A.K., Rigby, C.S., Ryan, R.M.: A motivational model of video game engagement. Rev. Gen. Psychol. **14**(2), 154–166 (2010)
49. Granic, I., Lobel, A., Engels, R.C.: The benefits of playing video games. Am. Psychol. **69**(1), 66–78 (2014)
50. Pallavicini, F., Ferrari, A., Mantovani, F.: Video games for well-being: a systematic review on the application of computer games for cognitive and emotional training in the adult population. Front. Psychol. **9**, 2127 (2018)
51. Green, C.S., Bavelier, D.: Action video game modifies visual selective attention. Nature **423**(6939), 534–537 (2003)
52. Clark, J.E., Lanphear, A.K., Riddick, C.C.: The effects of videogame playing on the response selection processing of elderly adults. J. Gerontol. **42**(1), 82–85 (1987)
53. Donchin, E.: The space fortress game. Acta Psychol. (Amst) **71**(1–3), 17–22 (1989)
54. Fleming, T.M., et al.: Serious games and gamification for mental health: current status and promising directions. Front. Psychiatry **7**, 215 (2016)
55. Boyle, et al.: An update to the systematic literature review of empirical evidence of the impacts and outcomes of computer games and serious games. Comput. Educ. **94**(C) 178–192 (2016)
56. Cherney, I.D., Bersted, K., Smetter, J.: Training spatial skills in men and women. Percept. Mot. Skills **119**(1), 82–99 (2014)
57. Parong, J., Mayer, R.E., Fiorella, L., MacNamara, A., Homer, B.D., Plass, J.L.: Learning executive function skills by playing focused video games. Contemp. Educ. Psychol. **51**, 141–151 (2017)
58. Toril, P., Reales, J.M., Mayas, J., Ballesteros, S.: Video game training enhances visuospatial working memory and episodic memory in older adults. Front. Hum. Neurosci. **10**, 206 (2016)
59. Schoene, D., Smith, S.T., Davies, T.A., Delbaere, K., Lord, S.R.: A stroop stepping test (SST) using low-cost computer game technology discriminates between older fallers and non-fallers. Age Ageing **43**(2), 285–289 (2014)
60. Tong, T., Chignell, M.: Developing a serious game for cognitive assessment. In: Proceedings of the Second International Symposium of Chinese CHI on - Chinese CHI 2014, pp. 70–79 (2014)
61. Tong, X., Gromala, D., Amin, A., Choo, A.: The Design of an Immersive Mobile Virtual Reality Serious Game in Cardboard Head-Mounted Display for Pain Management. In: Serino, S., Matic, A., Giakoumis, D., Lopez, G., Cipresso, P. (eds.) MindCare 2015. CCIS, vol. 604, pp. 284–293. Springer, Cham (2016). https://doi.org/10.1007/978-3-319-32270-4_29

62. Hagler, S., Jimison, H.B., Pavel, M.: Assessing executive function using a computer game: computational modeling of cognitive processes. IEEE J. Biomed. Heal. Inf. **18**(4), 1442–1452 (2014)
63. Chicchi Giglioli, I.A., de Juan Ripoll, C.., Parra, E., Alcañiz Raya, M.: EXPANSE: a novel narrative serious game for the behavioral assessment of cognitive abilities. PLoS One, **13** (11), e0206925 (2018)
64. Manera, V., et al.: Kitchen and cooking' a serious game for mild cognitive impairment and Alzheimer's disease: a pilot study. Front. Aging Neurosci. **7**, 24 (2015)
65. Martínez-Pernía, D., González-Castán, Ó., Huepe, D.: From ancient Greece to the cognitive revolution: a comprehensive view of physical rehabilitation sciences. Physiother. Theor. Pract. **33**(2), 89–102 (2017)
66. Lin, J.-H.T.: Fear in virtual reality (VR): fear elements, coping reactions, immediate and next-day fright responses toward a survival horror zombie virtual reality game. Comput. Human Behav. **72**, 350–361 (2017)
67. Pallavicini, F., Ferrari, A., Pepe, A., Garcea, G., Zanacchi, A., Mantovani, F.: effectiveness of virtual reality survival horror games for the emotional elicitation: preliminary insights using resident evil 7: biohazard. In: Antona, M., Stephanidis, C. (eds.) UAHCI 2018. LNCS, vol. 10908, pp. 87–101. Springer, Cham (2018). https://doi.org/10.1007/978-3-319-92052-8_8
68. Riva, G.: Is presence a technology issue? Some insights from cognitive sciences. Virtual Reality **13**(3), 159–169 (2009)
69. Pallavicini, F., et al.: What distinguishes a traditional gaming experience from one in virtual reality? An exploratory study. In: Ahram, T., Falcão, C. (eds.) AHFE 2017. AISC, vol. 608, pp. 225–231. Springer, Cham (2018). https://doi.org/10.1007/978-3-319-60639-2_23
70. Eggenberger, P., Schumacher, V., Angst, M., Theill, N., de Bruin, E.: Does multicomponent physical exercise with simultaneous cognitive training boost cognitive performance in older adults? A 6-month randomized controlled trial with a 1-year follow-up. Clin. Interv. Aging **10**, 1335 (2015)
71. Arbuthnott, K., Frank, J.: Trail making test, part B as a measure of executive control: validation using a set-switching paradigm. J. Clin. Exp. Neuropsychol. **22**(4), 518–528 (2000)
72. Sanchez-Cubillo, I., et al.: Construct validity of the trail making test: role of task-switching, working memory, inhibition/interference control, and visuomotor abilities. J. Int. Neuropsychol. Soc. **15**(03), 438 (2009)
73. Mondini, S.: Esame neuropsicologico breve 2 (ENB-2) : una batteria di test per lo screening neuropsicologico. Raffaello Cortina (2011)
74. Lezak, M.D.: Neuropsychological Assessment. Oxford University Press, Oxford (2004)
75. Anderson-Hanley, C., Maloney, M., Barcelos, N., Striegnitz, K., Kramer, A.: Neuropsychological benefits of neuro-exergaming for older adults: a pilot study of an interactive physical and cognitive exercise system (iPACES). J. Aging Phys. Act. **25**(1), 73–83 (2017)
76. Best, J.R.: Effects of physical activity on children's executive function: contributions of experimental research on aerobic exercise. Dev. Rev. **30**(4), 331–551 (2010)
77. Kline, R.B.: Principles and Practice of Structural Equation Modeling, 5th edn. The Guilford Press, New York (2011)
78. Gath, E.G., Hayes, K.: Bounds for the largest Mahalanobis distance. Linear Algebra Appl. **419**(1), 93–106 (2006)
79. Hu, L., Bentler, P.M.: Cutoff criteria for fit indexes in covariance structure analysis: conventional criteria versus new alternatives. Struct. Equ. Model. Multidiscip. J. **6**(1), 1–55 (1999)
80. Marsh, H.W., Hau, K.-T.: Assessing goodness of fit. J. Exp. Educ. **64**(4), 364–390 (1996)

81. Morin, A.J.S., Marsh, H.W., Nagengast, B.: Exploratory structural equation modeling. In: Hancock, G.R., Mueller, R.O., Charlotte, N.C. (eds.) Structural Equation Modeling: A Second Course, 2nd edn., pp. 395–436, Information Age Publishing, Inc (2013)

82. Mackinnon, D.P., Lockwood, C.M., Williams, J.: Confidence limits for the indirect effect: distribution of the product and resampling methods. Multivar. Behav. Res. **39**(1), 99 (2004)

83. Salthouse, T.A.: What cognitive abilities are involved in trail-making performance? (2011)

84. Miyake, A., Friedman, N.P., Emerson, M.J., Witzki, A.H., Howerter, A., Wager, T.D.: The unity and diversity of executive functions and their contributions to complex 'frontal lobe' tasks: a latent variable analysis. Cogn. Psychol. **41**(1), 49–100 (2000)

85. Lang, B.: Valve reveals top selling VR games on steam in 2018, Road to VR (2018). https://www.roadtovr.com/valve-reveals-top-selling-vr-games-on-steam-2018/?platform=hootsuite. Accessed 11 Jan 2019

86. Baniqued, P.L., et al.: Cognitive training with casual video games: points to consider. Front. Psychol. **4**, 1010 (2014)

87. Oei, A.C., Patterson, M.D.: Playing a puzzle video game with changing requirements improves executive functions. Comput. Hum. Behav. **37**, 216–228 (2014)

88. Takahashi, D.: The DeanBeat: How much did Red Dead Redemption 2 cost to make? (updated)—VentureBeat, Venture Beat (2018). https://venturebeat.com/2018/10/26/the-deanbeat-how-much-did-red-dead-redemption-2-cost-to-make/

89. Danielle, T.: GTA V: How Much Did It Cost to Make, Twinfinite (2017). https://twinfinite.net/2017/12/gta-v-cost-make-how-much/

Evaluation of Handwriting Skills in Children with Learning Difficulties

Wanjoo Park[1], Georgios Korres[1], Samra Tahir[2], and Mohamad Eid[1(✉)]

[1] Engineering Division, New York University Abu Dhabi,
Saadiyat Island, Abu Dhabi, United Arab Emirates
{wanjoo,george.korres,mohamad.eid}@nyu.edu
[2] Child Clinical Department, American Center for Psychiatry and Neurology,
Abu Dhabi, United Arab Emirates
s.tahir@americancenteruae.com

Abstract. Many children have physical, cognitive, motor, and other limitations that influence their ability to develop handwriting skills. Recently, haptic technology is gaining rising interest as an assistive technologies to improve the acquisition of handwriting skills for children with learning difficulties. In this paper, we introduce a method and an experimental protocol to evaluate the quality of handwriting for children with learning difficulties. We developed a copy work task comprising four categories of handwriting tasks, namely numbers, letter, shapes, and emoticons (a total of 32 tasks, covering low to high complexity handwriting tasks). Results demonstrated that shapes are more difficult to learn than emoticons, even though emoticons are more complex to construct. This is probably due to the fact that children are more familiar with emoticons than abstract shapes. Findings in this study are crucial for developing a longitudinal experimental study to evaluate the effectiveness of various haptic guidance methods to improve learning outcomes for children with learning difficulties.

Keywords: Haptic · Handwriting · Drawing, assistance · Learning difficulty

1 Introduction

Handwriting is a complex human activity that requires fine motor control, perceptual, and visual-motor integration skills [1, 2]. Early writing skills is the ability to create easy-to-read text with minimal physical and mental effort. Generally, the fluency of handwriting skills improves with age and education [3]. Children usually acquire these skills as they grow over a long term of repetitive training. Also, for typically growing children, handwriting become automatic and therefore, text generation does not interfere with creative thinking process [4]. However, it is reported that the difficulty of handwriting in school-aged children varies from 10% to 34% [5]. Hamstra-Bletz and Blote defined "dysgraphia"

© Springer Nature Switzerland AG 2019
M. Antona and C. Stephanidis (Eds.): HCII 2019, LNCS 11573, pp. 150–159, 2019.
https://doi.org/10.1007/978-3-030-23563-5_13

as a disturbance or difficulty in the creation of a written language [6]. Writing difficulties are officially diagnosed as a part of Developmental Coordination Disorder [7]. There are many assistive technologies to support students with writing difficulties, but they all have their own practical limitations [8]. Drawing skill is also considered to be one of the important factor to evaluate learning process. In order to support students with handwriting learning difficulties, a workbook was created that include shapes, emoticons, numbers and letters in order to provide a wide range of handwriting skills complexity (starting from simple tasks such as numbers or characters to more complex shapes and emoticons). We also developed a haptic-based handwriting training system (Fig. 1) to provide haptic guidance for improving handwriting skills acquisition. Our previous work reported significant improvement in the acquisition of handwriting skills for adults using various haptic guidance methods (full and partial guidance) [9]. In this study, we designed an experimental protocol for students with learning difficulties to investigate if haptic guidance may improve further the learning outcomes. Before proceeding with the experimental study, We must evaluate the children's performance with paper-based copy work to verify the experimental protocol.

2   Materials and Methods

2.1   Participants

Twenty children with learning difficulties are recruited for this study (14 males and 6 females; age range, 4–12, all met our inclusion/exclusion criteria). The inclusion criteria were: (a) children who struggle with letter, shape formation and have moderate, mild and borderline intellectual deficits, (b) children who suffer from visual spatial skill deficits. The exclusion criteria were: (a) children who have no intellectual disabilities, (b) children who demonstrate normal executive and marking memory skills. The study was approved by the Institutional Review Board for Protection of Human Subjects in the American Center for Psychiatry and Neurology (Project # 0017) and New York University Abu Dhabi (Project # 101–2016).

2.2   Design of the System

We have developed a haptic-based handwriting platform for physically guiding the children along a trajectory of handwriting task [9]. The platform has been extensively revised and calibrated, in collaboration with the therapists, for children with learning difficulties. A seven inch display was mounted on a base under the revised pen-shaped end-effector in order to act as a writing pad. Subsequently, the haptic device had to be calibrated with respect to the monitor so that the end-effector tip of the stylus matches accurately the trace path when a letter is written. The revised design of the end-effector stylus was more robust and ergonomically efficient in comparison to the previous one (verified through

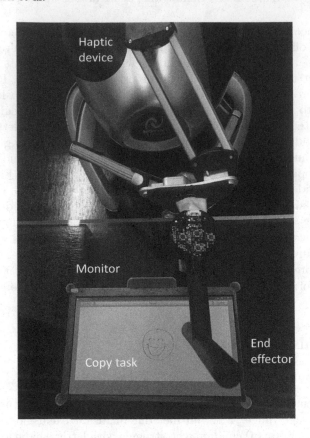

Fig. 1. Haptic handwriting assistance system.

several trials with the therapist). It provides a firm grip and thus a better coupling between the end-effector stylus and the Novint Falcon haptic device [10]. A firm coupling minimizes errors in haptic playback and thus provides accurate reconstruction of handwriting tasks. The revised experimental setup is presented in Fig. 1.

The handwriting system is capable of delivering three different modes of haptic guidance namely full guidance, partial guidance and disturbance guidance. In the full guidance mode which is described by Eq. 1, the force applied by the haptic device is derived from the maximum stiffness provided (K_{max}) times the point-to-point displacement (δu). For the partial guidance mode, a Proportional-Integral-Derivative (PID) controller was used as described in Eq. 2 whereas C_p, C_i and C_d are the gains for the proportional, integral and derivative components of the controller respectively. e(t) is the error set by the difference between current position (x_{cur}) and desired position (x_{des}). Motivated by Lee's work on motor learning through cognitive effort [11], we designed the disturbance haptic guidance mode so that the haptic device would cause the stylus end-effector to provide vibration patterns at strategic positions along the hand-

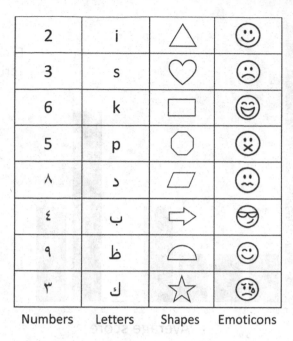

Numbers	Letters	Shapes	Emoticons

Fig. 2. Four types of task for the students' copy work.

writing trajectory, with the intention to increase the participants attention to the task at hand. This impulsive behaviourforce of the haptic device is randomly activated and deactivated by a set of predefined parameters while the task is performed. Finally, a No-Haptic guidance mode was also designed by driving the haptic device in high admittance in which there is no haptic feedback and the participant can freely move the end-effector stylus in any direction.

$$\mathbf{F}(t) = K_{max}\Delta\mathbf{u} \tag{1}$$

$$\mathbf{F}(t) = C_p\mathbf{e}(t) + C_i \int_{\Delta T} \mathbf{e}(t)dt + C_d\frac{de(t)}{dt}$$
$$\mathbf{e}(t) = \mathbf{x}_{cur} - \mathbf{x}_{des} \tag{2}$$

2.3 Experimental Tasks

We developed a copy work task comprising of numbers, letters, shapes, and emoticons. These four categories represent cognitive and academic tasks. In this task, we are able to evaluate visuomotor and fine motor skill in the ability of students to recognize tasks and copy works. We also designed various difficulties of tasks to evaluate the students' fine motor skills. Figure 2 shows 32 tasks for the students' copy work with four types of categories. In case of numbers and letters, we included Arabic numerals and letters, since these are handwriting tasks the children are currently learning. We selected not only simple shapes such

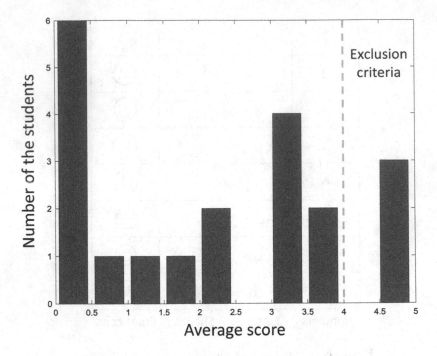

Fig. 3. Histogram.

as triangle and square, but also difficult shapes such as arrow and star. We have added emoticons, four positive and four negative emotional expressions, in the copy work to investigate the students' recognition and development depending on emotional expressions.

2.4 Evaluation of the Experimental Protocol

To verify the longitudinal experimental protocol, we asked the candidates to perform the 32 copy work task on a sheet of paper. Three therapists evaluated individual copy work tasks on a scale 0–5 points. The scores were calculated as the average of three experts' evaluations. We checked distribution of average score and differences between/within the four categories. We designed various difficulties, thus we investigated if there are significant differences of score among all copy work tasks. We also checked the correlations between age/gender and average scores.

3   Results

First of all, we analyzed histogram of average scores. Seven students achieved a very low score of 1 or less on average. four and six students achieved 1 to 2.5 and 3 to 4 points respectively, as shown in Fig. 3. Unexpectedly, three students achieved

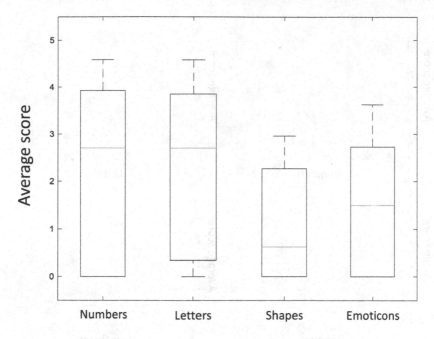

Fig. 4. Distribution of average scores according to four types of copy work.

a very high score of 4.5 or higher. They are excluded from the longitudinal study because they no longer have room for improvement.

Figure 4 shows the average of scores for the four categories of handwriting tasks. As expected, the average scores are higher for numbers and letters than shapes and emoticons, but these differences are not statistically significant. Figure 5 shows the distributions of the average scores of eight copy work tasks in each category. There are no significant difference among the eight copy work tasks in each category. We expected Arabic numbers and letters (tasks 5 to 8) to be a little more difficult than others. However, there are no significant differences. Also, we found no significant differences in the average scores of copy work of emoticons according to the positive and negative emotional expression.

We investigated significant difference in the average scores among the 32 copy work tasks. Figure 6 shows the comparison intervals of Kruskal-Wallis test, showing significant differences between the letter 'i' and the arrow/star shapes (Kruskal-Wallis test, $p < 0.01$; ad-hoc, Bonferroni). This result shows that we have designed a wide spectrum of handwriting tasks, varying from very difficult to very easy, and that is shown to be statistically valid.

We also investigated whether there is a correlation between average score and gender/age, but there is no significant correlation.

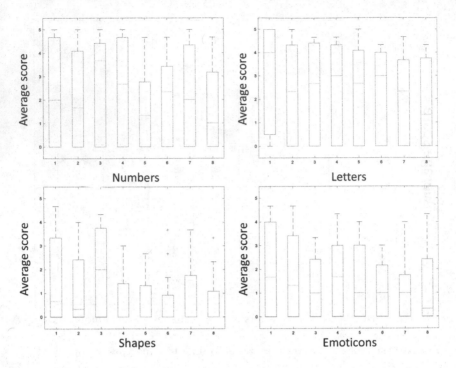

Fig. 5. Distribution of average scores in each category of copy work

4   Discussion

As shown in the histogram in Fig. 3, students with diagnosed learning difficulties showed significant differences in ability to copy work. It will be interesting to see how the improvements during the longitudinal training depending on their initial ability of the copy work. We also added this condition to the exclusion criteria for the longitudinal study because high scores with a score of 4 or higher have little room for improvement and do not need to be trained in copy work.

An interesting result of distribution analyzing among four categories of copy work is that the average scores of shape had the lowest scores. We expected the copy of emoticons to be the most difficult task since emoticons have far more complex construction trajectories. It is presumed that the face is a picture that young children draw well. Especially happy face is a psychologically familiar emoticon because it is their favorite emotion. It is likely that the familiarity, in addition to the complexity of the copy work task, plays a significant role in defining what makes a difficult task.

Although there was no significant differences among the four categories of copy work tasks and the eight tasks in each category, it is necessary to investigate their effects during the following longitudinal experiment. There could be significant improvements in the students' performance in relation to the assigned copy work task. In addition, it will be interesting to examine how task difficulty influences the development of handwriting skills for children with learning difficulties.

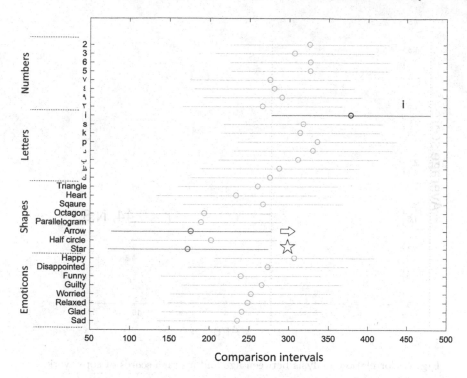

Fig. 6. Comparison intervals. The blue and red lines indicate that scores of the first letter, 'i' are higher than scores of the sixth and eighth shapes, arrow and star. (Kruskal-Wallis test, $p < 0.01$; ad-hoc, Bonferroni) (Color figure online)

In general, it is natural that the ability of copy work varies according to age, however we could not find a significant correlation of average scores of copy work according to age. Previous studies showed that implicit learning has not been correlated with age [12]. It is presumed that the copy work which is not familiar to the children with learning difficulties could not have a correlation with the score according to age because it is a new task requiring visual recognition, visual-motor function, and fine motor function. On the other hand, there is no significant correlation between age and copy work (Pearson correlation coefficient, $r = 0.44$, N.S.), but there is a trend to improve the score according to age as shown in Fig. 7. Therefore, the assignment of treatment/control groups for the following longitudinal study should be adjusted to match the age group balance.

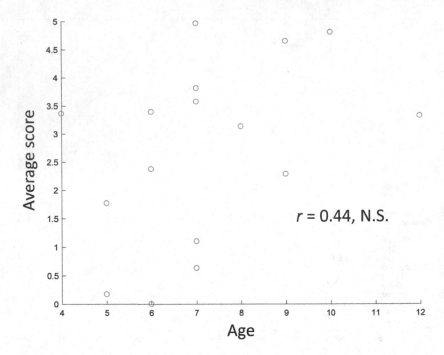

Fig. 7. Correlation analysis between age and average scores of copy work.

5 Conclusion

This study introduced a methodology and copy work for evaluating the handwriting skills of children with learning difficulties. The proposed method can be used to define appropriate tasks, depending on the complexity of the task and the abilities of the learner, to maximize the learning outcomes. This may also be used as a pre-test to place learners in different groups for comparative studies. Our immediate future work is to form balanced groups of students, based on their abilities, that will train with the haptic-based handwriting platform. We would like to study the effectiveness of various haptic guidance methods (partial, full and disturbance) towards improving the acquisition of handwriting skills.

Acknowledgments. This work has been supported by the ADEK Award for Research Excellence (AARE) 2017 program (project number: AARE17–080).

References

1. Bonny, A.: Understanding and assessing handwriting difficulties: perspective from the literature. Aust. Occup. Ther. J. **39**(3), 7–15 (1992)
2. Reisman, J.E.: Development and reliability of the research version of the minnesota handwriting test. Phys. Occup. Ther. Pediatr. **13**(2), 41–55 (1993)

3. Hamstra-Bletz, L., Blöte, A.W.: Development of handwriting in primary school: a longitudinal study. Percept. Motor Skills **70**(3), 759–770 (1990)
4. Scardamalia, M., Bereiter, C., Goelman, H.: The role of production factors in writing ability. What Writers Know: Lang. Process Struct. Written Discourse **3**, 173–210 (1982)
5. Smits-Engelsman, B.C., Niemeijer, A.S., van Galen, G.P.: Fine motor deficiencies in children diagnosed as DCD based on poor grapho-motor ability. Hum. Mov. Sci. **20**(1–2), 161–182 (2001)
6. Hamstra-Bletz, L., Blöte, A.W.: A longitudinal study on dysgraphic handwriting in primary school. J. Learn. Disabil. **26**(10), 689–699 (1993)
7. Diagnostic and Statistical Manual of Mental Disorders, 5th edn. DSM-5. American Psychiatric Association, Washington, DC (2013)
8. Kivisto, L.R.: The use of assistive technology in school-aged children with learning disorders. Electronic theses and Dissertations 7270 (2017)
9. Teranishi, A., Korres, G., Park, W., Eid, M.: Combining full and partial haptic guidance improves handwriting skills development. IEEE Trans. Haptics **11**(4), 509–517 (2018)
10. Martin, S., Hillier, N.: Characterisation of the novint falcon haptic device for application as a robot manipulator. In: Australasian Conference on Robotics and Automation (ACRA), pp. 291–292. Citeseer (2009)
11. Lee, T.D., Swinnen, S.P., Serrien, D.J.: Cognitive effort and motor learning. Quest **46**(3), 328–344 (1994)
12. Vinter, A., Perruchet, P.: Implicit learning in children is not related to age: evidence from drawing behavior. Child Dev. **71**(5), 1223–1240 (2000)

"Express Your Feelings": An Interactive Application for Autistic Patients

Prabin Sharma[1], Mala Deep Upadhaya[3], Amrit Twanabasu[3],
Joao Barroso[1,2], Salik Ram Khanal[1,2], and Hugo Paredes[1,2(✉)]

[1] Universidade de Trás-os-Montes e Alto Douro, Vila Real, Portugal
prabinent7@gmail.com, {jbarroso,salik,hparedes}@utad.pt
[2] INESC TEC, Porto, Portugal
[3] Kathmandu University, Dhulikhel, Nepal
maladeep.upadhaya@gmail.com, amrit.cas@gmail.com

Abstract. Much effort is put into Information Technology (IT) to achieve better efficiency and quality of expressing communication between autistic children with the surrounding. This paper presents an application that aims to help the autistic child to interact and express their feeling with their loved ones in easy manner. The major objective of the project is to connect autistic children with their family and friends by providing tools that enable an easy way to express their feeling and emotions. To accomplish this goal an Android app has been developed through which, autistic child can express their emotion based on emoji. Child's emotions are share by sending the emoji to their relatives. The project aims a high impact within the autistic child community by providing a mechanism to share emotions in an "emotionless world". The project was developed under the Sustainable Development Goal (SDG) 3: good health and well-being in the society by making the meaningful impact in the life of autistic child.

Keywords: Autism spectrum disorder · Android application · Emotion

1 Introduction

A challenge for parents with children with autism disorder is communication. The children cannot exchange information with their parents because of the delay in the development of language [1].

A first approach to fill the existing communication gap of children with autism was the Picture Exchange communication system. The system aimed to act as a bridge and connect the children with autism with other people. The main objective was to enhance the exchange information, by introducing a communication augmentation system [2].

Despite the efforts and the research in the last decades in augmented and alternative communication (AAC) systems there are still open issues. One of them is the expression of basic daily needs, required for an autonomous life that a person with autism usually cannot express, and therefore became dependent. To overcome this problem, we have developed an application which will help the users to express their feelings, so they can easily communicate using emoji. Although the target audience of

M. Antona and C. Stephanidis (Eds.): HCII 2019, LNCS 11573, pp. 160–171, 2019.
https://doi.org/10.1007/978-3-030-23563-5_14

the application is the child with autism, it can also be used by the elderly as well as speech impaired people who cannot express themselves for their needs.

In the proposed mobile application, we have used a picture (emoji) with their respective sounds. When the user touches the picture representing their needs or emotions, it will sound like the respective emotions or need so they will be confident about their needs and learn about their emotions and needs listening to the voice that was included in the application.

This paper is organized into seven sections: "Background", describing the current state of the art, including Express Feelings using mobile application; "Design and Implementation", explaining how the application works and its main features; "Experimental Design", describing the methods used to test the application, the type of audience, the evaluation variables (effectiveness, efficiency and user motivation); "Apparatus", describing the devices used to run the application; "Practical Results", describing the test users feedback, and "Conclusion", discussing the evaluation of the results obtained during the field testing.

2   Background

Autism is a Spectrum of disorders which is characterized by problem with social interaction and communication. People with such disorder shows respective and repetitive behaviors which makes them upset [3]. Usually it appears gradually, when the human development began to unfolds [4]. Autism was first recognized by Leo Kanner (1943) as a completely different disorder which was not described before under childhood psychosis. Later this disorder was named as Kanner's Autism. Autistic children have poor social and language skills with less interest in people and they are rigid up to their daily routines, as described by Kanner in 'Autistic disturbances in Affective Contact' [5].

According to the Centers for Diseases Control and Prevention (CDC), about one in every fifty-nine child have autism spectrum disorder [6]. Autism, or Autism Spectrum Disorder, is a combination of condition characterized by challenges to speech and nonverbal communication, skills and repetitive behaviors.

Autistic children have problems regarding understanding and expressing their emotions [7]. So, expressing their emotions through mobile application can be good medium to communicate with their family members.

Due to which autistic children are caught in an impasse of social isolation, studies show that intervention focused on teaching social interaction, interpersonal problem solving, and affective knowledge has resulted in positive social interaction with peers by triggering their ability to share experiences, emotion and cast interest [8]. So, a platform that can simplify expressing of their emotion, and the connection between child and parents will act as a communication bridge and break the communication barrier for autistic children. Currently there are some apps, available in the apps' stores that aim helping children to cope their behavior. A selection of these applications is presented in the next subsections:

2.1 Awesomely Autistic Test

The Awesomely autistic test application (Fig. 1) was developed by Harry Marcus Ltd, and aims to measure autistic traits. It is based on the Autism Quotient (AQ) created by the psychologist Simon Baron-Cohen and his team at Cambridge's Autism Research Center.

Fig. 1. Autism Quotient (AQ) result in Awesomely autistic test

The test contains 50 choice-based questions (with a Likert: slightly disagree, definitely disagree, definitely agree and slightly agree). Based on the answer it generates AQ and based on AQ it expresses the level of autistic in the test taker. Overall, this app provides a simple screening mechanism that could be used for further autism evaluation.

In Google play it has 5000 downloads and 4.0 stars by 30 persons [9]. It does not have iOS version. It has a file size of 5.24 MB in version 1.05.1 and latest updates was 22 Aug 2016.

2.2 Autism Therapy with MITA

The Autism therapy with MITA application (Fig. 2) provides visual puzzles that facilitates early development for children. It was developed by three Ivy League researchers to help children with autism to learn using pivotal response treatment (PRT) method. MITA stands for Mental Imaginary Therapy for Autism. According to the authors, the app provides puzzles that help the children to improve attention, visual skills, and languages.

It includes several games, namely; arithmetic and auditory memory games that train auditory working memory; and combing toys for developing spatial direction choice.

The application is organized by users' age categories. Three age groups are defined: Preschool (2-5); Children (6-12); and Adolescents (1317). In Apple App Store, it has not enough rating and reviews. Google Play Store reports 5000 download with rating of 4.6 (687 voters) [10]. This app also got the prize of Best Autism app for 2017 on the healthiness' list. The app was released on 27 Jul 2015 offered by ImagiRation LLC, Boston, MA 02135. The last update was on 25 Jun 2018, version is 4.0.5 with file size: 273 MB.

Fig. 2. Game play of Autism therapy with MITA

2.3   Autism Emotion

The Autism Emotion application (Fig. 3) shows a story line to choose an emotion from four given options. Based on the choice (happy, sad, proud and calm), it shows a photo and a sentence describing the photo. The consecutive five photos and in the last slide it has a music button to play the story about it. This app is developed by "modelmekids" that produces teaching tools for children with Autism, Asperger's Syndrome, PDD-NOS, and nonverbal learning disorder.

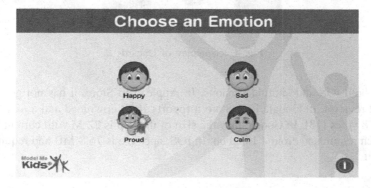

Fig. 3. Game play of Autism Emotion

This app is only compatible with iOS and has a rating of 3.7 stars with 30 votes [11]. First version of the app came to the store on August 10, 2012. The most recently updated was on Nov 16, 2017 version 4.0 and file size of 23.5 MB that is compatible with IOS 11.

2.4   Upcard

Autism children may not articulate word all the times, but they have a visual vocabulary. Based on this, Royden James created this app for his autistic son.

Upcard application (Fig. 4) uses the picture exchange communication system (PECS) [12], developed for people with cognitive, physical and communication disabilities. This app is made of "picture cards" that users can place in sequences to form simple statements and requests. Sign in and login validation is presented in the app. Through the app parents can make any plan, can keep timer for the child so on specific that message gets popup and on the other hand the autistic child can speak based on the cards or icon he selects from the permitted icons that parents has validated. It also has Flash card game based on the icons used before. So, the UI of the app is slick, but on usability of the app is hard for the children to understand. For the children, it is more time consuming and challenging at prior hand.

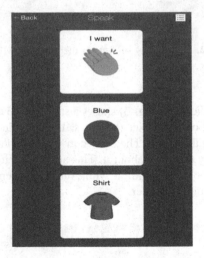

Fig. 4. Gameplay of Upcard

It has both iOS and android versions. In Apple App Store, it has not got enough rating and reviews. In Google Play Store it reports 1000 download and a rating of 3.9 stars with 8 votes [13]. In Google play the size of the app is 22 M with current version 0.7.3 which requires android 4.1 and up. In IOS, app size is 76.8 MB and requires IOS 9.0 and up.

2.5 Summary

The analysis of the applications allowed to identify their pros and cons, summarized in Table 1. We found the limitation of the existing apps are: difficulty and more time consuming in understanding the app structure are only designed for specific game propose.

The aforementioned app do not provide platform for children with autism to develop important social skills and express their feeling to their family and friends.

Table 1. Overview of the current applications (scale from — to +++)

Application	Design	Quantity of Ads	Memory
"Awesomely Autistic Test"	+	–	–
"Autism therapy with MITA"	+++	+	–
"Autism Emotion"	+	+	–

3 Design and Implementation

The study of the state of the art revealed the need for a general app to help autistic child express their feeling and speak through the app. The "Express Feelings" application was designed taking into account the following requirements:

- Have an intuitive design.
- Target demographic of people with Autism.
- Clear feedback of user actions.

The user interface was designed to be used by children with Autism, so well-known icons derived from emoticons (or emojis) were used to give visual representation of the user's emotion.

3.1 Implementation

The application was designed to work on Android devices with android version from Jelly Bean to Oreo without connecting to internet. To increase the user base, the app was made free and open source at GitHub [14].

3.2 Using the App

When the user starts the application, a short description of the app can be seen along with a set of instructions to use it.

After clicking "Get Started", a screen to save password is shown as illustrated in Fig. 5. The role of password to keep the settings unchangeable by the autistic user in course of usage.

On selecting a password, the user is allowed to select at most 5 contacts to send the text message by the autistic user as illustrated in Fig. 6 (a).

On selecting the contacts, the app is ready to use. The app shows two tabs titled with "Feelings" and "Send to". The "Feelings" tab shows a grid of icons representing various emotions the user may feel to express with clear distinction between the icons using white spaces. Nine general feelings expressing emoji are represented in the tab for fast and reliable emotion expression.

Fig. 5. Setting password for the for App

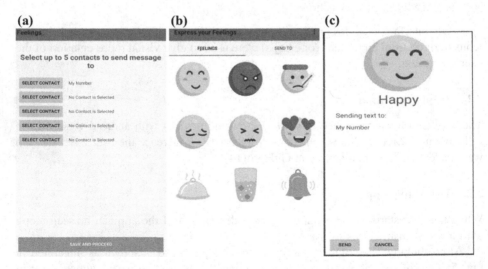

Fig. 6. (a) Page to select the Contacts (b) Grid of emotion icons (c) Dialog showing prompting the user to send the text.

The app consists of emoticon icons which represents certain emotion of the user which is shown in Fig. 6 (b). When the user taps on an emoticon icon, an audio representing the emotion is played. A dialog box appears which shows the details of the emoticon including a text to be sent to the listed contacts – Fig. 6 (c).

After clicking "Send", the user can send a predefined message to the selected list of contacts. If "Cancel" is pressed, then the dialog closes and the app resumes to show the list of emotions.

When the user clicks on an icon, respective speech that indicates that respective feelings is generated and a dialog appears along with the which shows the visual representation of the emotion, textual form of the emotion and the list of contacts who will get the message.

After clicking "Send", the user can send a predefined message to the selected list of contacts. If "Cancel" is pressed, then the dialog closes and the app resumes to show the list of emotions.

The "Send to" tab offers the ability to include or exclude previously selected users from getting the message.

A dotted vertical icon is shown at the top right corner of the application. Clicking this shows a menu to go to "Settings" of the app. Accessing "Settings" requires a password which was set by the user earlier. After setting correct password the user can edit contacts and password.

The State Chart Diagram, presented in Fig. 7, summarizes the application usage, describing all the states and important events.

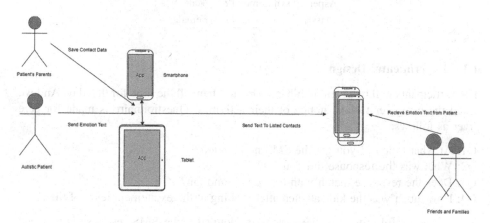

Fig. 7. State chart diagram

4   User Testing

With the help of user testing (UT), we study if the target group can use the proposed application in easy and effective way. The approach was a method of formative evaluation or think aloud as it serves as a "window on the soul", letting users do all the tasks they will be doing in the future through our application.

During the UT, we looked for critical incidents i.e. moments that strongly affect usability, task performance (efficiency or error rate) and came to following points:

4.1   Methods

User test were performed in two phases: using an application in real time with people with autism; and questionnaire. In the first phase, while the user is using the application, data is collected by a member of the research team.

To test the performance of the user a questionnaire was performed by the caregiver, considering the communication barriers with people with autism.

4.2   Participants

Four participants were chosen from a child care with autism people. The people or caregiver were asked to fill the authorization form prior to the data collection. These kids were chosen by Special education teacher at SERC School (Special Education and Rehabilitation Center for Disabled Children) in Kathmandu, Nepal.

Participants characterization is presented in Table 2.

Table 2. Participants

Participants	Type of Autism	Age	Gender	Remarks
P1	PDD	13	Male	With glasses
P2	Classic	9	Female	–
P3	Asperger syndrome	12	Male	–
P4	Classic	17	Female	–

4.3   Experimental Design

The participants will be given mobile phone with them all the time for that day. And we asked the caregiver to take notice of their activities. Questionnaire is made for care giver as follows:

Q1: At what time did you get the SMS notification?
Q2: What was the response did you get?
Q3: Does the response match with the actual emotion?
Q4: How much was the kid satisfied after looking at the excitement level of the user?

The successful criteria of the system are defined if the SMS response is matched with the response that actually user wants to deliver after the care giver approach to the user just after getting SMS.

4.4   Apparatus

For the user tests, we used a Mobile phone (Samsung Galaxy Note 9 sm-g960f). We choose this phone because we believe it could be easier to see all the emotion pictures easily for the user to interact with app more easily.

4.5   Results and Discussion

The data collection before testing an application is compared with the data collection after the testing the application. Specially, the usability of an application was tested with autism people, but questioner is asking with the caregiver or their parents and sometime with the doctor and nurses. The performance of the application or the impact of application was evaluated according to the behavior changes by the autism people.

The data collected with this application is described in Table 3.

Table 3. Representing the result of Questionnaire

Participants	Time	Response	Does the response match with the actual emotion?	Is user satisfied with the response?
P1	7:00 AM	Thirsty	Yes	Very satisfied
	8:00 AM	Hungry	Yes	Satisfactory
	10:00 AM	Help	Yes	Satisfied
	12:00 PM	Sad	Yes	Satisfied
	4:00 PM	Hungry	Yes	Satisfied
P2	8:00 AM	Hungry	Yes	Very satisfied
	9:00 AM	Help	Yes	Satisfactory
	12:00 PM	Thirsty	No	Unsatisfactory
	2:00 PM	Sad	Yes	Satisfactory
P3	5:00 AM	Help	Yes	Satisfactory
	8:00 AM	Hungry	Yes	Unsatisfactory
	11:00 AM	Help	Yes	Satisfactory
	3:00 PM	Angry	Yes	Satisfactory
P4	7:30 AM	Thirsty	Yes	Very satisfied
	9:00 AM	Hungry	Yes	Satisfied
	10:00 AM	Help	Yes	Unsatisfactory
	1:00 PM	Happy	Yes	Satisfied

Figure 8 shows the user satisfaction chart for using the application which was done when the care taker notices the emotion of the respective user after getting the SMS and dealing with the user afterward.

From the chart we can see that, Participant 1 gave 5 responses in which 50% of his response are very satisfying and other 50% was satisfactory. And 25% of the response of Participant P2 was very satisfying and 50% of the responses were satisfactory and remaining 25% was unsatisfactory. For Participant P3, 75% of the responses were satisfactory where 25% of the response was unsatisfactory. Whereas the 25% of the responses of Participant P4 was very satisfying and 50% of the responses were satisfactory and remaining 25% was unsatisfactory.

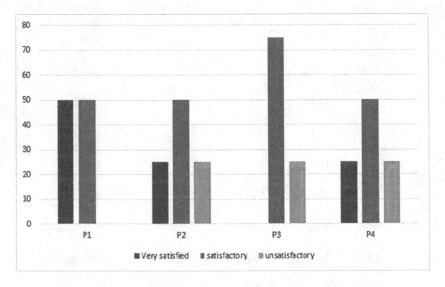

Fig. 8. Application satisfaction chart

5   Conclusion

In this paper, we presented an android application for children with autism to express their emotions to their parents and relatives. The application allows speech output by selection an emotion in the mobile application. Emotions are chosen from a preset of emoji as they can represent emotions in a user-friendly way. Moreover, emoji are a common and well accepted icon representation of emotions.

The result indicated that this application can be implemented to all kind of user having autism. Their response showed that they are interested to use the app as it acts as a bridge between the people. Globally, the obtained result indicates all kind of user were successful to use the app. But the user with pervasive developmental disorder was more satisfied than the user having Classic autism and Asperger syndrome.

For future work, we will try to add more emotions that users can give so that their all the needs can be solved and put the Native speech instead of English when they click the emotion.

Acknowledgements. This work is financed by National Funds through the Portuguese funding agency, FCT - Fundação para a Ciência e a Tecnologia within project: UID/EEA/50014/2019.

References

1. Biklen, D.: Communication unbound: autism and praxis. Harvard Educ. Rev. **60**, 291–315 (1990)
2. Schwartz, I., Garfinkle, A., Bauer, J.: The picture exchange communication system. Top. Early Child. Spec. Educ. **18**, 144–159 (1998). https://doi.org/10.1177/027112149801800305
3. Iowa Department of Education. http://educateiowa.gov/sites/files/ed/documents/Parent-Factsheets_April2010_Autism.pdf. Accessed 04 Apr 2019
4. Frith, U.: Autism: Explaining the Enigma, 2nd edn. Blackwell Publishing, Malden (2003)
5. Matson, J.: Clinical Assessment and Intervention for Autism Spectrum Disorders. Elsevier Science, Burlington (2011)
6. Data and Statistics on Autism Spectrum Disorder—CDC. Centers for Disease Control and Prevention. https://www.cdc.gov/ncbddd/autism/data.html. Accessed 04 Apr 2019
7. Gay, V., Leijdekkers, P.: Design of emotion-aware mobile apps for autistic children. Health Technol. **4**, 21–26 (2013). https://doi.org/10.1007/s12553-013-0066-3
8. Bauminger, N.: The facilitation of social-emotional understanding and social interaction in high-functioning children with autism: intervention outcomes. J. Autism Dev. Disord. **32**, 283–298 (2002). https://doi.org/10.1023/A:1016378718278
9. Google Play. https://play.google.com/store/apps/details?id=com.androidinlondon.autismtests. Accessed 04 Apr 2019
10. Google Play. https://play.google.com/store/apps/details?id=com.imagiration.mita. Accessed 04 Apr 2019
11. Autism Emotion. https://itunes.apple.com/us/app/autism-emotion/id550027186?mt=8. Accessed 04 Apr 2019
12. Picture Exchange Communication System (PECS). https://pecsusa.com/pecs/. Accessed 04 Apr 2019
13. Google Play. https://play.google.com/store/apps/details?id=com.james.upcard. Accessed 04 Apr 2019
14. GitHub. https://github.com/amrittb/express-your-feelings. Accessed 03 Apr 2019

The Design of an Intelligent LEGO Tutoring System for Improving Social Communication Skills Among Children with Autism Spectrum Disorder

Qiming Sun and Pinata Winoto[✉]

Wenzhou-Kean University, Wenzhou, Zhejiang, China
{sunq, pwinoto}@kean.edu

Abstract. A system intended to help children with autism spectrum disorder (ASD) to play with LEGO bricks is proposed. The system could provide step-by-step guidance to complete pre-defined task using interactive dialog strategy. A camera is used to capture the bricks, and an image recognition module is implemented to identify the color and size of each brick. The system can also detect a misplaced brick, provide guidance to reassemble it, and suggest a correct one. Dialog is generated using a speech synthesizer on a set of pre-defined statements used in daily social communication with the children. To further enrich the interaction between children and the system, teachers may intervene remotely in real time to alter existing statements in the system.

Keywords: LEGO therapy · Autism · Education software

1 Introduction

LEGO bricks are very popular among children, and for those with autism spectrum disorder (ASD), the bricks could be a good auxiliary toy to improve their social communication skills [1–3]. In this paper, we proposed a system to assist children with ASD playing with LEGO bricks, in which it could recognize a 2D structure ($n \times m$ bricks) and generate the whole layout through a camera automatically. More specifically, the system allows teachers to create a task (structure), tracks the bricks played by children in real time and give children feedback according to their built result. The proposed system can also reduce the teacher's workload and allow them to monitor more children simultaneously.

At the current stage, we have implemented a prototype for detecting 2D structures containing up to eleven bricks in three different colors and three different dimensions. We also made a model for the educators to design their unique brick structures for the children. The system can load a pre-defined (target) layout and then generate instructions for the users in both text and speech. In the future, we will expand the system to detect 3D structures with more brick colors and shapes. More interactive methods will also be introduced to facilitate the learning process.

© Springer Nature Switzerland AG 2019
M. Antona and C. Stephanidis (Eds.): HCII 2019, LNCS 11573, pp. 172–181, 2019.
https://doi.org/10.1007/978-3-030-23563-5_15

2   Related Work

2.1   LEGO Therapy for Children with ASD

Many studies have shown that early intervention could be beneficial for children with autism as reported in [4], which could be performed for children as young as 18 months age [5]. LEGO bricks, which are very popular for toddlers, have been used as a therapeutic medium to enhanced children's social communication skills [1–3]. In the playing process, children are assigned with various roles and expected to collaborate to build a structure following a set of clear instructions provided by their teacher. Given a clear and predictable task in the LEGO play, children with ASD would feel more comfortable compared to in a free-play one; hence, would be more likely to initiate communication with their peers.

Recently, technology-based intervention using LEGO play, including LEGO robots, have also been introduced to children with ASD [6, 7]. However, using LEGO in an intelligent tutoring system for improving their social communication skills is rare. In this project, we attempt to fill this void.

2.2   The Adoption of Learning Technology

The increasing number of children diagnosed with ASD has become an issue around the world, especially for the educators [8, 9]. For instance, in the U.S., the prevalence of children aged 8 years with autism was 6.7 per 1,000 in 2000, and increased to 14.7 per 1,000 in 2010 [10]. Since the increase would bring more pressure to the special-education teachers, many attempts have been proposed to reduce their workload through the introduction of various assistive learning technology, including self-monitoring systems [11], electronic AAC [12, 13], collaborative-learning applications [14, 15], and intelligent tutoring systems, as in our case here. In fact, some applications are proven effective in improving children's social communication skills. In addition, some researches have also demonstrated a better acceptance of electronic technologies to the youth with autism [16, 17], which also motivates us to implement the system.

3   Motivation and Design Rationale

Our proposed system is intended for a small classroom of 8 to 10 students (preschool children age 3 to 5) and a supervising teacher. During the activity, students will be divided into 4 to 5 pairs, and each pair will be assigned to collaboratively build a LEGO structure on a special desk equipped with the system. Ideally, all five systems could simultaneously notify the teacher any important progress made by their students, such as the completion time, frequency and type of mistakes, etc. The teacher, according to their students' learning progress, may (re-)design appropriate learning scheme for them. Each learning scheme may consist of a set of LEGO structures to be built within a certain duration (e.g. 30 min). The system, on the other side, may automatically tag each learning scheme with its average completion time, the frequency of mistakes and/or some common mistake(s) in building them, so that it can be evaluated/redesigned in the future.

Since many preschool children with ASD exhibit delay or deficit in gross and fine motor skill, and many comorbid with ADHD and/or dyspraxia, the instructions used in our system must be comprehensible for them, for instance, using simple words delivered in low-volume voice, avoiding sensitive colors (e.g. yellow), providing repetitive instructions, etc. Finally, the material must be affordable for a special-education (public) school; for example, the current prototype consists of a decent laptop with an ordinary webcam, and some LEGO Duplo bricks.

4   System Implementation

The test set of bricks is shown in Fig. 1. Eleven basic LEGO Duplo bricks with three different shapes (short, long, curved) and colors (red, green, blue) are used in the testing. Figure 2 shows a USB camera, connected to a computer, is put in front of the bricks to track the changes. In terms of the software, there are two separate parts for both teachers and children. Teachers can use an interactive program to create a unique structure of bricks (designer mode). The module designed for children can then load the structure and generate the guidance automatically (player mode).

Fig. 1. The blocks used in the test. (Color figure online)

Fig. 2. An example of LEGO structures being tested.

4.1 Brick Recognition Module

The proposed system uses several functions, including some from the OpenCV library, to process frames from the camera and generate contours. Figure 3 shows the flowchart of the image processing. First, the pixels are filtered based on their RGB values. Since only red, blue, and green colors appear in the test set, pixels containing other color will be replaced with a white one. Then, the derivatives on x- and y-axis are calculated using an extended Sobel operator. The two derivatives will be subtracted to sharpen the contours of bricks. Threshold and blur functions are also used to remove the noise in the frame. The resulted grayscale image will then be passed to OpenCV's findContours function. For each contour, a minimum rotated rectangle contains a closed contour is computed, and the program will try to figure out whether the contour belongs to a LEGO brick or not based on properties such as the rectangle's area, slope, and length-width ratio. The program also samples points inside a rectangle to determine the brick's color. Figure 4 gives the program's final label on one frame (the labels are not displayed in a released version). Some critical properties to recognize the bricks are introduced in Table 1.

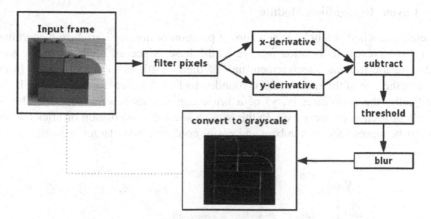

Fig. 3. The pre-processing of an input frame. (Color figure online)

Table 1. Some properties used to recognize the bricks.

Shape	Height	Length	Colors	# of bricks
Long brick	1	3	Red, blue, green	3
Short brick	1	1.5	Red, blue, green	6
Curved brick	1	2.25	Green	2

Fig. 4. The result of block recognition.

4.2   Layout Recognition Module

To determine whether a brick is at a correct position or not, we introduce a coordinate system to represent its position relative to the base. Since all bricks have the same height, we use it as the measurement unit of their length as shown in Fig. 4. In this example, the size of the short bricks is rounded to 1×2, hence, the long one is 1×4. Specifically, the coordinate (x, y) of a brick contains its row number and the left corner's projection on the x-axis. In this way, a specific composition of bricks on one side can be represented in numbers and easily compared with target layouts.

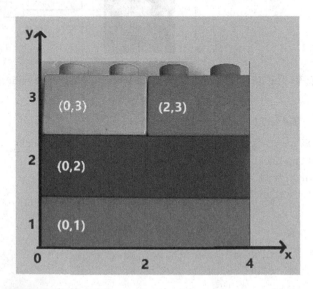

Fig. 5. The coordinate system used in our prototype to represent a brick layout. (Color figure online)

4.3   Interaction Module

The user interface is generated using *cvui*[1], a UI library built on OpenCV. Before children can use the system, their teacher must create a target layout (learning scheme) using the designer module (will be explained in the next section) and store it into the system. For the sake of maintainability, the target layout will be encoded in the format [size] [color] [x-value] to describe every block and saved in an unencrypted text file. For instance, a layout in Fig. 5 will be written as:

    long red 0
    long blue 0
    short green 0 short red 2

Using a plain text file, the system could easily generate textual (or speech) instructions for each building step, for instance, we could read the first line above, interpret and insert it into a textual template to generate a statement "put a [long red] brick []". The second line can be interpreted and used to generate statement "put a [long blue] brick [on top of the [long red] one]"; followed by "put a [short green] brick [on top of the [long blue] one]", and "put a [short red] brick [beside the [short green] one]", and so on. The system will allow teachers to change the textual templates to fit their students' language abilities.

When a wrong brick is placed, the system can also provide feedback and wait for the user to correct it. The checking process is done by comparing the output of *layout recognition module* with the *target layout* in real time.

To make the instructions more attractive and comprehensible, our system uses various text-to-speech modules, for example, a sentient child tone or male/female adult voice with pace appropriate for the target children. At the current stage, three types of voice feedback are provided in our system, some examples are shown in Table 2. The voice is generated using text to speech service from Baidu™.

Table 2. The voice feedbacks in the system.

Type	Example	Speed
Instruction	Please take out a red, long brick	Slow
Encouragement	Great!	Intermediate
Notification	All steps completed!	Intermediate

5   System Prototype

5.1   Designer Mode

Figure 6 shows the user interface of the basic designer module intended for creating learning schemes. Teachers can select a brick with specific shapes and colors on the right pane and then put it on the left pane (a 10 × 10 grid), layer by layer up to eleven

---

[1] https://github.com/Dovyski/cvui.

layers. After they finish designing, they can click 'Output Scheme' to save the target layout in the filesystem which can be retrieved in the future.

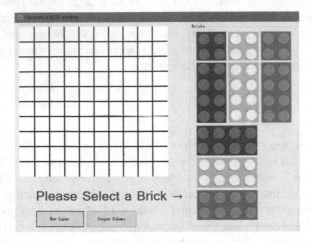

Fig. 6. The designer module.

5.2 Player Mode

Figure 7 demonstrates the user interface of the system in action. The main part on the interface is a camera view on the left pane, which will be used to track the building steps and provide visual notifications, if any.

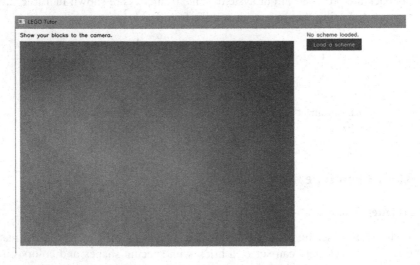

Fig. 7. The user interface of the player module.

Indeed, the camera must be adjusted manually to cover the working area or the system may not work properly. At this stage, no calibration is required; hence, a light color background (e.g. white) with bright ambient light is suggested for an optimal recognition. After adjusting the camera, the user may load a target layout, which be automatically shown in the right pane (see Fig. 8). In this example, the layout consists of six bricks, including a curved green brick on the top.

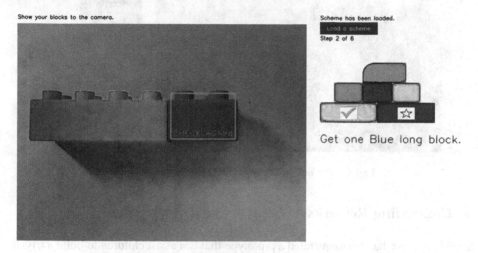

Fig. 8. The system is notifying a mistake. (Color figure online)

Instructions on what brick should be picked will be given in both textual (e.g. "Get one Blue long block" in Fig. 8) and auditory form (via computer speaker), because most young children may not read but their supervising teacher may read the message later (recall that a teacher may supervise five groups of students, hence, s/he may not remember the latest instruction of all five). During the building process, the system will monitor the whole structure and provide feedback accordingly, for example, in the right pane of Fig. 8 the correct red brick is ticked and the current target blue brick is marked with a yellow star. If a brick is incorrectly placed, the system will highlight the wrong brick in the camera view (left pane) and provide an instruction to check and replace it. In Fig. 8, the red long brick in the camera view is in a correct position, but the green short brick is incorrectly placed - it should be a blue long brick, instead. In this example, the mistake is marked clearly in the camera view along with a message 'check again!'.

When the target structure has been correctly built as shown in Fig. 9, all bricks in the right pane are ticked and the notification 'you have completed all of the tasks' is given to the user, also in both textual and auditory forms. In this case, the system may load a new layout, or the teacher may assign a new scheme for the children or exit the system.

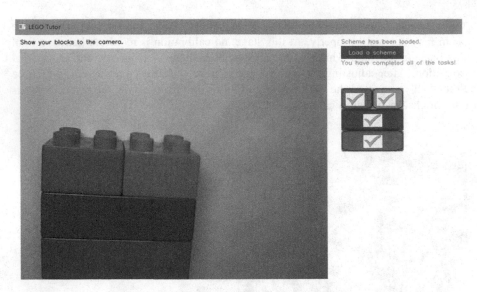

Fig. 9. The notification when all steps are completed.

6 Concluding Remarks

In this paper we have demonstrated a prototype that can assist children to build a trivial LEGO structure. The system enables teachers to design some target structures, assign one of them, and guide children to build it step by step. Currently, the system can recognize the brick structure from one side based on frames recorded by a webcam. Using the system, it may reduce teachers' workload, yet capable of tracking more children simultaneously. In addition, the dialog module for enhancing children's interaction with the system is under construction.

In the future, some improvements will be made in order to generate 3D structures to support a richer combination of LEGO bricks. Also, more instructions in speech will be added so that the children may focus more on their work instead of the screen.

Acknowledgement. This work is partially supported through research grant from Wenzhou-Kean University (WKU201718016 and WKU201718017). We wish to thank Tiffany Tang for her constructive comments.

References

1. LeGoff, D.B.: Use of LEGO© as a therapeutic medium for improving social competence. J. Autism Dev. Disord. **34**(5), 557–571 (2004)
2. LeGoff, D.B., Sherman, M.: Long-term outcome of social skills intervention based on interactive LEGO© play. Autism **10**(4), 317–329 (2006)
3. Owens, G., Granader, Y., Humphrey, A., Baron-Cohen, S.: LEGO® therapy and the social use of language programme: An evaluation of two social skills interventions for children with high functioning autism and Asperger Syndrome. J. Autism Dev. Disord. **38**, 1944 (2008)

4. Wagner, M., Marder, C., Blackorby, J., Cameto, R., Newman, L., Levine, P., Davies-Mercier, E.: The achievements of youth with disabilities during secondary school, a report from the national longitudinal transition study-2, SRI International, Menlo Park, CA (2003)
5. Dawson, G., Rogers, S., Munson, J., Smith, M., Winter, J., Greenson, J., Donaldson, A., Varley, J.: Randomize, controlled trial of an intervention for toddlers with autism: the early start Denver model. Pediatrics 125(1), e17–e23 (2010)
6. Huskens, B., Palmen, A., Van der Werff, M., Lourens, T., Barakova, E.: Improving collaborative play between children with autism spectrum disorders and their siblings: the effectiveness of a robot-mediated intervention based on Lego® therapy. J. Autism Dev. Disord. 45(11), 3746–3755 (2015)
7. Dorsey, R., Howard, A.M.: Examining the effects of technology-based learning on children with autism: a case study. In: Proceedings of the 11th International Conference on Advanced Learning Technologies, pp. 260–261. IEEE, USA (2011)
8. Loiacono, V., Valenti, V.: General education teachers need to be prepared to co-teach the increasing number of children with autism in inclusive settings. Int. J. Spec. Educ. 25(3), 24–32 (2010)
9. Blumberg, S.J., Bramlett, M.D., Kogan, M.D., Schieve, L.A., Jones, J.R., Lu, M.C.: Changes in prevalence of parent-reported autism spectrum disorder in school-aged US children, 2007 to 2011–2012, National Center for Health Statistics Reports 65, pp. 1–11 (2013)
10. Christensen, D.L., et al.: Prevalence and characteristics of autism spectrum disorder among children aged 8 years—autism and developmental disabilities monitoring network, 11 sites, United States, 2012. MMWR Surveill. Summ. 65(3), 1–23 (2018)
11. Bouck, E.C., Savage, M., Meyer, N.K., Taber-Doughty, T., Hunley, M.: High-tech or low-tech? Comparing self-monitoring systems to increase task independence for students with autism. Focus Autism Other Dev. Disabil. 29(3), 156–167 (2014)
12. Chien, M., et al.: iCAN: a tablet-based pedagogical system for improving communication skills of children with autism. Int. J. Hum.-Comput. Stud. 73, 79–90 (2015)
13. Ganz, J.B., Hong, E.R., Goodwyn, F.D.: Effectiveness of the PECS phase III app and choice between the app and traditional PECS among preschoolers with ASD. Res. Autism Spectr. Disord. 7(8), 973–983 (2013)
14. Parsons, S.: Learning to work together: designing a multi-user virtual reality game for social collaboration and perspective-taking for children with autism. Int. J. Child-Comput. Interact. 6, 28–38 (2015)
15. Winoto, P., Tang, T.Y.: Training joint attention skills and facilitating proactive interactions in children with autism spectrum disorder: a loosely coupled collaborative tabletop-based application in a Chinese special education classroom. J. Educ. Comput. Res. 57(1), 32–57 (2017)
16. MacMullin, J.A., Lunsky, Y., Weiss, J.A.: Plugged in: electronics use in youth and young adults with autism spectrum disorder. Autism 20(1), 45–54 (2016)
17. Putnam, C., Chong, L.: Software and technologies designed for people with autism: what do users want? In: Proceedings of the 10th International ACM SIGACCESS Conference on Computers and Accessibility, pp. 3–10. ACM, New York (2008)

An Augmented Reality-Based Word-Learning Mobile Application for Children with Autism to Support Learning Anywhere and Anytime: Object Recognition Based on Deep Learning

Tiffany Y. Tang[✉], Jiasheng Xu, and Pinata Winoto

Media Lab, Department of Computer Science, Wenzhou-Kean University,
Wenzhou, China
{yatang, xujias, pwinoto}@kean.edu

Abstract. An abundant earlier controlled studies have underscored the importance of early diagnosis and intervention in autism. Over the past several years, thanks to technological advances, we have witnessed a large number of technology-based teaching and learning applications for children with autism. Among them, augmented reality-based ones have gained much attention recently due to its unique benefits of providing multiple learning stimulus for these children via accessing a kinesthetic moving simply using a mobile device. Despite it, few have been developed for these young children in China, which motivates our study. In particular, in this paper, we present a mobile vocabulary-learning application for Chinese autistic children especially for outdoor and home use. The core object recognition module is implemented within the deep learning platform, *TensorFlow*; unlike other sophisticated systems, the algorithm has to run in an offline fashion. We conducted two small-scale pilot studies to assess the system's feasibility and usability with typically developing children, children with autism, their parents and special education teachers with very promising and satisfying results. Our studies did suggest that the downside of the application is the performance of the object-recognition module. Therefore, before we further examine the benefits of such AR-based learning tools in clinical settings, it is crucial to fine-tune the algorithm in order to improve its accuracy. Despite it, since the current literature of AR-technology on Chinese word-learning for children with special needs is still in its infancy, our studies offers early glimpse into the usefulness, usability and applicability of such AR-based mobile learning application, particularly to facilitate learning at anytime and anywhere.

Keywords: Autism · TensorFlow · Deep learning · Word-learning · Augmented reality · China · Children

1 Introduction

Autism spectrum disorder (ASD) is a neurodevelopmental disorder mainly characterized by repetitive and restricted behaviors and deficits in verbal communication, social interaction and emotion recognition [3, 21]. Children with ASD are less willing to

© Springer Nature Switzerland AG 2019
M. Antona and C. Stephanidis (Eds.): HCII 2019, LNCS 11573, pp. 182–192, 2019.
https://doi.org/10.1007/978-3-030-23563-5_16

communicate with others including in classrooms where education takes place; hence, compared to those typically developing (TD) children, they have more limited channels to acquire new knowledge. In addition, although autistic children demonstrated delays in expressive and receptive language, the extent of such delays largely varies across the population and contexts [18]. On the other hand, infant brains are very malleable, so early intervention which largely capitalize on the great potential of learning that an infant brain has could lead to positive effects in limiting some developmental impairments [7, 10, 17, 21], including early language and nonverbal skills [18]. Thanks to the technological advances, numerous technology-based intervention applications have been developed (see [19] for a brief discussion on these). However, an over-whelming number of these previous works have been focused on teaching individuals with autism social communication skills [20], where few addressed the feasibility and efficacy of early language intervention; in additions, many prior works targets learning in a more formal and established learning environment (such as classrooms, clinical centers, etc.).

On the other hand, AR uniquely combines multiple methods of instruction channels including static and dynamic visual stimuli and auditory stimuli via accessing a kinesthetic movement using a mobile phone; it allows users to interact with the real-world in an enhanced way. By learning in different ways according to learners' dif-ferent needs is paramount and has been emphasized and grounded in a foundational framework called Universal Design for Learning (UDL) theory [16] which has been served as a launching point for these AR-based learning tools. Over the past few years, there are a number of AR-based teaching tools; however, few have been built to target children with autism, which motivates our work.

In particular, in this paper we present a lightweight augmented reality-based (AR) mobile word-learning application which allows users to capture a photo where up-to-four objects can be recognized and spoken out in both English and Chinese. The core of the application is an offline deep-learning technique-based object recognition module which is capable of recognizing objects from any angle. Such unique feature offers superior learning opportunities for not only children with autism but also those with other special needs, which had been confirmed from two very small-scale pilot studies.

In particular, a small-scale feasibility and usability pilot test was conducted during a public show with typically development (TD) children and adults (including parents who tried our application with their young children), with very satisfying results. Based on their comments, we simplified the system design. We demonstrated this enhanced version in a special education school in one of the biggest cities in southern China, interviewed some teachers and let some children play with it; our very positive feed-backs provide us with valuable inputs to further adjusting the design of the system.

The organization of this paper is as follows. Previous works will be presented in Section Two, while system descriptions are shown in Section Three. Our observations and discussions in the two pilot studies are shown in Section Four. We conclude this paper by revealing the early yet valuable insights from children, their parents and special education teachers which can be used for further development of our system.

1.1 Motivation

Since the current literature of AR-technology on Chinese word-learning for children with special needs is still in its infancy, which thus offers limited insights into its therapeutic efficacy, feasibility and applicability of individualized intervention for autistic individuals, particularly children. Our works, although preliminary, particularly offers early insights into our understanding towards the usability and usefulness of such AR-based mobile learning application.

2 Previous Works on AR-Based Technology for Children with Autism

There exists an abundant previous work on the adoption of AR technology for therapeutically use and education, particularly for individuals with development disorders [15]; the major advantage of such an AR-enabled environment is that it highly facilitates the cognitive mapping of what is in users' prior knowledge with what they are observing in the real world [12]. Such authentic opportunities can thus promote knowledge transfer and offer more opportunistic learning. In this section, we focus on the use of the technology for tailored and personalized intervention for children with autism. The application of it for Chinese speaking autistic individuals is also discussed to motivate the development of our application.

The majority of AR-based applications had targeted intervention for enhancing children's social and communication skills. For example, [9] described an Object Identification System which allows teachers superimpose digital content on top of physical objects; a five-week study revealed that AR-based application could lead to increased sustained and selective attention of children with autism, and elicits positive emotions, which thus promoted engagement during therapies. However, since the application requires specially trained therapists, it could not easily be used outside the clinic, which thus restricted its usefulness, as most of autistic children's learning mainly occur outside a classroom. McMahon et al. [15] applied AR in teaching science vocabulary and strongly advocated the authentic opportunities enabled via AR for children with development disorder including autism. Improvement of attraction and enhanced social skills training had been observed in [5] where AR technologies had been used to visually conceptualize social stories for children with high-functioning autism. An AR-based application was also developed to train autistic children's emotion expression and social skills [4]. Enhancing pretend play had been the focus in [2, 8, 11], and results from these studies indicate that the AR-based technology offers superior advantages over other traditional intervention techniques; among these three studies, [8] focused on such AR-based play setting in a classroom. [1] implemented an audio-augmented paper which supports audio recording with standard sheets of paper in a storytelling activity; the unique feature of the tool is that it is built with tangible physical tools that can be shared between the therapist and the child. The AR-based Google Glass was studied in home-based social affective learning in children with autism [6]. In a series of studies at home with parents on facial affect recognition tasks, a reported increased eye contact and greater social acuity has led the researchers to support its therapeutic purposes [6]. Liu et al. [13] systematically explored, for the first time, the feasibility of autism-focused AR-based smart-glasses for training social communication and behavioral coaching and concluded that the AR-based can significantly increase children's engagement and fun, which thus might in turn improve the respective skills during intervention. However, a recent study on AR for social skills intervention failed to find significant improvement between groups [14]; further ecological studies are necessary. Despise it, due to the unique and immersive learning environment an AR-based system can offer, it would be more popular in future systems.

Of note, we failed to uncover any published English articles on AR-based application for Chinese speaking users, which is the motivation as well as the uniqueness of our works.

3 System Overview and Offline Object Recognition

3.1 The Offline Object Recognition Module Powered Within the TensorFlow Platform

The offline object recognition module was implemented within Google's TensorFlow[1] machine learning framework. Since our system aims at offering and facilitating teaching and learning at any time and any place (particularly in an outdoor setting and at home) without relying on online training, we did not modify the sample codes; instead, we took advantage of around one hundred already-trained models to cold start our system. We realized that in the absence of large amount of training data, such algorithm might suffer from inaccurate recognition which was constantly observed during our in-lab testing. To alleviate this problem, we added and integrated a small module to allow the user to correct the result manually; the discussion of it is beyond the scope of this paper.

At present, our system can recognize around one hundred typical daily objects in an offline fashion, which thus offers tremendous advantage particularly for rural users who have much less access to therapists and the internet [19].

3.2 System Development Phases and User Interfaces: System Versions Without Arduino Sensors

Our system went through several design iterations, which aims at facilitating the ease of use for autistic children and their loved ones, particularly in rural areas. Two types of the AR-based systems had been implemented, one with Arduino sensors and one without these sensors (target younger children). In this paper, we focus our discussion on the latter part. Figure 1 presented two sample user interfaces during these design iterations where the system can only recognize one object per use (the first and middle images in Fig. 1).

---

[1] https://www.tensorflow.org/tutorials/images/image_recognition.

Fig. 1. Three sample user interfaces in the system development iterations where more than one objects can be recognized (in the rightmost image, the system is seen to recognize four objects in one scene.

To differentiate and highlight the items in one scene, colored border will wrap the object and the colored border is matched to colored buttons; when a user presses a colored button, the Chinese word for the item corresponding to the colored button will be shown and spoken, as shown in Fig. 2. Notice in Fig. 2, the user targeted a photo in a browser, which shows that our system is capable of supporting learning anywhere and anytime for these children.

Fig. 2. A sample example of the word recognition where both Chinese and English words appear; and all the bicycles are highlighted.

Notice that as shown in Fig. 2, button design has been simplified by removing the texts and replacing them with simple buttons to facilitate younger autistic children with limited vocabulary.

4 Observations from Pilot Studies and Discussions

Since there is no prior works to draw in the design of such AR-based word-learning mobile application, we conducted two small pilot studies to obtain some feed-backs and inputs from TD children and their parents, children with autism and their loved ones and special education teachers. Main observations and discussions will be presented in the following sections.

4.1 Pilot Study One and Main Observations

The first small-scale usability testing is conducted with typical children and adults in a public show on the university campus. Another testing goal is to assess the accuracy of the offline object recognition algorithm and its portability. Figure 3 shows the moment when a young girl whose mother was seen by her side was shown the application; Fig. 4 illustrates a group of adults were demonstrated the offline object-recognition.

Fig. 3. A young girl was shown of the application recognizing an object by the researcher where her mother is seen at her back.

The simple and lightweight mobile application receives very good reviews from both adults and children, particularly for parents and young audiences who claimed that such AR-based applications are very rare in China; and children were observed to demonstrate high interest in trying it after several rounds. Parents are particularly satisfied with the audio module of the application which could facilitate teaching and learning at home and largely relieve them from repeated teaching.

Fig. 4. A group of adult attendees were shown of the application recognizing an object by the researcher.

Supervising, the accuracy of the application is satisfying, though a few items supplied by the audiences were wrongly recognized. Another main reason that the accuracy seems not be a big issue in this pilot study might be due to the limited time each child is playing with the application.

We hypothesize that objects that are used for teaching and learning with young children are typical ones which mostly can be accurately recognized by the algorithm. However, when children grow up with expanding vocabularies and sophisticated environment, the performance of the algorithm will be declined.

4.2   Pilot Study Two and Discussions

General Description and Goals. We conducted a second pilot test in a private special education center in Hangzhou, one of China's biggest cities. The main goals are two folds and same as those in study one.

Study Participants and Environment. Five children with two different age groups tried the application; one group consists of children under five years old; another is those between six to eight years old. Testing objects including a set of toy animals (see Figs. 5 and 6) we brought and those in the center. Besides children, accompanying parents and teachers had also tried the application and been interviewed by us.

General Observation with Children with Autism and Discussions. Overall, the group with younger children are observed to have difficulty in using the application correctly; they did show excited-ness and surprise after the application pronounces objects' names. However, they quickly lost interest in the app for hearing the voice several times. Comparably, the application is very well received among elder children who not only showed high interests in learning with application, but also were observed to use the application without any training and difficulty. Figure 7 shows one testing moment with a child on the animal toy group, while Fig. 8 shows testing on multiple cups.

Fig. 5. The toy animals brought to the center for children to try.

Feed-backs from Teachers. Very satisfying feed-backs had been received from teachers regarding its usefulness and usability. Specifically, they are especially satisfied with the advantages of technology-based application to significantly reduce their efforts: the repeatability of the application. Its ease of use and its audio features also attract their attention in that children with special needs could also learn the pronunciations. However, the performance of the offline algorithm is not satisfactory. For example, the application cannot recognize some books due to their seem-to-be strange and varied covers, shapes and colors. Teachers claimed that it would make an excellent learning companion if the performance could be significantly improved.

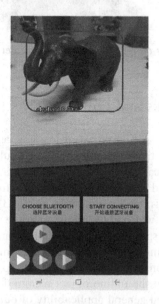

Fig. 6. The application can accurately recognize the elephant toy.

Fig. 7. The application is being tested among a child and his parent at a private education center.

Fig. 8. A boy was seen testing on multiple cups in the center with the research at a private education center.

5   Concluding Remarks and Future Works

We developed a lightweight AR-based word-learning application for children with autism to learn words at anytime and anywhere, particularly outside the classrooms where many AR-based applications targeting TD children works effectively.

Overall, the feasibility and usability results obtained from our two pilot studies are aligned with those in previous ones, particularly in [8, 13] that the application greatly attracts children's attention, which thus might promote learning at their own pace outside the classroom. Both special education teachers and parents' feed-backs highlighted the importance of learning while playing and learning at anytime and anyplace, not only for children with autism but those with other special needs.

However, the accuracy of the offline object recognition model significantly compromised the acceptability and general applicability of our application. That said, how to balance accuracy and lightweight is crucial. Our current thought is to integrate a reinforcement learning module to take user-inputs so as to further train the existing algorithm.

Despite it, to the best of our knowledge, our application, as the first few, offers valuable insights into the design of such mobile learning applications, particularly to facilitate learning at anytime and anywhere.

Acknowledgements. The authors gratefully acknowledge financial support from Zhejiang Provincial Natural Science Foundation of China (LGJ19F020001) and Wenzhou City Science and Technology Bureau (H20180001).

References

1. Alessandrini, A., Cappelletti, A., Zancanaro, M.: Audio-augmented paper for therapy and educational intervention for children with autistic spectrum disorder. Int. J. Hum Comput Stud. **72**(4), 422–430 (2014)
2. Bai, Z., Blackwell, A.F., Coulouris, G.: Using augmented reality to elicit pretend play for children with autism. IEEE Trans. Vis. Comput. Graph. **21**(5), 598–610 (2015)
3. Charman, T. Swettenham, J: Repetitive behaviors and social-communicative impairments in autism: implications for developmental theory and diagnosis. In: Burack, J.A., Charman, T., Yirmiya, N., Zelazo, P.R. (eds.) The Development of Autism: Perspectives from Theory and Research. Mahwah, New Jersey. Lawerence Eribaum Associates (2001)
4. Chen, C., Lee, I., Lin, L.: Augmented reality-based self-facial modeling to promote the emotional expression and social skills of adolescents with autism spectrum disorders. Res. Dev. Disabil. **36**, 396–403 (2015)
5. Chung, C.H., Chen, C.H.: Augmented reality based social stories training system for promoting the social skills of children with autism. In: Soares, M., Falcão, C., Ahram, T. (eds.) Advances in Ergonomics Modeling, Usability & Special Populations. Advances in Intelligent Systems and Computing, vol. 486, pp. 495–505. Springer, Cham (2016). https://doi.org/10.1007/978-3-319-41685-4_44
6. Daniels, J., et al.: Exploratory study examining the at-home feasibility of a wearable tool for social-affective learning in children with autism. npj Digit. Med. **1** (2018). Article no. 32
7. Dawson, G., et al.: Randomized controlled trial of the early start denver model: a developmental behavioral intervention for toddlers with autism: effects on IQ, adaptive behavior, and autism diagnosis. Pediatrics **125**(1), e17–e23 (2010)
8. Dragomir, M., Manches, A., Fletcher-Watson, S., Pain, H.: Facilitating pretend play in autistic children: results from an augmented reality app evaluation. In: Proceedings of the 20th International ACM SIGACCESS Conference on Computers and Accessibility (ASSETS 2018), pp. 407–409. ACM, New York (2018)
9. Escobedo, L., Tentori, M., Quintana, E., Favela, J., Garcia-Rosas, D.: Using augmented reality to help children with autism stay focused. IEEE Pervasive Comput. **13**(1), 38–46 (2014)
10. French L., Kennedy E.M.M: Annual research review: early intervention for infants and young children with, or at-risk of, autism spectrum disorder: a systematic review. J. Child Psychol. Psychiatry **59**(4), 444–456 (2018)
11. Hosseini, E., Foutohi-Ghazvini, F.: Play therapy in augmented reality children with autism. J. Mod. Rehab. **10**(3), 110–115 (2016)

12. Hwang, G.J., Wu, P.W., Chen, C.C., Tu, N.T.: Effects of an augmented reality-based educational game on students' learning achievements and attitudes in real-world observations. Interact. Learn. Environ. **24**(8), 1895–1906 (2016)
13. Liu, R., Salisbury, J.P., Vahabzadeh, A., Sahin, N.T.: Feasibility of an autism-focused augmented reality smartglasses system for social communication and behavioral coaching. Front. Pediatr. **5**, 145 (2017)
14. Lorenzo, G., Gómez-Puerta, M., Arráez-Vera, G., Lorenzo-Lledó, A.: Preliminary study of augmented reality as an instrument for improvement of social skills in children with autism spectrum disorder. Educ. Inf. Technol. **24**(1), 181–204 (2018)
15. McMahon, D.D., Cihak, D.F., Wright, R.E., Bell, S.M.: Augmented reality for teaching science vocabulary to postsecondary education students with intellectual disabilities and autism. J. Res. Technol. Educ. **48**(1), 38–56 (2016)
16. Smith, F.G., LeConte, P.: Universal design for learning. In: Encyclopedia of Distance Learning, vol. 4, pp. 1926–1928 (2004)
17. Su, M.S., Haga, C.: Effectiveness of cognitive, developmental, and behavioural interventions for autism spectrum disorder in preschool-aged children: a systematic review and meta-analysis. Heliyon **4**(9), e00763 (2018)
18. Szatmari, P., Bryson, S.E., Boyle, M.H., Streiner, D.L., Duku, E.: Predictors of outcome among high functioning children with autism and Asperger syndrome. J. Child Psychol. Psychiatry **44**(4), 520–528 (2003)
19. Tang, T.Y., Flatla, D.R.: Autism awareness and technology- based intervention research in China: the good, the bad, and the challenging. In: Proceedings of Workshop on Autism and Technology - Beyond Assistance & Intervention, in Conjunction with the CHI 2016 (2016)
20. Virnes, M., Kärnä, E., Vellonen, V.: Review of research on children with autism spectrum disorder and the use of technology. J. Spec. Educ. Technol. **30**(1), 13–27 (2015)
21. Zwaigenbaum, L., et al.: Early intervention for children with autism spectrum disorder under 3 years of age: recommendations for practice and research. Pediatrics **136**, S60–S81 (2015)

Design and Evaluation of Mobile Applications for Augmentative and Alternative Communication in Minimally-verbal Learners with Severe Autism

Oliver Wendt[✉], Grayson Bishop, and Ashka Thakar

Research Lab on Augmentative and Alternative Communication in Autism,
School of Communication Sciences and Disorders, University of Central Florida,
Orlando, FL, USA
oliver.wendt@ucf.edu

Abstract. One of the most significant disabilities in autism spectrum disorders (ASD) includes a delay in, or total lack of, the development of spoken language. Approximately half of those on the autism spectrum are functionally non-verbal or minimally verbal and will not develop sufficient natural speech or writing to meet their daily communication needs.

A suite of evidence-based mobile applications, SPEAKall!® and SPEAKmore!®, was developed to help these individuals achieve critical speech and language milestones. SPEAKall! and SPEAKmore! enable early language learning, facilitate natural speech development, enhance generalization skills, and expand social circles as students learn. These solutions grow with the learner, enabling better participation in school and community, thus reducing the lifetime cost of care while enhancing chances for classroom success.

Evidence generation for the newly created applications involved: (a) Single-subject experimental designs to evaluate treatment efficacy through repeated measurement of behavior and replication across and within participants; (b) quantitative electroencephalograms to gain information about brain functioning.

The comprehensive approach to evidence-generation facilitated adoption of SPEAKall! and SPEAKmore! in clinical practice. It also allowed identifying critical app features that enhance skill acquisition and contribute to treatment effectiveness.

Keywords: Autism spectrum disorders · Mobile technology · Augmentative and alternative communication

1 Communication Technology for Autism Spectrum Disorders

1.1 Communication Disorders in Autism

Individuals with autism spectrum disorder (ASD) experience a severe delay or atypical development in the area of communication. Sturmey and Sevin (1994) report that poor

M. Antona and C. Stephanidis (Eds.): HCII 2019, LNCS 11573, pp. 193–205, 2019.
https://doi.org/10.1007/978-3-030-23563-5_17

communication skills are hallmark symptoms included in most classification systems of ASD. Current diagnostic criteria for ASD are centered around a profound impairment in verbal and non-verbal communication used for social interaction (American Psychiatric Association 2013). The degree of this communication disorder can differ widely in individuals on the autism spectrum. Some learners develop speech and language slowly during the preschool years; it is estimated that about 50% reach phrased speech at the time of entering primary school (Anderson et al. 2007; Howlin et al. 2009). However, a proportion of about 30–50% demonstrate a severe lack in the acquisition of speech and language when making the transition to kindergarten settings (National Research Council 2001; Tager-Flusberg and Kasari 2013). These learners are often referred to as being "non-verbal" or only "minimally verbal" (Tager-Flusberg and Kasari 2013).

Because these learners do not develop sufficient natural speech or writing to meet their daily communication needs they often experience serious barriers for participation in education and society. Communicative abilities of these students may be limited to pre-intentional communication, such as reaching for a desired item, or communication may display intent through simple behaviors such as pointing (Yoder et al. 2001). If speech occurs it is typically limited to unusual or echolalic verbalizations (Paul 2005). Many become candidates for an area of clinical intervention that is known as augmentative and alternative communication (AAC).

1.2 Augmentative and Alternative Communication

Establishing even low levels of communication is an immediate and crucial need for learners with severe, minimally-verbal autism to take part in daily life. A common intervention approach to help these individuals successfully participate in communicative interactions is the use of augmentative and alternative communication (AAC). AAC is defined as the supplementation or replacement of natural speech and/or writing using aided and/or unaided strategies. Blissymbols, pictographs, Sigsymbols, tangible symbols, and electronically produced speech are examples of aided AAC. Manual signs, gestures, and body language are examples of unaided AAC. The use of aided symbols requires a transmission device, whereas the use of unaided symbols requires only the body (Lloyd et al. 1997). The most commonly applied AAC interventions for individuals on the autism spectrum include graphic symbols and speech-generating devices.

Graphic symbol sets and systems are a relatively modern AAC approach for individuals with ASD. As early as the 1980s, practitioners began to explore the potential benefits of graphic symbols because of their non-transient nature (e.g., Mirenda and Schuler 1988). Graphic symbol libraries can be constructed as sets or systems. Sets are collections of symbols that do not present with defined rules for their construction and expansion while systems are tied to an established repertoire of rules (see Lloyd et al. 1997). Graphic symbols most often seen in clinical practice for learners with ASD include: PCS, line drawings, colored photographs, Premack (all sets); Blissymbols, Orthography, Rebus (all systems) (Schlosser and Wendt 2008). Graphic symbol sets and systems that by their nature are more iconic (i.e., they present a better visual resemblance between symbol and referent) seem easier to be learned (Kozleski 1991).

Speech-generating devices (SGDs) represent another potentially viable option for minimally verbal learners with ASD. SGDs refer to a variety of high technology solutions including dedicated electronic communication devices, talking word processors, and handheld multi-purpose mobile devices (e.g., iPad®, iPod®, Android® tablets) loaded with AAC applications (apps). All of these solutions possess built-in technology that enables a learner to communicate via digitized and/or synthetic speech. Digitized speech is produced by using a recording of a human voice and transforming it into an electronic waveform. Digitized speech output can vary in its quality depending on the sampling rate used during the conversion process. SGDs and apps that have higher sampling rates typically produce higher quality speech output compared to those with lower sampling rates. Recording quality may also be influenced by noisy environments, equipment quality, age of the speaker, and characteristics of the speaker's natural voice (Drager and Fink 2012). Initial research into the potential benefits of SGDs indicates that these newer technologies may have advantages over non-electronic AAC strategies because they provide additional auditory stimuli for the learner via speech output, which can facilitate receptive and expressive language development (Romski et al. 2010).

1.3 New Mobile Technologies

When Apple introduced the iPad in 2010, the toolbox for AAC options changed significantly. Older forms of SGDs, the more expensive, dedicated devices, were no longer the preferred strategy for minimally verbal children with autism. The iPad allowed running various AAC apps to accomplish the same functions that many SGDs provide. AAC intervention via an iPad became more cost-efficient, more socially appealing, and more motivating (Wendt 2014). Because the iPad's functionality is broader than that of a dedicated device, it can be applied to a variety of activities that help children with autism beyond learning functional communication. In addition, using a tablet with an application for communication may be normalizing and less stigmatizing for a child with autism than a dedicated communication device. Lastly, the fact that the iPad provides many further opportunities for interacting with the device beyond communication purposes (i.e., games, video clips, photos, music, etc.) may provide very strong motivation for its use.

2 Design of AAC Applications for Intervention in ASD

2.1 The Rationale for an "Autism-Friendly" AAC Solution

When designing an AAC application for severe autism, it is critical to pay attention to the specific cognitive and sensory processing characteristics. In recent years, many AAC apps appeared on the market with the sole purpose of turning an iPad into a SGD and simply mimicking the SGD interface. The vast majority of these apps ignored particular behavioral and learning characteristics of individuals with autism. Such apps often came equipped with large sets of graphic symbols that when displayed on the iPad screen overwhelmed the learner with too many visual stimuli. An important sensory processing characteristic in autism is the difficulty to filter out salient and truly important incoming stimuli from a stimulus-rich environment (Minshew and Williams 2007). For the beginning communicator with autism, graphic symbols should be carefully selected and not be presented alongside other conflicting visual stimuli on the interface. Another less ideal feature of existing apps is a hierarchical organization of graphic symbol vocabulary. In order to retrieve a specific symbol, the learner first had to select from abstract and broader themes (e.g., "Food", "People", "Places", etc.). This retrieval strategy increased cognitive load and created a level of abstraction too difficult for individuals with severe autism. Previous apps also included graphic symbols that had very little resemblance with their referents and were not highly iconic, that is, their meaning was not easy to comprehend.

To address the need for an AAC solution that can target critical speech and language milestones while taking into account the learning and processing characteristics of individuals with autism, the research team developed a suite of mobile AAC applications, SPEAKall!® and SPEAKmore!®.

2.2   SPEAKall! for Initial AAC Intervention

SPEAKall! facilitates the development of functional communication skills, natural speech production, and social-pragmatic behaviors (see Fig. 1).

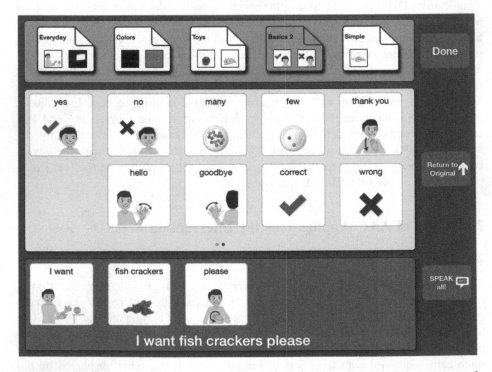

Fig. 1. An example for a communication display on the SPEAKall! application to teach functional communication skills. (Courtesy of the UCF Research Lab on AAC in Autism).

SPEAKall! provides a very intuitive interface for the non-verbal learner. The top portion of the screen features graphic symbols and pictures, and the bottom is a storyboarding strip where users can drag and drop symbols to create sentences. Learners with motor control difficulties can activate symbols with a simple touch gesture. The app also allows caretakers to add photos to the existing bank of symbols, creating relevant and recognizable content of the child's everyday life. Once a child creates a sentence using the symbols, they can push the "SPEAKall!" icon to hear the sentence out loud. Voice-output can be generated via pre-recorded utterances from caretakers or via Apple's synthetic Siri® voice.

2.3 SPEAKmore! for Advanced Language Training

SPEAKmore! teaches advanced language concepts and their generalization to grow vocabulary and extend complexity of utterances (see Fig. 2).

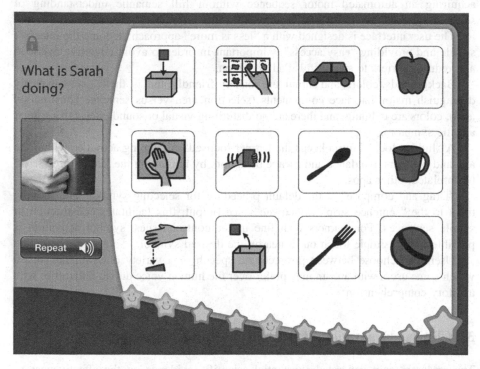

Fig. 2. An example for a 6 × 6 action-object matrix taught through the application "SPEAKmore!". (Courtesy of the UCF Research Lab on AAC in Autism).

SPEAKmore! addresses the advanced learner in need of focused language training. The app follows a matrix training format and has three components: (1) An assessment mode to conduct a brief assessment of expressive and receptive language. (2) An intervention mode to present stimuli, elicit responses from the learner, and engage in error correction as needed. Different language learning lessons can be programmed. (3) A generalization mode to test the learner's ability to extend the new skills learned in intervention to new stimuli.

2.4 Design Considerations

Development and clinical research on SPEAKall! and SPEAKmore! included direct experiences with end users and input from caretakers, clinicians, and teachers using a participatory design approach. This lead to identify subtle, but critical features that in turn influenced app design. This way, the research team was able to develop products that directly relate to the learning characteristics in autism and developmental disabilities. The following features are particularly noteworthy:

Randomization of symbol locations on the selection display forces the learner to truly look at the symbol and develop symbolic comprehension, an important precursor to language development; random placement of symbols prevents the learner from acquiring an automated motor sequence without full semantic understanding of selected symbols.

The user interface is designed with a "less is more" approach: keeping the interface simple and providing "easy access" is important in order to avoid cognitive overload and reduce barriers to successful iPad operation.

Backgrounds, colors, and stimuli are sensory-friendly, that is, the learner can easily distinguish major interface components (selection area versus sentence construction area), colors are calming, and there are no disturbing visual or sound effects that trigger autistic symptoms.

A "hidden lock" button keeps the learner focused on learning activities within the app and prevents avoidance and escape behaviors by not letting the learner browse off to unrelated, other apps.

"Drag and compose" is the default procedure for selecting symbols and moving them to the "sentence strip" to activate voice output; this facilitates construction of simple sentences. For learners with fine motor control issues, symbol activation is possible with a simple touch on or nearby the desired symbol.

Users can choose between pre-recorded speech or a variety of synthetic voices, which some users with autism may prefer over the human voice due to difficulties with auditory comprehension.

3 Clinical Evaluation of SPEAKall! and SPEAKmore!

The research team generated substantial scientific evidence on the effectiveness of SPEAKall! in addition to proof-of-concept data for SPEAKmore! Behavioral data and neurophysiological imaging data demonstrate treatment effects on developing functional communication skills and natural speech at a high level of scientific rigor.

3.1 Behavioral Evidence of Treatment Effects

A series of three different single-subject experiments were carried out to study different approaches to intervention with SPEAKall! Single-subject research designs are one of the most rigorous methods to examining treatment efficacy and are ranked equally to quasi-experimental group designs in evidence hierarchies for AAC (Schlosser and Raghavendra 2004). These designs are typically examining pre-treatment versus post-treatment performance within a small sample of participants (Kennedy 2005).

All study participants met the following criteria: (1) an official diagnosis of autistic disorder (all individuals were re-assessed by using the Childhood Autism Rating Scale, and the Autism Diagnostic Observation Schedule), (2) little or no functional speech operationalized as no more than 5 spoken words, (3) visual and auditory processing within normal levels, (4) adequate hand coordination for graphic symbol pointing on a tablet device, (5) understanding simple verbal commands (e.g., "Sit down") and responding to yes-no questions.

Target skills and dependent measures included (a) the number of correct requests during a 20-trials session; and (b) the numbers of non-intentional utterances versus intentional vocalizations or word approximations.

Experiment 1 sought to produce replication effects across settings when SPEAKall! intervention took place in clinic, school, and home environments (Wendt et al. 2013a; 2014b). Four students between 10–13 years of age were introduced to the intervention protocol; three of the four mastered all five phases, and one individual achieved mastery of phase 3. All students improved in their ability to produce natural speech: Two students made significantly more intentional vocalizations (e.g., first letter of intended word or a word approximation), and another two students developed sufficient natural speech to produce one-word or two-word spoken requests. For all four students, these intervention effects were replicable across the three settings.

Experiment 2 aimed to replicate the prior intervention results with a cohort of three adolescents and adults between 14–23 years while focusing on response generalization to untrained stimuli (Wendt et al. 2019). All participants improved in their requesting abilities and were able to generalize the newly learned skills to untrained items. Mixed results occurred for targeting natural speech production: one out the three participants achieved the ability to speak simple sentences, two remained minimally verbal, suggesting that the ideal time to start intervention on speech production is at a much earlier age.

Finally, Experiment 3 evaluated the effects of a parent-training protocol for SPEAKall! intervention (Wendt et al. 2013b; Wendt et al. 2014a). This study involved three families with children between 6–8 years of age. Specific emphasis was placed on response generalization to untrained items and procedural fidelity of parent implementation. Results obtained suggest that with proper supervision and training parents can carry-out SPEAKall! intervention. All participants mastered the five phases of the intervention protocol, systematically increased symbolic utterances and spontaneous communication. Response generalization was demonstrated by extending acquired requesting skills from snack to untrained toy stimuli. Two of three children ended up speaking simple sentences without the help of SPEAKall! or their iPad, one child with a dual diagnosis of autism and Down syndrome did not show gains in speech production.

Behavioral Evidence Example. Figure 3 below shows an example for the behavioral data for three participants from Experiment 2. These participants received SPEAKall! intervention to increase functional communication skills. Performance was measured on the ability to request desired items from baseline to intervention. As soon as intervention began, the participants started to show immediate improvement in requesting skills. This effect is replicated across the three participants, and is being maintained after intervention was completed. The figure shows the effects on developing requesting skills from baseline (pre-intervention) over the course of intervention. The horizontal axis displays the number of intervention sessions. The vertical axis displays the number of correct responses for requesting desired items.

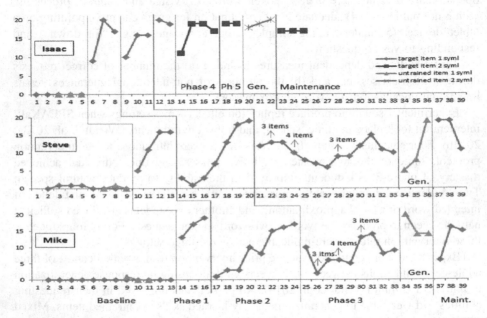

Fig. 3. Single-subject experiment showing SPEAKall! intervention effects on requesting skills replicated across three participants with severe autism, 14–23 years of age (Wendt et al. 2019).

3.2 Neurophysiological Evidence of Treatment Effects

An emerging clinical application of brain imaging in autism is to conduct a quantitative electroencephalogram (qEEG; Chan et al. 2007). A qEEG measures electrical activity produced by the brain and displays states of neural functioning in the form of a brain map. Such information allows one to pinpoint anomalies in brain function and to document neurophysiological changes over the course of intervention. This technique was used with some of the participants from Experiments 1–3 described above. The following is an example for one of those participants, a 14-year-old male with severe autism and no functional speech. Figure 4 shows the improvement of autistic brain activity from pre- to post-treatment (on the left and right hand sides, respectively).

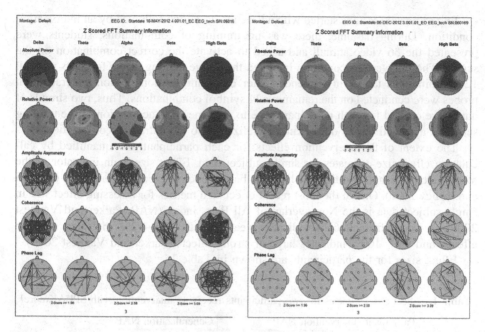

Fig. 4. Quantitative electroencephalogram (QEEG) shows resolving anomalies in electrical brain activity pre-and post-treatment for 14-year-old, male participant with severe autism (Courtesy of Autism Parent Care, LLC).

In the upper two rows of the brain maps, anomalies of electrical brain activity appear as red-shaded (too much activity) or blue-shaded (too little activity) areas. Improvement from pre-to post-intervention is demonstrated by a significant transformation of electrical activity to green-shaded areas (normal). The effect occurs on four major areas of neuro-cognitive functioning including emotion and sensation (theta), alertness (alpha), decision-making, information processing (beta), and agitation (high beta). The lower three rows show the ability of the autistic brain to establish connections between different cortical regions. Aberrant connections (over-connectivity) appear as red lines, lacking connections (under-connectivity) appear as blue lines; no lines indicate normal connectivity. Again, improvement is evident as red and blue lines significantly fade out over the course of intervention.

3.3   Proof-of-Concept Data for SPEAKmore!

Two participants, both 12 years old, diagnosed with severe autism and tested as minimally verbal on a standardized language assessment received matrix training through SPEAKmore! Both students were taught action-object symbol combinations on a 6 × 6 matrix as described earlier. The six actions included "point to", "drop", "take-out", "put-in", "shake", and "wipe"; the six objects included "ball", "cup", "spoon", "fork", "apple", and "car". From the total pool of 36 possible symbol combinations, the researchers created four different sets of three symbol combinations each that were actively taught. The remaining 24 combinations were tested for generalization effects.

Experimental sessions started with a baseline condition followed by an intervention condition. During baseline, there was no training of combinations, students were presented the 36 video stimuli and asked to activate the correct combination on the screen. The intervention condition involved the active training of the four sets with all 12 combinations taught in sequence. After each intervention session generalization probes were conducted on the remaining 24 symbol combinations. Thus, two strands of data were created for each participant, one to document the acquisition of the training sets, and a second one to document generalization to the untrained combinations.

The extent of the intervention effects for each participant was quantified by calculating effect size estimates for single-subject data. Effect size is an indicator of the magnitude of change or difference between baseline to intervention conditions (Beeson and Robey 2006). One of the most recent effect size metrics for assessing effect size in single-subject data is the Non-overlap of All Pairs index (NAP; Parker and Vannest 2009). A score from 0–65% indicates weak effects, 66%–92% indicates medium effects, and 93%–100% indicates large or strong effects (Parker and Vannest 2009).

Effect sizes for the participants are shown in Table 1:

Table 1. Effect size estimates for two participants receiving intervention with SPEAKmore!

Participant	Intervention NAP	Generalization NAP
Danny	92% (medium-strong effect)	92% (medium-strong effect)
Andy	100% (strong effect)	100% (strong effect)

These results suggest that the matrix training intervention implemented via the SPEAKmore! app created an effect large enough to help the participants acquire action-object symbol combinations and generalize to new combinations not previously taught.

4   Conclusions

Generating research evidence to document the effectiveness of newly developed mobile technology is critical for evidence-based practice and decision-making. Valuable clinical lessons were derived from these research experiences:

1. Only providing the learner with a sophisticated AAC solution will not automatically result in improved communication. A proper instructional or intervention approach build around the learning characteristics in ASD needs to accompany the technology.
2. One single best SGD or AAC app for autism does not exist. Learners with autism present with varying needs and learning profiles and should receive individualized AAC solutions.
3. Whenever possible, practitioners should pursue solutions that are grounded in strong research evidence and have documented empirical support. Evidence-based practice is an essential component of successful AAC interventions.

4. AAC intervention should not be restricted to one-on-one sessions between learner and clinician but involve a variety of other communication partners, and occur during the entire day.
5. AAC does not inhibit natural speech production but facilitates its acquisition process.
6. When high technology AAC solutions break down or run out of battery, clinicians need to have a back-up strategy for their learners (e.g., simple communication board, picture exchange book, etc.).
7. Technology can be become outdated or outgrown quickly; AAC interventions should be evaluated regularly and technology be modified to meet the changing needs of the learner.

If we remember these principles, we can unlock the communication abilities of our learners with severe autism, maximize the benefits of AAC intervention, and enable better participation across academic and social environments (see Fig. 5).

Fig. 5. Mother and young daughter engaging in social play facilitated by SPEAKall! (Courtesy of the UCF Research Lab on AAC in Autism).

References

American Psychiatric Association: Diagnostic and statistical manual of mental disorders (DSM-V), 5th edn. American Psychiatric Association, Washington (2013)
Anderson, D.K., Lord, C., Risi, S., DiLavore, P.S., Shulman, C., et al.: Patterns of growth in verbal abilities among children with autism spectrum disorder. J. Consulti. Clin. Psychol. **75**, 594–604 (2007)

Beeson, P.M., Robey, R.R.: Evaluating single-subject treatment research: Lessons learned from the aphasia literature. Neuropsychol. Rev. **16**, 161–169 (2006)

Chan, A.S., Sze, S.L., Cheung, M.: Quantitative electroencephalographic profiles for children with autistic spectrum disorder. Neuropsychology **21**, 74–81 (2007)

Drager, K.D., Finke, E.H.: Intelligibility of children's speech in digitized speech. Augmentative Altern. Commun. **28**, 181–189 (2012)

Howlin, P., Magiati, I., Charman, T.: Systematic review of early intensive behavioral interventions for children with autism. Am. J. Intellect. Dev. Disabil. **114**, 23–41 (2009)

Kennedy, C.H. (ed.): Single-case Designs for Educational Research. Pearson Education, Boston (2005)

Kozleski, E.: Visual symbol acquisition by students with autism. Exceptionality **2**, 173–194 (1991)

Lloyd, L.L., Fuller, D.R., Arvidson, H.H. (eds.): Augmentative and Alternative Communication: A Handbook of Principles and Practices. Allyn & Bacon, Needham Heights (1997)

Minshew, N.J., Williams, D.L.: The new neurobiology of autism: cortex, connectivity, and neuronal organization. Arch. Neurol. **64**(7), 945–950 (2007)

Mirenda, P., Schuler, A.: Augmenting communication for persons with autism: issues and strategies. Top. Lang. Disord. **9**, 24–43 (1988)

National Research Council: Educating Children with Autism. NRC, Washington, DC (2001)

Parker, R.I., Vannest, K.J.: An improved effect size for single case research: non-overlap of all pairs (NAP). Behav. Ther. **40**, 357–367 (2009)

Paul, R.: Assessing communication in autism spectrum disorders. In: Volkmar, F.R., Paul, R., Klin, A., Cohen, D. (eds.) Handbook of Autism and Pervasive Developmental Disorders, pp. 799–816. Wiley, Hoboken (2005)

Romski, M., et al.: Randomized comparison of augmented and non-augmented language interventions for toddlers with developmental delays and their parents. J. Speech, Lang. Hear. Res. **53**(2), 350–364 (2010)

Schlosser, R.W., Raghavendra, P.: Evidence-based practice in augmentative and alternative communication. Augmentative Altern. Commun. **20**, 1–21 (2004)

Schlosser, R.W., Wendt, O.: Augmentative and alternative communication intervention for children with autism: a systematic review. In: Luiselli, J.K, Russo, D.C., Christian, W. P. (eds.) Effective Practices for Children with Autism: Educational and Behavior Support Interventions that Work, pp. 325–389. Oxford University Press, Oxford (2008)

Sturmey, P., Sevin, J.A.: Defining and assessing autism. In: Matson, J.L. (ed.) Autism in Children and Adults: Etiology, Assessment, and Intervention, pp. 13–36. Brooks/Cole, Pacific Grove (1994)

Tager-Flusberg, H., Kasari, C.: Minimally verbal school-aged children with autism spectrum disorder: the neglected end of the spectrum. Autism Res. **6**, 468–478 (2013)

Wendt, O., Boesch, M.C., Hsu, N., Simon, K., Warner, K., Robertson, R.: Models of parent-implemented AAC intervention for children with severe autism. Paper presented at the Annual Convention of the American Speech-Language-Hearing Association (ASHA), Orlando (2014a)

Wendt, O., et al.: A series of experimental investigations on iPad-based AAC interventions for individuals with severe autism. Paper presented at the Communication Matters National Conference (ISAAC-UK). University of Leeds, UK (2013a)

Wendt, O., et al.: Effects of an iPad-enhanced augmentative and alternative communication intervention for children with severe autism, Manuscript in preparation (2014b)

Wendt, O., Hsu, N., Cain, L., Dienhart, A., Simon, K.: Experimental evaluation of a parent-implemented AAC intervention protocol for children with severe autism. Paper presented at the Annual Convention of the American Speech-Language-Hearing Association (ASHA) Chicago (2013b)

Wendt, O., Hsu, N., Dienhart, A., Cain, L.: Effects of an iPad-based speech generating device infused into instruction with the Picture Exchange Communication System (PECS) for young adolescents and adults with severe autism. Manuscript under review (2019)

Wendt, O.: Experimental evaluation of SPEAKall! an evidence-based AAC app for individuals with severe autism. Commun. Matters **28**, 26–28 (2014)

Yoder, P., McCathren, R.B., Warren, S.F., Watson, A.L.: Important distinctions in measuring maternal responses to communication in prelinguistic children with disabilities. Commun. Disord. Q. **22**, 135–147 (2001)

Multimodal Interaction

Principles for Evaluating Usability in Multimodal Games for People Who Are Blind

Ticianne Darin[1](✉), Rossana Andrade[2], and Jaime Sánchez[3]

[1] Virtual University Institute, Federal University of Ceará,
Humberto Monte, S/N, Fortaleza, Brazil
ticianne@virtual.ufc.br
[2] Department of Computer Science, Federal University of Ceará,
Humberto Monte, S/N, Fortaleza, Brazil
rossana@great.ufc.br
[3] Department of Computer Science, University of Chile,
Blanco Encalada 2120, Santiago, Chile
jsanchez@dcc.uchile.cl

Abstract. Multimodal video games designed for increasing cognition of people who are blind should be friendly and pleasant to use, instead of adding complexity to the interaction, leading people to acquire cognitive skills while interacting. There are specific issues that make multimodal usability evaluation different from the evaluation of traditional user interfaces in the context of improving cognition of people who are blind. In this context, identifying how well the Usability Evaluation Methods (UEM) meet the evaluation criteria to assess multimodal games for people who are blind is necessary. In this paper, we conducted an expert opinion survey to analyze how usability evaluation has been done by researchers and practitioners in this field. As a result, we propose the PrincipLes for Evaluating Usability of Multimodal Video Games for People who are Blind (PLUMB), a set of evaluation good practices that should be observed while planning the evaluation. This paper builds on the literature about how multimodal features affect people who are blind interaction with multimodal interfaces by focusing on their practical evaluation.

Keywords: Multimodal video games · Multimodal interfaces ·
Usability evaluation · Cognition · People who are blind

1 Introduction

Multimodal serious games are a type of multimodal interface application that have been widely used as tools to improve various cognitive skills in people who are blind, such as orientation and mobility, logical reasoning and collaboration [1–3]. Video games with this purpose should meticulously combine and integrate different sources of perceptual inputs [4], considering the limitations of the users as well as the cognitive goals that the game pursues [5]. Moreover, there is an increasing concern regarding the usability of the user interfaces and interaction in this type of games [6, 7]. There is a

© Springer Nature Switzerland AG 2019
M. Antona and C. Stephanidis (Eds.): HCII 2019, LNCS 11573, pp. 209–223, 2019.
https://doi.org/10.1007/978-3-030-23563-5_18

significant challenge in practice when considering the evaluation of such multimodal interactions, because they integrate the complexities related to games evaluation, multimodal interfaces evaluation, and cognition of people who are blind. Therefore, much better guidance and accumulated best practices are still needed [8]. Even though usability evaluation is the most frequent type of evaluation in the context of multimodal video games [5], recurrent situations jeopardize its effectiveness in the context of people with visual disabilities. For example, employing inappropriate usability methods and neglecting critical issues, such as the nature of the audience's limitations and whether the modalities offered by the game can support and help to enhance the desired cognitive skills [9].

The usability evaluation in this context should verify whether a multimodal gaming interface adapts to the needs and abilities of different users, as well as diverse contexts of use, considering their individual differences [8]. The consideration of such aspects leads researchers to identifying and fixing the relevant usability issues that affect people who are blind while playing multimodal video games. Considering the target users' specificities is necessary for the HCI evaluation of any system. However, it is a more sensitive matter when evaluating applications for people who are blind, because we cannot assume that all individuals with visual disabilities are identical, especially in different contexts and with specific cognitive development purposes [1, 2, 10]. For that reason, the evaluators must identify whenever is necessary to change or extend the interactive modalities and whether the customization options fit the actual user needs [10]. The nature of the interactions and modalities when carrying out interactive tasks should also be taken into consideration because it can make a considerable difference to usability whether a piece of abstract information has been represented in one or another modality, especially in for the perception and understanding of people who are blind [3, 10].

The importance of thinking about usability evaluation in this regard resides mainly in that the multimodal gaming interface should not add unnecessary complexity to the interaction of people with visual disabilities. On the contrary, such interfaces should be friendly and pleasant to use, supporting in different levels the diversity of the target users' abilities and disabilities [11, 12], and successfully leading them to acquire cognitive skills while interacting.

As part of a continuing effort to improve usability evaluation of multimodal video games for cognitive enhancement of people who are blind, we conducted an expert opinion survey [13]. It aimed to update the published literature review results [5, 9], by deepening and enriching the developing understandings of challenges and principles in this topic. Hence, this paper contributes with the literature by discussing how usability evaluation has been carried on by researchers and practitioners in this field and with the proposal of PrincipLes for Evaluating Usability of Multimodal Video Games for People who are Blind (PLUMB). PLUMB is a practical aid to help researchers and practitioners to properly plan and conduct usability evaluation of multimodal video games based on audio and haptics designed for people who are blind.

2 Background

2.1 Multimodal Interaction Characterization

Multimodal applications have evolved in a complex way regarding technological resources and interaction possibilities that might be offered, through processing multiple combined user input modes in a coordinated manner with multimedia output [14]. It is generally accepted that the multimodal interaction definition is built upon Norman's action cycle [15], using established findings on human-machine multimodal communication to orchestrate the fusion of multimodal inputs and the fission of multimodal outputs, resulting in an adequate outcome to the users, according to their context of use, and personal preferences and characteristics [16].

The applicability of multimodal interfaces in diverse areas is due to the range of possibilities brought by the combination of interaction modalities. An interaction modality can be seen as a communication channel related to the human senses or form of expression [17], describing an interaction technique that utilizes a particular combination of user ability and device capabilities [8]. The combination of such input and output channels and modes, in addition to the output modality selection based on context and user needs, turns the modeling of a multimodal application into a complex task [16]. As a result, researchers have put much effort in describing and modeling multimodal applications from both physical and conceptual point of view.

Considering the physical dimension of multimodal systems, the World Wide Web Consortium proposed a Multimodal Interaction Framework, emphasizing the interpretation and inner system layers [19]. Under a different perspective, the CASE model [20] describes four techniques to combine modalities at the integration engine level. However, for the purpose of considering the usability of such interfaces from a user-centered perspective, we are mostly interested in conceptual descriptions of the multimodal interaction, which can provide us with valuable insights on the human aspects that should be considered during the usability evaluation of multimodal interfaces.

Coutaz et al. [18] proposed the CARE properties (Complementarity, Assignment, Redundancy, and Equivalence) as a simple way of characterizing the features of the multimodal interaction that may occur between the interaction techniques available in a multimodal user interface, focusing on the modality combination possibilities at the user level. CARE properties can also be used to assess multimodal interfaces, considering the fusion and fission of information during the interaction. Each CARE property is represented in a formal expression based on state, goal, modality, and temporal relationship, and has a counterpart in CARE-like user properties.

Bongers & Van der Veer [17] introduced the Multimodal Interaction Space (MIS), a theoretical framework to place and described multimodal interactions, focusing on the space where the interaction occurs, in a human-centered approach. MIS is a descriptive framework for interaction styles and comprises levels, modes, and sensory modalities, but do not include physical interface input and output modalities. According to this approach, a multimodal interaction can be described in multiple layers, considering the user goal and intention, the formulation of tasks and subtasks, and finally the execution of such actions in the interface while receiving physical feedback and evaluating the outcome.

Based on an empirical study of multimodal usability [21], Bernsen (2010) proposed AMITUDE [10], a conceptual development-for-usability framework to express a model of use of the system under development, based on seven interaction aspects: Application type, Interaction, Task or domain, User, Environment of use, Modalities, and Device. According to the authors, each aspect addressed by AMITUDE must be taken into account when developing for usability. Therefore, while designing a multimodal application practitioners and researchers should provide detailed description of the intended application users, the tasks the application will support, the modalities and respective devices involved, as well as the environment in which the tasks will be carried on, and how each one of these aspects can affect the usability during user interaction.

2.2 Multimodal Interaction for People Who Are Blind

Receiving information in a multimodal way enables people who are blind to interact with real and virtual environments mainly through audio-based and haptic interfaces, which are capable of fomenting learning and cognition in this audience [8, 9]. The crucial features of the multimodal interaction in multimodal gaming interfaces in this context are Audio, Adaptation, Interaction Mode and Feedback, along with the cognitive aspects meant to be stimulated [5], in consonance with a motivating story [22]. The game input device - which may include a keyboard, natural language, force feedback devices, touchscreen, directional pads, and specific devices - determines the style of interaction available and influences the type of feedback the user receives after an interaction, which is usually a combination of haptic, sonorous and visual feedback [5]. Figure 1 summarizes the multiple aspects of the interface and interaction dimensions of multimodal games for people who are blind.

The different modalities combinations and choices affect the users' behavior towards the game and determine how cognitive processes are stimulated. For instance, audio and visual cues coordinated with haptic elements distributed in a virtual navigational environment can serve as references for orientation and mobility, as well as help people who are blind adopting and restructuring a mental model of spatial dimensions. The different types of audio cues can represent spatially and surroundings properties including location, size, distance, direction, separation and connection, shape, pattern, and movement; or be associated with each available object and action in the environment.

For that reason, the multimodal interaction provided by the game interface must adequate the use of modalities to the cognitive game goals, along with the game story, while providing the player with the proper interaction mode to develop the desired skills in a certain usage context.

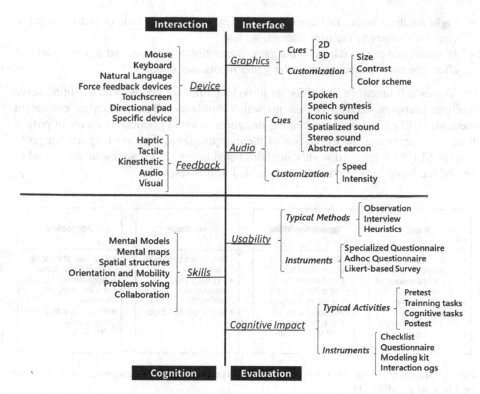

Fig. 1. Dimensions for the description of the key characteristics of the design and evaluation of multimodal video games for cognitive development of people who are blind [5]

In a previous work [23] we took the first step towards identifying the issues that may jeopardize the multimodal interaction of people who are blind, by defining a Standard List of Usability Problems (SLUP) for multimodal video games, which was submitted to the judgment of experts and end-users. In addition to the Overall Usability issues related to learnability, efficiency, satisfaction, and difficulties in handling the different modalities, SLUP details usability problems related to Audio, Adaptation, Interaction Mode and Feedback.

Being familiar with the types of issues that affect the interaction of people who are blind and with the main features of multimodal games can help researchers avoid these problems, as well as discover them more efficiently during usability evaluations. The final version of SLUP specifies:

- 15 audio issues that can affect iconic, spatialized and stereo sounds, iconic sounds and abstract earcons, or speech synthesis and spoken audio;
- Six customization issues that can be caused by the size, color scheme, or contrast of graphic elements, or even by the speed and intensity of sounds and voices;
- 13 issues related to the overall usability of the game, addressing learnability, satisfaction, errors, and efficiency;

- Eight feedback issues that can occur either in haptic (kinesthetic or tactile), aural or visual responses to the user interaction; and
- 19 issues that can be related to the game interaction techniques and devices, and that affect the user interaction with the game inputs and outputs.

The establishment of SLUP for multimodal games for people who are blind serve multiple purposes, such as comparing which problems different usability evaluation methods (UEMs) can disclose, helping designers to avoid predictable issues in project time, and supporting the evaluation of these applications. The most significant problems in SLUP are grouped in subcategories listed in Fig. 2. These results were used as one of the bases for the principles assembled in this research.

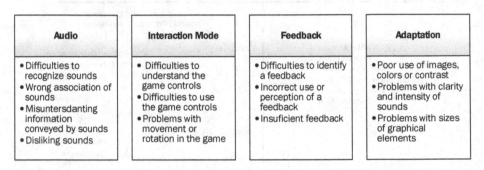

Fig. 2. Summary of the most frequent types of usability issues in multimodal interfaces for people who are blind [23]

3 Expert Opinion Survey

In previous studies [5, 9], we analyzed the usability evaluation performed in 17 multimodal games and 4 multimodal navigation virtual environments, in which we identified that practitioners and researchers often tend to administer informal usability evaluation methods in this field. To deepen our understanding of this issue, we created a 15-questions online questionnaire and emailed it to 16 international researchers that reported usability evaluation in the 21 applications previously identified. The survey aimed to update the published literature review results, as well as to deepen and to enrich the developing understandings of challenges and principles in this topic.

3.1 Methodology

Expert Opinion Surveys can be used to serve a variety of purposes and they result in predictions of how others will behave in a particular situation, according to persons with knowledge of the situation [13]. This technique can be used to assist in problem identification and in clarifying the issues relevant to a particular topic, by consulting individual experts [24]. Although it can be seen as a relatively informal technique, as individual expert opinion is not infallible. If a number of different experts provide the same feedback it is likely that real issues exist [24].

We sent personalized emails to the 16 identified authors. These emails included the title of the paper in which each researcher described a usability evaluation, a brief description of the ongoing study with a hyperlink to the survey, and an estimate of the time needed to fill out the questionnaire. We chose this approach because using personalized email in soliciting participation appears to be the most effective method to increase participation in surveys [25].

For two months, biweekly reminders were sent, until the response rate stopped improving. Researchers that had already responded the survey were contacted and kindly asked to remind their coauthors to answer our survey. About 56% percent of these researchers never answered, despite reminders and/or personalized emails to their coauthors. Hence, a response rate of 43, 75% (N = 7) was obtained, corresponding to seven authors responsible for nine of the previously analyzed papers.

Despite the small number of respondents, the strength of this exploratory approach resides in the fact that the respondent experts have a diverse background and work in different countries and research groups, which helps to avoid specific trends in their opinions. Among the respondents, there are researchers from North and South America, Europe and Asia, who answered the questionnaire independently. The literature also attests the fairness of the response rate obtained in this work. For instance, Rowe and Wright (2001) discuss principles for the use of expert opinions and state that the most relevant forecasts rely on unaided expert opinions. They state that, when conducting expert opinion surveys, researchers should obtain independent answers from between 5 to 20 experts [26]. In addition to that, according to Lazar, Feng & Hochheiser (2017), if the research goal is to gather requirements from domain experts, in-depth discussions with two or three motivated individuals can provide a wealth of data, which corroborates the adequacy of the approach this research purposes [27].

3.2 Overview of Survey Results

The participant's profiles are summarized in Table 1, based on the characterization of their research experience and practice on evaluating multimodal applications for people who are blind. The experts are from five different countries and work in different research groups, except for A3 and A6, who belong to the same University. Most of the experts are University professors and researchers (A2, A3, A4, A6, and A7).

A1 works with virtual environments, usability and accessibility at IBM Research and A5 investigates educational video games at the Organización Nacional de Ciegos Españoles (ONCE), a Spanish national nonprofit social corporation of public law. Some of the researchers had more than one paper identified in our previous literature review, such as A2 (3 papers) and A5, A6 and A7, with two papers each. In the survey, they answered specific questions regarding the types of usability instruments and methods administered in the evaluations described in each of these papers.

Regarding their research experience in the field, 43% have been researching applications designed for people who are blind for about four to six years (A4, A5, and A6), and 29% have been researching in this area for seven to nine years (A2 and A7). While one expert has been researching this field for up to ten years (A1), another one occasionally collaborates with researchers involving evaluation with people who are blind (A3). It is interesting to point out that neither the most experienced nor the occasional researcher are familiar with any particular models for the design and evaluation of multimodal applications, as well as most of the experts (57%).

The experts that reported being familiar with models for the design or usability evaluation of multimodal interaction mentioned the CARE properties, the CASE model, and the AMITUDE model. However, none of them based their evaluation instruments in any of these models. Instead, all the researchers use mostly ad-hoc instruments during the usability evaluation of multimodal applications for people who are blind. In this work, they are classified as ad-hoc any instruments generated by the authors, according to the specific goals of an ongoing evaluation, but not formally validated and often not reusable.

Table 1. Researchers' profiles and behavior towards multimodal usability evaluation

ID	Country	N# of papers on literature review	Research experience in the field	Familiarity with multimodal models	Types of usability instruments administered	Types of usability methods carried out
A1	USA	1	Up to 10 years	Unfamiliar with any particular models	Both validated and ad-hoc instruments	Both quantitative and qualitative methods
A2	Chile	3	7–9 years	Unfamiliar with any particular models	Ad-hoc instruments	Both quantitative and qualitative methods
A3	Spain	1	Occasional	Unfamiliar with any particular models	Ad-hoc instruments	Both quantitative and qualitative methods
A4	Ecuador	1	4–6 years	Unfamiliar with any particular models	Ad-hoc instruments	Both quantitative and qualitative methods
A5	Spain	2	4–6 years	CARE properties, CASE model	Ad-hoc instruments	Both quantitative and qualitative methods
A6	Spain	2	4–6 years	AMITUDE model, CASE model	Ad-hoc instruments	Both quantitative and qualitative methods
A7	Israel	2	7–9 years	CASE model	Both validated and ad-hoc instruments	Quantitative methods only

Although A7 is familiar with CASE model and uses validated instruments reportedly, his evaluations usually do not employ any instrument based on that model.

Instead, his research group developed and validated a number of quantitative specific instruments to measure usability in the specific context of their research. This information implies that their familiarity with formal models describing multimodal interaction models does not affect the conduction of usability evaluations.

All the researchers agreed that the use of both quantitative and qualitative methods is necessary to assess the interaction of people who are blind with multimodal interfaces. Although also agreeing with the use of quantitative and qualitative methods, A7 focus in the use of quantitative evaluation methods and instruments, as his research usually measures the cognitive impact of multimodal games on the intellect of people who are blind.

4 Challenges on Usability Evaluation of Multimodal Interaction with People Who Are Blind

When answering the survey, the experts provided a detailed discussion of the topics approached, revealing challenges and needs in the field of usability evaluation of multimodal video games for cognitive development of people who are blind. Some insights emerged from the experts' answers on their recent research work and evaluation practices, as well as challenges related to usability evaluation of multimodal video games for people who are blind. These results are further discussed in the remainder of this section.

4.1 Challenge 1: Lack of Guidance on the Conduction of Usability Evaluation

Overall, experts indicated that it is common to perform informal usability evaluation in this field, which usually consists in applying ad-hoc questionnaires or interviews after a gameplay session. According to them, it happens due to time or team issues, as well as for the need to perform multiple types of tests (e.g. performance and cognitive impact). In practice, little time is left for planning and conducting a usability evaluation because most of the project schedule accommodates the development and other types of tests.

In addition to that, usually, the team is unfamiliar with usability evaluation instruments or methods that offer useful specific support to this context. Even experienced researchers who are familiar with models for design and evaluation of multimodal applications (e.g. CARE, CASE, and AMITUDE) claim that the use of these models for multimodal usability evaluation is complicated and too laborious, mainly when performed by practitioners. They argue that, in this scenario, it seems more beneficial to create their own evaluation instruments that are easier to administer and context-specific.

Indeed, some usability aspects – efficiency and effectiveness, for instance – can be evaluated independently of the domain [28]. However, all the researchers agreed that a drawback of doing this is the lack of guarantee of meeting the user's needs, principally considering the game cognitive requirements. Ratifying this concept, researcher A3 highlighted that visual disabilities are very particular, and each user is a world, so it is hard to apply very general solutions for usability evaluation in this context.

The applicability of general UEMs to diverse areas is questionable when evaluating specific characteristics, because they may not be adequate for the new contexts of use, generating gaps in the evaluation [29]. In fact, all the expert researchers agreed that, according to their experience the multimodal inputs, the specificities of users' visual disabilities and the type of cognitive skills to be supported are specific characteristics that profoundly affect the usability evaluation of an application designed for people who are blind.

Hence, these characteristics have to be considered in any usability evaluation in this context. Researcher A6 further remarked that, while it is true that all of these aspects can have an impact on the usability evaluation, the most significant challenge in measuring usability is that the solutions that grant accessibility for some levels of visual impairment actually hinder their usability for other levels.

The discussion raised by the researchers pointed out that there is a need for practical guidance on usability evaluation in this field. It is necessary to ensure that multimodal game interaction and interface elements suit the game cognitive purposes and the people' characteristics, leading them to interact pleasantly and correctly while playing and learning. All the experts affirmed that they would welcome evaluation principles to assist researchers in choosing the methods that better fit their usability requirements to assess diverse aspects of multimodal interfaces in this context.

4.2   Challenge 2: Evaluate Multimodal Interaction While Considering the Cognitive Dimension

The usability evaluations described in the experts' papers showed that the crucial features to be evaluated during the interaction of people who are blind with multimodal gaming interfaces are audio, customization capability, interaction mode, and feedback (including audio and haptic stimulus). However, the experts highlighted that these aspects could not be evaluated apart from the game target cognitive aspects.

Despite that, they reportedly conduct usability evaluation more frequently than cognitive impact evaluation because the last one requires specialized people and procedures. All the experts agreed that it is not possible to assure that any multimodal application or game is capable of developing or enhance any cognitive skills in people with visual disabilities if an adequate cognitive impact evaluation is not conducted.

Having performed cognitive impact evaluation in this context several times, experts A2 and A7 suggested that the key for the success of this type of evaluation is the understanding of what data to collect from users, in order to compare, analyze and measure the skills and the development of the subjects. They recommend the use of tests based on experimental and control groups or two-sample test analysis based on a pretest-posttest of the same group. Besides, researchers were unanimous in their opinion that it is crucial to investigate how multimodal game elements can be meaningfully used to develop cognition in people who are blind.

4.3   Challenge 3: To Go Beyond Both Usability and Cognitive Impact Evaluation

Although researchers spoke in one voice regarding the need for sound usability evaluation of multimodal games for people who are blind, they also demonstrated concern about evaluating other aspects. According to the researcher A7, when a usability evaluation is applied to the users' context, the mental model and the cultural environment are also critical because the interface was created specifically for them. In other words, all the interaction is part of the design for people who are blind, including all their culture. Besides, A7 suggested that an equally important challenge could be these users' experience with the multimodal interface because the experience is more than usability.

For the experts, usability is clearly an important aspect for the game quality, but there are other aspects to consider related to pleasure-based human factors, such as the satisfaction of people who are blind, the multisensory aesthetic experience and their emotional response. In addition, they indicated that the behavior of people who are blind towards multimodal games should also be evaluated considering gameplay experience, social interaction, fun, and playability.

There is an opportunity for the academic community to take an active role in creating diverse types of evaluation instruments, as well as evaluating the effectiveness of the existing ones, in the context of multimodal video games for blind person's cognition enhancement.

5   Principles for Evaluating Usability of Multimodal Video Games for People Who Are Blind (PLUMB)

Seeking to meet the identified challenges regarding usability evaluation, a set of PrincipLes for Evaluating Usability of Multimodal Video Games for people who are Blind (PLUMB) were proposed. It is based on the analysis of the Expert Opinion Survey; on usability evaluation reported in the literature; and on the Standard List of Usability Problems (SLUP) in multimodal video games for people who are blind.

PLUMB is a practical aid to help researchers and practitioners to properly plan and conduct usability evaluation of multimodal video games based on audio and haptics, designed for enhancing and improving cognition in people who are blind. When asked about their opinion regarding the establishment of principles to assist researchers and practitioners in choosing the methods that better fit their usability requirements, to assess diverse aspects of multimodal interfaces in this context, all the authors agreed that it would be a useful aid to theirs and others' research. A6 commented that this outcome would fill significant gaps in their original research, while A7 stated that it would be useful to have a simple guideline on how to make a usability testing with people who are blind.

Inspired by the experts' comments on this topic and based on reviews of previous related studies and observations of current trends previously reported in this paper, PLUMB was created as a list of five principles for evaluating the usability of multimodal video games for people who are blind. The usability issues detailed in SLUP [23] include problems reported by users and issues pointed out by researchers. Considering that, and in addition to the discussion and results presented regarding the experience of researchers in this field, we propose that the usability evaluation of multimodal video games for developing cognition in people who are blind should observe the following principles.

1. **Be connected to the design process, in a formative way.** The identification of usability issues in this context should follow a "find-and-fix" approach, to ensure an interaction free of possibilities to distract the person who is blind from the game cognitive purposes. Hence, the usability evaluation should be planned focused on identifying usability problems before the game is completed. A formative evaluation during the game design process can maximize the chances of effecting change and implementing the usability recommendations.
2. **Combine quantitative and qualitative methods to provide a holistic view of the data.** This approach can help develop rich insights into phenomena of interest while evaluating multimodal games for people who are blind that cannot be fully understood using only a quantitative or a qualitative method. This principle aims to guarantee that usability evaluation uses multiple ways to understand the interaction and possible issues between the user who is blind and the multimodal gaming interface. Hence, data collection should involve any techniques available to researchers that allow at least two types of data (e.g., numerical and text), two types of data analysis (e.g., statistical and textual) and two types of conclusions (e.g., objective and subjective).
3. **Combine empirical and analytical methods to comprehend both the users and researchers' point of view.** This principle aims to improve the accuracy of the identification of usability issues sources in the gaming interface and interaction. The use of both empirical (test-based) and analytical (inspection-based) usability evaluation methods provides direct information about how people who are blind use the multimodal game and their exact issues with its interface, while also having usability specialists judging whether each interactive element follows the necessary usability principles.
4. **Include both users who are blind and with visual impairments, preferably in the real context of use.** This principle aims to guarantee that the usability evaluation considers the different issues that arise from the diversity of perception and behavior between people who still rely on visual residues and those who rely on hearing and touch only. Besides, the multimodal video game has to adequate the presentation of abstract information, feedback, and game stimulus to the real conditions where people who are blind interact with the game: in schools or at home, assisted by a tutor. It is important to notice that testing applications with blindfolded users is not the same, due to the very different mental models.
5. **Guarantee a combination of methods capable of analyzing**
   a. the user's perception of each interaction modality to execute specific tasks in the game;
   b. the user's understanding of the relationship between the modalities offered and the game tasks;
   c. the user's comprehension of the game goals and context, including the cultural and social context of the game narrative;
   d. the user's ability to perform the expected tasks in the game correctly, in a way that the planned cognitive skills can be exercised;
   e. whether the user can distinguish the diverse sonorous, visual and haptic feedbacks, associating them with the correct actions and objects in the game;

f. f.whether the user can combine modalities to achieve a goal in the game successfully;

g. whether the combination of modalities offered by the gaming interface and devices is adequate for executing the game tasks;

h. whether the modalities are appropriate to convey the information related to the game tasks;

i. whether the game devices offer the desirable support for the game modalities;

j. whether the modalities offered can ease the execution of a task;

k. whether the user can recognize visual, aural and haptic feedback in a game task;

l. whether the user can associate visual, aural and haptic feedback to a game task;

m. whether the user has a positive acceptance to the visual, aural and haptic feedback associated with the game tasks, objects, and instructions.

The principles for setting a usability evaluation environment and for choosing UEMs in this context should be used as a guide to help practitioners and researchers to employ the most appropriate UEMs to evaluate the required aspects of these games in a particular context. However, we highlight that PLUMB is not a closed list. It can be expanded and improved, particularly as more knowledge is produced on the suitability of usability evaluation methods in this field.

6 Conclusion

Identifying usability problems in serious multimodal video games designed with cognitive proposes for people who are blind matters because these issues will make them focus on the problems, distracting them from learning cognitive skills when interacting with the video game. However, the planning and conduction of usability field tests involving these users is not an easy task. For this reason, these tests are often conducted using inappropriate instruments, UEMs and procedures to their contexts, or even left aside. In this paper, we discussed the characterization of multimodal interaction under multiple points of views, considering how different modalities combinations and choices may affect the users and be considered to an adequate usability evaluation.

In addition, we discussed the current usability evaluation practices for multimodal games for people who are blind and pointed out some challenges regarding this field, based on research with experts in development and evaluation of such applications.

Finally, we argued that there are specific aspects of user interface and interaction of multimodal video games that should be considered for the evaluation of a multimodal game for people who are blind to identify and correct the relevant usability issues that affect these users. In a nutshell, an adequate usability evaluation of such applications should consider (i) what are the specific characteristics of the target users; and (ii) how the multimodal features affect the interaction of users who are blind with the interface. PLUMB was proposed to assist evaluators in considering these aspects in their practice.

Our expectation is that the future designs and evaluation processes of multimodal interactions to improve cognition of people who are blind take into consideration their broadly different abilities and disabilities and provide them with usable and pleasurable gaming interfaces, by considering our findings.

Acknowledgements. This paper is funded by Chilean FONDECYT Project #1150898; Basal Funds for Centers of Excellence, Project FB003, from CONICYT; Brazil LGE - Project Mobile, Tool & CAS with FCPC administration (Law 8248/1991). Rossana Andrade is a researcher fellow of CNPq Productivity (DT-2).

References

1. Espinoza, M., Sánchez, J., de Borba Campos, M.: Videogaming interaction for mental model construction in learners who are blind. In: Stephanidis, C., Antona, M. (eds.) UAHCI 2014. LNCS, vol. 8514, pp. 525–536. Springer, Cham (2014). https://doi.org/10.1007/978-3-319-07440-5_48
2. Sánchez, J., Sáenz, M.: Three-dimensional virtual environments for blind people. CyberPsychol. Behav. **9**(2), 200 (2006)
3. Sánchez, J., Saenz, M., Garrido, J.M.: Usability of a multimodal video game to improve navigation skills for blind people. ACM Trans. Access. Comput. **3**(2), 35–42 (2010). Article No. 7. Proceedings of the 11th international ACM SIGACCESS conference on Computers and Accessibility
4. Westin, T.: Game accessibility case study: terraformers – a real-time 3D graphic game. In: Proceedings of the 5th International Conference on Disability, Virtual Reality and Associated Technologies. ICDVRAT 2004, pp. 95–100, Oxford (2004)
5. Darin, T., Sánchez, J., Andrade, R.: Dimensions to analyze the design of multimodal videogames for the cognition of people who are blind. In: XIV Simpósio Brasileiro sobre Fatores Humanos em Sistemas Computacionais (IHC 2015), pp. 2–11 (2015)
6. Darin, T.: Towards a methodology to evaluate multimodal games for cognition in people who are blind. In: INTERACT 2015 Adjunct Proceedings of the 15th IFIP TC. 13 International Conference on Human-Computer Interaction, Bamberg, Germany, 14–18 September 2015, vol. 22, p. 61. University of Bamberg Press, September 2015
7. Sánchez, J., Viana, W., Darín, T., Gensel, J., Andrade, R: Multimodal interfaces for improving the intellect of the blind. In: TISE - XX Conferência Internacional sobre Informática na Educação, 2015, Santiago. Nuevas Ideas en Informática Educativa, vol. 11, pp. 404–413, Santiago (2015)
8. Turk, M.: Multimodal interaction: a review. Pattern Recogn. Lett. **36**, 189–195 (2014). https://doi.org/10.1016/j.patrec.2013.07.003. ISSN 0167-8655
9. Sánchez, J., Darin, T., Andrade, R.: Multimodal videogames for the cognition of people who are blind: trends and issues. In: Antona, M., Stephanidis, C. (eds.) UAHCI 2015. LNCS, vol. 9177, pp. 535–546. Springer, Cham (2015). https://doi.org/10.1007/978-3-319-20684-4_52
10. Bernsen, N.O., Dybkjær, L.: Multimodal Usability. Springer, London (2009)
11. Pagliano, P.J.: Using a Multisensory Environment: A Practical Guide for Teachers. David Fulton, London (2001)
12. Raisamo, R., Hippula, A., Patomaki, S., Tuominen, E., Pasto, V., Hasu, M.: Testing usability of multimodal applications with visually impaired people. IEEE Multimedia **13**(3), 70–76 (2006). https://doi.org/10.1109/MMUL.2006.68
13. Armstrong, J.S. (ed.): Principles of Forecasting: A Handbook for Researchers and Practitioners. Springer, Heidelberg (2001)

14. Oviatt, S.: Multimodal interfaces. Hum.-Comput. Interact. Handb.: Fundam., Evolving Technol. Emerg. Appl. **14**, 286–304 (2003)
15. Norman, D.A.: The Design of Everyday Things. Basic Book, New York (1988)
16. Dumas, B., Lalanne, D., Oviatt, S.: Multimodal interfaces: a survey of principles, models and frameworks. In: Lalanne, D., Kohlas, J. (eds.) Human Machine Interaction. LNCS, vol. 5440, pp. 3–26. Springer, Heidelberg (2009). https://doi.org/10.1007/978-3-642-00437-7_1
17. Bongers, B., van der Veer, G.C.: Towards a multimodal interaction space: categorisation and applications. Pers. Ubiquit. Comput. **11**(8), 609–619 (2007)
18. Coutaz, J., Nigay, L., Salber, D., Blandford, A., May, J., Young, R.M.: Four easy pieces for assessing the usability of multimodal interaction: the CARE properties. In: Nordby, K., Helmersen, P., Gilmore, D.J., Arnesen, S.A. (eds.) Human—Computer Interaction. IFIP Advances in Information and Communication Technology, pp. 115–120. Springer, Boston (1995). https://doi.org/10.1007/978-1-5041-2896-4_19
19. Larson, J.A., Raman, T.V., Raggett, D. (eds.) Multimodal interaction framework. W3C Note (2003). http://www.w3.org/TR/mmi-framework
20. Nigay, L., Coutaz, J.A.: Design space for multimodal systems: concurrent processing and data fusion. In: Proceedings of the INTERACT 1993 and CHI 1993 Conference on Human (1993)
21. Bernsen, N.O.: Modality theory: supporting multimodal interface design. In: Proceedings from the ERCIM Workshop on Multimodal Human–Computer Interaction, Nancy, pp. 13–23, November 1993
22. Allain, K., et al.: An audio game for training navigation skills of blind people. In: Sonic Interactions for Virtual Environments (SIVE), 2015 IEEE, pp. 1–4, IEEE (2015)
23. Darin, T., Andrade, R., Merabet, L.B., Sánchez, J.: Investigating the mode in multimodal video games: usability issues for people who are blind. In: CHI 2017 Extended Abstracts on Human Factors in Computing Systems, pp 2487–2495. ACM (2017)
24. Poulson, D., Ashby, M., Richardson, S.: USERfit: a practical handbook on user-centered design for assistive technology. In: Handbook Produced Within the European Commission TIDE Programme USER Project. HUSAT Research Institute, Loughborough (1996)
25. Holland, R., Smith, A., Hasselback, J., Payne, B.: Survey responses: mail versus email solicitations. J. Bus. Econ. Res. **8**(4), 95 (2010)
26. Rowe, G., Wright, G.: Expert opinions in forecasting: the role of the Delphi Technique. In: Armstrong, J. (ed.) Principles of Forecasting. International Series in Operations Research & Management Science, vol. 30, pp. 125–144. Springer, Boston (2001)
27. Lazar, J., Feng, J., Hochheiser, H.: Research Methods In Human-Computer Interaction. Morgan Kaufmann, Burlington (2017)
28. Frøkjær, E., Hertzum, M., Hornbæk, K.: Measuring usability: are effectiveness, efficiency, and satisfaction really correlated? In: Proceedings of the SIGCHI Conference on Human Factors in Computing Systems, pp. 345–352. ACM (2000)
29. Zaharias, P., Poylymenakou, A.: Developing a usability evaluation method for e-learning applications: beyond functional usability. Int. J. Hum.-Comput. Interact. **25**(1), 75–98 (2009)

A Low Resolution Haptic Interface
for Interactive Applications

Bijan Fakhri$^{(\boxtimes)}$, Shashank Sharma, Bhavica Soni, Abhik Chowdhury,
Troy McDaniel, and Sethuraman Panchanathan

Center for Cognitive Ubiquitous Computing,
Arizona State University,
Tempe, AZ, USA
{bfakhri,sshar139,bsoni,achowdh5,troy.mcdaniel,panch}@asu.edu

Abstract. This paper introduces a novel haptic interface for use as
a general-purpose sensory substitution device called the Low Resolution Haptic Interface (LRHI). A prototype of the LRHI was developed
and tested in a user study for its effectiveness in conveying information
through the sense of touch as well as for use in interactive applications.
Results are promising, showing that participants were able to accurately
discriminate a range of both static and dynamic haptic patterns using
the LRHI with a composite accuracy of 98.38%. The user study also
showed that participants were able to sucessfully learn to play a completely haptic interactive cat-mouse game with the device.

Keywords: Haptics · HCI · Sensory substitution ·
Assistive technology

1 Introduction

The Cornell Employment and Disability Institute estimates there are 7.3 million
adults in the U.S. who are blind [1]. Because most digital experiences involve
a screen, most remain inaccessible to people with severe visual impairments.
While there has been a recent push towards more inclusive design [2] and screen-
less interfaces [3], implementations are few and far between. A possible solution
lays in Sensory Substitution Devices (SSDs): having shown promise for making
visual experiences accessible [4]. Most SSDs though are purpose-built [5–8] and
are thus unfit for other uses. Towards rectifying this inequity, we introduce the
Low Resolution Haptic Interface (LRHI), a general-purpose haptic interface for
sensory substitution that abstracts 2D haptic patterns into "Haptic Images".
The LHRI can be controlled using our library: https://github.com/bfakhri/lrhi.

2 Related Work

Dr. Bach-y-Rita showed with his Tactile-to-Vision Sensory Substitution (TVSS)
device [4] that individuals who are blind are capable of understanding simple

© Springer Nature Switzerland AG 2019
M. Antona and C. Stephanidis (Eds.): HCII 2019, LNCS 11573, pp. 224–233, 2019.
https://doi.org/10.1007/978-3-030-23563-5_19

visual scenes with the aid of a sensory substitution device (SSD). His device utilized 400 solenoid actuators placed on a user's back that were controlled by a camera. Users felt the images captured by the camera on their backs, and with training were able to distinguish a variety of common objects. The successor to the TVSS was the BrainPort, which used ETVSS (electro-tactile visual sensory substitution) to first augment a user's sense of balance in order to regain autonomy [9] and later as an alternative means to vision [10], similar to the original TVSS. SSDs are a great alternative for circumventing the loss of a sensory modality to devices requiring surgical procedures such as the Cochlear Implant (CI) [11] and retinal prosthesis [12] as these procedures are expensive and invasive.

Examples of more modern SSDs include the Social Interaction Assistant [5] and the VibroGlove [6] where facial expressions are identified by the system and relayed to the user via haptics. SSDs that make use of the auditory modality have also been developed such as KASPA (Kay's Advanced Spatial Perception Aid) [13], the Sonic Pathfinder [14], and the EyeCane for virtual environments [15] and real environments [16]. More generally, SSDs towards general vision substitution such as the "vOICE" [17,18] and EyeMusic [19] abstract images into tones or musical notes and instruments to convey visual information. Unfortunately, the usability of auditory SSDs for vision substitution is limited as they obstruct a valuable sensory modality (hearing) which is often counterproductive to SS [6]. Alternatively, haptic SSDs allow the interface to work without obstructing modalities that are often also in use while taking part in typical daily tasks. Because most SSDs are purpose-built they are difficult to modify for other uses, effectively presenting large design costs to researchers and engineers.

3   Low Resolution Haptic Interface

Building a general-purpose haptic sensory substitution device required a standardised and general interface. Towards this, we propose that haptic patterns be abstracted into "haptic images", which are essentially two dimensional arrays of haptic intensities and frequencies $i, f = H[x, y]$ analogous to how a visual image can be modeled as a 2D array of color intensities $r, g, b = V[x, y]$ (RGB model) where x, y are discrete coordinates relating to space. A series of haptic images can thus convey moving patterns over time similar to how a series of images becomes a video. The LRHI is a system that communicates using "haptic images" and converts them into tactile representations. The LRHI consists of a *computing platform* which sends haptic images to be displayed, a *controller* which intereprets the haptic images and converts them into anolog signals, and a *display* which converts the analog signals into vibrotactile actuation. Figure 1 shows a block diagram of the LRHI.

The *computing platform* can be any USB enabled computer: its role is to generate the haptic images in a digital and abstract form. The computing platform may take on a variety of roles in generating the haptic images. In sensory substition applications for instance, the computing platform converts images from a video stream into haptic images and sends them to the controller. The actual

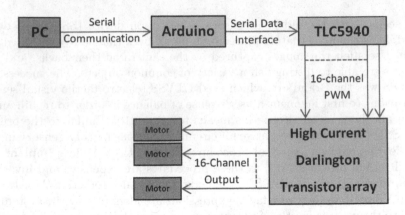

Fig. 1. Block diagram of the LRHI. In red *Computing Platform*. In green *Controller*. In blue *Display* (Color figure online)

algorithms for conversion are left up to the designers. In our incarnation of the LRHI, it communicates with the *controller* by sending 4 × 4 8-bit haptic images.

The *controller* consists of an Arduino microcontroller, TLC5940 analog to digital converter, and a collection of high-current Darlington Transistor Arrays. The Arduino accepts the haptic image and using the TLC5940 converts the haptic image into 16 analog electrical signals (8-bit PWM). These are transmitted to the transistor arrays where the signals are amplified and made suitable to drive the *display*. A full version of the LRHI would allow haptic images to specify not only an intensity but also a vibration frequency for each actuator on the display.

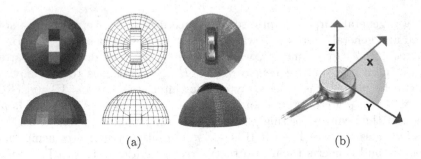

(a)   (b)

Fig. 2. Motor Housing: (a) digital and 3d-printed models (b) vibration axis

Our prototype of the *display* consists of a 4 × 4 array of pancake motors housed in custom 3D printed mounts that orient the motors orthogonal to the user's back. The housing is shown in Fig. 2a. This accomplishes two objectives: first, the vibration axis is made perpendicular to the user's back (illustrated in Fig. 2b). Second, the contact point is made smaller. These two objectives increase the perceived intensity of the vibrations which is especially important when the user is wearing thick clothing. The motors and housing are mounted on accoustic foam to provide a maleable surface that adheres to a user's back and simultaneously transmits minimal intermotor vibration. The haptic display is shown in Fig. 3.

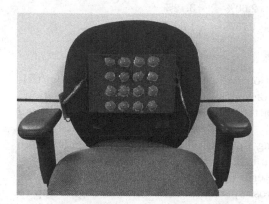 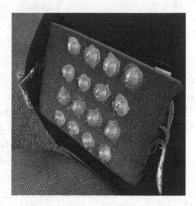

Fig. 3. 4 × 4 haptic display mounted on an office chair

The haptic display consumes 50 mA in an idle state with a maximum consumption of 412 mA when all motors are at full power (energy consumption summarized in Table 1). During the non-interactive portion of the user study, the LRHI had a mean power consumption of 0.73 W. During the interactive portion of the study the LRHI showed a mean power consumption of 0.56 W.

Table 1. LRHI energy table

	Current (A)	Power (W)
Idle	0.05	0.17
Single motor	0.10	0.33
1/4 motors	0.19	0.63
2/4 motors	0.28	0.92
3/4 motors	0.34	1.12
4/4 motors	0.41	1.35

4   User Study

In order to assess the LRHI's potential as an SSD, we performed a preliminary user study with 8 participants to explore its ability to convey information through haptics. The study consisted of a non-interactive and an interactive

Table 2. Results for non-interactive phase

Non-interactive	Repeat %	Error %
Phase 1	1.13%	0.92%
Phase 2	1.46%	0.83%
Phase 3	2.60%	3.12%
Total	1.73%	1.62%

component. The non-interactive portion consisted of 3 phases wherein participants were introduced to a finite set of haptic patterns during "familiarization" (being exposed to each individual pattern only once) and were asked to recall those patterns during "testing".

During the non-interactive testing portion of the study participants were given the option to repeat the pattern if they were not confident in their assessment. Phase 1 consisted of static patterns (Top Left, Bottom Right, etc). Phase 2 consisted of patterns that vary across space and time (Left to Right, Top to Bottom, etc). Phase 3 is similar to Phase 2, but users were asked to recall how fast the pattern was displayed (Left to Right - Fast, Top to Bottom - Slow, etc) in addition to the original pattern identity (Left to Right). The patterns increased in complexity in each subsequent phase, beginning with simple single-motor patterns to patterns that move through space and time. Participants were given the option to repeat a pattern if they were not confident in their initial assessment. The patterns for each phase are illustrated and described in Appendix A.

In order to assess the LRHI's potential in interactive environments, we designed a completely haptic, cat-mouse game to play (illustrated in Fig. 7). The user plays as a cat, and the goal is to find a mouse. The cat is presented on the haptic display as a solid vibration, while the mouse is a pulsing vibration. Participants used a computer-mouse to control the position of the cat on the haptic display, leading it towards the mouse - the goal being to catch the mouse as quickly as possible. The duration between the beginning of the game and capturing the mouse was recorded, each participant playing 60 games in total (results shown Table 3). Increasing performance in this game (decreasing game time) was intended to show that participants were in fact able to learn to use the LRHI to interact with dynamic environments.

5   Results

For the non-interactive portion of the study, participants were able to identify the patterns with considerable accuracy. Phase 1, which included static patterns only did not significantly differ in accuracy over Phase 2 (dynamic patterns). Only when participants were asked to discern both the pattern and the speed at which it was presented did performance suffer slightly. Results are compiled in Table 2 - participants were able to achieve an aggregate accuracy of 98.38%.

Table 3. Results for interactive phase

Interactive	Avg	StdDev
Total	4.81 s	2.99 s
First 3rd	6.64 s	3.23 s
Last 3rd	4.40 s	2.78 s

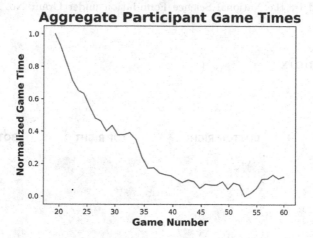

Fig. 4. Normalized mean game times

For the interactive portion of the study (illustrated in Fig. 7), participants were able to capture the mouse in 4.81 s on average, and showed a significant performance increases the longer they played. A comparison of the first third of the gaming session (first 20 games) and the last third as well as total performance can be seen in Table 3. Figure 4 illustrates the participants' performance over time.

6   Conclusion and Future Work

In conclusion, we have introduced the LRHI, a general haptic interface for sensory substitution applications and have performed a preliminary user-study to show its efficacy in conveying information through the sense of touch as well as an interface to interactive environments. The LRHI shows promise as a general purpose sensory substitution device because we were able to show that users learn as they play the cat-mouse game, successfully navigating a virtual, non-visual, and interactive environment solely through their sense of touch. For future work we intend to augment the LRHI with peripheral controls and test its effectiveness in visual-to-haptic sensory substitution in 3D environments (both virtual and real-world).

Awknowledgements. The authors thank Arizona State University and the National Science Foundation for their funding support. This material is partially based upon work supported by the National Science Foundation under Grant No. 1069125 and 1828010.

A   Appendix

See Figs. 5 and 6.

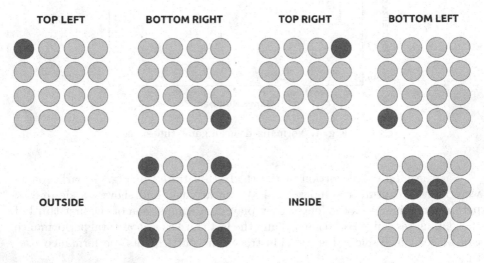

Fig. 5. User Study Patterns for Phase 1: Patterns in this phase do not change over time. They remain static. A red motor in this illustration represents a motors on at full power, while a blue motor represents a motor that is completely inactive (off). Participants were asked to recall what pattern they are experiencing after being subjected to the pattern for 1 s. The participant's accuracy was recorded. (Color figure online)

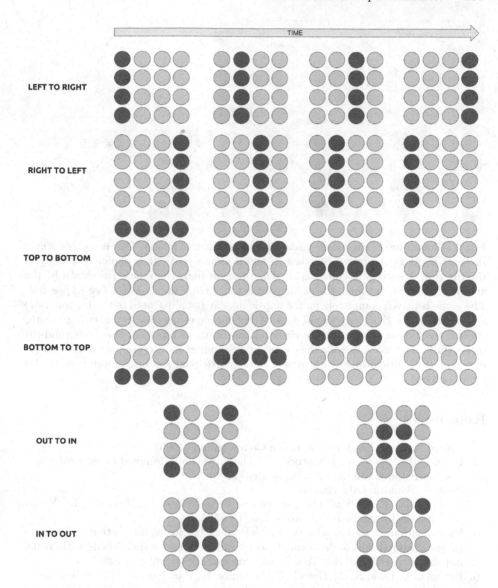

Fig. 6. User Study Patterns for Phases 2 and 3: Patterns in these two phases were dynamic, meaning they evolve over time. A red motor in this illustration represents a motors on at full power, while a blue motor represents a motor that is completely inactive (off). The illustration shows how the patterns evolve over time (left to right). The first four patterns (Left-to-Right, Right-to-Left, Top-to-Bottom, and Bottom-to-Top) have 4 states while the last two (Out-to-In and In-to-Out) only have two states. In Phase 2, the patterns lasted a total of 1 s, while in Phase 3 they lasted either 0.6 s or 1.4 s depending on the speed. In Phase 2 participants were asked to recall what pattern they had experienced and in Phase 3 they were asked what pattern they had experienced as well as how fast the pattern was presented (Slow, Fast). The participant's accuracy was recorded. (Color figure online)

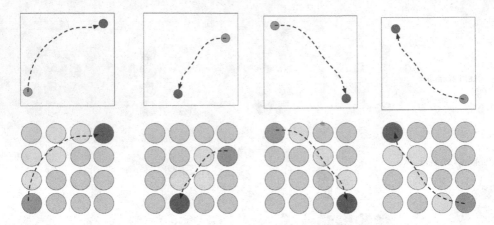

Fig. 7. Illustration of cat-mouse game for The interactive Phase: This is a game where the user plays a cat (green) that has to find a mouse (red). It is depicted visually in the top row. A successful game occurs when the cat finds the mouse (as shown by the arrows). The game is timed: the participant is told to find the mouse as fast as possible. The game is played completely on the haptic display (as illustrated in the bottom row) - the user is not given any visual cues. A red motor represents a "pulsating" motor while a green motor is a statically vibrating motor, these represent the mouse and cat respectfully. The user controlls the cat using a computer mouse peripheral and the cat moves on the haptic display with respect to the participant's mouse movements. (Color figure online)

References

1. United Nations: Disability Statistics Compendium (1990)
2. Rose, D.H., Meyer, A., Hitchcock, C.: The universally designed classroom: accessible curriculum and digital technologies (2005)
3. Bala, S.: Somatic Labs (2018)
4. Bach-y-Rita, P., Collins, C.C., Saunders, F.A., White, B., Scadden, L.: Vision substitution by tactile image projection [18] (1969)
5. Panchanathan, S., Chakraborty, S., McDaniel, T.: Social interaction assistant: a person-centered approach to enrich social interactions for individuals with visual impairments. IEEE J. Sel. Top. Signal Process. **10**(5), 942–951 (2016)
6. Krishna, S., Bala, S., McDaniel, T., McGuire, S., Panchanathan, S.: VibroGlove: an assistive technology aid for conveying facial expressions. In: CHI '10 Extended Abstracts on Human Factors in Computing Systems, pp. 3637–3642 (2010)
7. Eagleman, D.: Plenary talks: a vibrotactile sensory substitution device for the deaf and profoundly hearing impaired. In: 2014 IEEE Haptics Symposium (HAPTICS), p. xvii (2014)
8. Novich, S.D.: Sound-to-Touch Sensory Substitution and Beyond (2015)
9. Bach-y Rita, P., Danilov, Y., Tyler, M., Grimm, R.J.: Late human brain plasticity: vestibular substitution with a tongue BrainPort human-machine interface. Plasticidad y Restauracion Neurologica **4**(1–2), 31–34 (2005)
10. Nau, A., Bach, M., Fisher, C.: Clinical tests of ultra-low vision used to evaluate rudimentary visual perceptions enabled by the BrainPort vision device. Transl. Vis. Sci. Technol. **2**(3), 1 (2013)

11. Merzenich, M.M., Michelson, R.P., Pettit, C.R., Schindler, R.A., Reid, M.: Neural encoding of sound sensation evoked by electrical stimulation of the acoustic nerve. Ann. Otol. Rhinol. Laryngol. **82**(4), 486–503 (1973)
12. Caspi, A., Dorn, J.D., McClure, K.H., Humayun, M.S., Greenberg, R.J., McMahon, M.J.: Feasibility study of a retinal prosthesis: spatial vision with a 16-electrode implant. Arch. Ophthalmol. **127**(4), 398–401 (2009)
13. Kay, L.: A sonar aid to enhance spatial perception of the blind: engineering design and evaluation. Radio Electron. Eng. **44**(11), 605–627 (1974)
14. Heyes, A.D.: The sonic pathfinder: a new electronic travel aid. J. Vis. Impair. Blind. **78**(5), 200–2 (1983)
15. Maidenbaum, S., Levy-Tzedek, S., Namer-Furstenberg, R., Amedi, A., Chebat, D.R.: The effect of extended sensory range via the eyecane sensory substitution device on the characteristics of visionless virtual navigation. Multisens. Res. **27**(5–6), 379–397 (2014)
16. Chebat, D.R., Maidenbaum, S., Amedi, A.: Navigation using sensory substitution in real and virtual mazes. PLoS ONE **10**(6), 1–18 (2015)
17. Meijer, P.B.: An experimental system for auditory image representations. IEEE Trans. Biomed. Eng. **39**(2), 112–121 (1992)
18. Ward, J., Meijer, P.: Visual experiences in the blind induced by an auditory sensory substitution device. Conscious. Cogn. **19**(1), 492–500 (2010)
19. Abboud, S., Hanassy, S., Levy-Tzedek, S., Maidenbaum, S., Amedi, A.: EyeMusic: introducing a 'visual' colorful experience for the blind using auditory sensory substitution. Restor. Neurol. Neurosci. **32**(2), 247–257 (2014)

A Fitts' Law Evaluation of Hands-Free and Hands-On Input on a Laptop Computer

Mehedi Hassan[1]($\boxtimes$), John Magee[2], and I. Scott MacKenzie[1]

[1] York University, Toronto, ON, Canada
mhassan@eecs.yorku.ca
[2] Clark University, Worcester, MA, USA

Abstract. We used the Fitts' law two-dimensional task in ISO 9241-9 to evaluate hands-free and hands-on point-select tasks on a laptop computer. For the hands-free method, we required a tool that can simulate the functionalities of a mouse to point and select without having to touch the device. We used a face tracking software called *Camera Mouse* in combination with dwell-time selection. This was compared with three hands-on methods, a touchpad with dwell-time selection, a touchpad with tap selection, and face tracking with tap selection. For hands-free input, throughput was 0.65 bps. The other conditions yielded higher throughputs, the highest being 2.30 bps for the touchpad with tap selection. The hands-free condition demonstrated erratic cursor control with frequent target re-entries before selection, particularly for dwell-time selection. Subjective responses were neutral or slightly favourable for hands-free input.

Keywords: Hands-free input · Face tracking · Dwell-time selection · Fitts' law · ISO 9241-9

1 Introduction

Most user interfaces (UIs) for computers and mobile devices depend on physical touch from the user. For instance, a web page on a laptop computer's screen requires a mouse or touchpad for pointing and selecting. Most UIs also require a keyboard to enter text. In this paper, we explore pointing and selecting without using a physical device. Our ultimate goal is to test the hands-free system for accessible computing.

We are particularly interested in methods that do not require specialized hardware, such as eye trackers. Our focus is on methods that use inexpensive built-in cameras, either on a laptop's display or in a smartphone or tablet. Tracking a body position, perhaps on the head or face, is easier than tracking the movement of a user's eyes, which undergo rapid jumps known as saccades [12]. The smoother and more gradual movement of the head or face, combined with the

M. Antona and C. Stephanidis (Eds.): HCII 2019, LNCS 11573, pp. 234–249, 2019.
https://doi.org/10.1007/978-3-030-23563-5_20

ubiquity of front-facing cameras on today's laptops, tablets, and smartphones, presents a special opportunity for users with motor disabilities. Such users desire access to the same wildly popular devices as used by non-disabled users.

Magee et al. [16] did similar research with a 2D Fitts' law task, but we present a modified approach herein. We present and evaluate a hands-free approach, comparing it with hands-on approaches, and provide the results of a comparative evaluation. The hands-free method uses camera input combined with dwell-time selection. The hands-on methods use camera or touchpad input combined with tapping on the touchpad surface for selection. For camera input, we used *Camera Mouse*, described below. We evaluated the participants on a completely hands-on method with pointing and selecting with the touchpad, a partially hands-free method with pointing with *Camera Mouse* and selecting with touchpad, and finally a completely hands-free method with pointing and selecting with *Camera Mouse* only. These methods make our user study relevant for people with partial or complete motor disabilities. Although this experiment only had participants with no motor disabilities, in future we intend to do case studies with disabled participants as well.

We begin with a review of related work, then describe the use of Fitts' law and ISO 9241-9 for evaluating point-select methods. This is followed with a description of our system and the methodology for our user study. Results are then presented and discussed followed by concluding remarks and ideas for future work. Our contribution is to provide the first ISO-conforming evaluation of hands-free input on a laptop computer using a built-in webcam.

2 Related Work

Research on hands-free input methods using camera input is now reviewed. The review is organized in two parts. First, we examine research not using Fitts' law and follow with research where the experimental methodology used Fitts' law testing.

2.1 Research Not Using Fitts' Law

Roig-Maimó et al. [19] present *FaceMe*, a mobile head tracking interface for accessible computing. Participants were positioned in front of a propped-up *iPad Air*. Via the front-facing camera, a set of points in the region of the user's nose was tracked. The points were averaged, generating an overall head position which was mapped to a display coordinate. *FaceMe* is a picture-revealing puzzle game. A picture is covered with a set of tiles, hiding the picture. Tiles are turned over revealing the picture as the user moves her head and the tracked head position passes over tiles. Their user study included 12 non-disabled participants and four participants with multiple sclerosis. All non-disabled participants could fully reveal all pictures with all tile sizes. Two disabled participants had difficulty with the smallest tile size (44 pixels). *FaceMe* received a positive subjective rating overall, even on the issue of neck fatigue.

Roig-Maimó et al. [19] described a second user study using the same participants, interaction method, and device setup. Participants were asked to select icons on the *iPad Air*'s home screen. Icons of different sizes appeared in a grid pattern covering the screen. Selection involved dwelling on an icon for 1000 ms. All non-disabled participants were able to select all icons. One disabled participant had trouble selecting the smallest icons (44 pixels); another disabled participant felt tired and was not able to finish the test with the 44 pixel and 76 pixel icon sizes.

Gips et al. [7] developed the *Camera Mouse* input method that we included in our evaluation. *Camera Mouse* uses a camera to visually track a selected feature of the body. The feature could be the nose or, for example, a point between the eyebrows. During setup, the user adjusts the camera until their face is centered in the image. Upon clicking on a face feature, *Camera Mouse* begins tracking and draws a 15 × 15 pixel square centered at the clicked location. This location is output as the "mouse position". Camera images are processed at 30 frames per second. The tracked location moves as the user moves their head. No user evaluation was presented in this initial paper on *Camera Mouse*.

Cloud et al. [3] conducted an experiment with *Camera Mouse* that tested 11 participants, one with severe physical disabilities. The participants were tested on two applications, *EaglePaint* and *SpeechStaggered*. *EaglePaint* is a simple painting application that uses a mouse pointer. *SpeechStaggered* allows users to spell words and phrases by accessing five boxes that contain the English alphabet. Measurements for entry speed or accuracy were not reported; however, a group of participants wearing glasses showed better performance than a group not wearing glasses.

Betke et al. [1] describe further advancements with *Camera Mouse*. They compared different body features for robustness and user convenience. Twenty participants without physical disabilities were tested along with 12 participants with physical disabilities. Performance was tested on two applications, *Aliens Game*, which is an alien catching game requiring movement of the mouse pointer, and *SpellingBoard*, a typing application where entry involved selecting characters with the mouse pointer. The non-disabled participants showed better performance with a normal mouse than *Camera Mouse*. Nine of the 12 disabled participants showed eagerness in continuing to use the *Camera Mouse* system.

Magee et al. [17] present *EyeKeys*, a gaze detection interface which exploits the symmetry between the left and right eyes to determine whether the user's gaze direction is center, left, or right. They developed a game named *BlockEscape* for a quantitative evaluation. *BlockEscape* presents horizontal black bars with gaps in them. The bars move upward on the display. The user controls a white block which is moved left and right, and aligned to fall through a gap in the black wall to the wall below, and so on. If the block reaches the bottom of the display, the user wins. If the block is pushed to the top of the screen, the game ends. Three input methods were compared: *EyeKeys* (eyes), camera mouse (face tracking), and the keyboard (left/right arrow key). The win percentages were 100% (keyboard), 83% (*EyeKeys*), and 83% (*Camera Mouse*).

2.2   Research Using Fitts' Law

Magee et al. [16] did a user study using an interactive evaluation tool called *FittsTaskTwo* [13, p. 291]. *FittsTaskTwo* runs on a laptop computer and implements the two-dimensional (2D) Fitts' law test in ISO 9241-9 (described in the following section). The primary dependent variable is *throughput* in bits per second, or bps. They also used *Camera Mouse* configured with two selection methods: 1000 ms dwell-time and *ClickerAID*. *ClickerAID* generates button events by sensing an intentional muscle contraction from a piezoelectric sensor contacting the user's skin. The sensor was positioned under a headband, making contact with the user's brow muscle. A third baseline condition used a conventional laptop computer touchpad. In a user study with ten participants, throughputs were 2.10 bps (touchpad), 1.28 bps (*Camera Mouse* with dwell-time selection), and 1.43 bps (*Camera Mouse* with *ClickerAID*). For the *Camera Mouse* conditions, participants indicated a subjective preference for *ClickerAID* over dwell-time selection.

Magee et al. [16] included a follow-on case study with a patient affected by the neuromuscular disease Friedreich's Ataxia. Throughputs were quite low at 0.49 bps (*Camera Mouse* with dwell-time selection) and 0.45 bps (*Camera Mouse* with *ClickerAID*). To accommodate the patient's motor disability, the dwell time was increased to 1500 ms.

Cuaresma and MacKenzie [4] designed an experimental application named *FittsFace*, which is similar to *FittsTaskTwo*, except it runs on Android devices and uses facial sensing and tracking for input (instead of touch). A user study with 12 participants evaluated two navigation methods (positional, rotational) in combination with three selection methods (smile, blink, dwell). Positional navigation with smile selection was best in terms of throughput (0.60 bps) and movement time (4383 ms). Positional navigation with smile selection and positional navigation with blink selection had similar error rates, about 11%. Ten of the 12 participants preferred positional navigation over rotational navigation. Seven out of the 12 participants preferred dwell-time selection.

Roig-Maimó et al. [18] conducted a target selection experiment using a variation of the *FaceMe* software described above. As their motivation was to test target selection over an entire display surface by head-tracking, they used a non-standard Fitts' law task: The targets were positioned randomly during the trials. The mean throughput was 0.74 bps. They also presented design recommendations for non-ISO tasks which include keeping amplitude and target width constant within each sequence of trials and using strategies to avoid reaction time.

Hansen et al. [8] described a Fitts' law experiment using a head-mounted display. They compared three pointing methods (gaze, head, mouse) in combination with two selection methods (dwell, click). The hands-free conditions are therefore gaze or head pointing combined with dwell-time selection. Dwell time was 300 ms. In a user study with 41 participants, throughputs were 3.24 bps (mouse), 2.47 bps (head-pointing), and 2.13 bps (gaze pointing). Gaze pointing was also less accurate than head pointing and the mouse.

3   Evaluation Using Fitts' Law and ISO 9241-9

Fitts' law – first introduced in 1954 [6] – is a well-established protocol for evaluating target selection operations on computing systems [2,11]. This is particularly true since the mid-1990s with the inclusion of Fitts' law testing in the ISO 9241-9 standard for evaluating non-keyboard input devices [9,10,20]. The most common ISO evaluation procedure uses a two-dimensional task with targets of width W arranged in a circle. Selections proceed in a sequence moving across and around the circle (see Fig. 1). Each movement covers an amplitude A, the diameter of the layout circle. The movement time (MT, in seconds) is recorded for each trial and averaged over the sequence of trials.

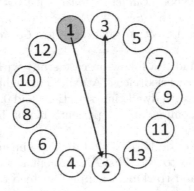

Fig. 1. Two-dimensional Fitts' law task in ISO 9241-9.

The difficulty of each trial is quantified using an index of difficulty (ID, in bits) and is calculated from A and W as

$$ID = \log_2\left(\frac{A}{W} + 1\right). \tag{1}$$

The main performance measure in ISO 9241-9 is throughput (TP, in bits/second or bps) which is calculated over a sequence of trials as the ID-MT ratio:

$$TP = \frac{ID_e}{MT}. \tag{2}$$

The standard specifies calculating throughput using the effective index of difficulty (ID_e). The calculation includes an adjustment for accuracy to reflect the spatial variability in responses:

$$ID_e = \log_2\left(\frac{A_e}{W_e} + 1\right) \tag{3}$$

with

$$W_e = 4.133 \times SD_x. \tag{4}$$

The term SD_x is the standard deviation in the selection coordinates computed over a sequence of trials. For the two-dimensional task, selections are projected onto the task axis, yielding a single normalized x-coordinate of selection for each trial. For $x = 0$, the selection was on a line orthogonal to the task axis that intersects the center of the target. x is negative for selections on the near side of the target center and positive for selections on the far side. The factor 4.133 adjusts the target width for a nominal error rate of 4% under the assumption that the selection coordinates are normally distributed. The effective amplitude (A_e) is the actual distance traveled along the task axis. The use of A_e instead of A is only necessary if there is an overall tendency for selections to overshoot or undershoot the target (see [14] for additional details).

Throughput is a potentially valuable measure of human performance because it embeds both the speed and accuracy of participant responses. Comparisons between studies are therefore possible, with the proviso that the studies use the same method in calculating throughput. Figure 2 is an expanded formula for throughput, illustrating the presence of speed and accuracy in the calculation.

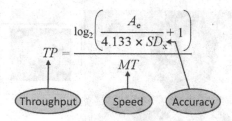

Fig. 2. The calculation of throughput includes speed and accuracy.

Our testing used *GoFitts*[1], a Java application which incorporates *Fitts Task-Two* and implements the 2D Fitts' law task described above. *GoFitts* includes additional utilities such as *Fitts Trace* which plots the cursor trace data captured during trials.

4   Method

The goal of our user study was to empirically evaluate and compare two pointing methods (touchpad, *Camera Mouse*) in combination with two selection methods (tap, dwell). The hands-free method combines *Camera Mouse* with dwell-time selection. A 2D Fitts' law task was used with three movement amplitudes combined with three target widths.

We recruited 12 participants. Nine were male aged 23–33 and three were female aged 23–29. All participants were from the local university community. None had prior experience using *Camera Mouse*.

---

[1] http://www.yorku.ca/mack/GoFitts.

4.1   Apparatus

An Asus *X541U* laptop was used as hardware. Both the built-in touchpad and the webcam provided input, depending on the pointing method. The touchpad was configured with the medium speed setting ("5") and with single-tap selection enabled.

The laptop's webcam provided images to *Camera Mouse*, as described under Related Work. Both horizontal and vertical sensitivity were set to medium. Although *Camera Mouse* can generate click events upon hovering the mouse cursor for a certain dwell time, this feature was not used since *GoFitts* provides dwell-time selection (see below).

Camera Mouse was setup to track the participant's nose. An example of the initialization screen is shown in Fig. 3. The experiment tasks were presented using *GoFitts*, described earlier. The 2D task was used with 11 targets per sequence. Three amplitudes (100, 200, 400 pixels) were combined with three target widths (20, 40, 80 pixels) for a total of nine sequences per condition.

Fig. 3. *Camera Mouse* initialization screen.

Selection was performed by the *GoFitts* software (not *Camera Mouse*). For dwell-time selection, a setting of 2000 ms was used. This somewhat long value was chosen after considerable pilot testing as it provided a balance between good selection and avoiding inadvertent selections.

Selection occurred after the cursor entered and remained in the target for 2000 ms. Errors were not possible. Visual feedback on the progress of the dwell timer was provided as a rotating arc inside the target. See Fig. 4.

During dwell-time selection, if the cursor exited the target before the timeout, the timer was reset. When the cursor next entered the target, the software logged a "target re-entry" event.

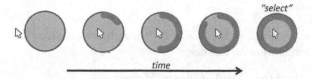

Fig. 4. Visual feedback indicating the progress of the dwell timer.

For tap selection, participants were instructed to perform a single-tap with their finger on the touchpad surface.

4.2   Procedure

Participants were welcomed into the experiment. We explained the experiment to each participant and made them aware of the purpose of it. To make participants comfortable with the setup of the experiment and *Camera Mouse*, practice trials were allowed until they felt comfortable with the interaction.

Participants were instructed to select targets as quickly and accurately as possible, but at a comfortable pace. For each sequence, they were to proceed from the first to last target without hesitation. Between sequences, they could pause at their discretion. Figure 5 shows a participant doing the experiment task (a) using the touchpad with tap selection and (b) using *Camera Mouse* with dwell-time selection. At the end of the experiment, participants provided feedback on a set of questions. They were asked about their preferred combination of pointing method and selection method. They also provided feedback on two 5-point Likert scale questions for physical fatigue and the overall rating of the hands-free phase.

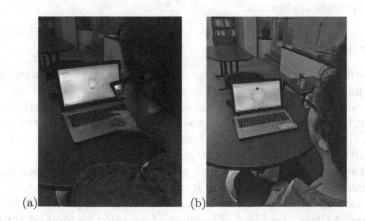

Fig. 5. Participant doing the experiment task (a) touchpad + tap selection (b) *Camera Mouse* + dwell-time selection.

4.3 Design

The experiment was a $2 \times 2 \times 3 \times 3$ within-subjects design. The independent variables and levels were as follows:

- Pointing method (touchpad, *Camera Mouse*)
- Selection method (tap, dwell)
- Amplitude (100, 200, 400 pixels)
- Width (20, 40, 80 pixels)

The primary independent variables were pointing method and selection method. Amplitude and width were included to ensure the conditions covered a range of task difficulties. The result is nine sequences for each test condition with *IDs* ranging from $\log_2(\frac{100}{80} + 1) = 1.17$ bits to $\log_2(\frac{400}{20} + 1) = 4.39$ bits. For each sequence, 11 targets appeared.

The dependent variables were throughput (bps), movement time (ms), error rate (%), and target re-entries (TRE, count/trial). There were two groups for counterbalancing, one starting with the touchpad and the other starting with *Camera Mouse*.

The total number of trials was 4752 ($= 2 \times 2 \times 3 \times 3 \times 11 \times 12$).

5   Results and Discussion

Results are presented below organized by dependent variables. For all dependent variables, the group effect was not statistically significant ($p > .05$). This indicates that counterbalancing was effective in offsetting learning effects.

Cursor trace examples, Fitts' law regression models, and a distribution analysis of the selection coordinates are also presented. Statistical analyses were done using the *GoStats* application.[2]

5.1 Throughput

Pointing with the touchpad and *Camera Mouse* had mean throughputs of 1.70 bps and 0.75 bps, respectively. The effect of pointing method on throughput was statistically significant ($F_{1,10} = 117.8, p < .0001$). Clearly, doing the experiment task with *Camera Mouse* was more difficult than with the touchpad. Of course, there is no expectation that hands-free point-select interaction would compete with hands-on point-select interaction.

During pointing with the touchpad, selecting with tap and dwell had mean throughputs of 2.30 bps and 1.10 bps, respectively. While pointing with *Camera Mouse*, selecting with tap and dwell had mean throughputs of 0.85 bps and 0.65 bps, respectively. See Fig. 6. The effect of selection method on throughput was statistically significant ($F_{1,10} = 93.0, p < .0001$). The lowest throughput of 0.65 bps was for the *Camera Mouse* with dwell-time selection – hands-free

---

[2] http://www.yorku.ca/mack/GoStats.

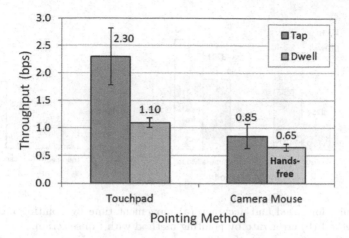

Fig. 6. Throughput (bps) by selection method and pointing method. Error bars show ±1 *SD*.

interaction. This value is low, but is expected given the pointing and selection methods employed. Throughput values in the literature are generally about 4–5 bps for the mouse [20, Table 4]. Other devices generally fair poorer with values of about 1–3 bps for the touchpad or joystick. Throughputs <1 bps sometimes occur when testing unusual cursor control schemes or when engaging participants with motor disabilities [4,5,15,18].

The closest point of comparison is the work of Magee et al. [16] who also used *Camera Mouse* with dwell-time selection. They obtained a throughput of 1.28 bps, about 2× higher than the value reported above. The biggest contributor to the difference is probably their use of a 1000-ms dwell-time, compared to 2000 ms herein. All else being equal, an increase in dwell-time yields an increase in movement time which, in turn, decreases throughput (see Eq. 2). Other points of distinction are their use of an external webcam (our apparatus used the laptop's built-in web cam) and having dwell-time selection provided by *Camera Mouse* (vs. *GoFitts* in our study). It is not clear how these differences might impact the value of throughput, however.

5.2   Movement Time and Error Rate

Since throughput is a composite measure combining speed and accuracy, the individual results for movement time and error rate are less important. They are briefly summarized below. See Fig. 7.

The effects on movement time were statistically significant both for pointing method ($F_{1,10} = 395.1, p < .0001$) and for selection method ($F_{1,10} = 93.0, p < .0001$). Note in Fig. 7a the long movement time of 5012 ms for *Camera Mouse* with dwell-time selection. As errors were not possible with dwell-time selection, the long movement time is likely caused by participants having difficulty

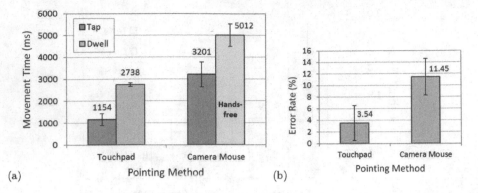

Fig. 7. Results for speed and accuracy (a) movement time by pointing method and selection method (b) error rate by pointing method with tap selection.

maintaining the cursor inside the target for the required dwell-time (2000 ms). This point is examined in further detail below in the analyses for target re-entries.

Figure 7b only shows the results by pointing method using tap selection, since errors were not possible for dwell-time selection. The effect of pointing method on error rate was statistically significant ($F_{1,10} = 67.3, p < .0001$).

5.3 Target Re-entries (TRE)

The grand mean for target re-entries (TRE) was 0.21 re-entries per trial. The implication is that for approximately one in every five trials the cursor entered the target, then left and re-entered the target. Sometimes this occurred more than once per trial.

Pointing with the touchpad and *Camera Mouse* had mean TREs of 0.09 and 0.31, respectively. So, TRE was about 3× higher for *Camera Mouse*. The effect of pointing method on TRE was statistically significant ($F_{1,10} = 10.2, p < .01$).

During pointing with the touchpad, selecting with tap and dwell had mean TREs of 0.08 and 0.11, respectively. While pointing with *Camera Mouse*, selecting with tap and dwell had mean TRE of 0.18 and 0.46, respectively. See Fig. 8. The effect of selection method on TRE was statistically significant ($F_{1,10} = 27.7, p < .0005$). Further discussion on target re-entries continues below in an examination of the trace paths for the cursor during pointing.

5.4 Cursor Trace Examples

The high value for TRE with *Camera Mouse* warrants further investigation. This was done by examining the cursor trace files generated by *GoFitts*. Cursor movements were relatively clean for the touchpad pointing method. It was a different story for *Camera Mouse*, however, where some erratic cursor movement patterns were observed. For comparison, Fig. 9 provides two examples. Both are

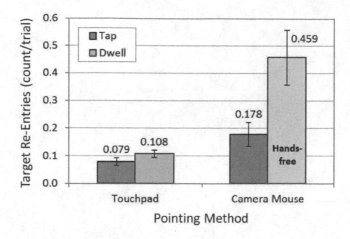

Fig. 8. Target re-entries (count/trial) by selection method and pointing method. Error bars show ±1 *SE*.

for dwell-time selection with $A = 200$ pixels and $W = 20$ pixels. Figure 9a is for pointing with the touchpad, while Fig. 9b is for pointing with *Camera Mouse*.

It is evident that the cursor movement paths were more direct for the touchpad (Fig. 9a) than for *Camera Mouse* (Fig. 9b). In fact, the difference is dramatic, as seen in the call-out in Fig. 9b. This particular trial had $MT = 14199$ ms with six target re-entries. Clearly, the participant had considerable difficulty keeping the cursor inside the target for the 2000-ms interval required for dwell-time selection.

5.5   Fitts' Law Models

To test for conformance to Fitts' law, we built least-squares prediction equations for each test condition. The general form is

$$MT = a + b \times ID \tag{5}$$

with intercept a and slope b. See Table 1. The most notable observation in the table is the very high intercepts for the dwell models. Ideally, intercepts are 0 (or ≈ 0) indicating zero time to complete a task of zero difficulty, which has intuitive appeal. However, large intercepts occasionally occur in the literature. A notable case is the intercept of 1030 ms in Card et al.'s Fitts' law model for the mouse [2, p. 611].

Scatter plot and regression line examples are seen in Fig. 10 for pointing using *Camera Mouse*. Although the tap model provides a good fit ($r = .9622$, Fig. 10a), the dwell-time model is a much weaker fit ($r = .8627$, Fig. 10b). Behaviour was clearly more erratic in the *Camera Mouse* + dwell condition.

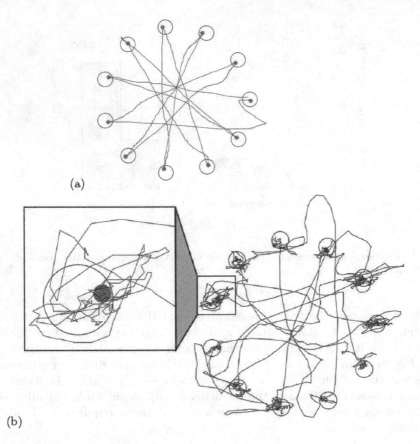

(a)

(b)

Fig. 9. Cursor trace examples for dwell-time selection with $A = 200$ pixels and $W = 20$ pixels. The pointing methods are (a) touchpad and (b) *Camera Mouse*. See text for discussion.

Table 1. Fitts' law models

Condition	Intercept, a (ms)	Slope, b (ms/bit)	Correlation (r)
Touchpad + tap	699.2	171.4	.9124
Touchpad + dwell	2270.5	176.6	.9653
Camera Mouse + tap	404.5	1055.1	.9622
Camera Mouse + dwell	1135.9	1462.7	.8627

5.6   Distribution of Selection Coordinates

The calculation of throughput uses the effective target width (W_e) which is computed from the standard deviation in the selection coordinates for a sequence of trials (see Eq. 4). There is an assumption that the selection coordinates are normally distributed. To test the assumption, we ran normality tests on the

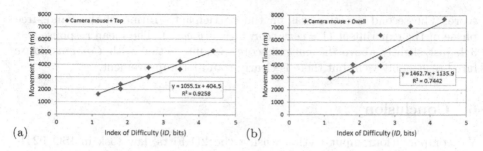

Fig. 10. Example Fitts' law models for *Camera Mouse*. Selection using (a) tap or (b) dwell.

Table 2. Lilliefors normality test on selection coordinates by trial sequence

Condition	Sequences	Normality hypothesis	
		Rejected	Not-rejected
Touchpad + tap	108	25	83
Touchpad + dwell	108	3	105
Camera Mouse + tap	108	15	93
Camera Mouse + dwell	108	7	101
Total	432	50	382

x-selection values, as transformed onto the task axis. A test was done for each sequence of trials. We used the Lilliefors test available in *GoStats*. The results are seen in Table 2.

As seen in Table 2, the user study included 432 sequences of trials (12 participants × 2 pointing methods × 2 selection methods × 3 amplitudes × 3 widths). Of these, 382, or 88.4%, had selection coordinates deemed normally distributed. Thus, the assumption of normality is generally held. The best results in Table 2 are for dwell-time selection; however, this is expected since all the selection coordinates were inside the targets. For some reason, the touchpad with tap selection had 25 of 108 sequences (23.1%) with selection coordinates considered not normally distributed.

5.7   Participant Feedback

Participants were asked to provide feedback on the experiment and indicate their preferred test condition. Eight of 12 participants chose the touchpad with tap selection as their preferred test condition. *Camera Mouse* with tap selection was preferred by two of the 12 participants. *Camera Mouse* with dwell-time selection and touchpad with dwell-time selection were preferred by one participant each.

Participants also provided responses to two 5-point Likert scale questions. One question was on the participant's level of fatigue with *Camera Mouse* (1 = *very low*, 5 = *very high*). The mean response was 2.4, closest to the *low* score. The second question was on the participant's rating of the hands-free phase of the experiment ($1 = very\ poor$, $5 = very\ good$). The mean response was 3.4, just slightly above the *normal* score. So, interaction with *Camera Mouse* fared reasonably well, but there is clearly room for improvement.

6   Conclusion

We compared four input methods using the 2D Fitts' law task in ISO 9241-9. The methods combined two pointing methods (touchpad, *Camera Mouse*) with two selection methods (tap, dwell). Using *Camera Mouse* with dwell-time selection is a hands-free input method and yielded a throughput of 0.65 bps. The other methods yielded throughputs of 0.85 bps (*Camera Mouse* + tap), 1.10 bps (touchpad + dwell), and 2.30 bps (touchpad + tap).

Cursor movement was erratic with *Camera Mouse*, particularly with dwell-time selection. This was in part due to the long 2000 ms dwell-time employed. Participants gave the hands-free condition a neutral, or slightly better than neutral, subjective rating.

For future work, we plan to extend our testing to different platforms. Effort to port *Camera Mouse* to mobile devices is on-going. We are also planning to test with disabled participants and with different age groups.

References

1. Betke, M., Gips, J., Fleming, P.: The camera mouse: visual tracking of body features to provide computer access for people with severe disabilities. IEEE Trans. Neural Syst. Rehabil. Eng. **10**(1), 1–10 (2002). https://doi.org/10.1109/TNSRE.2002.1021581
2. Card, S.K., English, W.K., Burr, B.J.: Evaluation of mouse, rate-controlled isometric joystick, step keys, and text keys for text selection on a CRT. Ergonomics **21**, 601–613 (1978). https://doi.org/10.1080/00140137808931762
3. Cloud, R.L. Betke, M., Gips, J.: Experiments with a camera-based human-computer interface system. In: Proceedings of the 7th ERCIM Workshop on User Interfaces for All, UI4ALL, pp. 103–110. European Research Consortium for Informatics and Mathematics, Valbonne, France (2002)
4. Cuaresma, J., MacKenzie, I.S.: FittsFace: exploring navigation and selection methods for facial tracking. In: Antona, M., Stephanidis, C. (eds.) UAHCI 2017. LNCS, vol. 10278, pp. 403–416. Springer, Cham (2017). https://doi.org/10.1007/978-3-319-58703-5_30
5. Felzer, T., MacKenzie, I.S., Magee, J.: Comparison of two methods to control the mouse using a keypad. In: Miesenberger, K., Bühler, C., Penaz, P. (eds.) ICCHP 2016. LNCS, vol. 9759, pp. 511–518. Springer, Cham (2016). https://doi.org/10.1007/978-3-319-41267-2_72
6. Fitts, P.M.: The information capacity of the human motor system in controlling the amplitude of movement. J. Exp. Psychol. **47**, 381–391 (1954). https://doi.org/10.1037/h0055392

7. Gips, J., Betke, M., Fleming, P.: The camera mouse: preliminary investigation of automated visual tracking for computer access. In: Proceedings of RESNA 2000, pp. 98–100. Rehabilitation Engineering and Assistive Technology Society of North America, Arlington (2000)
8. Hansen, J.P., Rajanna, V., MacKenzie, I.S., Bækgaard, P.: A Fitts' law study of click and dwell interaction by gaze, head and mouse with a head-mounted display. In: Proceedings of the Workshop on Communication by Gaze Interaction - Article no. 7. ACM, New York (2018). https://doi.org/10.1145/3206343.3206344
9. ISO: Ergonomic requirements for office work with visual display terminals (VDTs) - part 9: Requirements for non-keyboard input devices (ISO 9241-9). Technical report, Report Number ISO/TC 159/SC4/WG3 N147, International Organisation for Standardisation (2000)
10. ISO: Evaluation methods for the design of physical input devices - ISO/TC 9241–411: 2012(e). Technical report, Report Number ISO/TS 9241–411:2102(E), International Organisation for Standardisation (2012)
11. MacKenzie, I.S.: Fitts' law as a research and design tool in human-computer interaction. Hum.-Comput. Interact. **7**, 91–139 (1992). https://doi.org/10.1207/s15327051hci0701_3
12. MacKenzie, I.S.: An eye on input: research challenges in using the eye for computer input control. In: Proceedings of the ACM Symposium on Eye Tracking Research and Applications - ETRA 2010, pp. 11–12. ACM, New York (2010). https://doi.org/10.1145/1743666.1743668
13. MacKenzie, I.S.: Human-Computer Interaction: An Empirical Research Perspective. Morgan Kaufmann, Waltham (2013)
14. MacKenzie, I.S.: Fitts' law. In: Norman, K.L., Kirakowski, J. (eds.) Handbook of Human-Computer Interaction, pp. 349–370. Wiley, Hoboken (2018). https://doi.org/10.1002/9781118976005
15. MacKenzie, I.S., Teather, R.J.: FittsTilt: the application of Fitts' law to tilt-based interaction. In: Proceedings of the 7th Nordic Conference on Human-Computer Interaction - NordiCHI 2012, pp. 568–577. ACM, New York (2012). https://doi.org/10.1145/2399016.2399103
16. Magee, J., Felzer, T., MacKenzie, I.S.: Camera mouse + ClickerAID: dwell vs. single-muscle click actuation in mouse-replacement interfaces. In: Antona, M., Stephanidis, C. (eds.) UAHCI 2015. LNCS, vol. 9175, pp. 74–84. Springer, Cham (2015). https://doi.org/10.1007/978-3-319-20678-3_8
17. Magee, J.J., Scott, M.R., Waber, B.N., Betke, M.: EyeKeys: a real-time vision interface based on gaze detection from a low-grade video camera. In: Computer Vision and Pattern Recognition Workshop at CVPRW 2004, pp. 159–159. IEEE, New York (2004). https://doi.org/10.1109/CVPR.2004.340
18. Roig-Maimó, M.F., MacKenzie, I.S., Manresa, C., Varona, J.: Evaluating Fitts' law performance with a non-ISO task. In: Proceedings of the 18th International Conference of the Spanish Human-Computer Interaction Association, pp. 51–58. ACM, New York (2017). https://doi.org/10.1016/j.ijhcs.2017.12.003
19. Roig-Maimó, M.F., Manresa-Yee, C., Varona, J., MacKenzie, I.S.: Evaluation of a mobile head-tracker interface for accessibility. In: Miesenberger, K., Bühler, C., Penaz, P. (eds.) ICCHP 2016. LNCS, vol. 9759, pp. 449–456. Springer, Cham (2016). https://doi.org/10.1007/978-3-319-41267-2_63
20. Soukoreff, R.W., MacKenzie, I.S.: Towards a standard for pointing device evaluation: perspectives on 27 years of Fitts' law research in HCI. Int. J. Hum.-Comput. Stud. **61**, 751–789 (2004). https://doi.org/10.1016/j.ijhcs.2004.09.001

A Time-Discrete Haptic Feedback System for Use by Persons with Lower-Limb Prostheses During Gait

Gabe Kaplan[✉], Troy McDaniel, James Abbas, Ramin Tadayon, and Sethuraman Panchanathan

Arizona State University, Tempe, AZ 85281, USA
gbkaplan@asu.edu

Abstract. Persons with lower-limb amputations experience limited tactile knowledge of their prostheses due to the loss of sensory function from their limb. This sensory deficiency has been shown to contribute to improper gait kinematics and impaired balance. A novel haptic feedback system has been developed to address this problem by providing the user with center of pressure information in real-time. Five piezoresistive force sensors were adhered to an insole corresponding to critical contact points of the foot. A microcontroller used force data from the insole to calculate the center of pressure, and drive four vibrotactile pancake motors worn in a neoprene sleeve on the medial thigh. Center of pressure information was mapped spatially from the plantar surface of the foot to the medial thigh. Human perceptual testing was conducted to determine the efficacy of the proposed haptic display in conveying gait information to the user. Thirteen able-bodied subjects wearing the haptic sleeve were able to identify differences in the speed of step patterns and to classify full or partial patterns with $(92.3 \pm 2.6)\%$ and $(94.9 \pm 2.1)\%$ accuracy respectively. The results suggest that the system was effective in communicating center of pressure information through vibrotactile feedback.

Keywords: Haptic feedback · Lower-limb rehabilitation · Prosthesis

1 Introduction

Skin and muscle receptors in the leg and foot provide able-bodied humans with crucial sensory information for balance and movement control. In lower-limb amputees however, this information is either missing or incomplete. In order to compensate for the partial loss of somatosensory function, people with lower-limb prostheses rely on haptic feedback from the interface between the socket and the residual limb to recognize gait-phase events and monitor balance [1]. This is problematic because the areas of the residual limb that directly interface with the socket are prone to painful skin irritation, which greatly degrades the haptic feedback users receive [2].

Without somatosensory input from the prosthetic limb, persons with lower-limb prostheses experience difficulty monitoring and correcting their center of gravity [3]. This manifests in measurable gait deficiencies such as impaired balance [1], asymmetrical walking patterns [4], larger stride-width than able-bodied humans [4],

M. Antona and C. Stephanidis (Eds.): HCII 2019, LNCS 11573, pp. 250–261, 2019.
https://doi.org/10.1007/978-3-030-23563-5_21

increased risk of falling [5], and decreased balance confidence [6]. These problems not only contribute to the difficulty of using a prosthesis in everyday life, but can result in the development of musculoskeletal diseases over time (*e.g.* osteoarthritis in the unamputated leg) [7].

Augmented sensory feedback systems have been established as useful tools to non-invasively replace missing somatosensory information in persons with lower-limb amputations. These systems generally fall into two categories: therapy aids, and assistive technologies. Therapy aids are used during rehabilitation sessions, and work by informing a patient when a specific biomechanical variable falls outside of a preset range [8–10]. Assistive technologies encode a specific biomechanical variable into stimulus patterns. A hypothetical example of an assistive technology for persons with lower-limb protheses would be a system that communicates ground-reaction forces through patterns of electrical stimulation on the thigh. Assistive feedback systems are intended to enable users to incorporate the feedback into their body control schemes, with the end goal of promoting more physiologically sound gait. These systems function through sensory substitution, a process in which the brain processes information from an alternative receptor in place of information that is normally transmitted through an intact sensory organ [11].

Several studies have found promising results from sensory feedback systems intended for daily use. Crea *et al.* [12] developed a system based on piezoresistive force sensors worn on the plantar surface of the foot, and vibrotactile actuators worn on the thigh. This system provided time-discrete vibrotactile stimuli to communicate gait-phase transitions. The efficacy of this system was verified through testing with able-bodied subjects. Fan *et al.* [13] proposed a similar system that communicated plantar ground-reaction forces through four pneumatically controlled balloon actuators worn around the thigh. Again, the efficacy of this sensory feedback scheme was verified through testing with able-bodied subjects. Lastly, Sabolich *et al.* [3] developed a device that encoded ground-reaction forces into continuous electrical stimuli displayed on the thigh via electrodes worn within the socket of the prosthesis. Results of testing showed significant improvements in weight distribution, balance time, and step length symmetry in both trans-tibial and trans-femoral amputees.

This study presents a novel haptic feedback system designed for daily use. The device consists of a custom force-sensing insole, a belt-worn microcontroller, and a linear array of vibrotactile motors worn on the thigh. The system communicates movement of the anteroposterior center of pressure (COP) to the user in real-time through time-discrete haptic feedback on the medial thigh.

To the best of our knowledge, this device is the first sensory feedback system to communicate movement of the anteroposterior COP through time-discrete vibrotactile stimuli. Time-discrete stimuli were used for two primary reasons: to increase the intelligibility of the stimuli, and to avoid some of the disadvantages of time-continuous stimulation. Previous studies supported that a time-discrete approach helped subjects perceive feedback patterns as *rhythms* [12, 14]. Additionally, time-discrete stimulus patterns have been shown to aid users in integrating the information into their body control scheme [15–17]. Furthermore, time-continuous stimulation can be perceived as aggravating by the user, and is likely to lead to habituation [12, 18]. A time-discrete approach was implemented to avoid these drawbacks.

This paper presents the design of the haptic feedback system, and verification of the stimulus approach through perceptual testing with 13 able-bodied subjects.

2 System Architecture

In this section, we detail the modules of the proposed haptic feedback system. In particular, we describe the hardware portion (sensorized insole, microcontroller, and haptic sleeve) and the feedback algorithm.

2.1 Hardware Design

The haptic feedback system is composed of a sensorized insole, a microcontroller in a belt-mounted enclosure, and a linear array of vibrotactile pancake motors, as detailed in Fig. 1.

The sensorized insole design was inspired by previous work by Rana [19], Al-Baghdadi et al. [20], Howell et al. [21], and Ferenczl et al. [22]. A flat foam insole was fitted with five piezoresistive force sensing resistors (FSRs; Tekscan A201) in positions corresponding to critical contact points on the foot: hallux, first and fifth metatarsal heads, and heel. Each FSR was calibrated to characterize the voltage-force relationship individually. An x, y coordinate system was imposed on the insole, and sensors were assigned coordinates corresponding to their individual positions. In this coordinate system, the x plane corresponded with the coronal plane, and the y plane corresponded with the sagittal plane. Analog voltage readings from the five FSRs were taken as inputs to the microcontroller, and used to calculate the COP.

A one-dimensional array of vibrotactile eccentric rotating mass pancake motors (tactors) was used to apply feedback. The tactors were 10 mm in diameter, and had a height of 3 mm. The tactors were soldered to a WS2801 flexible LED strip, and 5 V was applied to all tactors. The tactor array was driven by the microcontroller. Individual tactors were placed 5.1 cm apart, a greater distance than the 4.3 cm two-point discrimination threshold for touch on the thigh [23]. The vibrotactile array was mounted in a neoprene thigh wrap, and worn so that the array ran proximal to distal on the medial thigh. The medial thigh was chosen to display the stimuli based on previous research, which indicated that it is more sensitive to vibrotactile stimulation than the lateral or anterior thigh [24].

The current system was designed for laboratory testing with able-bodied subjects. The final version of this system will be fully integrated into prosthetic limbs, with FSRs mounted directly on the prosthetic foot and vibrotactile motors embedded in the socket liner. The processing unit may be incorporated into the prosthesis, or worn on the waist. Integrating the system into prostheses will ensure that users' gait will be unencumbered by the device, thereby improving the usability of the system.

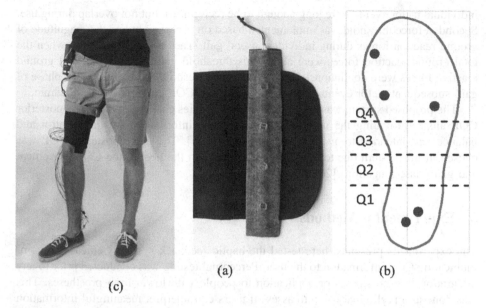

(a)   (b)

(c)

Fig. 1. (a) Haptic feedback system as worn by an able-bodied subject. The system consists of a sensorized insole, wearable processing unit, and thigh-mounted one-dimensional vibrotactile array. (b) Detailed view of the vibrotactile motors mounted within the neoprene thigh wrap. (c) Diagram of the sensorized insole with FSR placement indicated by red dots. Quadrants 1–4 are as labeled. (Color figure online)

2.2 COP Calculation and Feedback Strategy

COP was calculated in the sagittal plane using the following weighted average equation,

$$COP_y = \frac{\sum_{i=1}^{5} F_i y_i}{\sum_{i=1}^{5} F_i} \tag{1}$$

where F_i was the force registered by the ith sensor, and y_i was the y position of the ith sensor, and COP_y was the COP in the anteroposterior plane. Feedback control was handled by the microcontroller. The plantar surface of the insole was divided into four quadrants, as shown in Fig. 1. Each quadrant directly corresponded to one of the four tactors. Movement of the COP during the stance phase was communicated to the user though geographically mapped haptic feedback on the thigh. The feedback scheme during a normal step is as follows: When the COP is in the heel region during the heel-strike, the most proximal tactor fires. As the COP continues to progress through the quadrants from the heel to toe region, corresponding tactors fire. Finally, when the COP enters the final quadrant during toe-off the most distal tactor is fired.

Two strategies were employed to ensure that tactor events only occurred during the stance phase of the gait cycle, and stimulus patterns remained time-discrete. First, individual tactors would fire for 100 ms. This time interval was chosen so that the individual tactor events were long enough to be recognized, but not overlap during use. Second, a force threshold was implemented based on user weight and the magnitude of ground reaction forces during individual users' gait. Tactors were only fired when the total ground reaction forces were above this threshold value. While the total ground reactive forces were continuously above the force threshold during the stance phase of gait, subsequent tactor events were triggered by the COP entering a new quadrant.

The proposed system was designed to communicates changes in the anteroposterior COP alone. Providing the user with only essential information for gait control and balance was intended to avoid sensory overload [25]. The anteroposterior COP was chosen in this system due to the relationship between the heel-to-toe COP movement and gait phase transitions [26, 27].

3   Experimental Methods

The experiments presented here tested the haptic feedback system's efficacy in communicating COP information to the user. Perceptual testing was conducted prior to any integration into prostheses or application to people with lower-limb prostheses. This was done as a preliminary step to assess if users can interpret meaningful information from the haptic patterns.

All tests were conducted with able-bodied subjects who were seated wearing the haptic sleeve. Step patterns were generated based on hypothetical patterns that an end-user would receive during gait. These patterns were applied to the subjects by the experimenter via a PC serial connection with the microcontroller. Response times (RTs) were obtained using a timer integrated into the GUI used by the experimenter. The timer was started when a stimulus pattern was administered and stopped by the experimenter when the subject spoke their response. All subjects participated in both Experiments 1 and 2 in order.

3.1   Experiment 1: Classification of Step Pattern Speed

Subjects. For this experiment, 13 able-bodied subjects (7 male and 6 female) between ages 20 and 32 were recruited to participate. The subjects had no known sensory impairments.

Experimental Setup. The haptic sleeve described in Sect. 2.1 was used. Subjects were seated for the duration of the experiment. The vibrotactile motors in the sleeve fired sequentially and administered 100 ms pulses.

The stimulus patterns in this experiment focused on classification of step pattern speed. Stimulus patterns were developed to represent slow, medium, and fast steps. Stance times were selected to be 850 ms for the Slow pattern, 700 ms for Medium, and 550 ms for Fast based on laboratory testing with one subject. Tactors were activated sequentially at evenly spaced time intervals through each pattern.

Procedure. Subjects were instructed to wear shorts or pants, and the haptic sleeve was worn on top of the clothing. The sleeve was positioned on the subjects so that the vibrotactile motors ran along the medial thigh, and the most distal tactor (tactor #4) was approximately 4 cm above the knee. Subjects were instructed to rest their forearms on the arms of the chair to prevent the perception of additional sensory information from the haptic sleeve through the forearms.

Once situated in the apparatus, subjects were familiarized with the three step patterns (Slow, Medium, and Fast). The experimenter spoke the name of each pattern, then applied each stimulus in order with approximately 5 s between stimuli. This procedure was performed four times for each subject. The training phase consisted of 18 trials, where each pattern was administered 6 times in random order. Subjects were instructed to guess what each pattern was, and feedback was given to correct or confirm their responses. Like the training phase, the testing phase consisted of 18 trials, however no feedback was provided. Correct and incorrect responses were recorded, along with sensorimotor reaction time (time between application of the stimulus and subject response).

3.2   Experiment 2: Classification of Full or Partial Step Patterns

Subjects. The 13 subjects that participated in Experiment 1 also participated in Experiment 2.

Experimental Setup. The apparatus in Experiment 2 was identical to that of Experiment 1, however the stimulus patterns were different. The stimulus patterns in Experiment 2 were as follows: Full, Partial-123, and Partial-234. In each of these patterns, tactors were active for 100 ms periods, and there were 100 ms pauses between each tactor firing. The Full pattern consisted of tactors #1–4. Partial-123 recruited tactors #1–3, and Partial-234 recruited tactors #2–4.

Full and partial step patterns were related to changes in the movement of the COP in normal and abnormal gait. The ability to recognize these differences is the basis of understanding changes in gait phase through the haptic feedback system. For example, if a person were to take a physiologically-sound step on flat ground, the COP would progress from the heel to the toe, resulting in a Full pattern. If a person were to take a step and not complete the toe-off gait phase, this would result in a partial pattern similar to Partial-123. Similarly, if a person were to take a step and land flat footed, omitting the heel strike phase, it would result in a pattern similar to Partial-234.

Procedure. In Experiment 2, subjects were familiarized with the stimulus patterns that were tested. The experimenter spoke aloud each of the stimulus names, then applied each pattern sequentially. This was repeated four times for each subject. The subjects then went through a training phase, which consisted of 18 trials with each stimulus presented 6 times in random order. Feedback was given by the experimenter to confirm or correct subjects' responses. Testing was performed after the training phase, also with 18 stimuli presented. The procedure for testing was identical to the training procedure, however no feedback was given by the experimenter. Correct and incorrect responses, and sensorimotor reaction times were recorded.

4 Results

4.1 Classification of Step Pattern Speeds

The classification accuracy from Experiment 1 was averaged across all subjects and summarized in Fig. 2. The overall classification accuracy for Experiment 1 was $(92.3 \pm 2.6)\%$. A one-way ANOVA was performed on the classification accuracy data for all three patterns, and the results indicated that there were no significant differences in accuracy between any of the stimulus patterns ($F(2) = 1.80$, $p = 0.18$). The distribution of subject responses is shown in the confusion matrix in Table 1. Out of all incorrect responses, only one was not from a directly adjacent speed pattern.

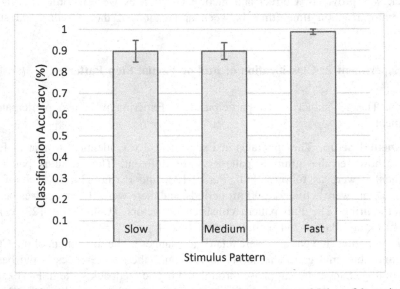

Fig. 2. Classification accuracy for Experiment 1. Error bars represent 95% confidence intervals.

Table 1. Confusion matrix for Experiment 1 showing classification accuracy. Rows represent the stimulus provided, and columns represent subject responses.

	Slow	Medium	Fast
Slow	70	7	1
Medium	3	70	5
Fast	0	1	77

The response time averaged across all stimulus patterns was 1.25 ± 0.04 s. A one-way ANOVA was performed on the data from the three stimulus groups, which indicated that there were significant differences between the RTs of at least 2 of the groups. *Post hoc* paired t-tests with Bonferroni Correction were performed between each of the three stimulus groups. Significant differences ($F(2) = 13.7$, $p = 3.80E-05$)

were found between RTs associated with Fast $(1.0 \pm 0.03$ s) and Medium $(1.41 \pm 0.08$ s) patterns, and Fast and Slow $(1.32 \pm 0.05$ s) patterns.

4.2 Recognition of Full and Partial Patterns

The classification accuracy in Experiment 2 was averaged across all subjects, as summarized in Fig. 3. The overall classification accuracy was $(94.9 \pm 2.1)\%$. There were no significant differences in classification accuracy across the three stimulus patterns, as indicated by the results of a one-way ANOVA $(F(2) = 1.97, p = 0.155)$. The subjects' performance in Experiment 2 is detailed in the confusion matrix in Table 2.

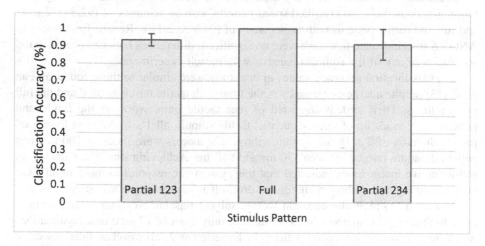

Fig. 3. Classification accuracy for Experiment 2. Error bars represent 95% confidence intervals.

Table 2. Confusion matrix for Full and Partial Pattern testing. Rows represent the stimulus given, and columns represent subject answers.

	Full	Partial 123	Partial 234
Full	78	0	0
Partial 123	0	73	5
Partial 234	1	6	71

RTs averaged across all stimulus patterns were found to be 1.32 ± 0.03 s. A one-way ANOVA was performed on the RT data from all three stimulus patterns which supported that there were no significant differences in RT between any of the patterns $(F(2) = 0.208, p = 0.81)$.

5 Discussion

Artificial sensory feedback systems have great potential for improving the mobility of people with lower-limb prostheses by compensating for lost somatosensory function. For an artificial sensory feedback system to be useful in dynamic situations such as gait, it is essential that the feedback patterns can be classified accurately and rapidly by the user [28]. Data from the two perceptual tests indicated that potential feedback patterns from the proposed system can be classified accurately by the subjects. Furthermore, results showed that RTs were relatively quick and consistent. Low RTs suggest that classification was not cognitively demanding, but further work must be done to verify this.

Classification accuracy was very high in both experiments, despite relatively brief training periods. Subjects classified speed patterns with an accuracy of $(92.3 \pm 2.6)\%$, and full and partial patterns with an accuracy of $(94.9 \pm 2.1)\%$. Results from one-way ANOVA tests indicated that there were no significant differences between classification accuracies of any of the stimulus patterns within each experiment.

The classification accuracy values in our study were similar to those found by Fan et al. [13] despite a large discrepancy in the time each tactile unit was fired and the full pattern times. Their system consisted of four tactile units worn on the thigh, and patterns were made up of three sequential tactile stimuli, all 1 s in duration for a total pattern duration of 3 s. In our system, vibrotactile motors were fired for 100 ms, and pattern durations ranged between 550 ms and 850 ms. Achieving similar accuracy with shorter active motor times indicated that this system can respond to more rapid COP changes during gait without a significant drop-off in classification accuracy.

In addition to high classification values, subject misclassifications were generally close to the correct responses. In Experiment 1, only 1 out of 17 total misclassifications was from a non-adjacent speed pattern. In Experiment 2, all misclassifications came when Partial patterns were administered, and only 1 of the 12 total misclassifications was incorrectly identified as the Full pattern. This indicated that when the subjects failed to correctly classify the stimulus patterns, their responses tended to be close to the correct responses. When the complete system is implemented, users will encounter stimulus patterns with much more subtle differences than the ones tested here. The ability to identify exactly how the COP is progressing in all cases is ideal, but being able to roughly interpret the movement of the COP and understand how it compares to stimulus patterns from other steps may be sufficient. Future perceptual testing will examine the minimum perceivable difference between stimulus patterns to find how similar patterns can be while still remaining consistently classifiable.

Subject response times were relatively low across all experiments. The overall RTs were 1.25 ± 0.04 s for Experiment 1, and 1.32 ± 0.03 s for Experiment 2. Within Experiment 1, there were significant differences between the average RTs to Fast and Medium, and Fast and Slow patterns. In both cases, the Fast RT was significantly lower than the other RT. Factors that could have lead to this discrepancy were pattern length, and the method of recording RT values. Developing a system in which the subject responded and stopped the timer by pressing a button may have decreased the observed

RTs, as response modality has been demonstrated to have a significant effect on observed RT in previous studies [28].

This initial study featured a small subject pool, and varied tactor contact due to differences in subjects' clothing and thigh length. Despite these factors, the system was shown to perform well, as evidenced by the high classification accuracies and low RTs observed across all subjects in both experiments. These initial results supported that the proposed system was able to sufficiently communicate data about step speed and full or partial steps to the user, and that this feedback strategy may be effective in dynamic conditions such as gait.

6 Conclusion

In this paper, we have presented a novel haptic feedback system to deliver antero-posterior COP information to persons with lower-limb prostheses in the goal of substituting for missing somatosensory information useful for gait. In addition to the description of the hardware and stimulus algorithm, perceptual experiments were performed with able-bodied subjects to determine the efficacy of the proposed feedback system in communicating COP information to the user. From the results of these experiments, it was ascertained that the subjects could accurately recognize differences in the speed of step patterns, and between full and partial step patterns. Future studies will proceed in two primary directions: testing the efficacy of the haptic feedback system in communicating COP data based on challenging conditions for gait (uneven terrain, sloped surfaces, stairs), and testing subjects' ability to classify step patterns during gait.

Acknowledgements. The authors thank Arizona State University and the National Science Foundation for their funding support. This material is partially based upon work supported by the National Science Foundation under Grant No. 1828010.

References

1. Lamoth, C.J.C., Ainsworth, E., Polomski, W., Houdijk, H.: Variability and stability analysis of walking of transfemoral amputees. Med. Eng. Phys. **32**, 1009–1014 (2010). https://doi.org/10.1016/j.medengphy.2010.07.001
2. Klute, G.K., Kantor, C., Darrouzet, C., Wild, H., Wilkinson, S., Iveljic, S., Creasey, G.: Lower-limb amputee needs assessment using multistakeholder focus-group approach. J. Rehabil. Res. Dev. **46**, 293 (2009). https://doi.org/10.1682/JRRD.2008.02.0031
3. Sabolich, J.A., Ortega, G.M.: Sense of feel for lower-limb amputees. JPO J. Prosthet. Orthot. **6**, 36–41 (1994). https://doi.org/10.1097/00008526-199404000-00003
4. Jaegers, S.M.H.J., Arendzen, J.H., de Jongh, H.J.: Prosthetic gait of unilateral transfemoral amputees: a kinematic study. Arch. Phys. Med. Rehabil. **76**, 736–743 (1995). https://doi.org/10.1016/S0003-9993(95)80528-1
5. Kulkarni, J., Toole, C., Hirons, R., Wright, S., Morris, J.: Falls in patients with lower limb amputations: prevalence and contributing factors. Physiotherapy **82**, 130–136 (1996). https://doi.org/10.1016/S0031-9406(05)66968-4

6. Miller, W.C.: Speechley MR balance confidence among people with lower-limb amputations. Phys. Ther. (2002). https://doi.org/10.1093/ptj/82.9.856

7. Burke, M.J., Roman, V., Wright, V.: Bone and joint changes in lower limb amputees. Ann. Rheum. Dis. **37**, 252–254 (1978). https://doi.org/10.1136/ard.37.3.252

8. Wu, S.W., et al.: Torso-based tactile feedback system for patients with balance disorders. In: 2010 IEEE Haptics Symposium, pp. 359–362 (2010). https://doi.org/10.1109/haptic.2010.5444630

9. Chow, D.H.K., Cheng, C.T.K.: Quantitative analysis of the effects of audio biofeedback on weight-bearing characteristics of persons with transtibial amputation during early prosthetic ambulation. J. Rehabil. Res. Dev. **37**, 255–260 (2000)

10. Redd, C.B., Member, S., Bamberg, S.J.M., Member, S.: A wireless sensory feedback device for real-time gait feedback and training. IEEE/ASME Trans. Mechatron. **17**, 425–433 (2012). https://doi.org/10.1109/TMECH.2012.2189014

11. D'Alonzo, M., Cipriani, C.: Vibrotactile sensory substitution elicits feeling of ownership of an alien hand. PLoS One **7** (2012). https://doi.org/10.1371/journal.pone.0050756

12. Crea, S., Cipriani, C., Donati, M., Carrozza, M.C., Vitiello, N.: Providing time-discrete gait information by wearable feedback apparatus for lower-limb amputees: usability and functional validation. IEEE Trans. Neural Syst. Rehabil. Eng. **23**, 250–257 (2015). https://doi.org/10.1109/TNSRE.2014.2365548

13. Fan, R., et al.: A haptic feedback system for lower-limb prostheses. IEEE Trans. Neural Syst. Rehabil. Eng. **16**, 270–277 (2008). https://doi.org/10.1109/TNSRE.2008.920075

14. Roerdink, M., Lamoth, C.J., Kwakkel, G., van Wieringen, P.C., Beek, P.J.: Gait coordination after stroke: benefits of acoustically paced treadmill walking. Phys. Ther. **87**, 1009–1022 (2007). https://doi.org/10.2522/ptj.20050394

15. Cipriani, C., Segil, J.L., Clemente, F., Weir, R.F., Edin, B.: Humans can integrate feedback of discrete events in their sensorimotor control of a robotic hand. In: Proceedings of the 8th International conference on Electronics, Computers and Artificial Intelligence ECAI 2016, vol. 232, pp. 3421–3429 (2017). https://doi.org/10.1109/ecai.2016.7861192

16. Edin, B.B., Ascari, L., Beccai, L., Roccella, S., Cabibihan, J.J., Carrozza, M.C.: Bio-inspired sensorization of a biomechatronic robot hand for the grasp-and-lift task. Brain Res. Bull. **75**, 785–795 (2008). https://doi.org/10.1016/j.brainresbull.2008.01.017

17. Johansson, R.S., Edin, B.B.: Predictive feed-forward sensory control during grasping and manipulation in man. Biomed. Res. - Tokyo **14**, 95 (1993). https://doi.org/10.1006/abbi.1994.1154

18. Hahn, J.F.: Vibrotactile adaptation and recovery measured by two methods. J. Exp. Psychol. **71**, 655–658 (1966)

19. Rana, N.K.: Application of force sensing resistor (FSR) in design of pressure scanning system for plantar pressure measurement. In: 2009 International Conference on Computer and Electrical Engineering, ICCEE 2009, vol. 2, pp. 678–685 (2009). https://doi.org/10.1109/iccee.2009.234

20. AL-Baghdadi, J.A.A., Chong, A.K., Milburn, P.D.: Fabrication and testing of a low-cost foot pressure sensing system. In: Proceedings of the 2nd International Conference on Industrial Application Engineering 2015, pp. 246–253 (2015). https://doi.org/10.12792/iciae2015.046

21. Howell, A.M., Kobayashi, T., Hayes, H.A., Foreman, K.B., Bamberg, S.J.M.: Kinetic gait analysis using a low-cost insole. IEEE Trans. Biomed. Eng. **60**, 3284–3290 (2013). https://doi.org/10.1109/TBME.2013.2250972

22. Ferenczi, D.C., Jin, Z., Chizeckf, H.J., Boulevard, E.: Estimation of center-of-pressure during gait s i, pp. 3–4 (1993)

23. Bach-y-Rita, P., Kercel, W.S.: Sensory substitution and the human-machine interface. Trends Cogn. Sci. **7**, 541–546 (2003)
24. Wentink E, et al.: Vibrotactile stimulation of the upper leg: effects of location, stimulation method and habituation. In: IEEE Engineering in Medicine and Biology Society, pp. 1668–1671 (2011). ROBAR project-Evaluation and implementation of arm support in rehab View project Reflex-leg View project. https://doi.org/10.1109/iembs.2011.6090480
25. Kristjánsson, Á., et al.: Designing sensory-substitution devices: principles, pitfalls and potential 1. Restor. Neurol. Neurosci. **34**, 769–787 (2016). https://doi.org/10.3233/RNN-160647
26. Grundy, M., Tosh, P.A., McLeish, R.D., Smidt, L.: An investigation of the centres of pressure under the foot while walking. J. Bone Jt. Surg. Br. **57**, 98–103 (1975)
27. Kawamura, K., Tokuhiro, A., Takechi, H.: Gait analysis of slope walking: a study on step length, stride width, time factors and deviation in the center of pressure. Acta Med. Okayama **45**, 179–184 (1991). https://doi.org/10.18926/AMO/32212
28. Sharma, A., Torres-Moreno, R., Zabjek, K., Andrysek, J.: Toward an artificial sensory feedback system for prosthetic mobility rehabilitation: examination of sensorimotor responses. J. Rehabil. Res. Dev. **51**, 907–917 (2014). https://doi.org/10.1682/JRRD.2013.07.0164

Quali-Quantitative Review of the Use of Multimodal Interfaces for Cognitive Enhancement in People Who Are Blind

Lana Mesquita[1]([⊠]) and Jaime Sánchez[2]

[1] Universidade Federal do Ceará (UFC), Master and Doutorate in Computer Science (MDCC), Fortaleza, CE, Brazil
lanabeatriz.mesquita@gmail.com
[2] Department of Computer Science, Universidad de Chile, Santiago, Chile

Abstract. Visual disability has a major impact on people's quality of life. Although there are many technologies to assist people who are blind, most of them do not necessarily guarantee the effectiveness of the intended use. As part of research developed at the University of Chile since 1996, we investigated the interfaces for people who are blind regarding a gap in cognitive impact. We first performed a systematic literature review concerning the cognitive impact evaluation of multimodal interfaces for people who are blind. Based on the papers retrieved from the systematic review, a high diversity of experiments was found. Some of them do not present the data results clearly and do not apply a statistical method to guarantee the results. We conclude that there is a need to better plan and present data from experiments on technologies for cognition of people who are blind. Moreover, we also performed a Grounded Theory qualitative-based data analysis to complement and enrich the systematic review results.

Keywords: Cognitive evaluation · Impact evaluation · Multimodal interfaces · Blind people

1 Introduction

In 2018, the "World Health Organization" [24] informs that 1.3 billion people are estimated to have visual disabilities worldwide, of which 36 million are blind. In face of this huge amount, visual disability has a significant impact on the quality of life of people, including their ability to study, work and to develop personal relationships. In this aspect, technologies have been designed to assist people

Supported by Research Project of the National Fund for Science and Technology, Fondecyt-Chile #1150898, "Knowing with Multimodal Interfaces in Blind Learner"; as well as being part of the Basal Funds for Centers of Excellence, FB0003, CONICYT-Chile; and Fundação Cearense de Apoio ao Desenvolvimento (Funcap), that financed the master scholarship of author Lana Mesquita.

M. Antona and C. Stephanidis (Eds.): HCII 2019, LNCS 11573, pp. 262–281, 2019.
https://doi.org/10.1007/978-3-030-23563-5_22

who are blind to support daily life activities. These technologies work as aids to facilitate their independence, autonomy, and safety. Thus, such technologies help to improve the quality of life of people with visual disabilities and could stimulate and develop several skills, such as cognitive skills.

Even though there is technology specialized for people who are blind (e.g., serious game [3]), they are still using applications that are similar to older applications for the sighted population. For example, Battleship was one of the earliest games designed as a computer game with its release in 1979 [7]. AudioBattle-Ship, a version for both blind children and sighted playing together came in 2004 [18]. In general, the people who are blind have particular human-computer interaction needs, and the user interface should be suitable for them [19].

Considering these aspects, there are many efforts to develop accessible multimodal interfaces for people with visual disabilities, especially in multimodal games [3]. Despite this effort and in contrast to the visual interface evolution of games and applications for sighted people, interfaces for people who are blind explore other ways to interact with the user. In general, technologies for people with visual disabilities combine different sources of perceptual inputs and outputs. Although multimodal interfaces could help to improve the learning skills of people with visual disabilities, most of these technologies have been not wholly validated. Mostly, they remain in the prototype phase without being integrated into the people's everyday life [6].

This paper is a continuation of the research developed and reported in the proposal [13] and of an initial version of the systematic review [14] yet without including the snowballing process.

2   Background

The impact evaluation of software could use several evidence-based methods. This work focuses on the experiments. To treat the cognitive impact evaluation is necessary to comprehend the experiment design in both cognitive psychology and software engineering areas. The two main theoretical background bibliographies used in this work are the books "Cognitive Psychology" [23] and "Experimentation in Software Engineering" [26]. In this multidisciplinary context, the next two subsections are dedicated to explain each point of view and highlight the differences.

2.1   Design Experiment in Software Engineering

The experiment process includes several steps: Scoping; Planning; Operation; Analysis and interpretation; Presentation and package [26]. This process provides a high level of control, which uses a formal, rigorous and controlled investigation. These steps were used to analyze the data of experiments in the methodology proposed. The main concepts involved in the experiment, shown in Table 1, are used to understand how the cognitive impact is evaluated.

Table 1. The main concepts of experimental design [26]

Concepts	Description
Measure	A mapping from the attribute of an entity to a measurement value
Instrumentation	The instruments for an experiment are of three types, namely objects, guidelines and measurement instruments
Dependent variables	The dependent variables are those we want to see the effect; the independent variables are those controlled and manipulated
Independent variables	The independent variables are those controlled and manipulated
Factors	The independent variables which the experiment changes to verify the effect. Treatment is one value of a factor

2.2 Design Experiment in Cognitive Psychology

The research methods in cognitive psychology focus on describing particular cognitive phenomena such as how people preconceive notions regarding what they may find while gathering data [23]. The characteristics used in this work of controlled experiments to explore cognitive phenomena are based on Sternberg and Sternberg [23].

Regarding the experiment concepts, the variables of the experiment are (i) independent variables, that are individually manipulated, or carefully regulated, by the experimenter; or (ii) dependent variables, that are outcome responses, the values of which depend on how one or more independent variables influence or affect the participants in the experiment [23]. This literature also presents the concepts of (iii) irrelevant variables, which affect the outcomes (dependent variable) when manipulated; (iv) control variables, which are held constant; and (v) confounding variables, which affect the dependent variables without be controlled or manipulated and should be avoided.

Independent and dependent variables must be chosen with great care, because what is learned from an experiment will depend almost exclusively on the variables one chooses to isolate from the often complex behavior one is observing [23]. The authors suggest two dependent variables that are used in cognitive psychological research: percent correct (or its additive inverse, error rate) and reaction time. These measures are popular because they can tell the investigator, respectively, the accuracy and speed of mental processing [23].

Among the myriad possibilities for independent variables are characteristics of the situation, of the task, or of the participants [23]. For example, characteristics of the situation may involve the presence versus the absence of particular stimuli or hints during a problem-solving task, as virtual versus real navigation [11]. Characteristics of the task may involve reading versus listening to a series of words and then responding to comprehension questions. Characteristics of the participants may include age differences, differences in educational status, or differences based on test scores. Characteristics of the participant are not easily manipulated experimentally due to the ethical regulation.

3   Systematic Literature Review (SLR)

In contrast to an ad-hoc literature review, the Systematic Literature Review (SLR) is a methodologically rigorous analysis and study of research results. To achieve our goal, the main research question for this first part of the study was: How is the cognitive impact evaluated on multimodal interfaces for people who are blind? For a better understanding, a second goal question was formulated "What are the challenges regarding the impact evaluation on multimodal interfaces for people who are blind on this scenario?".

The process of a SLR includes three main phases: planning the review; conducting the review and reporting the review [10]. Each of these stages has a qualitative methodological design that aims to offer a better specification and evolution in the development of the SLR. Figure 1 presents the SLR process adopted in this study by using a UML language[1] for Activity Diagram. The process suits the guidelines from [10]. The next subsections describe the planning (the study selection criteria, the research sources selected) and the conducting phase (the search process, the data extraction form fields and the studies quality evaluation). The entire process was stored in an excel worksheet available online[2].

During all the SLR process, we used the tool StArt ("StArt", 2016) and the software Microsoft Excel[3] as a support to create the protocol, apply the filters, select the papers and show the results. We organized all references on software Mendeley[4]. As the papers retrieved from PubMed Central are in MEDLINE format, we developed the tool Medline2bibtex[5]. It works as a parser to permit the list to be read by both StArt and Mendeley.

3.1   Planning: Definition of the Protocol

In the planning phase, we defined a review protocol that specifies the research question being addressed and the methods that will be used to perform the review [10].

Sources Selection. The first suggested digital libraries as sources are: *ACM Digital Library*; *Engineering Village*; *IEEE Xplore*; *Scopus*; *Science Direct*; *Springer Link*; *PubMed*; *Web of Science*; *Google Scholar*, that includes the leading conferences and journals from Computer Science.

---

[1] UML - http://www.uml.org/.
[2] https://www.dropbox.com/s/4wiiqnwqyd3cquw/SystematicLiteratureReviewv5.xlsm?dl=0.
[3] Microsoft Excel - https://products.office.com/pt-br/excel.
[4] Mendeley - https://www.mendeley.com/.
[5] Medline2Bibtex - https://github.com/lanabia/Medline2Bibtex.

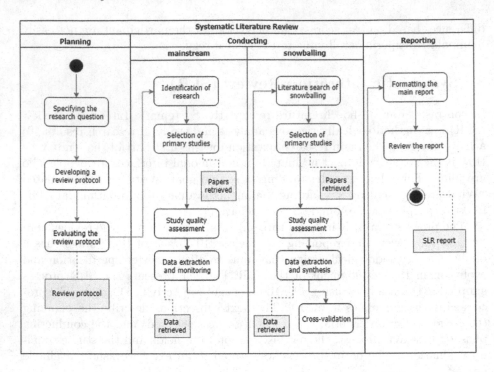

Fig. 1. Systematic literature review process (based on [4])

The bases chosen are the primary research bases for scientific articles in the research area or the bases that index them. Since Scopus, which index ACM Digital Library and IEEE Xplore, is based on the same database of ScienceDirect. As the research includes visual disabilities, we expected to find some works in PubMed database. The final list of sources to SLR is *Scopus, Springer, Web of Science* and *PubMed Central.* We also researched in the *Journal of Visual Impairment & Blindness* that addresses a variety of topics related to visual impairment. The research in this journal changes a little the string applied to get their proceedings, that are in the scientific base *PubMed Central.*

Studies Initial Selection. The initial selection occurs by applying a search string into in each source. Before the final version of the string, we simulated many other strings in searching the best solution and results with the principal papers of the area. The final string defined was the string bellow.

```
(
    (
    (evaluat* OR assessment)
    AND ((cognitive OR psychomotor OR physical OR emotional)
            (NEAR/10 (impact or effectiveness)))
    )
    AND (design OR development)
    AND (blind* OR "visually impaired" OR "visial disability")
    AND (multimodal OR hapatic OR audio OR auditory OR vibrotactile)
```

```
     AND interface
 )
```

Study Selection Criteria. The inclusion and exclusion criteria are according to the goal of the SLR. Table 2 presents the inclusion *(I)* and exclusion *(E)* selection criteria. To be accepted, a scientific paper must cover all inclusion criteria and none exclusion criterion.

Table 2. Inclusion and exclusion criteria

Code	Inclusion criteria	Code	Exclusion criteria
I.01	The study has the technology for people who are blind or visually impaired	E.01	Title and abstract out the search criteria (I.02, I.03, I.04)
I.02	The study evaluates the technology by using some approach that involves the user	E.02	Entire text out the search criteria (I.02, I.03, I.04)
I.03	The study evaluates the technology by using a method to evaluate the cognitive impact	E.03	The document is a book, a congress' abstract, an extended abstract, a poster, an oral communication, proceedings of a conference, a seminar, a research plenary, a dictionary or an encyclopedia
		E.04	The study must be published after 1998 and written in English

The technologies defined in the *I.01* criterion include mobile application, computer software, IoT systems, virtual environments or a video game with multimodal interfaces. Also, we accepted technologies that are not specifically for people who are blind or visually impaired with the goal of expanding the results, but the studies present the technology focused on users with visual disabilities. We excluded from all technologies that uses Sensory Substitution Devices (SSD) [16], which substitutes a sense as sensory augmentation, bionic eyes, retinal visual prosthesis, cortical implants and others. This definition is important to plan the methodology proposed and to delimit the focus.

We defined the studies type in the *E.03*. This criterion excludes all studies type different from primary studies that present technology for people who are blind and its evaluation. We accepted articles, conference papers, short papers, and book sections. This criterion includes documents that have the minimum information to understand the evaluation. We did not cover books because the information is dispersed inside them.

The $E.04$ criterion defines the scientific articles must be in English, because it is the mandatory language for the main events and scientific journals in the search area. And they must be published between 1st January 1998 and 2nd August 2017. The year 1998 was a milestone due to the paper [12] which concerns with 3D acoustic interfaces for blind children and is the first study known.

3.2 Conducting

Search Process. The conducting phase starts with the initial search in the scientific bases proposed. The string was applied on the metadata of papers, which includes abstract, index terms, and bibliographic citation data (such as document title, publication title, etc.).

The first filter excluded papers duplicated and document types out the scope due their format ($E.03$). The second filter identified which paper is in and out the scope by reading their titles and abstracts ($E.01$). A lot of papers were excluded in the first filter because the scientific base PubMed Central (PMC) brings a lot of medical papers focused on disease effectiveness and specific medical statements. Even though the area of this study is computer science, we decided to insert the PMC in the bases' list due to the nature of the subject.

Next, in the third filter, we evaluated each retrieved paper in its entirety ($E.02$). If necessary, besides the entire text, we searched more about the technologies and processes described, as project and institutional websites, videos, newspaper articles and others. In the fourth filter, we searched and compared the experiments to find the same experiment described in two or more papers. This occurs when the experiment is not the primary goal of the paper, and more than one paper cites the experiment methodology and results according to the paper goal.

Once we have chosen the select papers, we extracted all data required to achieve the objective. The organization of the data generates data synthesis. The main reason for withdrawing papers in the last filter was the evaluation performed is out the search and at most times related to the system performance, e.g., sensor performance evaluation.

In face the final papers selected, we make a snowballing approach for an opportunistic search for other relevant papers. Snowballing is a manual search using the reference and citations (known as backward and forward snowballing) list of a paper or the citations to the paper select aiming to identify additional papers [25]. Thus, we performed other interaction among the forward and backward snowballing list to select the papers reading the title firstly and abstract and after the whole text.

In the snowballing process, the acceptance rate was high compared to the mainstream search. This acceptance rate is due to that the articles are closely related to the theme of tools for people who are blind and usually use similar processes of validation building.

Figure 2 resumes the conducting phase results. The spreadsheet[6] details the complete list of selected articles and the process.

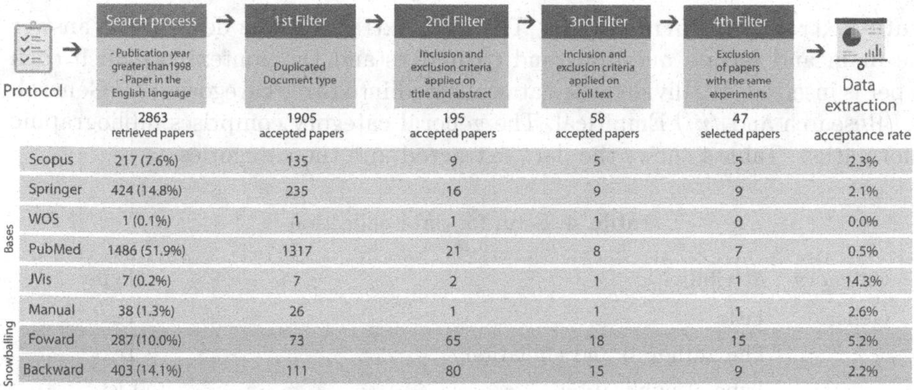

Protocol	Search process → - Publication year greater than 1998 - Paper in the English language 2863 retrieved papers	1st Filter → Duplicated Document type 1905 accepted papers	2nd Filter → Inclusion and exclusion criteria applied on title and abstract 195 accepted papers	3nd Filter → Inclusion and exclusion criteria applied on full text 58 accepted papers	4th Filter → Exclusion of papers with the same experiments 47 selected papers	Data extraction acceptance rate
Scopus	217 (7.6%)	135	9	5	5	2.3%
Springer	424 (14.8%)	235	16	9	9	2.1%
WOS	1 (0.1%)	1	1	1	0	0.0%
PubMed	1486 (51.9%)	1317	21	8	7	0.5%
JVis	7 (0.2%)	7	2	1	1	14.3%
Manual	38 (1.3%)	26	1	1	1	2.6%
Foward	287 (10.0%)	73	65	18	15	5.2%
Backward	403 (14.1%)	111	80	15	9	2.2%

Fig. 2. Filters in the systematic review process

Studies Quality Evaluation. Each paper was qualified into the following quality checklist to assess the studies and measure the weight of each study found on the results (Table 3).

Table 3. Quality assessment form

	Quality question
Q1	Was it possible to extract all data regarding the data the Key features in multimodal interfaces? (-0.1 pt per missing input; min value: -4.1 pts)
Q2	Is there a complete description of how the evaluation has been applied? (1.0 pt per complete input in Empirical category; max value: 8,0 pts)
Q3	Are the groups of participants in the experiment randomly assigned? (0.5 pt)
Q4	Is the description of the impact evaluation understandable? (0.5 pt)
Q5	Does the article present different evaluation types of the proposal? (1 pt)
Q6	How many experiments does the paper present? (0.5 pt per experiment, if more than two experiments)
Q7	Is the goal of the evaluation cleared defined? (0.5 pt)
Q8	Is the hypothesis (null and alternative) explicitly described in the study? (0,5 pt)

---

[6] https://www.dropbox.com/s/lp7ehmxqvwem8ln/Systematic%20Review%20-%20Mesquita%20L.%20-%20HCII%202019.xlsm?dl=0.

The results of quality form range from 52% to 96% of expected quality, with average of 74%. The expected quality means the maximum points in each question and none misses in the first question.

Data Extraction Form Fields. The data extraction was designed to answer the main and second questions and to understand the context in which each paper is inserted. We divided the data collected into three categories: *(i)* General, *(ii)* Research and *(iii)* Empirical. The general category comprises bibliographic information. Table 4 shows the data extracted and the categories.

Table 4. Form for data collection

Category	Attribute	Type
General	Title	Text
	The author(s) and affiliation	Text
	Type of publication	List
	Year of publication	Number
	Research type	List
	Empirical Methods classification	List
	Technologies the paper presents for people who are blind	Text
Research	Key features in multimodal interface	Text
	Other evaluations	Text
Empirical	Sample	Text
	Instruments	Text
	Variables in the experimental design	Text
	Statistical methods used	Text
	Tasks defined	Text
	Investigation cost	Text
	Ethical concepts treated	Text

The general category comprises bibliographic information and classifies the papers. We classified the experiment of a scientific paper into two classifications: research type, based on [15], which can be validation research, evaluation research, solution research, philosophical research, opinion paper or experience papers. The empirical method classification is based on the classification of [1]. This classification aims to confirm the papers retrieved are in the search focus, since we look for papers in the "evaluation research" type and that are "Experiments". Although, we retrieved one paper as a Case Study. For this one, we consider only the experiment data. All papers retrieved are in Evaluation Research category.

The research category comprises the classification that fits the technology presented in the key features of multimodal interfaces for the cognition of people who are blind [3]. This classification is divided into 4-dimension: Interface,

Interaction, Cognition, and Evaluation; and it is applied to video games and virtual environments. For our purpose, we classified only in the interaction, interface and cognition dimensions. We covered in the classification more than video games and virtual environments, since we also found these features present in the technologies selected. The research category also shows other strategies used to evaluate, as usability evaluation.

The empirical category provides information specifically about how the empirical method that evaluates the impact of the cognitive impact. This category will be explored in the results section.

4   Grounded Theory

The Grounded Theory [5] analysis was performed to enhance and strengthen the findings of the SLR regarding cognitive evaluation concepts used in the context of this search: multimodal interfaces for people who are blind. The data gathered from the literature became the population used in the analysis. We aimed to analyze in the deepest level of generating theory but as the authors state: knowledge and understanding take many forms [2]. The Grounded Theory analysis was supported by the MAXQDA12[7] tool in all process, which is composed of the following steps: planning, data collection, coding and reporting results.

In Grounded Theory analysis, we used the method to analyze in detail four items from Empirical category: Hypothesis, Variables, Measures, and Tasks. The final map produced by the Selective Coding phase represents the main idea of the cognitive evaluation in the context of this study and connect all elements.

The planning step aims to identify the area of interest and the research question that will drive the work. In our case, the area of interest is the Cognitive Impact Evaluation. In the current research, the Grounded Theory analysis suits well some characteristics of cognitive impact experiment extracted from the systematic review, as it assists in the interpretation and clarification of the results found.

In the data collection step, we prepared an Excel spreadsheet with data from the SLR. We import the data extraction form from each experiment into the MAXQDA. In this way, each experiment is a document in the MAXQDA analysis. All data from experiments are imported as variables (59 variables), that could be used to quantify your qualitative analysis results or to add additional information to pieces of data. These one are already imported as excerpts coded as the empirical categories. The data retrieved is organized and modified from the paper to answer the data form. Although, in some experiments, we added some excerpt from the paper to facilitate the coding step on that experiment.

The coding step, the heart of the Grounded Theory analysis, is composed of (i) Open Coding, (ii) Axial Coding, and (iii) Selective Coding. On this step, we extract concepts from raw data and relate them to each other until reaching a core concept [27]. In our case, we pursued to extract and relate some concepts

---

[7] MAXQDA - https://www.maxqda.com/.

the experiments of cognitive impact evaluation. After considering all possible meanings and examining the context carefully produced 91 codes and 808 tags. Figure 3 shows the 6 top categories of the codes.

Explicit Hypothesis	14
Explicit Variable	11
Groups	44
Measures	231
Tasks	152
Variables	356

Fig. 3. Open coding

In the open coding, constant comparative analysis is a regular procedure to execute it. Whenever coming across another excerpt that seemed to talk about the same concept or shared a common attribute, these were grouped together into the same code.

Axial Coding is stepping to relate concepts to each other [27]. With this step, the fractions of data from the open coding can be reassembled and organized into the categories and subcategories with their descriptions, properties or dimensions. The Axial Code produced 95 codes and 603 tags.

The Selective Coding merge all concepts grounded in the process and others captured in the SLR. As a result, we produced maps of concepts and a map to ground the theory of the cognitive impact evaluation in multimodal interfaces for people who are blind or visually impaired.

5   General Results

Despite we considered as the final result of the SLR the papers selected in the fourth filter, in this analysis of general data, we consider the 58 papers selected in the third filter. Regarding the empirical method, almost all papers use an Experiment as an empirical method (21 papers). Only 1 paper present a Case Study as an evaluation, that justifies the choice due to the small number of participants [21]. This information was encountered in the text explicit or it was deducted from the details presented.

Among the papers selected in third filter, 30 are conference papers, 27 are scientific journals, and one is book sections. The main conferences where the papers selected are published are ACM SIGACCES with 6 papers, ICDVRAT with 5 papers and UAHCI with 3 papers. The leading journal with three papers retrieved is the International Journal on Disability and Human Development.

6   Research Results

About other Strategies of the evaluation applied, many papers apply another approach to evaluate other criteria not covered in our study. Usability evaluation is the most used assessment besides impact evaluation.

6.1   Classification

The result of this classification is shown in Fig. 4. As we can see, the Keyboard is the main mode to interact with the technologies analyzed. It is an instrument of interaction already common as input device [18]. The keyboard does not generate more complexity and expenses, like the Novint Falcon device, that is used in [20]. The Novint Falcon is one of the Force Feedback Device that promotes a Tactile and Kinesthetic Feedback. Mouse and Natural language are less used. The mouse is replaced by buttons or other specific devices for a better interaction, as shown in [22]. The most common Feedback is the Sonorous and the main Audio Interface used is the Iconic Sound, which are sounds associated with each available object and action in the environment [20]. These characteristics are important to understand which kind of interfaces are assessed and how the impact is evaluated on them.

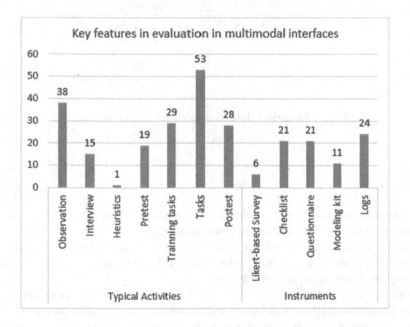

Fig. 4. Key features in multimodal interfaces (evaluation)

7   Empirical Data Results

The Empirical category aims to answer the research question of the SLR in detail. The data retrieved ground the comprehension of the main research question. Thus, we searched the main aspects concerning the empirical method applied. Notice that in this category, we analyze the 47 papers selected on the fourth filter that brings 52 experiments process since five papers have two different experiments. In addition to the SLR (SLR) method, we used the Grounded Theory method to analyze the empirical data. With this combination of methods, we constructed a concept based on cognitive assessment data. Figure 5 shows the data were chosen to be retrieved and which method had been primarily used.

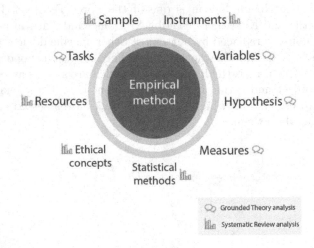

Fig. 5. Empirical data extracted from experiments

7.1   Sample

In the sample data analysis, we looked for the sampling strategy and the samples description, including the number of participants (per condition) and the kind of participants (e.g., computer science students). The sample information retrieved from each experiment are divided into 4 categories: number of users, blindness level, gender distribution and age range[8].

From the number of users to the onset age of blindness, there are many sample combinations in the selected experiments. The sample choice includes previous experience required to do the task and others characteristic controlled in the experiment, like disabilities, the onset of blindness, the etiology of a visual

---

[8] Code Map of sample - https://www.dropbox.com/s/ulijj6x74u2lurc/sample_information.png?dl=0.

impairment or the presence of another disability. The quantity of users varies as shown in Fig. 6. Most of the experiments (22) are applied to 6 to 10 users. One experiment does not inform the number of users. One experiment [8] expresses the small number of the sample (4 blind participants) as a limitation/implication of the research. Moreover, the paper talks about the limited access to people who are blind, and those who use a smartphone are even more difficult to find.

There is no a guideline about the number of users on an experiment sample in the context of this research. The tradeoff between limitations and quality of the research should define the number of users. The planning phase must propose a plausible solution with the group of participants that are sought to the experiment have a reasoned result, preferably based on statistical data.

The rage of age varies a lot among the experiments. To found a pattern, we adopted age groups based on indications of the Brazilian Child and Adolescent Statute (ECA) and the Brazilian Statute of the Elderly, which defines four groups: child (under age12), adolescent (between age 12 and under 21) , adult (between age 21 and 59) and elderly (over 60 years old). Figure 6 also shows the distribution among age groups.

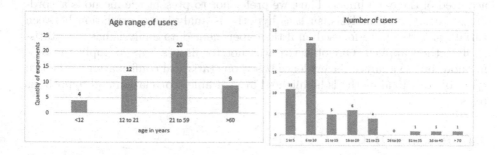

Fig. 6. Age distribution by groups

The gender distribution in samples is equilibrated in most cases in the 36 papers that describes this information. The mean of gender proportion, among all samples, is 47% for women and 53% for men.

The distribution between the blindness level is varied. There are experiments where the sample is all formed by people who are blind, and there are samples formed only by people who are blindfolded. The blindness level distribution between the samples is an essential information in the context of the experiments studied here. Although 9 experiments do not inform the proportions of blind, low vision, sighted and blindfolded.

7.2 Instruments

The instruments used in the evaluation had the objective to identify some user ability controlled on the Experiment (as an independent variable), e.g., the mathematics knowledge test or is used to guide the evaluation process, e.g., observation guideline to assess O&M skills. There were 37 experiments that described the instruments used. Among the instruments used, there are 5 Likert-based surveys, 20 checklists (which include guidelines and specific tests), 15 questionnaires (which include surveys), 8 modeling kits (which are manual instruments, as pen and papers or bricks), and 9 logs (which include, in addition to the system log, the video and audio logs). None instrument has been used in more than one study experiment from our list. We speculated this is due to the distinct natures of the skills evaluated.

7.3   Statistical Methods

As a result, in 3 experiments the statistical method was not described or explicit in the text, neither in the references. T-test, which uses statistical concepts to reject or not a null hypothesis, is the most used, followed by ANOVA (one, two and three-way) and Person's Correlation, that is used to analyze the variation between groups.

The statistical method used depends on the research goal and the data acquired of the experiment. Thus, we prefer not to present the methods according to the type of analysis, such as hypothesis analysis or correlation between variables. Instead, we focused on if the experiment data analysis has any statistical method applied. Our objective was not to discuss how to apply the best method, but to analyze the evidence based on statistical confirmations. The two types of statistical methods highlighted are variables correlation and hypothesis tests.

7.4   Resource Data

In general, the papers do not present resource data of the experiments in the text. 19 experiments describe some information about the resource, as time-period, costs or human resources. Among this information, the time-period presented varies from 2 days to 6 months with the mean of 3 months-period. There were 4 experiments that explain some information about cost, two of them chose some technology due to their low cost. One experiment says pay $25 per hour to participants, and another pays $60 for 3-hour sessions. One paper describes their team to apply the experiment. Almost always there is little or no information about the resources, this can hinder the replication of the experiment, as well as it lacks evidence to the descriptive text about adequacy, limits, qualities, costs and associated risks.

7.5   Ethical Concepts

The ethical concepts also are not well covered in the papers; even it is an essential step to produce an experiment with people who have disabilities [26]. Among papers that threats the ethical concepts, 8 papers mentioned signing consents, two of them also applies to stop rules to enforcing ethical concept; 6 papers present the ethics council approved by legal institutions; two experiments express some information concern the user safety. It should be considered the local laws of ethical concepts and also the published year for each experiment.

7.6   Hypothesis and Groups

Hypothesis is the basis for the statistical analysis of an experiment. The hypothesis is the core of the experiment design. Seeking to understand the hypothesis, we selected which experiments have explicit in the text the Hypothesis and which of them test it. Even if the paper explains the research question or the goal of the experiment in the text, we did not count them. Wohlin divides the Hypothesis formulation from the Goal definition as two different phases of the experiment process. The two codes derived from this category is Explicit hypothesis and hypothesis well explicit. The "Hypothesis well explicit" (4 papers) is coded when the paper presents the hypothesis structured very well, as the experiments from "Model of Cognitive Mobility for Visually Impaired" (11 papers).

7.7   Variables

Almost no paper presents the variables explicitly in the text as dependent variables and independent variables. Thus, along with the text, we looked for these variables to understand how the experiment is designed. After the grounded theory analysis process, we produced three categories of dependent variables and three categories of independent variables. In searching this, we can see that the variables are strongly related to the experiment hypothesis or research objective. As a result, the Code Map[9] presents the code map generated and next we explain each variable structure.

As cognitive impact evaluation, the dependent variable is related to cognition processes. The main category of dependent variables is the task performance that has many ways to measure (explored in the Sect. 7.9). We coded as skills all dependent variables that work with a cognitive skill affected by some independent variable. Problem-solving skill stands out in the experiments. Only two papers use some user emotion, as user opinion.

The independent variables are related to domain knowledge. The independent variables encountered are characteristics of the situation, the participants and the tasks. Generally, the characteristics of the situation and tasks are factors, while participant characteristics are variables controlled, as etiology of blindness or age of participants. Nevertheless, usually, the levels of blindness are factors in the experiments used to divide the sample into groups.

Beyond the independent and dependent variables, we encountered in the grounded data other topics about the variables. There were 11 experiments with their variables well described in a separated section in the text. There were 8 experiments that work with a control group to compare with the treatment. Also,

---

[9] Code Map of variables: https://www.dropbox.com/variables_code_map.png.

three experiments stand out concern about the confounding variables. Report these topics show more rigor in the controlled experiments, moreover it is not a usual practice.

7.8 Tasks

The tasks explored in the experiments are related to the technology assessed. The Code Map have five categories[10]. The two main categories grounded in the experiment data are: Maps and Problem-Solving, already discussed in the Variables section. There were 31 experiments that work with maps applications aiming to develop Orientation & Mobility skills and other related skills. Not all presented in detail the tasks and measures, but we divide the Maps category into: *(i)* location (indoor or outdoor maps), *(ii)* number of spaces (one or more spaces), *(iii)* known spaces (new or public spaces), *(iv)* type (virtual or real map tasks), *(v)* tasks that the participant need to reproduce a map using some modeling kit, as bricks, and *(vi)* tasks that use the clock [17]. Notice that the same task of the experiment could have more than one code.

7.9 Measures

The measures are defined according to the variables of the experiment. We found five categories in the Measures code map grounded in data of the experiments[11], that are measures based on *(i)* technology, *(ii)* activity type, *(iii)* tests, *(iv)* user behavior, and *(v)* task performance. The categories "activity type" and "task performance" highlight in front of the others because they are the most used. The "activity type" depends on the technology domain and the variables defined. For example, one measure defined to a map task is the obstacle detection. The task performance measures are the most important cognitive measures because they can be used in several domains, for example, time duration and error rate.

8 Theory of Cognitive Impact Evaluation

After organizing the core category and related the concepts of Cognitive Impact Evaluation, the data was arranged to observe the excerpts related to each concept. The relations found were identified by analyzing patterns and implicit meaning between codes. Once the concepts are related they were organized into a theoretical framework [2], meaning the representation of concepts, together with their definitions and relations among them. Figure 7 presents the cognitive impact evaluation central idea, considering the context of multimodal interfaces for people who are blind or visually impaired.

---

[10] Code Map of tasks: https://www.dropbox.com/tasks_code_map.png.

[11] Code Map of measures - https://www.dropbox.com/measures_code_map.png.

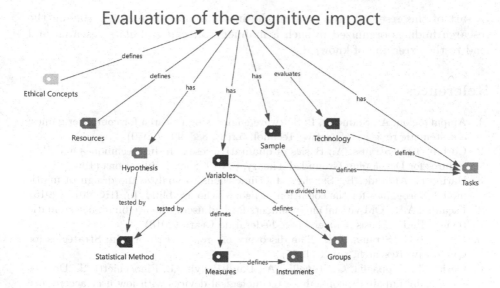

Fig. 7. Central idea of cognitive impact evaluation for blind users

9   Conclusion

The applications for people who are blind have many needs due to the target audience and unique characteristics related to multimodal interfaces. Moreover, a lot of the applications for people who are blind or visually impaired aims to improve a cognitive skill, such as cognitive enhancement in O&M, wayfinding, and navigation skills, and thus supporting the user in daily lives. Among so many evidence-based methods, the experiments to evaluate the cognitive impact of an application or a technology present faults in the experiment process.

This perception is based on data of experiments retrieved from the scientific papers found in the SLR which report the experiment, and the empirical study cannot be distinguished from its reporting [26]. The experiments that do not report well their characteristics become weak in one of the main points of the experiment: the repeatability. If the differences in factors and settings are well documented and analyzed, more knowledge may be gained from replicated studies [26].

The importance of the experiment is to consider the formalism in terms of the cognition assumptions underpinning that knowledge. This concern introduces challenges in concepts and formalism of the area of cognition that must be well studied and understood within the experiment. Thus, it is important to involve other disciplines to test an in-depth hypothesis, so as to guarantee rigor in the result, as indicated by [9]. It states that without the link from theory to the hypothesis, empirical results cannot contribute to a broader body of knowledge.

We expected to have created a bibliographic review of the cognitive impact evaluation based on the steps of the SLR approach. It organizes the state-of-the-art of this research area making the information summarized. Having the research findings organized in such a manner that can stimulate research and lead to the extension of knowledge.

References

1. Ampatzoglou, A., Stamelos, I.: Software engineering research for computer games: a systematic review. Inf. Softw. Technol. **52**(9), 888–901 (2010)
2. Corbin, J.M., Strauss, A.: Basics of qualitative research. In: Techniques and Procedures for Developing Grounded Theory, p. 379. Sage Publications (1998)
3. Darin, T., Andrade, R., Sánchez, J.: Dimensions to analyze the design of multimodal videogames for the cognition of people who are blind. In: IHC 2013 (2015)
4. Façanha, A.R.: OM virtual environments for end-users who are blind: a systematic review. Ph.D. Thesis, Universidade Federal do Ceará (2018)
5. Glaser, B.G., Strauss, A.L.: The discovery of grounded theory. In: Strategies for Qualitative Research, vol. 1 (1967)
6. Goria, M., Cappaglia, G., Tonellia, A., Baud-Bovyb, G., Finocchietti, S.: Devices for visually impaired people: high technological devices with low user acceptance and no adaptability for children. Neurosci. Biobehav. Rev. **69**, 79–88 (2016)
7. Hinebaugh, J.P.: A Board Game Education. Rowman & Littlefield Education (2009)
8. Hossain, G., Shaik, A.S., Yeasin, M.: Cognitive load and usability analysis of R-MAP for the people who are blind or visual impaired. In: Proceedings of the 29th ACM International Conference on Design of Communication, pp. 137–143. ACM Press, New York (2011)
9. Kitchenham, B.B.A., et al.: Preliminary guidelines for empirical research in software engineering. IEEE Trans. Softw. Eng. **28**(8), 721–734 (2002)
10. Kitchenham, B., Charters, S.: Guidelines for performing systematic literature reviews in software engineering. Engineering **45**(4ve), 1051 (2007)
11. Lahav, O., Mioduser, D.: Haptic-feedback support for cognitive mapping of unknown spaces by people who are blind. Int. J. Hum Comput Stud. **66**(1), 23–35 (2008)
12. Lumbreras, M., Sánchez, J.: 3D aural interactive hyperstories for blind children. Virtual Reality **3**(4), 119–128 (1998)
13. Mesquita, L., Sánchez, J.: In search of a multimodal interfaces impact evaluation model for people who are blind. In: 16th Brazilian Symposium on Human Factors in Computing Systems (IHC 2017), Joinville, Brazil (2017)
14. Mesquita, L., Sánchez, J., Andrade, R.M.C.: Cognitive impact evaluation of multimodal interfaces for blind people: towards a systematic review. In: Human-Computer Interaction International Conference (HCII), pp. 4–6 (2018)
15. Petersen, K., Feldt, R., Mujtaba, S., Mattsson, M.: Systematic mapping studies in software engineering. In: 12th International Conference on Evaluation and Assessment in Software, pp. 1–10 (2008)
16. Pissaloux, E., Velazquez, R., Hersh, M., Uzan, G.: Towards a cognitive model of human mobility : an investigation of tactile perception for use in mobility devices. J. Navig. **70**(1), 1–17 (2017)
17. Sánchez, J., Espinoza, M., Garrido-Miranda, J.M.: Videogaming for wayfinding skills in children who are blind. In: International Conference on Disability, Virtual Reality and Associated Technologies - ICDVRAT, Laval (2012)

18. Sánchez, J., Baloian, N., Hassler, T.: Blind to sighted children interaction through collaborative environments. In: de Vreede, G.-J., Guerrero, L.A., Marín Raventós, G. (eds.) CRIWG 2004. LNCS, vol. 3198, pp. 192–205. Springer, Heidelberg (2004). https://doi.org/10.1007/978-3-540-30112-7_16
19. Sánchez, J., Darin, T., Andrade, R.M.C., Viana, W., Gensel, J.: Multimodal interfaces for improving the intellect of the blind. In: XX Congreso de Informática Educativa - TISE, vol. 1, pp. 404–413 (2015)
20. Sánchez, J., de Borba Campos, M., Espinoza, M., Merabet, L.B.: Audio haptic videogaming for developing wayfinding skills in learners who are blind. In: IUI/International Conference on Intelligent User Interfaces, pp. 199–207 (2014)
21. Sánchez, J., Flores, H., Sáenz, M.: Mobile science learning for the blind. In: Extended Abstracts on Human Factors in Computing Systems, pp. 3201–3206. ACM Press, Florence (2008)
22. Shafiq, M., et al.: Skill specific spoken dialogues based personalized ATM design to maximize effective interaction for visually impaired persona. In: Marcus, A. (ed.) DUXU 2014. LNCS, vol. 8520, pp. 446–457. Springer, Cham (2014). https://doi.org/10.1007/978-3-319-07638-6_43
23. Sternberg, J.R., Sternberg, K.: Cognitive Psychology. Cengage Learning, Wadsworth (2011)
24. WHO: Blindness and vision impairment (2018)
25. Wohlin, C.: Guidelines for snowballing in systematic literature studies and a replication in software engineering. In: Proceedings of the 18th International Conference on Evaluation and Assessment in Software Engineering - EASE 2014, pp. 1–10 (2014)
26. Wohlin, C., et al.: Experimentation in Software Engineering. Kluwer Academic Publishers, Norwell (2000)
27. Strauss, C.: Basics of Qualitative Research: Techniques and Procedures for Developing Grounded Theory, 3rd edn., vol. 36, no. 2. SAGE Publications Inc., Thousand Oaks (2010)

Statistical Analysis of Novel and Traditional Orientation Estimates from an IMU-Instrumented Glove

Nonnarit O-larnnithipong[1]([⊠]), Neeranut Ratchatanantakit[1],
Sudarat Tangnimitchok[1], Francisco R. Ortega[2], Armando Barreto[1],
and Malek Adjouadi[1]

[1] Department of Electrical and Computer Engineering,
Florida International University, Miami, FL, USA
{nolar002,nratc001,stang018,barretoa,adjouadi}@fiu.edu
[2] Department of Computer Science,
Colorado State University, Fort Collins, CO, USA
fortega@colostate.edu

Abstract. This paper outlines the statistical evaluation of novel and traditional orientation estimates from an IMU-instrumented glove. Thirty human subjects participated in the experiment by performing the instructed hand movements in order to compare the performance of the proposed orientation correction algorithm with Kalman-based orientation filtering. The result of two-way multivariate analysis of variance indicates that there is no statistically significant difference in the means of the orientation errors: Phi ($F(1, 580) = .080; p = .777$), Theta ($F(1, 580) = 2.556; p = .110$) and Psi ($F(1, 580) = .049; p = .825$) between the orientation correction algorithm using the gravity and magnetic North vectors (GMV) and the correction using Kalman-based orientation filtering (KF). The different hand poses have a statistically significant effect on the orientation errors: Phi ($F(9, 580) = 129.555; p = .000$), Theta($F(9, 580) = 85.109; p = .000$) and Psi ($F(9, 580) = 134.474; p = .000$). The effect of the two algorithms on the orientation errors is consistent across the different hand poses.

Keywords: Inertial measurement unit · Gyroscope drift ·
Orientation correction algorithm · Bias offset error ·
Quaternion correction using gravity vector and magnetic North vector ·
3D hand orientation tracking

1 Introduction

There have been multiple studies on human hand motion tracking for the development of the human-computer interfaces. Each system employs different types of sensing units to detect the hand orientation and finger configurations based on the applications. The systems for hand-shape and gesture recognition [7,9] utilize

© Springer Nature Switzerland AG 2019
M. Antona and C. Stephanidis (Eds.): HCII 2019, LNCS 11573, pp. 282–299, 2019.
https://doi.org/10.1007/978-3-030-23563-5_23

a number of accelerometers to determine the hand orientation from the acceleration measurements. Some technologies [4,6] use magnetic sensors to determine the movement of the thumb and fingers by sensing the change in magnetic field.

We have proposed and developed [16,17] an IMU-instrumented glove to determine the orientation of the human hand as an alternative approach for human-computer interaction. The IMU-instrumented glove utilizes inertial measurement unit (IMUs), which consist of accelerometer, magnetometer and gyroscope within a single module.

The gyroscope is an inertial measurement device, which generates its output signal as a measurement of angular velocity. Then, the orientation can simply be determined by mathematical integration with respect to time. Nevertheless, the Microelectromechanical Systems (MEMS) gyroscope may produce a non-zero erroneous output signal even when the sensor is not in motion. This type of error is called "bias offset error", which is a common error in MEMS inertial sensors that severely disrupts the orientation tracking and causes a major problem called "drift" in several applications [3,18].

One of the common ways to improve orientation tracking and eliminate gyroscope drift in MEMS inertial measurement units is Kalman-based orientation filtering [13,19]. Kalman-based orientation filtering is capable of determining the orientation estimates based on the measurements, state variable model, statistical modeling of noise, and other uncertainties. Kalman filtering has been used in numerous applications, especially in navigation systems for aircraft and autonomous vehicles [8,12]. However, several studies [2,10] utilize sensor fusion approaches to determine the orientation estimates by combining the measurements from gyroscopes, accelerometers and magnetometers. Since, modern MEMS IMUs contain gyroscopes, accelerometers and magnetometers in a single module, IMUs then become a valid option in terms of cost, dimension and integrability to be used for determining orientation in human-computer interaction applications.

This paper outlines the statistical analysis of the effect of novel and traditional orientation estimates on the orientation errors in the form of Euler Angles (Phi, Theta and Psi) from an IMU in a glove. The orientation correction algorithm using the gravity and magnetic North vectors was performed in real-time while the hand movement was being performed. The traditional orientation estimates were obtained from the on-board Kalman-based orientation filtering of the IMU.

2   Methodology and Materials

2.1   IMU-Instrumented Glove

For this evaluation, the "3-Space embedded" inertial measurement unit from Yost Labs, as shown in Fig. 1(a) is used. The Yost Labs 3-Space sensor is a commercial-grade Attitude and Heading Reference System (AHRS), consisting of tri-axial accelerometer, gyroscope and magnetometer. The module integrates on-board Kalman-based orientation filtering and can be connected to the host

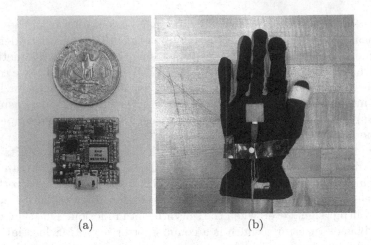

(a)　　(b)

Fig. 1. (a) Yost Labs 3-Space^{TM} sensor compared with a quarter (b) IMU-instrumented glove.

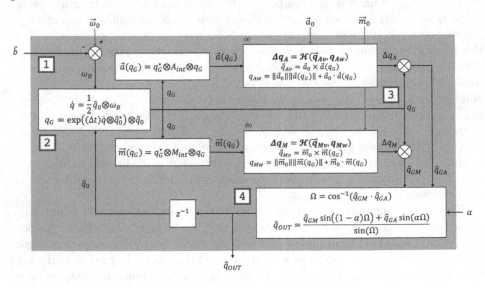

Fig. 2. Block diagram of the orientation correction algorithm using the gravity vector and the magnetic North vector correction (GMV).

computer via a USB cable. The Yost Labs 3-Space sensor was attached on a glove, at the back of the hand (as shown in Fig. 1(b)), in order to measure the angular velocity and acceleration of the hand motion. In addition, the module can determine the direction of the magnetic field with respect to the sensor's body frame. These three quantities are used to calculate the orientation estimates using the proposed algorithm.

2.2 Orientation Correction Using the Gravity and Magnetic North Vectors

For this evaluation, we compared the performance of human hand orientation estimation obtained from the traditional method (Kalman-based orientation filtering) with our proposed orientation estimation method, which utilizes the sensor fusion approach. The approach initially involves the dead-reckoning process, which is the calculation of a current orientation based upon previously-determined angle and measured angular velocity from the MEMS gyroscope in the Yost Labs 3-Space sensor. The proposed orientation estimation method also includes the correction of the orientation estimates, utilizing the measurements of acceleration due gravity from the accelerometer and the direction of magnetic North from the magnetometer, both integrated in the same sensor module.

A novel orientation correction algorithm was previously proposed and implemented [15–17] to improve the orientation estimation. Each step of the orientation correction algorithm using the gravity and magnetic North vectors, labeled by a number in Fig. 2, is described as follows:

(1) Prediction of the Bias Offset Error. A gyroscope should generate zero angular velocity output when the sensor module is not in motion. However, a commercial-grade MEMS gyroscope can generate an erroneous output signal that deviates from zero, called "the bias offset error", which yields an unacceptable orientation error, called "drift". The bias offset error in a MEMS gyroscope changes randomly through time and consists of both deterministic and stochastic components [1]. Therefore, the bias offset error should be determined and updated whenever the IMU is in a static period. The bias offset error of the gyroscope reading ($\hat{b}$) is determined by averaging five consecutive gyroscope samples, in which their magnitudes are smaller than the pre-defined thresholds.

By subtracting the bias offset error ($\hat{b}$) from the raw gyroscope reading (ω_0), the unbiased angular velocity (ω_B) is obtained as described in Eq. 1. The unbiased angular velocity is then used to calculate the dead-reckoning orientation estimates in the next step.

$$\omega_B = \omega_0 - \hat{b} \tag{1}$$

(2) Estimation of a Quaternion Orientation. For the purpose of orientation correction and to avoid the gimbal lock problem, quaternion notation will be used to describe the orientation in this work. A quaternion is a 4-dimensional quantity, consisting of one real part and three imaginary parts. It is a common representation for the rotation of an object in computer graphics [11].

The unbiased angular velocity (ω_B) obtained from the previous step is converted into a quaternion domain as the quaternion rate ($\dot{q}$), as shown in Eq. 2, where $\hat{q}_0$ is the quaternion estimation from the previous iteration of the algorithm. At the beginning, $\hat{q}_0$ is initialized as $[0, 0, 0, 1]$ to indicate zero-degree rotation for all three axes.

$$\dot{q} = \frac{1}{2} \hat{q}_0 \otimes \omega_B \tag{2}$$

$$q_G = exp((\Delta t)\dot{q} \otimes \hat{q}_0^*) \otimes \hat{q}_0 \tag{3}$$

Equation 3 implements the dead-reckoning process, in which the quaternion q_G represents the estimated orientation (current orientation) calculated from the quaternion rate $\dot{q}$ (known angular velocity in quaternion domain) and the quaternion estimation from the previous iteration $\hat{q}_0$ (previously-determined orientation). Here Δt is the sampling interval of the data received from the IMU. This estimated quaternion (q_G) can be used to describe the approximate orientation of the IMU. However, the removal of the bias offset error from gyroscope reading can be over- or under-compensated. Consequently, a small amount of drift is still present and it is corrected by using the gravity and magnetic North vectors in the next step.

(3) Quaternion Correction. The acceleration due to gravity of the Earth always has its direction pointing towards the Earth's center. The estimated orientation q_G in Eq. 3 obtained from the previous step, which can approximately describe the orientation of the IMU with respect to the Earth frame, can also be used to transform the gravity vector in the Earth frame into the gravity vector referenced in the IMU's body frame, namely the *calculated gravity vector* ($\boldsymbol{a}(q_G)$), by using Eq. 4. In this equation, $\boldsymbol{A}_{int}$ is the initial measured gravity vector in the Earth frame. Since the *calculated gravity vector* ($\boldsymbol{a}(q_G)$) is referenced in the IMU's body frame, it can be used to compare with the *measured gravity vector* ($\boldsymbol{a}_0$) from the accelerometer, which ideally measures only the acceleration due to gravity (in the IMU's body frame) when the IMU is in a static period.

$$\boldsymbol{a}(q_G) = q_G^* \otimes \boldsymbol{A}_{int} \otimes q_G \tag{4}$$

The angular difference between the *calculated gravity vector* ($\boldsymbol{a}(q_G)$) and the *measured gravity vector* ($\boldsymbol{a}_0$) represents the error of orientation estimation, which can be determined in the form of a quaternion denoted by Δq_A, as shown in Eq. 5, where $\boldsymbol{q}_{Av} = \boldsymbol{a}_0 \times \boldsymbol{a}(q_G)$, and $q_{Aw} = \|\boldsymbol{a}_0\|\|\boldsymbol{a}(q_G)\| + \boldsymbol{a}_0 \cdot \boldsymbol{a}(q_G)$. Equation 6 describes the correction of the orientation estimates q_G by Δq_A since the product of 2 quaternions implies compounding of their rotations. The estimated quaternion $\hat{q}_{GA}$ yields the orientation estimates fully corrected by the gravity vector.

$$\Delta q_A = \mathbb{H}(\boldsymbol{q}_{Av}, q_{Aw}) \tag{5}$$

$$\hat{q}_{GA} = q_G \otimes \Delta q_A \tag{6}$$

The correction of the orientation estimate based on the magnetic North vector follows a similar principle. The direction of the Earth's magnetic field is pointing from the geographic south pole to the geographic north pole and is constant at a particular point on the Earth surface. The estimated orientation q_G in Eq. 3 obtained from the dead-reckoning process, which can approximately describe the orientation of the IMU with respect to the Earth frame, can also be used to transform the magnetic North vector in the Earth frame into into the magnetic North vector referenced in the IMU's body frame. This yields the *calculated magnetic North vector* $(\boldsymbol{m}(q_G))$, using Eq. 7, where $\boldsymbol{M}_{int}$ is the initial measured magnetic North vector in the Earth frame. Since the *calculated magnetic North vector* $(\boldsymbol{m}(q_G))$ is referenced in the IMU's body frame, it can be used to compare with the *measured magnetic North vector* $(\boldsymbol{m}_0)$ from the magnetometer readings.

$$\boldsymbol{m}(q_G) = q_G{}^* \otimes \boldsymbol{M}_{int} \otimes q_G \qquad (7)$$

The angular difference between the *calculated magnetic North vector* $(\boldsymbol{m}(q_G))$ and the *measured magnetic North vector* $(\boldsymbol{m}_0)$ can also represent the error of orientation estimation, which can be determined in the form of a quaternion denoted by Δq_M, as shown in Eq. 8, where $q_{Mv} = \boldsymbol{m}_0 \times \boldsymbol{m}(q_G)$, and $q_{Mw} = \|\boldsymbol{m}_0\|\|\boldsymbol{m}(q_G)\| + \boldsymbol{m}_0 \cdot \boldsymbol{m}(q_G)$. Equation 9 describes the correction of the orientation estimates q_G by Δq_M. The estimated quaternion $\hat{q}_{GM}$ yields the orientation estimates fully corrected by the magnetic North vector.

$$\Delta q_M = \mathbb{H}(q_{Mv}, q_{Mw}) \qquad (8)$$

$$\hat{q}_{GM} = q_G \otimes \Delta q_M \qquad (9)$$

(4) Adaptive Quaternion Interpolation. The last step of the orientation correction algorithm is to calculate the final quaternion estimate from the orientation estimates fully corrected by the gravity vector $(\hat{q}_{GA})$ and the orientation estimates fully corrected by the magnetic North vector $(\hat{q}_{GM})$, based on the current conditions of the IMU. When the IMU is in rapid motion, the measurement from accelerometer includes an acceleration due to linear motion. Therefore, the accelerometer cannot be used as a reference for orientation correction. In contrast, the magnetometer measurement is not affected by the rapid movement of the hand (but it could be affected by local distortions of the magnetic field). To calculate the final orientation estimates $(\hat{q}_{OUT})$, Spherical Linear Interpolation (SLERP) [5] is adapted as an approach to determine the interpolated quaternion between two quaternions by using an adaptive weight (α), as shown in Eq. 10.

$$\hat{q}_{OUT} = \frac{\hat{q}_{GM}sin((1 - \alpha)\Omega) + \hat{q}_{GA}sin(\alpha\Omega)}{sin(\Omega)} \qquad (10)$$

$$\Omega = cos^{-1}(\hat{q}_{GM} \cdot \hat{q}_{GA}) \qquad (11)$$

The adaptive weight α, ranging from 0 to 1, is a measure of the "stillness" of the IMU and can be used to linearly interpolate the two quaternions. When the IMU is in a static period, the value of $\alpha = 1$, making the final estimated quaternion orientation $(\hat{q}_{OUT})$ match the orientation estimates using only the gravity vector correction $(\hat{q}_{GA})$. When the sensor is in rapid motion, the value of α drops towards zero. Thus, the final estimated quaternion orientation $(\hat{q}_{OUT})$ is the interpolated quaternion, which tends towards the orientation estimates using only the magnetic North vector correction $(\hat{q}_{GM})$. The final estimated quaternion orientation $(\hat{q}_{OUT})$ is then used in the dead-reckoning process for the next iteration of the algorithm.

3 Implementation

3.1 Evaluation Protocol

Thirty human subjects were recruited to participate in the evaluation experiment. Each of the subjects wore the IMU-instrumented glove on his/her left hand and sat on a chair, facing towards a computer screen. A rectangular frame with guiding position and orientation markers was placed in between the subject and the computer screen. The testing environment was set up as shown in Fig. 3. Each subject was asked to perform specific hand movements as instructed on the computer screen. This instructed hand movement guide was created within a virtual 3D environment using Unity. The virtual 3D environment consists of a 3D left-hand model and the model of rectangular frame which is identical to the physical frame placed in between the subject and the computer screen. The model of the rectangular frame in the 3D environment helped the test subjects to easily understand and perform the requested hand movements correctly. The hand movement guide was displayed as an animation of the 3D hand model movement, visualizing a sequence of 10 pre-defined movements of the 3D hand (as shown in Fig. 4). The subjects were asked to replicate the orientation of each virtual hand pose shown on the screen while wearing the IMU-instrumented glove. The angular velocity, acceleration and the direction of magnetic field from an inertial measurement unit (IMU) attached at the back of the hand, were recorded. The proposed orientation correction algorithm, described in Sect. 2, was implemented to calculate the estimated orientation. The orientation estimates obtained with the orientation correction algorithm using the gravity and magnetic North vectors (GMV) were compared to quaternion outputs from the Kalman-based orientation filtering, streamed directly from the Yost Labs 3-Space sensor module (KF).

3.2 Experiment Procedure

Each subject was asked to wear the IMU-instrumented glove on his/her left hand and orient the hand to match the hand orientation in Pose 1 of Fig. 4. Then, The experimenter recorded the initial orientation of the hand by clicking the button "Mark this orientation". After that, the experimenter clicked on the button "Show hand movement" in order to display the animation of the 3D hand movement, stopping at the next hand pose that the subject had to achieve. The subject then moved his/her hand to match the orientation as requested on the screen. The experimenter clicked on the button "Mark this orientation" to record the current orientation of the subject's hand. The procedure was repeated until all 10 poses of the hand movement had been performed by the subject. At the end of the session, the subject removed the IMU-instrumented glove and answered some questions regarding gender, age, and his/her dominant hand.

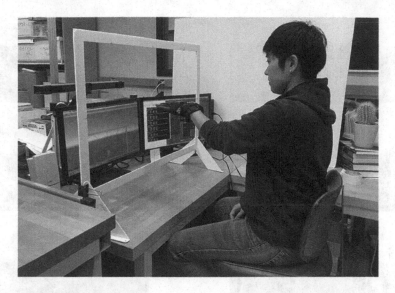

Fig. 3. The testing environment set up to evaluate orientation correction algorithms for IMU-instrumented glove

4   Results and Discussion

Thirty test subjects participated in the experiment to evaluate the performance of the orientation correction algorithms. Among the thirty subjects, there were 20 male and 10 female healthy participants. Only one participant was left-handed. Each test subject was asked to wear the IMU-instrumented glove and move his/her hand to match the 10 pre-defined poses of the hand movement. The orientation correction algorithm using the gravity and magnetic North vectors (GMV) was implemented to determine the orientation estimates of the hand, while the corrected orientation using Kalman-based orientation filtering (KF) was being streamed directly from the Yost Labs 3-Space sensor module. For the orientation output from both orientation correction algorithms (GMV and KF), the orientation errors in the form of Euler angles (Phi, Theta and Psi) were calculated based on the assumption that the test subjects were able to exactly match the instructed hand poses, which were considered as "ground truth". The orientation errors were used to evaluate the performance of the orientation correction algorithms.

A total of 600 data points of the orientation errors for each of the three Euler angles (30 test subjects × 10 hand poses × 2 algorithms) were recorded and analyzed using SPSS. The experiment was designated to predominantly test for the effects of the two different orientation correction algorithms on the mean of orientation errors. In addition, the effect of the 10 hand poses on the mean of orientation errors was also investigated. Table 1 shows the marginal means and the standard deviations of the orientation errors for each of the two algorithms, computed by averaging across the 10 hand poses. The means of the orientation

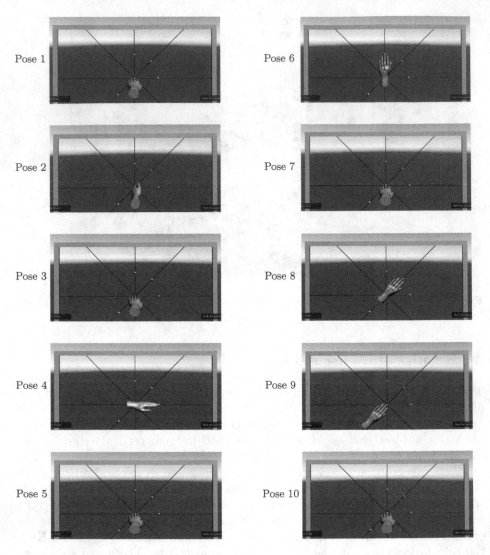

Fig. 4. The sequence of the 3D hand model movement (poses 1 to 10).

errors for the orientation correction algorithm using the gravity and magnetic North vectors (GMV) are slightly different from those of the correction using Kalman-based orientation filtering (KF), in all three Euler angles. The marginal means and the standard deviations of the orientation errors for each of the 10 hand poses averaging across the two algorithms, are also calculated and shown in Table 2. The means of the orientation errors are noticeably different across all poses in all three Euler angles. To test for the significance of the effects of the two factors (type of algorithm and hand poses) on three dependent variables (Euler angles: Phi, Theta and Psi), two-way multivariate analysis of variance

Table 1. Marginal means and standard deviations of the orientation errors for each algorithm

Dependent variable	Algorithm	N	Mean (deg)	Std. deviation
Phi	GMV	300	5.240	4.975
	KF	300	5.171	5.151
Theta	GMV	300	4.646	4.560
	KF	300	4.258	4.365
Psi	GMV	300	4.854	5.230
	KF	300	4.799	5.219

(two-way MANOVA) is suggested [14] as an appropriate statistical test model. The null hypotheses for two-way MANOVA are described below.

1. There is no difference in the means of orientation errors between the two algorithms.
2. There is no difference in the means of orientation errors for all hand poses.
3. There is no interaction between the two factors (type of algorithm and hand poses).

The orientation error data was tested using the General Linear Model (multivariate) in SPSS with .05 level of significance. The result of two-way multivariate analysis of variance (two-way MANOVA) for the effects of orientation correction algorithms and hand poses on the mean of the orientation errors is shown in Table 3. It is found that there is no statistically significant difference in the means of the orientation errors: Phi ($F(1, 580) = .080; p = .777$), Theta ($F(1, 580) = 2.556; p = .110$) and Psi ($F(1, 580) = .049; p = .825$) between the orientation correction algorithm using the gravity and magnetic North vectors (GMV) and the correction using Kalman-based orientation filtering (KF). The estimated marginal means plots of the orientation errors for the two algorithms across 10 different hand poses in Figs. 5, 7 and 9 also show that the means of the orientation errors for both algorithms are very similar across the 10 hand poses in all three Euler angles. This result indicates that the proposed orientation correction algorithm (GMV) is capable of estimating the orientation of the human as well as the traditional Kalman-based orientation filtering (KF). However, for some hand poses, both algorithms produced large orientation errors. Large orientation errors are present in Poses 6, 9 and 10 of the estimated marginal means plot for Phi (Fig. 5); Poses 2 and 4 of the estimated marginal means plot for Theta (Fig. 7); and Poses 8 and 9 of the estimated marginal means plot for Psi (Fig. 9). Since both algorithms produced large orientation errors in these poses, it raises the question whether the errors are possibly caused by the effect of the hand poses.

The test for the effect of hand poses on the orientation errors in Table 3 shows p-values smaller than .05 (p=.000 for all three Euler angles), indicating that different hand poses have a statistically significant effect on the orientation errors:

Table 2. Marginal means and standard deviations of the orientation errors for each hand pose

Dependent variable	Pose	N	Mean (deg)	Std. deviation
Phi	1	60	.335	.324
	2	60	3.824	2.805
	3	60	2.750	1.634
	4	60	6.469	3.789
	5	60	1.978	1.633
	6	60	11.219	4.191
	7	60	1.978	2.336
	8	60	12.015	3.630
	9	60	9.796	4.614
	10	60	1.689	1.337
Theta	1	60	.317	.335
	2	60	10.017	5.114
	3	60	2.153	1.470
	4	60	11.546	4.926
	5	60	2.806	2.387
	6	60	4.266	2.587
	7	60	2.493	1.816
	8	60	3.669	2.719
	9	60	4.108	2.734
	10	60	3.145	2.104
Psi	1	60	.493	.453
	2	60	5.078	3.937
	3	60	2.295	1.764
	4	60	2.818	2.352
	5	60	1.892	1.511
	6	60	4.935	3.747
	7	60	2.549	2.118
	8	60	10.901	5.266
	9	60	14.759	3.521
	10	60	2.545	2.101

Phi ($F(9, 580) = 129.555; p = .000$), Theta($F(9, 580) = 85.109; p = .000$) and Psi ($F(9, 580) = 134.474; p = .000$). The result provides a strong indication that hand poses affect the orientation errors for all three Euler angles. This can be interpreted that both algorithms could possibly produced less orientation errors. The marginal means plots of the orientation errors in Phi for 10 different hand poses across the two algorithms in Fig. 6 shows three plots (Poses 6, 8 and 9) with large orientation errors corresponding to the large errors observed in Fig. 5.

Table 3. Result of two-way multivariate analysis of variance (two-way MANOVA) test for the effects of orientation correction algorithms and hand poses on the mean of the orientation errors

Source	df	Dependent variable	Mean square	F statistic	Sig
Algorithm	1	Phi	.700	.080	.777
		Theta	22.560	2.556	.110
		Psi	.447	.049	.825
Pose	9	Phi	1133.721	129.555	.000
		Theta	751.284	85.109	.000
		Psi	1225.674	134.474	.000
Algorithm*Pose	9	Phi	5.984	.684	.724
		Theta	3.384	.383	.943
		Psi	.579	.064	1.000
Error	580	Phi	8.751		
		Theta	8.827		
		Psi	9.115		

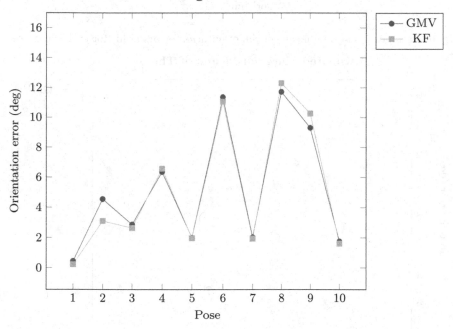

Estimated Marginal Means of Phi

Fig. 5. Estimated marginal means of the orientation errors (Phi) for each algorithm.

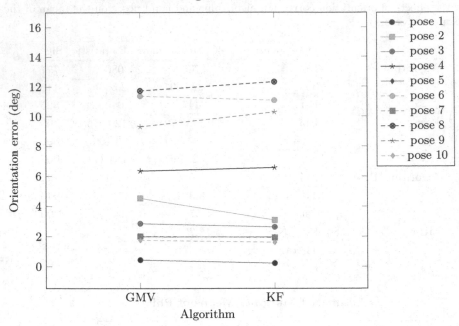

Fig. 6. Estimated marginal means of the orientation errors (Phi) for each hand pose.

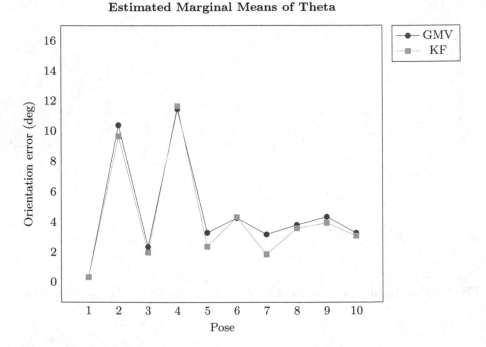

Fig. 7. Estimated marginal means of the orientation errors (Theta) for each algorithm.

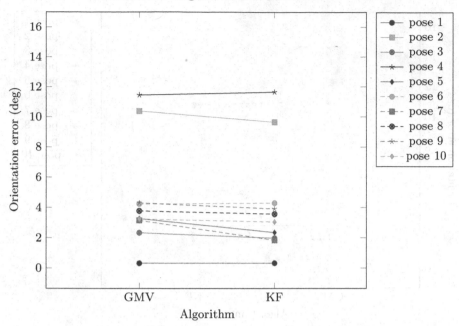

Fig. 8. Estimated marginal means of the orientation errors (Theta) for each hand pose.

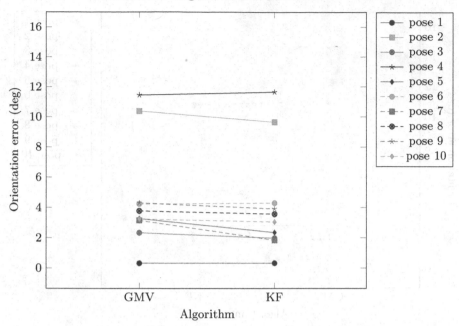

Fig. 9. Estimated marginal means of the orientation errors (Psi) for each algorithm.

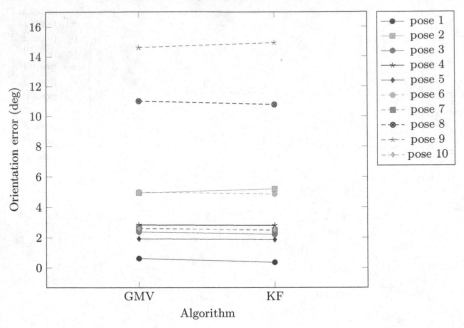

Fig. 10. Estimated marginal means of the orientation errors (Psi) for each hand pose.

The marginal means plots of the orientation errors in Theta for 10 different hand poses across the two algorithms in Fig. 8 also shows two plots (Poses 2 and 4) with large orientation errors that correspond to the large errors observed in Fig. 7. Similarly, the marginal means plots of the orientation errors in Psi for 10 different hand poses across the two algorithms in Fig. 10 has two plots (Poses 8 and 9) with large orientation errors corresponding to the large errors observed in Fig. 9. The test for the interaction between the type of algorithm and hand poses (row Algorithm*Pose) in Table 3 shows that there is no statistically significant interaction effect on the orientation errors: Phi $(F(9, 580) = .684; p = .724)$, Theta$(F(9, 580) = .383; p = .943)$ and Psi $(F(9, 580) = .064; p = 1.000)$. This means that the effect of algorithms is consistent across the different hand poses. Equivalently, the effect of the different hand poses on the orientation errors is consistent for both algorithms.

We speculate that large orientation errors recorded for some specific poses could be possibly caused by using the orientation of the instructed hand poses as the ground truth to calculate the orientation errors. Perhaps the algorithms estimated the actual orientation of the hand which might have been different from the orientation of the instructed hand poses. It is possible that these large errors could have been caused by the difficulties of the test subjects in trying to match the instructed hand orientations, particularly for poses that are unnatural and difficult to achieve. In the sequence of the hand movements (Fig. 4), the test subjects were asked to rotate their hands into 45-degree orientation for poses 8 and 9, and had to align their hands in parallel with the plane of the guiding frame for poses 4 and 6. In these difficult poses, the orientation errors can become large in one or more Euler angles. In the future, we will seek to test the glove with a position-measurement system that might have higher, controlled accuracy (e.g., laser-based) to use its measurements as the effective "ground truth" for each hand pose. THis might remove the uncertainty we faced in this experiment about the cause of larger errors yielded (by both orientation estimation methods) for some specific poses.

5 Conclusion

Thirty human subjects participated in the evaluation of novel and traditional orientation correction algorithms for an IMU-instrumented glove. Each subject was asked to move his/her hand to match the instructed hand poses. The orientation correction algorithm using the gravity and magnetic North vectors was implemented to determine the orientation estimates. The orientation errors in the form of Euler angles (Phi, Theta and Psi) were calculated as the departures from the orientation of the instructed hand poses. The orientation errors are compared with those of the correction using Kalman-based orientation filtering. The statistical analyses show that there is no statistically significant difference in the means of the orientation errors in Phi, Theta and Psi between the orientation correction algorithm using the gravity and magnetic North vectors and the correction using Kalman-based orientation filtering. The test for the effect of hand poses on the orientation errors indicates that different hand poses have a statistically significant effect on the orientation errors in Phi, Theta, and Psi. There is no statistically significant interaction effect on the orientation errors, implying that the effect of the two algorithms is consistent across the different hand poses. Even though the orientation errors for the novel and traditional orientation correction algorithms are not significantly different, the proposed orientation correction algorithm has one key advantage over the correction using Kalman-based orientation filtering: The proposed orientation correction algorithm handles acceleration measurements and magnetometer measurements separately during the execution of the algorithm. This allows the control parameters that define the weight for both measurements to be changed dynamically, which gives flexibility in defining the final orientation estimate according to different operational circumstances.

Acknowledgments. This research was supported by National Sciences Foundation grants HRD-0833093 and CNS-1532061 and the FIU Graduate School Dissertation Year Fellowship awarded to Dr. Nonnarit O-larnnithipong.

References

1. Aggarwal, P.: MEMS-Based Integrated Navigation. Artech House, Norwood (2010)
2. Bachmann, E.R., Duman, I., Usta, U.Y., McGhee, R.B., Yun, X.P., Zyda, M.J.: Orientation tracking for humans and robots using inertial sensors. In: Proceedings of the 1999 IEEE International Symposium on Computational Intelligence in Robotics and Automation, CIRA 1999, pp. 187–194. IEEE (1999)
3. Borenstein, J., Everett, H.R., Feng, L., Wehe, D.: Mobile robot positioning sensors and techniques. Naval Command Control and Ocean Surveillance Center RDT and E Division, San Diego, CA (1997)
4. Chen, K.-Y., Lyons, K., White, S., Patel, S.: uTrack: 3D input using two magnetic sensors. In: Proceedings of the 26th Annual ACM Symposium on User Interface Software and Technology, pp. 237–244. ACM (2013)
5. Dam, E.B., Koch, M., Lillholm, M.: Quaternions, Interpolation and Animation, vol. 2. Datalogisk Institut, Kbenhavns Universitet (1998)
6. Fahn, C.-S., Sun, H.: Development of a fingertip glove equipped with magnetic tracking sensors. Sensors **10**(2), 1119–1140 (2010)
7. Hernandez-Rebollar, J.L., Lindeman, R.W., Kyriakopoulos, N.: A multi-class pattern recognition system for practical finger spelling translation. In: Proceedings of the Fourth IEEE International Conference on Multimodal Interfaces, pp. 185–190. IEEE (2002)
8. Kelly, A.: A 3D state space formulation of a navigation Kalman filter for autonomous vehicles. Technical report, Carnegie-Mellon University, Robotics Institute, Pittsburgh, PA (1994)
9. Kim, J.-H., Thang, N.D., Kim, T.-S.: 3-D hand motion tracking and gesture recognition using a data glove. In: IEEE International Symposium on Industrial Electronics, ISIE 2009, pp. 1013–1018. IEEE (2009)
10. Kong, X.: INS algorithm using quaternion model for low cost IMU. Robot. Auton. Syst. **46**(4), 221–246 (2004)
11. Kuipers, J.B.: Quaternions and rotation sequences: a primer with applications to orbits, aerospace, and virtual reality. Princeton University Press, Princeton, N.J. (1999). ID: 024971770; Includes bibliographical references (pp. 365–366) and index
12. Lefferts, E.J., Markley, F.L., Shuster, M.D.: Kalman filtering for spacecraft attitude estimation. J. Guidance Control Dyn. **5**(5), 417–429 (1982)
13. Marins, J.L., Yun, X., Bachmann, E.R., et al.: An extended Kalman filter for quaternion-based orientation estimation using MARG sensors. In: Proceedings of the 2001 IEEE/RSJ International Conference on Intelligent Robots and Systems, vol. 4, pp. 2003–2011. IEEE (2001)
14. Montgomery, D.C.: Design and Analysis of Experiments. Wiley, Hoboken (2017)
15. O-larnnithipong, N., Barreto, A.: Gyroscope drift correction algorithm for inertial measurement unit used in hand motion tracking. In: 2016 IEEE SENSORS, pp. 1–3 (2016)
16. O-larnnithipong, N., Barreto, A., Tangnimitchok, S., Ratchatanantakit, N.: Orientation correction for a 3D hand motion tracking interface using inertial measurement units. In: Kurosu, M. (ed.) HCI 2018. LNCS, vol. 10903, pp. 321–333. Springer, Cham (2018). https://doi.org/10.1007/978-3-319-91250-9_25
17. O-larnnithipong, N., Barreto, A., Ratchatanantakit, N., Tangnimitchok, S., Ortega, F.R.: Real-time implementation of orientation correction algorithm for 3D hand motion tracking interface. In: Antona, M., Stephanidis, C. (eds.) UAHCI 2018. LNCS, vol. 10907, pp. 228–242. Springer, Cham (2018). https://doi.org/10.1007/978-3-319-92049-8_17

18. Sukkarieh, S., Nebot, E.M.: A high integrity IMU/GPS navigation loop for autonomous land vehicle applications. IEEE Trans. Robot. Autom. **15**(3), 572 (1999)
19. Yun, X., Lizarraga, M., Bachmann, E.R., McGhee, R.B.: An improved quaternion-based Kalman filter for real-time tracking of rigid body orientation. In: Proceedings of the 2003 IEEE/RSJ International Conference on Intelligent Robots and Systems, (IROS 2003), vol. 2, pp. 1074–1079. IEEE (2003)

Modeling Human Eye Movement Using Adaptive Neuro-Fuzzy Inference Systems

Pedro Ponce[1], Troy McDaniel[2(✉)], Arturo Molina[1], and Omar Mata[1]

[1] Tecnologico de Monterrey, 14380 Mexico City, Mexico
{pedro.ponce,armolina,omar.mata}@tec.mx
[2] School of Computing, Informatics and Decision Systems Engineering,
Arizona State University, Tempe, AZ 85281, USA
troy.mcdaniel@asu.edu

Abstract. The eye's muscles are difficult to model to build an eye prototype or an interface between the eye's movements and computers; they require complex mechanical equations for describing their movements and the generated voltage signals from the eye are not always adequate for classification. However, they are very important for developing human machine interfaces based on eye movements. Previously, these interfaces have been developed for people with disabilities or they have been used for teaching the anatomy and movements of the eye's muscles. However, the eye's electrical signals have low amplitude and sometimes high levels of noise. Hence, artificial neural networks and fuzzy logic systems are implemented using an ANFIS topology to perform this classification. This paper shows how the eye's muscles can be modeled and implemented in a concept prototype using an ANFIS topology that is trained using experimental signals from an end user of the eye prototype. The results show excellent performance for prototype when the ANFIS topology is deployed.

Keywords: Human eye movement · Eye muscles · Artificial neural networks · Fuzzy logic

1 Introduction

Movement of the human eye is caused by the action of six extra ocular muscles. Because of the eye's geometry and nature of movement, its mechanical analysis is complex and difficult to model. Similarly, understanding how these movements occur is challenging. In 1975, David A. Robinson wrote about the mechanics of ocular movement, describing the joint action of the extra ocular muscles [1]. Robinson used the mechanical properties of the extra ocular muscles to describe movement using force-balancing mechanical equations. Using this method, it's possible to describe how the force is distributed between the different muscles to reach the horizontal, vertical, and torsion movements depending on the length and innervations of each muscle. A physical model that can fully reproduce the ocular movements still does not exist; virtual simulations are instead used by specialists to gain a better comprehension of how ocular movements occur [2]. Artificial intelligence techniques exist that allow

© Springer Nature Switzerland AG 2019
M. Antona and C. Stephanidis (Eds.): HCII 2019, LNCS 11573, pp. 300–311, 2019.
https://doi.org/10.1007/978-3-030-23563-5_24

linguistic rules to be transformed into an adequate control that leads to the creation of an approximated model close to reality. In this work, we propose a model for describing the human eye's movement to allow a clear understanding of the joint action of the muscles involved in ocular movement with didactic purposes in areas such as medicine and ophthalmology.

2   General Aspects

2.1   Description of the Human Eye's Movements

The human eye is our vision organ, and it is located in the orbital cavities. These cavities are bone structures which contain the eyeballs, the extra ocular muscles, nerves, blood vessels, fat, and most of the lacrimal apparatus. The eyeball covers one third of this cavity and consists of five sixths spherical segments, one comprising the fifth posterior parts and one comprising the sixth fore part. Ocular movement is mainly due to the conjunct action of six extra ocular muscles mentioned below: superior rectus, inferior rectus, medial rectus, lateral rectus, and superior and inferior oblique. All these muscles except for the inferior oblique have their origin at a common point called the Annulus of Zinn.

These muscles come in pairs, and therefore every pair has a common plane. This plane is formed by joining the midpoints of the origin of the muscles to the midpoints in the insertions of the eyeball tendons. The plane of the medial rectus and the lateral rectus muscles is a horizontal plane which divides the eye in two parts. The plane of the superior rectus and inferior rectus is a vertical plane that has a 23° angle with the fixation line when the eyes are looking to the front, as shown in Fig. 1. The oblique muscles have a plane that has a 51° angle with the fixation line when the eye is looking in the same position as mentioned [2].

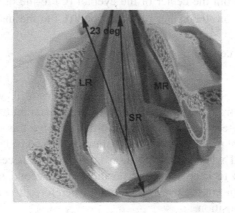

Fig. 1. Superior and inferior rectus angle with the plane

Ocular movement occurs on three axes (X, Y, Z) better known as Fick's axes [3] as shown in Fig. 2. The movements generated by the extra ocular muscles are purely rotational due to the action of the ligaments that attach the orbital cavity with the eye. Translational movements are very limited, avoiding the eye to get out of position. Nevertheless, it is considered that the center of rotation is always inside a zone called the centroid. In an adult eye, the center of rotation is located approximately 13.5 mm behind the cornea and 1.6 mm from the eye's nasal side [4].

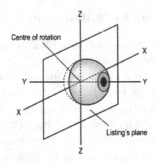

Fig. 2. Fick's axes

2.2 Vector Analysis of the Human Eye

θ Horizontal component of ocular rotation
ϕ Vertical component of ocular rotation
ψ Ocular torsion component
O   Origin of each muscle
R   Insertion point of each muscle in the eyeball
$\vec{O}$ Vector that goes from the center of the eyeball (C) to the origin (O)
$\vec{r}$ Vector that goes from the center of the eyeball (C) to the insertion point (R)
$\vec{F}$ Force vector
T   Point where the force acts
$\vec{m}$ Action unit vector

As shown in the Fig. 3, vector $\vec{F}$ begins to act from T, which is a muscular trajectory which goes from the insertion point to the separation of itself; therefore, the force is acting completely on the muscle. In order to solve this model, there are two main problems: the first problem is the innervations which consider the assumption that the eye is normal and when a new position is given it is necessary to find the six muscular innervations to keep it in that position. The second problem is when the innervations are known but the eye is abnormal. Therefore, it is necessary to find the deviation of the eye position.

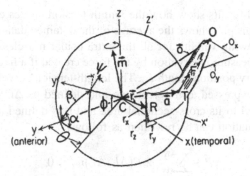

Fig. 3. Vector analysis of the eye movements

If the eye is moving in a horizontal position with an angle θ, in a vertical position with an angle φ, and with a torsion ψ, the eye will remain in this position as long as the sum of forces acting on it are zero. For determining the forces or moments, it is necessary to consider the insertion and origin points of each muscle; these follow a path through the eyeball's surface ending at T. The exerted force of the muscle is directed from T to O. Therefore, the moment created by this force is a vector perpendicular to the plane formed by C, T, and O with its direction following the right-hand rule. Vector $\vec{m}$ is directed in the same direction of this axis and it is extremely important because its components m_x, $m_y y m_z$ indicate the relative quantities of force distribution needed for acting on the eyeball vertically, horizontally and in torsion.

Vector $\vec{m}$ indicates the direction of the moment which has a magnitude F. Thus, the moment created by each muscle is $F\vec{m}$. There are seven moments in total, one for each extra ocular muscle and a passive moment $\vec{P}$ created for considering the muscular tissues that act for returning the eye to a neutral position close to the primary position. The force equilibrium equation results in the following:

$$\vec{P} + F_1\vec{m_1} + F_2\vec{m_2} + \ldots + F_6\vec{m_6} = 0 \tag{1}$$

As mentioned, the force on each muscle in a stationary state depends only on its length L and its innervation I, so it can be expressed as F (L, I). Robinson [1] describes how this relationship is measured experimentally (Fig. 4):

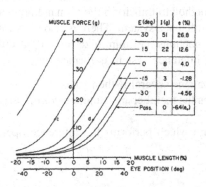

Fig. 4. Length-tension plot for different innervation levels

The experimental results show how the length-tension curves change for different innervation levels. This plot shows the average of the obtained data in which the curves were normalized allowing validity for all the extra ocular muscles.

It is possible to measure innervation by the force created if a fixed length is chosen, in this case the primary position length L_p. The length-tension curves are determined by some change in L_p expressed as a percentage and defined as Δl. The resistance of a muscle is proportional to its cross-sectional area and it is defined as λ. Therefore, the force equilibrium equation can be rewritten as follows:

$$\vec{P} + \sum_{i-1}^{6} \lambda_i F(\Delta l_i, \Delta I_i)\vec{m_i} = 0 \tag{2}$$

2.3 Artificial Intelligence Systems

The problem with modeling human eye movement is that it requires expressing many equations that could not be understandable by a field expert, like an ophthalmologist, for validation. Hence, we cannot obtain detailed information about the real eye movements or most of the times a precise model may not exist or is very difficult to model. For this, we use artificial intelligence because it facilitates the translation of a series of equations to linguistic rules that are given by the expert on the field whose knowledge and experience can explain the model [5]. In this work, we used a combination of fuzzy logic and neural networks to make an ANFIS system (Adaptative-Network-Based Fuzzy Inference System) because the complexity of most of the biological systems makes traditional quantitative approaches of analysis difficult [5].

Fuzzy Logic

Conventional mathematics tools, like differential equations, are not adequate for modeling systems that are not well defined or are with uncertainty. For this reason, fuzzy logic is useful for modeling systems using if-then rules because they can model qualitative aspects of reasoning and human knowledge without the need of quantitative analysis. Two approaches were created for fuzzy systems, the Mamdani and the Takagi-Sugeno. These models were explored systematically by Takagi and Sugeno, who found many applications in control, prediction, and inference. The Mamdani model is a fuzzy linguistic model that is focused on the interpretability, resulting in a fuzzy set. The Takagi-Sugeno model focuses on accuracy, resulting in crisp values [6].

There is no standard method to transform the human knowledge and experience into rules and databases of an inference system. There is also no completely effective method to tune the membership functions and minimize the output error or maximize its performance index. A fuzzy inference system is formed by five functional blocks [7, 8]:

- A rule base which contains several if-then rules.
- A database that defines the membership functions of the fuzzy sets used in the fuzzy rules.
- A decision-making unit which performs the inference operations on the rules.

- A fuzzification interface which transform the crisp inputs into degrees of match with the linguistic values.
- A defuzzification interface which transforms the fuzzy results of the inference into a crisp output.

Artificial Neural Networks

The theory and model of neural networks was inspired by the structure and function of the human nervous system where the neuron is the fundamental element. The neurons of the nervous system communicate to each other using signals. These signals are received by the dendrites and the cellular body, in which they are combined and emitted as output signals. The information is distributed by the axon to the axon's terminals to be transmitted again to a new set of neurons. A similarity is created between biological neurons and artificial neurons is which the signals that enter the synapses are the inputs of the neuron, which are weighted depending on the strength of each synapse and can be attenuated or simplified through a parameter called synapse associated weight. The signals that pass through the synapses can be excitatory or inhibitory. The neuron is excited with a positive weight and inhibited with a negative weight; if the weight is zero, the synapse is lost.

Neural Networks have big advantages in the creation of intelligent systems. Their resemblance to the human brain allows them to learn from experience, generalize from previous cases, abstract essential characteristics, among others. Below are some of their advantages [9]:

- *Adaptive Learning*. The capacity to learn to complete tasks based on training or initial experience. The weights of the interconnections are adjusted to obtain the desired results. Learning continues after initial training. An adequate structure and learning algorithm should be chosen to achieve the desired capabilities.
- *Auto organization*. Through adaptive learning, networks classify the information they receive and subsequently modify their own organization during learning. This capability provides the faculty to react to information or situations not previously seen.
- *Failure Tolerance*. When there is a failure in the network, the network should be resilient and immediately begin reorganizing to recover; much like the human brain rewires and builds new pathways.
- *Real-Time operation*. Neural networks can be run in parallel for efficiency; therefore, machines are often designed with special hardware in mind for efficiency.
- *Easy implementation with existing technology*. The ease of training, testing, and verification allow neural networks to be implemented with low cost hardware.

ANFIS Systems

ANFIS systems, or Adaptive-Network-Based Fuzzy Inference Systems, combine the use of Artificial Neural Networks with Fuzzy Logic techniques to make an artificial inference system. Adaptive networks, as its name implies, are neural network structures that contain nodes that can be adapted. The adaptability relies on the fact that the output depends on parameters that belong to that specific node that change depending on the rule of learning, minimizing the error. Thus, an adaptive network is a multilayer network with forward connection in which each node carries out a function in the input signal and on the parameters.

Inside the network, the adaptive nodes are represented by squares. Circular nodes do not have parameters because they are not adaptive; their values are fixed. The set of parameters in an adaptive network is the union of all parameters in each node. To obtain the desired output in response to the input, the parameters are updated depending on the training data. The architecture is functionally equivalent to a fuzzy rule base of type Sugeno allowing it to tune the existing rules with a learning algorithm based on the collection of data from the training, optimizing the rule base and adapting to its environment. ANFIS systems use neural networks to optimize the coefficients of the membership functions instead of modifying the weights of the matrix W of the neural network [10, 11].

Unlike using only an artificial neural network to generate a solid knowledge base, ANFIS can construct an input-output mapping based on human knowledge and stipulated input-output data pairs. As ANFIS uses linguistic rules to generate its knowledge, its behavior becomes clearer, so it stops being a "black-box" [8].

3 Implementation

3.1 LabVIEW

The system used for this project is an ANFIS with two inputs and one output. The inputs are coordinates in the X and Y axes in which the eye moves. Three triangular membership functions where used for each axis. To train the system, we researched the percentage of elongation of each muscle for each movement. From these facts, we chose to normalize the minimum value that represents the maximum contraction and maximum values to represent the maximum elongation of each muscle fixing the biggest value as the unity value. With the use of the *Anfiseditor* library from Matlab software, an ANFIS network was created fixing the number and form of the membership functions. The method used was backpropagation with a maximum of 5% error allowed. The results show nine rules with equal number of output functions of Singleton type, where using the mass center equation, we obtained a crisp output value. The system was programmed in National Instruments LabVIEW software due to its easy connection to acquisition cards and its graphical programming language.

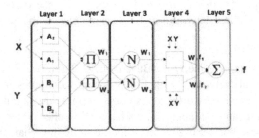

Fig. 5. Block diagram of the ANFIS system

The layer 1 from Fig. 5 shows the membership functions, which are adaptable. Layer 2 and 3 have the suggested rules to obtain the minimum membership as well as its proper normalization, obtain the respective weights. In Layer 4, we have the adaptable output functions that allow its sum to give us the function of the desired muscle. In Layer 5, the functions are summed, making its behavior as a Sugeno type system.

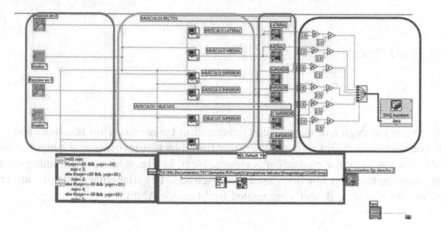

Fig. 6. Block diagram of the general program for the right eye

In the first block of Fig. 6, the position in degrees is obtained; X ranges from $-30°$ to $30°$, and Y from $0°$ to $30°$. The next block includes the programming for each muscle. The next step is a block that outputs a crisp value for each muscle. In the last block, the outputs are merged and sent to the acquisition card where 0 volts indicates completely closed and 5 volts completely opened.

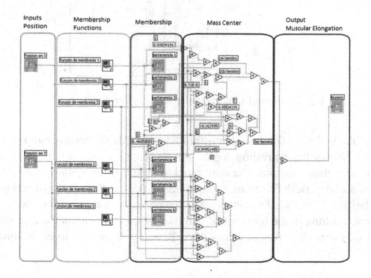

Fig. 7. Software block of the artificial muscular elongation

In the Fig. 7 the first block shows the square which receives position values. The second block shows the membership functions, three for X and three for Y. In the third block, a fuzzy numerical value is generated according to the belonging of the degree of membership in each of the membership functions. In the fourth block, the mass center is calculated by adding and multiplying the constant values generated by Matlab during the training phase using the following equations:

$$O_{2,i} = w_i = \mu A_i(x) * \mu B_i(y)$$ (3)

where $i = 1, 2$, and

$$O_{5,1} = \frac{\sum_i w_i f_i}{\sum_i w_i}$$ (4)

where w_i is the degree of membership generated in the membership functions and f_i is the constant output function used. In the fifth block, the numerical output which represents the muscular elongation is shown. Next, in Fig. 8. the programming of the triangular membership functions used are described. The triangular functions are created by the values α, β, and γ, generated by Matlab's ANFIS.

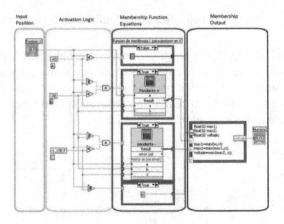

Fig. 8. Control of a triangular membership function

The first block obtains the desired position which needs the membership value. The second block shows the activation logic, which activates each of the equations to calculate the triangular function. The third block returns a 0 when the value is less than the values of α and γ. Both functions in the middle calculate their positive and negative slope. In the fourth block, the membership degrees are compared by choosing the higher degree resulting in the fuzzy value. All membership functions were generated in the same way, changing only the values of α, β, and γ according to each function.

3.2   Physical Implementation

A model of the right eye with a 5:1 scale was built using a white acrylic sphere of 120 mm diameter for the eye and a transparent resin box for the structure for simulating the eye cavity, as shown in Fig. 9. To make the muscles of the eye, an air muscle was made; when these muscles are filled with air they expand and when the air is removed, they contract. To move the muscles, we used proportional air valves from Dynalloy which are based on Flexinol. These valves can gradually open or close the stem caps to produce proportionally controlled air flow. While heated with internal resistance, the Flexinol contracts and opens the cap; however, as it opens, air begins to flow through and cool the same wire. The equilibrium between the electrical input and the mass of air entering the valve determines the aperture size and air flow.

Fig. 9. Physical eye model

According to the software and for simplicity of design, we chose to use the control signal for the muscles from 0 to 5 volts to produce 0 to 1 A which assured proper functioning of the valves. We created proportional conversion modules to accomplish this. Each module has a voltage to current converter with load to ground plus a power stage using transistors TIP 41 and TIP 42 for amplifying the current in a push-pull array, an op amp for compensation of the 0.7 volts drop of the emitter followers, and another op amp for impedance coupling and protection of the acquisition cards.

4   Results

Eye movement was controlled by software, in which the movements of the eye are selected by the rotational coordinates of the axes X and Y. However, to demonstrate the usability, we tested the eye's model movement by imitating the movement of a real eye. For this, an electrooculography circuit was implemented. The signals were acquired by electrodes connected as shown in Fig. 10.

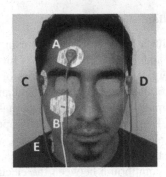

Fig. 10. Position of electrodes

The movement of the eye is measured in degrees from its initial position to the desired position. To characterize the movement of the model, the degrees moved and time taken were measured for each movement. The measurements where made with the eye in its center position, as the initial position, and to the up, left, and right position; also, from the up position to the left and right. The time was measured in seconds and the velocity as the ratio of degrees and time. The results are shown in Table 1.

Table 1. Model measurements

Origin	Destiny	Degrees [°]	Time [s]	Velocity [°/s]
Center	Right	5	0.8	6.3
Up	Right	4	1.1	3.6
Up	Left	10	1.5	6.7
Center	Left	11	1.4	7.9
Center	Up	15	1.6	9.4

Compared to the human eye, the model is 45 times slower due to its mechanical limitations. These limitations include the valve's response time, the program processing time, the acquisition cards voltage response time, and the valves closed state to fully flowing state. The model is limited in the down motion because of the physical restrictions, achieving the 6 of the 9 basic eye positions.

This interface demonstrates to be simple and yet reliable to monitor the movement of the muscles in the human eye. Thus, the interface can be an advantage in many applications where the eye movement is tracked like in computer games [12] where it can be used as a joystick to navigate through a level, or to help severely disabled people [13] to help them control a computer, or a way to measure spatial attention [14] for therapies and studies. and even helping studies of visual marketing [15].

5 Conclusions

Representing human eye movements accurately using vector or mechanical analysis results is difficult because of the complexity of the human vision system. Nevertheless, it turns out to be easier to understand how eye movement is produced with the use of everyday language as "too much" or "too little". With use of artificial intelligence techniques, such movements can be represented as a simple and easy to understand process, achieving good approximation of the real movement without the use of a mathematical model. Artificial intelligence techniques can be used as the link between physiology and engineering. With the use of this system, medical students can comprehend in a more interactive and didactic way such complexities. The muscles were a challenge since real muscles can be contracted and elongated. The design of the air muscles can elongate easily; but the contraction was made by rubber bands which not always applied the same force. This model proved to be a good alternative for emulating the eye's movement for potential applications of algorithms designed for robotic systems.

References

1. Robinson, D.: A quantitative analysis of extraocular muscle cooperation and squint. Invest. Ophthalmol. **14**, 801–825 (1975)
2. Kowler, E.: Eye movements: the past 25 years. Vis. Res. **51**, 1457–1483 (2011)
3. Robert, S.: Ocular torsion and th function of the vertical extraocular muscles. Am. J. Ophtalmol. **79**, 292–304 (1975)
4. Fry, G.: The center of rotation of the eye. Am. J. Optom. **39**, 581–595 (1962)
5. Maysam, F., Diedrich, G., et al.: Survey of utilisation of fuzzy technology in medicine and healthcare. Fuzzy Sets Syst. **120**, 331–349 (2001)
6. Zimmermann, H.: Fuzzy Control, Fuzzy Set Theory – and Its Applications. Springer, Heidelberg (1996). https://doi.org/10.1007/978-94-010-0646-0
7. Sugeno, M.: An introductory survey of fuzzy control. Inf. Sci. **36**, 59–83 (1985)
8. Jang, J.R.: ANFIS: adaptive-network-based fuzzy inference system. IEEE Trans. Syst. Man Cybern. **23**(3), 665–685 (1993)
9. Haykin, S.: Neural Networks and Learning Machines. Prentice Hall, Upper Saddle River (2008)
10. Nauck, D., Klawonn, F., Kruse, R.: Foundations of Neuro-Fuzzy Systems. Wiley, Hoboken (1997)
11. Jang, J.-S.: ANFIS: adaptive-network-based fuzzy inference system. IEEE Trans. Syst. Man Cybern. **23**, 665–685 (1993)
12. Lin, C., Huan, C., et al.: Design of a computer game using an eye-tracking device for eye's activity rehabilitation. Opt. Lasers Eng. **42**, 91–108 (2004)
13. Norris, G., Wilson, E.: The eye mouse, an eye communication device. In: Proceedings of the IEEE 23rd Northeast Bioengineering Conference (2002)
14. Daniel, C., Michael, J.: Eye tracking: research areas and applications. In: Encyclopedia of Biomaterials and Biomedical Engineering, pp. 573–582 (2004)
15. Wedel, M., Pieters, R.: Eye tracking for visual marketing. Found. Trends Mark. **1**(4), 231–320 (2008)

Creating Weather Narratives

Arsénio Reis[1]([✉]) [ID], Margarida Liberato[2,3] [ID], Hugo Paredes[1] [ID],
Paulo Martins[1] [ID], and João Barroso[1] [ID]

[1] INESC TEC, University of Trás-os-Montes and Alto Douro,
Vila Real, Portugal
{ars,hparedes,pmartins,jbarroso}@utad.pt
[2] Instituto Dom Luiz, Faculdade de Ciências, Universidade de Lisboa,
Lisbon, Portugal
mrl@utad.pt
[3] Escola de Ciências e Tecnologia,
Universidade de Trás-os-Montes e Alto Douro, Vila Real, Portugal

Abstract. Information can be conveyed to the user by means of a narrative, modeled according to the user's context. A case in point is the weather, which can be perceived differently and with distinct levels of importance according to the user's context. For example, for a blind person, the weather is an important element to plan and move between locations. In fact, weather can make it very difficult or even impossible for a blind person to successfully negotiate a path and navigate from one place to another. To provide proper information, narrated and delivered according to the person's context, this paper proposes a project for the creation of weather narratives, targeted at specific types of users and contexts. The proposal's main objective is to add value to the data, acquired through the observation of weather systems, by interpreting that data, in order to identify relevant information and automatically create narratives, in a conversational way or with machine metadata language. These narratives should communicate specific aspects of the evolution of the weather systems in an efficient way, providing knowledge and insight in specific contexts and for specific purposes. Currently, there are several language generator' systems, which automatically create weather forecast reports, based on previously processed and synthesized information. This paper, proposes a wider and more comprehensive approach to the weather systems phenomena, proposing a full process, from the raw data to a contextualized narration, thus providing a methodology and a tool that might be used for various contexts and weather systems.

1 Introduction

Weather narration for the blind is an extremely challenging objective as the blind can't visually check or confirm the information. Methodologically and technically there are other contexts for which this project proposal can be of great interest as well. The automatic construction of weather narratives will be developed with a strong focus on previously studied techniques for the narration to blind users of scenes and real-world dynamics in real time, as well as those techniques used in other projects,. Most of the time, computer interfaces for the visually impaired are based on narratives of the

M. Antona and C. Stephanidis (Eds.): HCII 2019, LNCS 11573, pp. 312–322, 2019.
https://doi.org/10.1007/978-3-030-23563-5_25

information to be transmitted to the user. A simple example is Microsoft Narrator, which narrates the computer screen content, allowing visually impaired users to access Microsoft Windows. In other more complex systems dedicated to blind users narration play a very important role, as it defines how information is presented and communicated to the user. In the systems BLAVIGATOR [1] and C4BLIND [2] an electronic white cane was developed, in collaboration with the University of Texas (UT) (the project C4BLIND UTAPEXPL/EEISII/0043/2014). This cane uses several sensors, characteristic scene elements, and numerical models to detect the surrounding environment and to position the user in a specific context of the scene, as well as to identify the elements of the scene. With this knowledge, it is possible to assist the blind user in specific actions, e.g., navigate to a location, identify an object, etc. In order to effectively assist the user, it is necessary to extract meaningful information out of this knowledge, according to the user's intention and current context, which is communicated to the user as a narrative. This use case of a narrative to inform a blind user is particularly important, as it attributes to the narration the role of replacing the sense of sight.

In a broader sense, and including other types of information, namely big data from scientific activities, there has been a great progress in the data acquisition, processing, and visualization technologies, making large amounts of data available to the general public, e.g. data from: atmospheric activity, financial markets, consumer marketing, sports analysis, genome, etc. These data and technologies are very useful for science, providing rich and detailed information, though a simple and eloquent narrative is desirable in some contexts and for some purposes.

The observation and recording of weather phenomena generate large amounts of data, covering long periods of time and geographic areas. This data is used to build numerical models, which together with advanced computation techniques, provide forecasts of the evolution of weather systems. This activity is essential to understand our planet, as well as to plan human activities with greater predictability.

2 Literature Review

2.1 Natural Language Narration

The narration approach draws on the experience that the team at INESC TEC acquired with the development of the previous projects funded by FCT, SmartVision (PTDC/EIA/73633/2006) and Blavigator (RIPD/ADA/109690/2009) [1], and C4Blind (UTAPEXPL/EEISII/0043/2014) [2], funded by the UT Austin Portugal Program. Considering the visually impaired and the dynamics of the environments, the static description (narratives) of urban environments are not good enough, so the project developed a mobile digital platform, using noninvasive natural interfaces, well suited to meet user needs. A user-centric design methodology was used, in partnership with the ACAPO (Association of Blind and Partially Sighted Portugal) to conduct extensive tests [3, 4].

The Natural Language Generation (NGL) is included in the artificial intelligence (IA) research field, as a computational linguistics subfield, with the main objective of producing understandable texts in human language, by using computer systems algorithms. These algorithms will input some form of nonlinguistic information and use knowledge about language and application domain in order to automatically produce a text message. This field is characterized by a variety of theoretical approaches, which has the effects of fragmenting the community and reducing communication and cooperative research efforts. Still, several reference architectures have been proposed for NGL systems and the technology is mature enough to be introduced in commercial products [5, 6]. Some NGL systems have been created to produce textual weather reports, from simulation forecast models and human annotated forecast data. An excellent example is the system used by the Canadian Weather Services since 1993, which produces text reports from numerical weather simulation data, annotated by a human forecaster [7–9]. An automated narrator system can be based on a NGL architecture, as proposed by Reiter and Dale [6].

Currently, the Apache OpenNLP library is a machine-learning based toolkit for the processing of natural language text that supports the most common NLP simple and advanced tasks [10]. The OpenNLP CCG Library is an open source natural language processing library written in Java, which provides parsing and realization services based on Mark Steedman's Combinatory Categorial Grammar (CCG) formalism [11]. The library makes use of multimodal extensions to CCG developed by Jason Baldridge [12].

2.2 Advanced Computing and E-Science

Weather systems data analysis, information extraction and visualization are resource-demanding activities regarding data and computation management. The Texas University (TU), Texas Advanced Computer Center (TACC), provides a high-performance computing (HPC) cyberinfrastructure, designed to cope with the most demanding research projects. To do so, it has developed the Agave Platform, which provides an ecosystem of APIs, software development kits and tools to power cyber-infrastructure at the national research level and provide the foundation for the next generation of science gateways. Agave aims to reduce the development burden on science gateway engineers, as well as individual scientists making use of computational resources, by providing a set of flexible, scalable software components that solve common problems across hybrid cloud and high-performance computing. At the heart of the platform is a set of APIs for managing the systems, applications, jobs and metadata involved in any computational experiment. The primary goal of Agave is to accelerate the development of web-enabled science projects and, thereby, drastically reduce the time of discovery [13–15].

2.3 Weather Systems Phenomena

Extratropical cyclones, atmospheric fronts and frontal systems are central components of weather (and hence climate) over much of the world. These frequent phenomena (every few days over many extratropical regions) are associated with the day-to-day weather conditions, including among others, precipitation, dramatic changes in temperature and wind (direction and speed) and extreme events [16]. In fact, wind extremes and heavy precipitation events occurring in the winter over land in the midlatitudes are almost always associated with extratropical cyclones (e.g. [17–19]). It is well known that the Azores Archipelago, due to its location, is prone to the occurrence of these extreme phenomena and associated hazards [19, 20].

With the development of computer technology and its rapid adoption by the atmospheric and climate sciences, it became clear that comprehensive cyclonic and frontal analysis would benefit greatly from automatic, computer analysis [16, 21–23]. Algorithms on data provided on a four-dimensional structured grid for the efficient detection and tracking of features in spatiotemporal atmospheric data have and continue to be developed at an increasing level of complexity. These may include the precise localization of the occurring genesis, lysis, merging and splitting events, and may allow access to their natural and socioeconomic impacts. However, given the difficulties of arriving at an accepted definition of these weather systems, clearly designing a numerical scheme from 3D analyses to identify cyclones, fronts and frontal zones or atmospheric rivers is a task of even greater difficulty and is still an important research theme [24].

Long-term climate datasets are critical both for understanding climate variations and evaluating their simulation in climate models. Since the 1990s, major national and international efforts have led to the creation of climate datasets, with reconstructions of meteorological and oceanographic fields (see, e.g. [25–30]), called retrospective analyses or 're-analyses', which may span extended periods of time. Examples are re-analyses covering the whole 20th century (The Twentieth Century Reanalysis Project; 20CR or the ECMWF EU-funded ERACLIM project; ERA20C reanalysis) or the Last Millennium climate Reanalysis (LMR) project.

3 Plan and Methods

3.1 The Plan

To develop the project, we intent to execute 3 tasks, where task 1 and task 2 have a high degree of overlapping, as represented in the Timeline figure.

In a first phase, and as part of tasks 1 and 2, the weather system's related information will be researched, including the following:

Identification of the weather phenomena to narrate and to work on, including fronts, atmospheric rivers, and extratropical cyclones; identification of the available data sources and their parameters; identification of the information extraction methodologies for the selected phenomena. At the same time, some narrative format mockups will be studied and tested, in order to design the narrative styles to be developed later. These mockups will be tested with an audience, as defined by the target audiences for the proof-of-concept demonstrators. Also, in this first phase, and in parallel with the previously described work, an operation model will be designed to retrieve and process data, considering the data sources and the data characteristics.

On a second phase, overlapping task 2, the work will be focused on designing the pipeline model and designing the specific components to be developed later.

On a third phase, overlapping task 2, the focus will be on software development, including the development of the necessary components the development of the NLG, and the individual test of these elements.

On a fourth phase, overlapping task3 and a part of task 2, the elements will be combined and tested, as a full featured system pipeline and probably some adjustments and software refactoring will also be executed.

A fifth, and final phase, overlapping task3 and a part of task 2, will focus on assembling the system in specific configurations and data sources, in order to produce the proof-of-concept narratives demonstrators.

3.2   The Methods

Supporting the Full Narration Process

The system to support the process will be designed as a pipeline, constituted of several functional components, properly arranged for each specific narrative type. This approach was chosen in order to have functional independence and isolation of the several components, as well as to keep the system as flexible as possible. This solution provides a framework on which different developers can add components to the systems, as well as use the existing ones in combinations that might suite their specific needs: type of narrative, raw data, weather phenomena, etc. In other previous work, this narration paradigm has been useful to design interactions with users [31–33].

The system's components will be developed as software parts that can be independently designed, implements and tested. This degree of flexibility and the exploratory character of the project are well suited to justify an agile approach, regarding the software development methodology, which will be our main choice.

The final system assembly will be tested in the Agave system [8] at the UT TACC. The weather systems data is particularly demanding in terms of storage, bandwidth and processors. TACC and Agave are the perfect matches to deal with these big data science requirements.

Natural Language Narrative Construction

The narratives of weather phenomena, in natural language, will be produced by a system component of the type "Narration composition", as described in task 2. This component will be implemented as a simple Natural Language Generator (NLG), to be developed, using the OpenNLP CCG Library. The NGL will have a classic pipeline design, for which a set of document plans will be designed, according to a communicative goal for each specific context. These plans will define the information to be included in the text narrative, as well as provide a structure for a coherent text message. They will be further processed on a microplanning stage, where the plans for the individual sentences will be defined. A final stage of linguistic realization will render a text document, thus producing a text narrative meaningful for a specific context, as previously defined.

Weather Systems Phenomena Data Extraction and Classification

The IDL team has extensive experience in developing and using such objective algorithms, namely for detecting and tracking extratropical cyclones [18, 22] and atmospheric rivers [23]. Thus, extratropical cyclones (ECs), associated frontal systems and atmospheric rivers (ARs) will be the weather systems used for this exploratory project. These algorithms are mostly applied to large gridded data sets that are produced by climate models (e.g. multidecadal period of atmospheric and ocean reanalysis data at different resolutions from several data centres [25]).

Extratropical Cyclones (ECs)

The structure and evolution of ECs, as viewed from the surface, has been described throughout the 20th century by the development of refined conceptual models (namely, the Norwegian, Shapiro and Keyser, cyclone models). Today three-dimensional conceptual models of extratropical cyclones provide a framework for understanding their dynamical evolution (see the annex for details). Detailed analysis of the structure and evolution of individual extratropical cyclones suggest that although there is no universal lifecycle of extratropical cyclones, some general cyclone characteristics can be identified [28, 30].

Atmospheric Rivers (ARs)

ARs are relatively narrow and elongated filaments of high-water vapor transport, with their occurrence generally interpreted as large atmospheric water vapor transport events in the extra-tropics [26]. This phenomenon is associated with tropical moisture exports and occurs often in combination with the passage of extratropical cyclones [19]. The warm, moist air in the cyclones' warm sector is swept up by the advancing cold front, leading to a filament of high specific moisture content, which is transported northward at the basis of the warm conveyor belt ([19], Fig. 5). As the cyclones travel poleward, such bands of high humidity may trace back over large distances, extending from regions of high sea surface temperature in the subtropics into the midlatitudes, leading to strong precipitation along its path (e.g., storm "Xynthia") [18]. Such structures transport more than 90% of the total midlatitude vertically-integrated water vapor [26] and can lead to intense precipitation over different continental regions due to its interaction with the topography [28]. Recently, an international agreement has been reached regarding the relationships between ARs, warm conveyor belts, and tropical moisture exports [26]. The term warm conveyor belt refers to the zone of dynamically uplifted heat and vapor transport close to a midlatitude cyclone. This vapor is often transported to the warm conveyor belt by an AR and was earlier brought poleward to the extra-tropics by tropical moisture exports ([26], Fig. 1). The uplift associated with the warm conveyor belt typically leads to heavy rainfall. This generally marks the downwind end of an AR, unless the AR has experienced orographic uplift earlier on, causing rainout over mountain areas.

The impacts of ARs have been analyzed in detail in Western Europe and revealed the importance of ARs in extreme weather events, both for case studies and from a climatological perspective [23].

An objective identification of ARs has been developed by the IDL team by means of the IVT computed with both the NCEP–NCAR and ERA Interim datasets [23].

Weather Fronts

Another important feature associated with cyclones and heavy precipitation events and its impacts are weather fronts. An objective method, the Thermal Method [29] is used to detect and depict weather fronts over a gridded map. The method is based on the choice of a Thermal Frontal Parameter (θe), directly related with the location and intensity of the frontal zone. The link between frontal precipitation and extreme weather events will also be addressed by objectively evaluating the proportion of precipitation collocated to the fronts.

Conceptual Models

All these conceptual models are widely used in educational meteorology courses and textbooks throughout the world to illustrate the basic structure and evolution of weather systems, such as extratropical cyclones and atmospheric rivers. These conceptual models have been the basis for objective detection and tracking methods developed by IDL researchers [21–23], which will be used here in this exploratory project.

To describe the structural evolution of ECs or ARs, a combination of surface observations, satellite imagery, radar data and model output are needed. These allow meteorologists to identify and describe the evolution of cyclonic flows such as the warm and cold conveyor belt flows and the dry intrusion. Atmospheric parameters such as horizontal wind (u, v), pressure velocity (ω), temperature and geopotential may be obtained with a 0.125° longitude × 0.125° latitude grid resolution available at 37 isobaric levels from 1000 to 1 hPa, at time intervals of 6 h or more.

The Project Proposal

This exploratory project will be based mainly on reconstructions of meteorological and oceanographic fields, meaning different available reanalysis datasets (e.g. 6hourly ERA-Interim 19792017 reanalysis over the Northern Hemisphere with a 0.75 degree of horizontal resolution; [25] and references therein). However, the methodology will be developed for any gridded data, so that it will also be tested for atmospheric and ocean reanalysis data at different resolutions from several data centres – ECMWF, National Centers for Environmental Prediction–National Center for Atmospheric Research (NCEP–NCAR), or the Japan Meteorological Agency (JMA) as well as for simulations from six Coupled Model Intercomparison Project Phase 5 (CMIP5) global climate models (GCMs) to quantify possible changes during the current century, with emphasis on the Atlantic Ocean.

This project will be aimed at building a general tool which will include future developments expected as reanalysis datasets become more diverse (atmosphere, ocean and land components), more complete (moving towards Earth-system coupled reanalysis), more detailed, and of longer timespans.

4 Tasks

The project is divided into 3 straightforward tasks.

4.1 Task 1 - Identification and Definition of Contexts and Phenomena

In this task we will identify and define the phenomena for which the narratives will be created, including atmospheric rivers, fronts, and extratropical cyclones, as well as the contexts for those narratives. A narrative provides insight and meaning to data in a specific context. Three user scenarios (contexts) will be adopted to develop the narration process. This set represents three very different types of narration, regarding the target audience and the message format:

- Weather phenomena description, in a natural and conversational language, targeted at the general public. This narrative will provide a description or explanation of a phenomena as it is represented by the data at an exact moment and location.
- Weather system's data description, as a metadata classification of the phenomena data, targeted at machine systems. The main goal is to use the narration process as a method to classify and index the original data, thus creating the metadata to be used by other machine systems, e.g., indexing and search engines systems, machine leaning systems, image analysis, etc.
- Weather phenomena progression, in a natural and scientific language, targeted at climate science students. This narrative describes the progression of a phenomena during its occurrence, as it develops over time and location. It provides a didactic storyline that can be used in teaching or science dissemination, as a solo message or as part of a multimedia message, including video and computer-generated graphics.

According to the previously described contexts, a set of weather phenomena will be chosen for narration. The phenomena will be characterized regarding the detailed methods, including algorithms and criteria that might be used to identify their occurrence and development in the original climate and weather data. These methods will later be used to extract information from the data, which will be rendered and formatted according to the context and target usage, creating a meaningful message.

4.2   Task 2 - Development of the Narration Process

In this task the narration process will be developed, which will be applied to three demonstration scenarios, according to the previously described contexts.

The process will be designed as a pipeline process, in which several components will be combined in order to have a full functional process, starting with the raw data and concluding with a narrative. The types of components will be:

- Data biding, to connect and bind to a raw data source;
- Data processing, to process raw data from several sources and create usable datasets;
- Information extraction, to search and extract information from data;
- Narration composition, to create a narrative from the extracted information, giving it meaning according to a context.

In this task several components of the previously defined types will be developed, which will then be assembled in distinct configurations, in order to create three narratives with different characteristics, which will act as proofs of concept for this project of weather system's narration.

The narratives will be created according to the phenomena and context previously defined in task 1.

The pipeline design approach provides flexibility and segregation in the different phases of the process by providing a framework for the independent development and lifecycle management of components regarding the following: distinct data sources, raw data formats, usable data sets, weather systems phenomena, and narration type.

4.3   Task 3 – Integration, Tests and Evaluation

In this task, the previously developed components will be integrated, by assembling them into several pipeline configurations, in order to evaluate the complete narration process, regarding its functionality as a fully automatic narrative creation system. First, the components will be individually tested, as independent functional units, after which they will be assembled and tested as a full functional system. To produce the final narratives, the system will be assembled in distinct pipeline configurations, using the necessary components, according to the desired outcome.

The ultimate evaluation of the system will be carried out by assessing the narratives produced, according to the three contexts defined in task 1 and later implemented in task2. In fact, the quality of these narratives will be an indicator of the overall quality and validity of the project's proof-of-concept.

The final narratives deliverables to assess are the following: a weather phenomenon description, in a natural and conversational language; a weather system's raw data metadata classification; and a weather phenomenon progression, in natural and scientific language.

Besides the evaluation of the final outcomes, other aspects of the project will also be evaluated regarding the functionality, sustainability and reproducibility of the automatic weather systems narration process. In fact, the process is heavily based on data and processing capacity, which can be provided by science infrastructures, but after having validated the automatic weather systems narration process, the next step should be to tailor and streamline the process for some marketwise application segments, in order to have the same positive outcomes, but consuming a reasonable amount of resources, accordingly to each application segment.

5   Conclusion

This exploratory project follows up the work started by CE4Blind project, extending the experience of the team to the domain of climate change. The proposal extends to supercomputing, the Big Data analytics domain and climate and atmospheric sciences.

The final goal is to combine and study the areas of weather systems, data computation, narrative creation, and human computer interface for the blind. A major milestone is the development of a working prototype, which may be tested and demonstrated outside the lab.

Acknowledgements. This work was supported by the project "WEx-Atlantic - Weather Extremes in the Euro Atlantic Region: Assessment and Impacts" (PTDC/CTA-MET/29233/2017) funded by Fundação para a Ciência e a Tecnologia, Portugal (FCT) and Portugal Horizon2020.

References

1. Fernandes, H., Sousa, A., Paredes, H., Filipe, V., Barroso, J.: Feature detection applied to context-aware blind guidance support. In: Antona, M., Stephanidis, C. (eds.) UAHCI 2015. LNCS, vol. 9178, pp. 129–138. Springer, Cham (2015). https://doi.org/10.1007/978-3-319-20687-5_13

2. Fernandes, H., Costa, P., Paredes, H., Filipe, V., Barroso, J.: Integrating computer vision object recognition with location based services for the blind. In: Stephanidis, C., Antona, M. (eds.) UAHCI 2014. LNCS, vol. 8515, pp. 493–500. Springer, Cham (2014). https://doi.org/10.1007/978-3-319-07446-7_48

3. Costa, P., Barroso, J., Fernandes, H., Hadjileontiadis, L.J.: Using Peano–Hilbert space filling curves for fast bidimensional ensemble EMD realization. J. Adv. Sig. Process. 1 (2011). https://doi.org/10.1186/168761802012181

4. Fernandes, H., Conceição, N., Paredes, H., Pereira, A., Araújo, P., Barroso, J.: Providing accessibility to blind people using GIS. Univers. Access Inf. Soc. 11(4), 19 (2011). https://doi.org/10.1007/s1020901102557

5. Mellish, C., Scott, D., Cahill, L., Paiva, D., Evans, R., Reape, M.: A reference architecture for natural language generation systems. Nat. Lang. Eng. 12(1), 1–34 (2006)

6. Reiter, E., Dale, R., Feng, Z.: Building Natural Language Generation Systems, vol. 33. Cambridge University Press, Cambridge (2000)

7. Goldberg, E., Kittredge, R., Driedger, N.: FoG: a new approach to the synthesis of weather forecast text. IEEE Expert (Special Track on NLP) (1994)

8. Goldberg, E., Driedger, N., Kittredge, R.I.: Using natural language processing to produce weather forecasts. IEEE Expert 9(2), 4553 (1994)

9. Ramos Soto, A., Bugarín, A.J., Barro, S., Taboada, J.: Linguistic descriptions for automatic generation of textual short term weather forecasts on real prediction data. IEEE Trans. Fuzzy Syst. 23, 4457 (2015). ISSN 10636706

10. Apache OpenNLP (2017). https://opennlp.apache.org

11. Open CCG (2017). http://openccg.sourceforge.net/

12. Baldridge (2017). http://www.jasonbaldridge.com/

13. Dooley, R., Hanlon, M.R.: Recipes 2.0: building for today and tomorrow. Concurr. Comput.: Pract. Exp. 27(2), 258–270 (2015)

14. Wang, L., Buren, P.V., Ware, D.: Architecting a Distributed Bioinformatics Platform with iRODS and iPlant Agave API. In: 2015 International Conference on Computational Science and Computational Intelligence (CSCI), Las Vegas, NV, pp. 420–423 (2015). https://doi.org/10.1109/csci.2015.121

15. Gesing, S.: et al.: Gathering requirements for advancing simulations in HPC infrastructures via science gateways. Futur. Gener. Comput. Syst. http://doi.org/10.1016/j.future.2017.02.042. Accessed 14 Mar 2017, ISSN 0167739X

16. Simmonds, I., Keay, K., Bye, J.A.T.: Identification and climatology of southern hemisphere mobile fronts in a modern reanalysis. J. Clim. (2012). https://doi.org/10.1175/jclid1100100.1

17. Liberato, M.R.L., Pinto, J.G., Trigo, I.F., Trigo, R.M.: Klaus an exceptional winter storm over Northern Iberia and Southern France. Weather 66, 330334 (2011). https://doi.org/10.1002/wea.755

18. Liberato, M.L.R., et al.: Explosive development of winter storm Xynthia over the subtropical North Atlantic Ocean. Nat. Hazards Earth Syst. Sci. 13, 22392251 (2013). https://doi.org/10.5194/nhess1322392013

19. Dacre, H.F., Clark, P.A., Martinez Alvarado, O., Stringer, M.A., Lavers, D.A.: How do atmospheric rivers form? Bull. Amer. Meteor. Soc. 96(8), 12431255 (2015). https://doi.org/10.1175/BAMSD1400031.1

20. Liberato, M.L.R.: The 19 January 2013 windstorm over the north Atlantic: largescale dynamics and impacts on Iberia. Weather Clim. Extrem. 56, 16–28 (2014)

21. Trigo, I.F.: Climatology and interannual variability of storm tracks in the EuroAtlantic sector: a comparison between ERA40 and NCEP/NCAR reanalyses. Clim. Dyn. 26, 127–143 (2006)

22. Neu, U., et al.: IMILAST—a community effort to intercompare extratropical cyclone detection and tracking algorithms. Bull. Am. Meteor. Soc. **94**(4), 529–547 (2013). https://doi.org/10.1175/BAMSD1100154.1

23. Ramos, A.M., Trigo, R.M., Liberato, M.L.R., Tome, R.: Daily precipitation extreme events in the Iberian Peninsula and its association with atmospheric rivers. J. Hydrometeorol. **16**, 579–597 (2015)

24. Limbach, S.: Software tools and efficient algorithms for the feature detection, feature tracking, event localization, and visualization of large sets of atmospheric data. PhD thesis, University of Mainz, 256 p. (2013)

25. Dee, D.P., Balmaseda, M., Balsamo, G., Engelen, R., Simmons, A.J., Thépaut, J.N.: Toward a consistent reanalysis of the climate system. Amer. Meteorol. Soc., Bull (2014). https://doi.org/10.1175/BAMSD1300043.1

26. Dettinger, M., Ralph, F.M., Lavers, D.: Setting the stage for a global science of atmospheric rivers. Eos Trans. AGU **96** (2015), https://doi.org/10.1029/2015eo038675

27. Ramos, A.M., Tomé, R., Trigo, R.M., Liberato, M.L.R., Pinto, J.G.: Projected changes in atmospheric rivers affecting Europe in CMIP5 models. Geophys. Res. Lett. **43**, 9315–9323 (2016). https://doi.org/10.1002/2016GL070634

28. Ramos, A.M., et al.: Atmospheric rivers moisture sources from a Lagrangian perspective. Earth Syst. Dynam. **7**, 371384 (2016). https://doi.org/10.5194/esd73712016

29. Shemm, S., Rudeva, I., Simmonds, I.: Extratropical fronts in the lower troposphere global perspectives obtained from two automated methods. Q. J. R. Meteorol. Soc. **141**(690), 1686–1698 (2015). https://doi.org/10.1002/qj.2471

30. Ulbrich, U., Leckebusch, G.C., Pinto, J.G.: Extratropical cyclones in the present and future climate: a review. Theor. Appl. Climatol. **96**, 117131 (2009). https://doi.org/10.1007/s0070400800838

31. Rocha, T., Fernandes, H., Reis, A., Paredes, H., Barroso, J.: Assistive platforms for the visual impaired: bridging the gap with the general public. In: Rocha, Á., Correia, A.M., Adeli, H., Reis, L.P., Costanzo, S. (eds.) WorldCIST 2017. AISC, vol. 570, pp. 602–608. Springer, Cham (2017). https://doi.org/10.1007/978-3-319-56538-5_61

32. Reis, A., Barroso, I., Monteiro, M.J., Khanal, S., Rodrigues, V., Filipe, V., Paredes, H., Barroso, J.: Designing autonomous systems interactions with elderly people. In: Antona, M., Stephanidis, C. (eds.) UAHCI 2017. LNCS, vol. 10279, pp. 603–611. Springer, Cham (2017). https://doi.org/10.1007/978-3-319-58700-4_49

33. Gonçalves, C., Rocha, T., Reis, A., Barroso, J.: AppVox: an application to assist people with speech impairments in their speech therapy sessions. In: Rocha, Á., Correia, A.M., Adeli, H., Reis, L.P., Costanzo, S. (eds.) WorldCIST 2017. AISC, vol. 570, pp. 581–591. Springer, Cham (2017). https://doi.org/10.1007/978-3-319-56538-5_59

RingBoard 2.0 – A Dynamic Virtual Keyboard Using Smart Vision

Taylor Ripke, Eric O'Sullivan, and Tony Morelli$^{(\boxtimes)}$

Central Michigan University, Mount Pleasant, MI 48859, USA
{ripkeltj,osullle,morella}@cmich.edu.com

Abstract. Computers have evolved throughout the digital era becoming more powerful, smaller, and cheaper. However, they are still lacking basic accessibility features that appeal to all users. They can be controlled with your voice and eye movement, but there is still much work to be done. This paper presents RingBoard 2.0, a dynamic virtual keyboard that uses computer vision to recognize and track hand movements and gestures. It allows for basic input to a computer using a web camera. This application was built to provide additional accessibility features for those who experience tremors or limited motor capability in their hands, which make it difficult to interact with a standard keyboard and mouse. At the core, it is built to recognize any form of a hand and can accurately track it, regardless of sporadic movement. This paper is an extension of previous work describing touch input for a computer using the HP Sprout [2].

Keywords: Accessibility · Vision · Touch · Keyboard · Tracking

1 Introduction

The QWERTY keyboard patented by Sholes in 1878 is the defining keyboard still used by many electronic devices, including smart phones and computers [1]. It provides a convenient way to quickly transcribe manuscripts into print. The advent of the digital era brought significant change to how information is conveyed and produced; however, it was rapidly leaving many individuals behind. Although revolutionary for its time, there is little information regarding additional accessibility features for those with cognitive or motor impairments. Those individuals face similar challenges today when interacting with modern technology, especially a keyboard and mouse which require precise motor control and concentration.

This paper introduces additional enhancements to RingBoard, an application designed by Wojcik et al. [2] as shown in Fig. 1. RingBoard was designed for use "with a personal computer such that a person with a mobility disability who cannot utilize a standard physical keyboard would be able to better interact with a standard computer" [2]. The updated version integrates advanced computer vision algorithms to track the user's hand while interacting with the system. These additions facilitate those who may have difficultly pressing on the mat with fingers due to limited motor control or loss of limb. The system can adapt to virtually any input, compensating for a limited range of motion or tremors. The paper will detail the algorithms and provide a comparison between different forms of input for the system.

© Springer Nature Switzerland AG 2019
M. Antona and C. Stephanidis (Eds.): HCII 2019, LNCS 11573, pp. 323–333, 2019.
https://doi.org/10.1007/978-3-030-23563-5_26

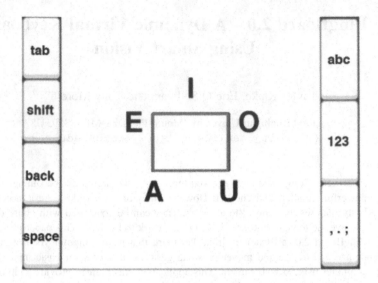

Fig. 1. RingBoard.

2 Related Work

Individuals with physical disabilities such as Cerebral Palsy, Spina Bifida, and tremors may have difficulty interacting with a traditional computer keyboard and mouse. These users may take longer to complete typing tasks and may experience various errors such as holding a key too long, pressing adjacent keys, missing keys, failing to hold two keys simultaneously, pressing the wrong key, or pressing more than one key at a time on accident (Trewin 1999). It is possible to provide accommodations for some of the errors by adjusting settings on the computer. Common solutions include "adjusting the repeat time of a key, looking for sticky keys, and using CAPS Lock [2].

As stated previously, users with limited mobile capabilities may find it difficult to interact with a standard keyboard. People who have limited capabilities in their hands may be able to use a computer keyboard keyguard, which is specifically designed for users with limited motor control. The design allows the user to increase typing accuracy and help stabilize fingers on the right keys [3]. Additional technologies include a no hands mouse controlled by foot movement and keyboard control through foot pedals, which allow the user to press important buttons such as shift, ctrl and alt [3].

AbleNet provides a variety of tools, such as switches, that can be used with a computer to provide users with the capability to interact with the system. Some popular switch devices include the Big Red, which is a large, red button users can press, and the Blue2 Bluetooth Switch, which gives users the capability to press two large buttons [4]. The devices can be used to work with existing applications but are also well-suited to be custom programmed to fit the user's needs.

Another possible solution to allow the user to interact with the system if they have limited control in their hands and arms is to use technology that tracks the position of the head relative to the screen. For example, the SmartNav 4: AT provides a hands-free mouse solution that allows the user to control their computer using only head movements. This solution provides a way to provide mouse input, as well as keyboard input through the use of a virtual keyboard. It even supports switch or foot pedal input to provide additional input [5]. Earlier versions of similar systems exist, which include printing a copy of a keyboard onto a piece of paper and attaching a laser to the user's head. A computer vision algorithm interfaced with a web camera was used to determine where the laser was on the keyboard and translate it into corresponding text input [8]. The second version of RingBoard presented in this paper was motivated from this implementation.

However, if the user has very limited motor capabilities, another possible solution is Eyegaze Edge, which is a eye-operated communication and control system that uses advanced image processing techniques to track the position of the eye relative to the screen at 60 frames per second, providing accuracy up to a 1/4 inch or less [6]. Users are able to type on a keyboard and generate speech. To click a button, the user waits a specified amount of time, such as half a second [6].

Similar systems also use a graphical keyboard that "allows the user to change the size and location of virtual onscreen keyboard buttons" [9]. These onscreen keyboards have also been used with specification applications, not just customizable character input. For example, one solution was to populate an onscreen keyboard with commonly type phrases specific to the active application [10]. The underlying system can determine a user's commonly typed phrases and "prepopulate the keyboard with user specific common words and phrases" [7].

Although this research is directly applicable for computers, it may also be possible to apply these accessibility features to mobile computers with the addition of a Bluetooth touch keyboard. Previous work has shown that users are okay using an additional keyboard when interacting on a mobile device [8]. In the upcoming section discussing future work, additional accessibility features are being considered that take advantage of predictive text. For example, Gkoumas implemented a solution to predictive typing where the keyboard enlarges letters it thinks are likely to be next [9]. After some training to understand an individual's diction, it may be possible to implement a similar approach on RingBoard.

Furthermore, additional research concerning virtual keyboards has been done by considering the position of the hand [10]. Similarly, as described in this paper, RingBoard is actively involved in tracking the user's hand and using it as input to the keyboard. Rashid proposed an additional enhancement by showing the benefits of having the virtual keyboard be relative to where the user starts typing, rather than having the keys remain in the same location on mobile devices [11]. This idea was considered when implementing the first iteration of RingBoard by having the secondary keys appear in the direction of the user's hand movement [2].

3 Motivation

As previously described, the current iteration of RingBoard was motivated by previous work done in [2]. It is developed on an HP Sprout computer running Windows 10 utilizing a touch sensitive mat, touch screen, and overhead camera and projector.

The projector displays a virtual keyboard on the touch mat that the user can interact with using their fingers. In the original version, the best results were achieved using fingers for touch input rather than using a fist. Due to the inherit design of the HP Sprout, it is designed to ignore the fist when users are interacting with the touch mat. This is inconvenient for people interacting with the system who have limited precision of movement in their arms and hands.

The current iteration of RingBoard is designed to be more accessible for individuals with limited motor capabilities. To overcome the challenges of fist-based touch input, computer vision algorithms were implemented to track the user's hand. As discussed in the upcoming sections, this type of input provided a convenient solution applicable to all users, compensating for sporadic movement or limited hand functionality.

4 Using the Keyboard

Interacting with the visual keyboard is quite easy. When the application is launched, a black screen will appear for a few seconds while the system calibrates. During this time, the camera is being prepared and the background image is captured. It is essential that only the mat is present during these few seconds, as other objects, such as hands or supplies, can interfere with subsequent processing. Next, the application will wait for the user to put their hand into the frame. It will pause for five seconds, allowing the user to position their hand. The keyboard will then become active and the user can interact with the different keys and buttons.

As described in [2], the keyboard displays a set of primary keys as shown in Fig. 1. These primary keys compose the default layout of the system. To access additional letters, the user needs to move their hand between two letters to display a menu of secondary keys as shown in Fig. 2. To return to the previous menu, the user must either keep their hand stationary for five seconds or increase the surface area of their hand by opening it, the details of which are discussed in the following section.

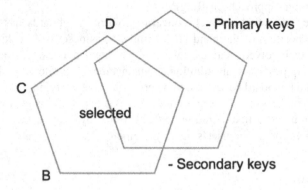

Fig. 2. Secondary keys.

Between each button press, a three-second delay is implemented to prevent the user from accidently pressing a button while moving their hand to a new location. For example, moving from the ring of letters to a button for punctuation may accidently trigger a letter, which is an inconvenience to the user. To reset the keyboard, the user can either keep their hand stationary for seven seconds or open their hand. The details of the implementations of both techniques are detailed in the following section.

5 Implementation

As stated previously, RingBoard provides two forms of input: touch and vision. This section will discuss the intricacies of how the application detects and tracks a user's hand or limb, while compensating for spurious muscle movement. The underlying vision algorithms are designed to be accessible to all individuals, as long as they can move their arm within view of the camera.

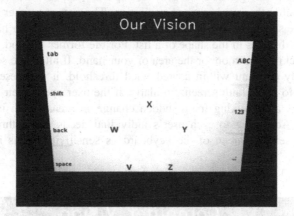

Fig. 3. Background image

The application was developed in Unity and programmed in C#. It is designed to run on the HP Sprout, although it can be modified to work on a system with multiple monitors and a downward facing camera. The Sprout has an interactive touch pad, which also serves as an additional screen, however the touch pad is not utilized and only acts as a background. Any relatively clear surface will work. The current version projects the display on the bottom screen, directly below the overhead camera. The Sprout's overhead camera can be accessed and controlled directly in Unity with the assistance of OpenCV. The OpenCV module provides an easy way to perform fundamental image processing tasks in Unity without having to run a separate program that interfaces with Unity through TCP.

$$I_n[x, y] = |I_c[x,]y - I_b[x, y]| > T \tag{1}$$

Eq. 1 Background Subtraction

The application begins by capturing an image of the touch pad without the user's hand present initially. This is done to calibrate the application for background subtraction. The image captured is taken at an angle due to the Sprout's design. A comparison can be seen by comparing Fig. 1 to Fig. 2. Furthermore, the image captures the sides of the touch mat, including the desk that it is sitting on. This interferes with image processing and is removed by isolating the interest area with black polygons as shown in Fig. 2.

During runtime, the application captures an image and compares it against the static background image to determine if an object is present. To reduce additional unwanted noise the images were converted to grey. Equation 1 depicts the equation used for background subtraction. The result of the background subtraction image is a binary image indicating areas where a change in pixel intensity was greater than the threshold T as shown in Fig. 3. I_n is the new image produced by taking the absolute value of the current image subtracted from the background image within a given threshold to produce a binary image.

As mentioned in the previous section, an important aspect of the keyboard is the ability to reset it after the user has made a selection. The user can either keep their hand stationary for five seconds or rapidly increase the surface area of their hand by opening it, assuming their hand is in the shape of a fist. For the former method, the application in constantly tracking and monitor the area of your hand. If after five seconds the area remains relatively constant within a predefined threshold, it will reset the keyboard bringing it back to the default screen. Similarly, if the user opens their hand suddenly, the application is also looking for a sudden change in area, which is also a predefined threshold. According to each user's individual needs, these thresholds can be modified in the settings menu of the keyboard as sensitivity levels may vary from person to person.

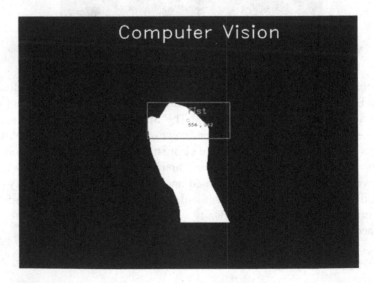

Fig. 4. Binary image with transform

OpenCV implements a findContours module that "retrieves contours from the binary image," which is useful for "shape analysis and object detection and recognition" [7]. For each of the detected contours, its area was computed and used to find the largest contour, which we assume to be the hand. To speed up the algorithm, only contours with an area >T (we used 5000), were considered. The process of finding the area of a contour is computed using image moments. Image moments are useful for describing properties of the contour, including the calculate center and area. Equations 2 and 3 are derived from OpenCV's documentation found in [7]. The [x, y] coordinates computed from the largest contour were assumed to be the hand.

$$\bar{x} = \frac{M_{10}}{M_{00}}, \; \bar{y} = \frac{M_{01}}{M_{00}} \tag{2}$$

Eq. 2 Centroid [7]

$$Area(M_{00}) = \sum \sum x^0 y^0 I[x, y] \tag{3}$$

Eq. 3 Contour Area (Adapted from [7])

This solution works assuming that only part of the hand is present at a given time. In cases where the user extends their hand to span the entire vertical space of the mat, using the x,y coordinates provides through image moments fails. Equation 2 computes the centroid, which is the center coordinates of the contour. Thus, when the hand extends the entirety of the hand, the coordinates will be centered around the wrist. To compensate, a simple transformation function was applied. The result of the image tracking can be seen in Fig. 4.

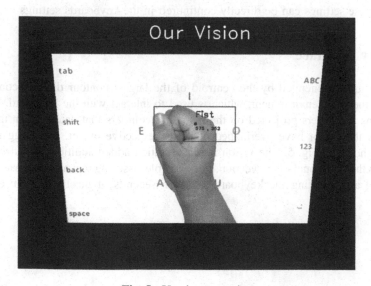

Fig. 5. User's perspective

In certain situations, or behavioral input, the centroid coordinates may be inconsistent, generating sporadic movement that in turn triggers incorrect input to the keyboard. To counteract this behavior and provide stability for those with muscle tremors, a trivial averaging function is applied to the input. The number of subsequent images used to calculate the average can be changed to determine how much averaging should be provided to minimize sporadic behavior. This was also useful to eliminate inconsistencies during image processing, which can be affected by lighting conditions or new stimuli.

The actual centroid position is proportional to the amount of area present. More area will affect how far the centroid is translated in the vertical direction. For example, when the user has their hand near the top of the mat, the corresponding centroid will also be at the top. We assume the approximate area for a hand at the top of the mat, which came about to be about 80,000 pixels. Using this information, we compute the area once the hand is visible as a proportion of the total pixels and apply it to the y-coordinate of the centroid. A quadratic expression can also be substituted for increased accuracy. As the area of the hand gradually increases, we want it to affect the y-coordinate greater once it has surpassed the center of the image. The current implementation uses a quadratic transform.

As discussed previously, the keyboard pauses for three seconds between keypresses to assure that the user does not accidently hit another button if they are moving their hand across the screen. At any screen in the application, the user has the ability to reset the keyboard back to the default screen by either keeping their hand stationary for seven seconds or by opening their hand. The first method is computed by looking at the area over a period of time. If the area did not change within a given threshold, the keyboard will automatically reset. Similarly, if the area of the hand suddenly changes by a decent amount, such as when opening their hand, it will also reset the keyboard. Both of these settings can be directly configured in the keyboards settings.

6 Unity Interface

The coordinates generated by the centroid of the largest contour directly control the cursor in the unity environment, which is used to interact with the keyboard. Each of the buttons and letters projected on the mat have colliders that detect when the coordinates of the cursor have overlapped onto the respective object, thus triggering the event as shown in Fig. 5. The vision implementation added additional challenges not present in the traditional touch version. For example, assuring that a button was pressed only once and resetting the keyboard after ten seconds of inactivity were essential features.

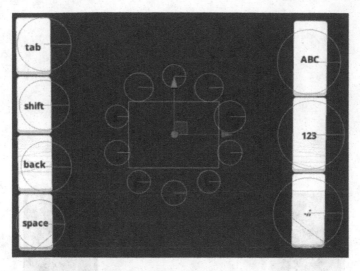

Fig. 6. Colliders

7 Practical Applications

The current version of the keyboard provides the user the capability to type text and control the mouse by tracking the location of the hand or arm. The initial release provides users the functionality of a real keyboard, including ways to type letters, numbers, and special characters. Additional buttons such as tab, shift, back, and space were also included for a more streamlined experience. The keyboard currently only supports these primary buttons as a method of input. Other buttons, such as the function keys or volume control, will be added in the future.

In addition, the current version allows the user to control the mouse. Currently, if the user stops moving the mouse, the program will delay for a specified amount of time, such as half a second, and then perform a double click action. This is useful if the user wants to open an application. The keyboard, which is displaying on the bottom screen, has an application button which provides a list of common applications that can launched by moving the mouse over the text, as shown in figure. The interface, which is running as a Unity application, communicates with a python script over TCP that is responsible for launching and putting the application into focus. Once in focus, the user can interact with the application similar to how they would with a standard mouse. An example of the system running can be seen in Fig. 6 (Fig. 7).

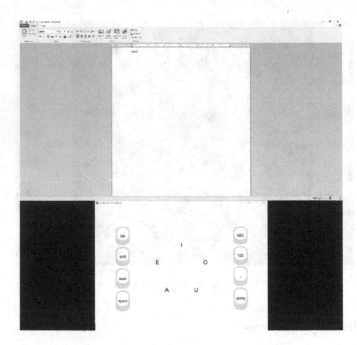

Fig. 7. Controlling external applications

8 Future Work

Many additional accessibility features will be added to RingBoard in the near future. The first will be the integration with the Windows OS. The Unity application will connect to a script responsible for controlling the computer based on the users input. For example, if the user presses the 'mouse' button, the user will be able to control the cursor on the top screen, allowing them to interact with the computer. The coordinates are mapped and sent using TCP to a script responsible for setting the position of the cursor in real time.

Furthermore, the user will have the capability to interact with common applications such as web-browsers and text-based entry applications. Current work at the time of publishing has implemented the ability for the user to type the name of the application in RingBoard, which is sent to the script responsible for spawning an instance of the application and focusing the window on the top screen. From there, the user can interact with the application using the mouse and keyboard.

Improvements to the tracking process will be continuously updated allowing for smoother control of the application while interacting with the keyboard. We are currently developing methods to provide a seamless interaction with Windows. Common challenges include handling situation requiring double-clicking events and handling multiple instances of the same application. Our goal is to have the application run at boot, which will automatically display the keyboard once the application reaches the login screen.

In addition, another area of improvement may be the addition of predictive text. Overtime the keyboard should learn the user's commonly typed phrases and be able to suggest them as additional options to the letters presented. Thus, the users would only have to navigate to the corresponding text to complete the word or phrase. These features are still being evaluated for usability and accessibility.

RingBoard provides an accessible way for those with muscular or cognitive, and verbal disabilities to interact with technology. Although current systems provide accessibility features such as text-to-speech and visual enhancements, such as color contrast and magnifying, there has not been much research in areas for manual input to a computer. RingBoard seeks to provide an accessible solution that can be easily used by all.

References

1. Sholes, C.S.: Type-Writing Machine. US Patent 207, 559, filed March 8, 1875, issued August 27 (1878)
2. Wojcik, B., Morelli, T., Hoeft, B.: RingBoard – a dynamic virtual keyboard for fist based text entry. The Journal on Technology and Persons with Disabilities (2018)
3. Fentek Industries: Computer Keyboard Keyguard Products. http://www.fentek-ind.com/Keyguard.htm#.Wy_SCqdKiUk. Accessed 23 June 2018
4. Ablenet. https://www.ablenetinc.com/. Accessed 23 June 2018
5. SmartNav by NaturalPoint: SmartNav 4: AT Overview. http://www.naturalpoint.com/smartnav/products/4-at/. Accessed 23 June 2018
6. LC Technologies, Inc.: Communicate with the world using the power of your eyes. http://www.eyegaze.com/eye-tracking-assistive-technology-device/. Accessed 22 June 2018
7. OpenCV. Web. Structural Analysis and Shape Descriptors (2018). https://docs.opencv.org/2.4/modules/imgproc/doc/structural_analysis_and_shape_descriptors.html?highlight=findcontours#findcontours
8. Ahsan, H., et al.: Vision based laser controlled keyboard system for the disabled. In: Proceedings of the 7th International Symposium on Visual Information Communication and Interaction. ACM (2014)
9. Missimer, E.S., et al.: Customizable keyboard. In: Proceedings of the 12th International ACM SIGACCESS Conference on Computers and Accessibility. ACM (2010)
10. Norte, S., Fernando, G.L.: A virtual logo keyboard for people with motor disabilities. In: ACM SIGCSE Bulletin, vol. 39, no. 3 (2007)
11. Wandmacher, T., et al.: Sibylle, an assistive communication system adapting to the context and its user. ACM Trans. Access. Comput. TACCESS 1(1), 6 (2008)
12. Armstrong, P., Wilkinson, B.: Test entry of physical and virtual keyboards on tablets and the user perception. In: Proceedings of the 28th Australian Conference on Computer-Human Interaction. ACM (2016)
13. Gkoumas, A., Komninos, A., Garofalakis, J.: Usability of visibly adaptive smartphone keyboard layouts. In: Proceedings of the 20th Pan-Hellenic Conference on Informatics. ACM (2016)
14. Yin, Y., et al.: Making touchscreen keyboards adaptive to keys, hand postures, and individuals: a hierarchial spatial abckoff model approach. In: Proceedings of the SIGCHI Conference on Human Factors in Computing Systems. ACM (2013)
15. Rashid, D.R., Smith, N.A.: Relative keyboard input system. Proceedings of the 13th International Conference on Intelligent User Interfaces. ACM (2008)

Introducing Pneumatic Actuators in Haptic Training Simulators and Medical Tools

Thibault Sénac[1], Arnaud Lelevé[1(✉)], Richard Moreau[1], Minh Tu Pham[1], Cyril Novales[2], Laurence Nouaille[2], and Pierre Vieyres[2]

[1] Univ Lyon, INSA Lyon, Laboratoire Ampère (UMR 5005), 69621 Lyon, France
`arnaud.leleve@insa-lyon.fr`
[2] Univ. Orléans, INSA-CVL, PRISME EA 4229, Bourges, France
`cyril.novales@univ-orleans.fr`

Abstract. Simulators have been traditionally used for centuries during medical training as the trainees have to improve their skills before practicing on a real patient. Nowadays mechatronic technology has open the way to more evolved solutions enabling objective assessment and dedicated pedagogic scenarios. Trainees can now practice in virtual environments on various body parts, with current and rare pathologies, for any kind of patient (slim, elderly ...). But medical students need kinesthetic feedback in order to get significant learning. Gestures to acquire vary according to medical specialties: needle insertion in rheumatology or anesthesia, forceps installation during difficult births ... Simulators reproducing such gestures require haptic interfaces with a variable rendered stiffness, featuring commonly called Variable Stiffness Actuators (VSA) which are difficult to embed with off-the-shelf devices. Existing solutions do not always fit the requirements because of their significant size. In contrast, pneumatic technology is low-cost, available off-the-shelf and has a better mass-power ratio. Its main drawback is its non-linear dynamics, which implies more complex control laws than with electrical motors. It also requires a compressed air supply. Ampère research laboratory has developed during the last decade haptic solutions based on pneumatic actuation, applied on a birth simulator, an epidural needle insertion simulator, a pneumatic master for remote ultrasonography, and more recently a needle insertion under ultrasonography simulator. This paper recalls the scientific approaches in the literature about pneumatic actuation for simulation and tools in the medical context. It is illustrated with the aforementioned applications to highlight the benefits of this technology as a replacement or for an hybrid use with classical electric actuators.

Keywords: Haptic training simulation · Pneumatic control · Medical robotics

© Springer Nature Switzerland AG 2019
M. Antona and C. Stephanidis (Eds.): HCII 2019, LNCS 11573, pp. 334–352, 2019.
https://doi.org/10.1007/978-3-030-23563-5_27

1 Introduction

Simulators have been traditionally used for centuries during medical training as the trainees have to improve their skills before practicing on a real patient. Classical simulators feature training boxes, manikins, animals and corpses. Although these former simulators have been used for years, nowadays mechatronic technology has open the way to more evolved solutions enabling objective assessment and dedicated pedagogic scenarios. This evolution attempts to satisfy the need for efficient teaching tools invoked by public institutes such as the French H.A.S. [15] (*Haute Autorité de la Santé*), which asks to "never do [it] the first time on a patient". Trainees can now practice in virtual environments on various body parts, with current and rare pathologies, for any kind of patient (slim, elderly ...). With these simulators, trainees can repeat a procedure several times without getting short of supplies (as in the case of surgery on corpses), get rapid and objective feedback from the simulator, and determine which skills they need to improve [43]. Medical students particularly need some haptic feedback to get significant learning, especially for the gestures that require kinesthetic feeling [35] (kinesthesia is the ability to sense the movement and position of the body limbs). Thanks to kinesthesia, haptic training, and in particular kinesthetic focused training, extends vision based training. Some haptic simulators are now available on the market and are used in various medical sectors [7].

Gestures to acquire vary according to medical specialties: needle insertion in rheumatology or anesthesia, forceps installation during difficult births ... Such gestures require haptic interfaces with a variable rendered stiffness which is sometimes difficult to reproduce with off-the-shelf devices (this kind of actuator is commonly called Variable Stiffness Actuator - VSA). Solutions have been proposed in the literature [8,16,22]. However, these solutions, based on electric or magneto-resistant fluids, do not always fit the training requirements because of their significant size. Unlike electric actuators, pneumatic cylinders are low-cost off-the-shelf components and are easy to mechanically embed. They also have a better mass-power ratio and they naturally provide a passive compliance thanks to the pressurized air contained inside their two chambers. Their drawbacks are their non-linear dynamics, which require more complex control laws than with electrical motors. They also require a compressed air supply. By modifying the pressure level in the actuator chambers, one can control the pneumatic stiffness in real-time over a large range of values starting from 0.1 N/mm, according to Takaiwa *et al.* in [44]. For instance, Semini *et al.* [38] introduced a VSA based on position-controlled hydraulic cylinders.

Position or force control of pneumatic cylinders is nowadays well mastered in the Fluid Power community as detailed in [7,36]. However, their stiffness control was less evolved at the start of our research. This is why we have progressively introduced compliant pneumatic control laws and, by the way, widened the variety of actuation solutions for haptic interfaces. At first, in 2013, Abry *et al.* introduced in [4] a control law to handle the global closed loop compliance of a pneumatic cylinder using a backstepping approach. In this case, high frequency disturbances are absorbed by the natural compliance of the pressurized chambers while the control loop adjusts its stiffness. Starting from these works, Senac et al. designed a syringe simulator [39] and Herzig *et al.* extended the initial model to introduce a two degrees-of-freedom (DOF) pneumatic robot [19].

Several applications have been designed, based on these works. Concerning haptic medical training, we designed a birth simulator (introduced in Sect. 3.1), an epidural needle insertion simulator (see Sect. 3.2) and we are working on a needle insertion under ultrasonography simulator (not introduced in this paper as this work is in progress [5]). Concerning supporting medical tools, we are also working on an haptic bilateral pneumatic control and remote haptic control of an ultrasonography probe (see Sect. 3.3).

This paper recalls the scientific approaches in the literature about pneumatic actuation in Sect. 2 and illustrates, in Sect. 3, their application in the aforementioned simulators and medical tools.

2   Control Laws for Compliant Pneumatic Actuators

Many robotic applications require an interaction between a human and the end-effector of a robot, for instance, for human rehabilitation, for haptic interfaces, or for prosthetic devices. When these interactions occur, most of the time, a compliant behavior of the robot is required in order to avoid human injuries or to avoid damaging the robot itself. Nevertheless, these robots have also to be stiff for some tasks requiring precision. Haptic interfaces are robots (mechatronic systems) with this kind of properties, in order to interact with a simulated tool in a virtual environment or a teleoperated system. However, in practice, when the interface requires to be realistic (as close as the real tool it simulates), off the shelf haptic interfaces are not always suitable [24]. For practical reasons, in commercial simulators, electric actuators are commonly used, in order to reproduce the force feedback mimicking the response of the human body behavior to medical tool interaction. Indeed, the control laws for electric actuators are quite well mastered and easy to set up. However they have some drawbacks such as a low power to weight ratio and difficulties to provide at the same time a high torque at high speed, and mechanical limitations in their backdrivability. This limits their performances to render a variable stiffness. To ensure a compliant behavior of a robot, various Variable Stiffness Actuators (VSA) or Variable Impedance Actuators (VIA) have been developed during last decades. These actuators allow the equilibrium position and the stiffness to be tuned independently. Van Ham *et al.* present a state of the art in the design of VSA in [48]. Most of these actuators are designed with two internal motors and passive compliant elements. An advantage of this design is that the control of the the position and stiffness is obtained by controlling the positions of two electric motors. The main drawbacks of this kind of solution are their high cost as two electric actuators are needed to control only one DOF, and their limited stiffness range due to the use of passive stiffness components [21].

Another approach to obtain a compliant behavior for the robots is based on control strategies such as stiffness control [37], impedance control [20] or hybrid force position control [17]. Most of these strategies have been developed for electromechanical actuated robots. The disadvantages of the electromechanical actuation are that, in order to implement these control strategies, a force/torque sensor is needed. This sensor is required to measure the interaction force between the robot and the environment, which implies knowing in advance where this interaction occurs. Moreover, these sensors are often expensive and fragile. When force/torque sensors are not used, the actuators have to be backdrivable which means reducing gear ratios and, consequently, the torque or force range of the robot.

On the opposite side, pneumatic actuators are quite adequate to reproduce human body behavior as they provide a natural compliance thanks to the air compressibility in their chambers. They are commonly used as bi-stable position actuators but they also can be considered as Variable Stiffness Actuators (VSA) as their stiffness can be tuned by modifying the pressure in both chambers so that they can instantaneously react to stimuli without requiring a fast control loop. Unfortunately, their control is more complex than electric actuators as the air compressibility induces non-linear behaviors. The recent development of servovalves and modern robust non-linear control laws based on sliding mode [47] and backstepping [23] allowed the development of position or force controller. A state of the art in compliant control for pneumatic cylinder is provided in [46], most of them are based on a sliding mode controller and need two proportional servovalves for a single cylinder. Thus, since pneumatic cylinders are inexpensive and have a good power to weight ratio, there has been a recent surge of interest for this technology.

Hereinafter, we present a model of a pneumatic actuator and a control law which objective is to control its stiffness.

2.1   Servo+Cylinder Modeling

Figure 1 shows the actuator often used in our applications: an Airpel® (Airpot® Corp.). We fit it with sensors (pressure, position, and force) and supply it with air by means of a Festo® *MPYE-5-M5-010-B* proportional servovalve. We usually embed our control laws in a *dSPACE®* *1104* control board which acquires in real time sensor signals and generates the servovalve control signals. The code is generated with *Matlab/Simulink®* which is suitable for control prototyping. A schematic of the whole architecture is shown in Fig. 2.

Fig. 1. Airpel M16D100D model

The actuator model can be obtained using two physical laws: the pressure dynamics of the chambers and the fundamental mechanical relation. The pressure evaluation of the chambers with variable volumes is obtained with the following assumptions [6]:

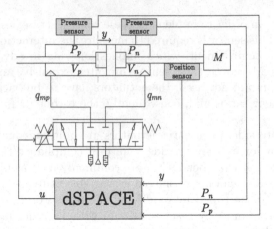

Fig. 2. Electropneumatic system

- air is a perfect gas and its kinetic energy is negligible in the chamber,
- the pressure and the temperature are homogeneous in each chamber,
- the evolution of the gas in each chamber is polytropic and is characterized by coefficient k,
- the temperature variation in chambers is negligible with respect to the supply temperature,
- the temperature in each chamber can be considered equal to the supply temperature,
- the mass flow rate leakages are negligible,
- the supply and exhaust pressures are constant.

A fourth order state model of the pneumatic actuator can be written:

$$
\begin{cases}
\dfrac{dy}{dt} = \dot{y} \\[2mm]
\dfrac{d\dot{y}}{dt} = \dfrac{1}{M}\left(P_p S_p - P_n S_n - b\dot{y} - F_{pext} - F_{st}\right) \\[2mm]
\dfrac{dP_p}{dt} = \dfrac{krT_a}{V_p(y)}\left(q_{mp} - \dfrac{P_p}{rT_a}S_p\dot{y}\right) \\[2mm]
\dfrac{dP_n}{dt} = \dfrac{krT_a}{V_n(y)}\left(q_{mn} + \dfrac{P_n}{rT_a}S_n\dot{y}\right)
\end{cases}
\tag{1}
$$

where the model variables and[1] parameters are listed in Table 1.

As ultra-low friction pneumatic cylinders are used, the stiction force F_{st} can be considered as negligible. Concerning servovalve dynamics, it can be neglected

---

[1] It has to be noticed that this model is available also for a single rod cylinder. In the case of a double rod cylinder, $S_p = S_n = S$ so $F_{pext} = 0$.

Table 1. Model parameters

Symbol	Description	Unit
y	Piston position	m
l	Stroke of the rod	m
$\dot{y}$	Piston velocity	m/s
M	load mass	kg
S_p	Chamber P section area	m^2
S_n	Chamber N section area	m^2
P_p	Pressure inside chamber P	Pa
P_n	Pressure inside chamber N	Pa
b	Viscous friction coefficient	N.s/m
F_{st}	Stiction force	N
F_{pext}	Force applied by the atmospheric pressure on the cylinder piston[3]	N
r	Perfect gas constant	J/(kg.K)
T_a	Temperature of the supply air	K
$V_p(y)$	Chamber P volume at position y	m^3
$V_n(y)$	Chamber N volume at position y	m^3
q_{mp}	Mass flow rate entering the chamber P	kg/s
q_{mn}	Mass flow rate entering the chamber N	kg/s

compared to the actuator ones. The pneumatic effort F_{pneu} corresponds to the pressure difference in both chambers: $F_{pneu} = P_p S_p - P_n S_n$.

Two separate stiffnesses are defined: the closed-loop stiffness $K_{cl} = -\dfrac{\partial \Sigma F}{\partial y}$ (where ΣF is the sum of every force applied on the cylinder rod), and the pneumatic stiffness, denoted $K_{pneu} = k\left(\dfrac{P_p S_p}{L_p(y)} + \dfrac{P_n S_n}{L_n(y)}\right)$ (where $L_p(y) = (\dfrac{l}{2} + y)$ and $L_n(y) = (\dfrac{l}{2} - y)$). The closed-loop stiffness denotes the electropneumatic actuator ability to reject a disturbance force, which, in the context of haptic interfaces, corresponds to the force F_e applied by the user. The pneumatic stiffness is a state of the electropneumatic actuator [40]. It illustrates the global pressurization in both chambers. This state can be controlled in order to optimize air consumption or to reduce air leakage.

We showed in [19] that, instead of controlling each mass flow rate entering each chamber, it is easier to define two virtual flow rates: the active mass flow rate, denoted q_{mA}, and the pressurization mass flow rate q_{mT}, using the reversible "A-T transform":

$$\begin{bmatrix} q_{mA} \\ q_{mT} \end{bmatrix} = \frac{l}{2} \begin{bmatrix} \dfrac{1}{L_p(y)} & -\dfrac{1}{L_n(y)} \\ \dfrac{1}{L_p(y)} & \dfrac{1}{L_n(y)} \end{bmatrix} \begin{bmatrix} q_{mp} \\ q_{mn} \end{bmatrix} \tag{2}$$

2.2 Position and Stiffness Control Based on Backstepping

Position or force control of pneumatic cylinders is well mastered nowadays in the fluid power community, as detailed in [7,36]. The control law exposed here (see Fig. 3) is based on a non-linear method: backstepping position control synthesis with a gain tuning strategy to control the stiffness [3]. It simultaneously controls both the actuator position and the pneumatic stiffness (it can also be used to control the actuator closed-loop stiffness [4]). A significant advantage of this control law is the use of only one servovalve per cylinder, in comparison to other ones which require two of them. The performance of this control law has been successfully compared in [18] with a "classical" impedance control law introduced in [27]. Controlling this stiffness allows to simulate different human organ behaviors such as rigid ones (bones) and soft ones (kidney ...). Indeed, most of the soft tissues in biomedical field are modeled with non constant or nonlinear stiffness, this is why real time tuning the closed-loop stiffness of the haptic simulators is needed. These controllers are also suitable for a medical simulator when the haptic interface is linked to a complex simulation model with numerous deformable objects. Indeed, in this specific case, the simulation software which has to compute the reference forces (or/and position) for the haptic interface, needs time to converge and cannot compute at a sufficiently high rate for an haptic control.

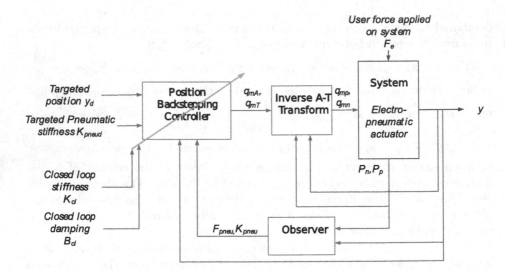

Fig. 3. Backstepping position control scheme

The state model of the cylinder introduced in (1) can be modified to include F_{pneu} and K_{pneu}. A new state model can thus be written as:

$$\begin{cases} \dfrac{dy}{dt} = \dot{y} \\[2mm] \dfrac{d\dot{y}}{dt} = \dfrac{1}{M}\left(F_{pneu} - F_{pext} - F_{st} - b\dot{y}\right) \\[2mm] \dot{F}_{pneu} = B_1 q_{mA} - \dot{y}K_{pneu} \\[2mm] \dot{K}_{pneu} = \dfrac{A_1 y\dot{y}K_{pneu} - A_2\dot{y}F_{pneu} - B_2 y q_{mA} + B_3 q_{mT}}{L_p(y)L_n(y)} \end{cases} \tag{3}$$

where $A_1 = 2(k+1)$, $A_2 = k(k+1)$, $B_1 = \dfrac{2krT_a}{l}$, $B_2 = \dfrac{2k^2 rT_a}{l}$ and $B_3 = k^2 rT_a$

The backstepping control law provides the two desired virtual flow rates q_{mA} and q_{mT} and ensures the system stability. q_{mA} allows to track a desired position y_d whereas q_{mT} allows to track a desired pneumatic stiffness evolution K_{pneud}. The expressions of q_{mA} and q_{mT} derived from the backstepping method are detailed in [18].

2.3 Experimental Results

Figure 4 shows the results of an experiment performed with the aforementioned controller. Figure 4a shows the disturbance force applied on the cylinder rod (in blue) and the values of the desired closed-loop stiffness (in red). Figure 5b shows the actuator response to these disturbances where y_d is the constant desired position, y is the measured position and $y_k = y_d + \dfrac{F_e}{K_{cl}}$ is computed to show the expected position of the cylinder when a disturbance is applied. One can observe a good tracking between y and y_k which means that the system reacts correctly to a position and stiffness references whatever disturbance. Experiments with a moving position reference have also been successfully performed. It demonstrates that it is possible to realize an actuator tracking a desired position which stiffness is tunable in real time.

We extended this strategy to multiple DOF systems in [19]. We applied it on a two-active-DOF pneumatic robot which is a part of the haptic interface embedded in the childbirth simulator presented in Sect. 3.1. In [19], the response to an external disturbance force and a strategy to ensure a desired closed-loop stiffness by tuning some gains are discussed. Simulation results and a comparison with a classical linear impedance controller without force sensor illustrate the interest of this approach.

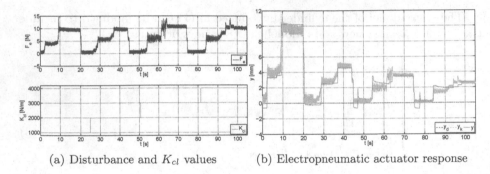

(a) Disturbance and K_{cl} values   (b) Electropneumatic actuator response

Fig. 4. Closed-loop stiffness tuning with the backstepping position controller (Color figure online)

3   Applications

This section illustrates how compliant pneumatic actuators are applied into medical tools and simulators.

3.1   Birthsim: A Simulator to Train on Difficult Childbirths

Since 2004, in order to train students in obstetrics and midwives, particularly for difficult childbirths, scientists have been proposing simulators more evolved than traditional passive manikins. For instance Lapeer *et al.* chose to develop an augmented environment to simulate obstetric forceps delivery [26]. Their simulator enables users to visualize their instruments inside the maternal pelvis. There is however no haptic interface to train on forceps extraction. Sielhorst et al. [41] and Abate *et al.* [1] have worked on delivery simulators coupling augmented reality with an haptic interface. The first one is based on a 6 DOF industrial robot to control the head trajectory. This kind of robot is easy to use but the range of the available forces is oversized for this kind of application. The second one uses an industrial exoskeleton haptic device, coupled with virtual reality goggles, which allows a perfect immersion of users and a good haptic feedback. However, as the user is the only one person immersed in the simulation, it is not appropriate for a team work practice. Some simulators are commercially available and one of the most advanced is the NOELLE simulator [12]. It is dedicated to team training but it is not adapted to evaluate a practitioner specific gestures.

During the past few years, Ampère laboratory has developed a childbirth simulator (called BirthSim) [42] which compensates for the limitations of existing ones. It has been designed especially to train and evaluate the instrumental delivery gestures. It consists of anthropomorphic models of maternal pelvis and fetal head. Obstetrics forceps are instrumented to measure their displacements [31]. Several scenarios have been implemented on the BirthSim simulator. Trainees can thus proceed to a risk free training on forceps blade placement and forceps extraction. A visualization interface is also available to let the trainee to see inside the pelvis and improve his gestures while using forceps.

The BirthSIM simulator now has two pneumatic actuated DOFs (cylinders 1 and 2 in Fig. 5a) to render the translation of the head of the fetus in the vertical symmetry plane and the efforts involved during deliveries. The rotation of the fetal head around the longitudinal axis of cylinder 2 is electrically controlled. A 2 DOF backstepping position control has been implemented, with gain tuning, in order to control the closed-loop stiffness and damping, as introduced earlier. Details and experimental results are available in [19].

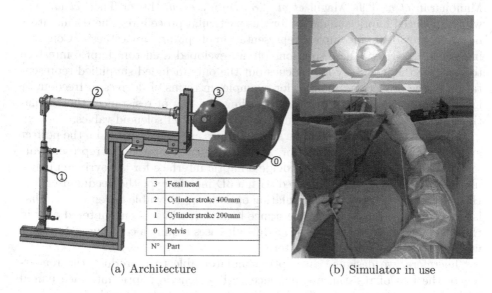

N°	Part
3	Fetal head
2	Cylinder stroke 400mm
1	Cylinder stroke 200mm
0	Pelvis

(a) Architecture   (b) Simulator in use

Fig. 5. BirthSim: difficult chilbirth haptic simulator

3.2   Perisim: Epidural Needle Insertion Simulator

Epidural anaesthesia, despite being a commonly used medical gesture performed mainly during childbirth delivery, remains a very complex and hard to learn gesture [25]. This is mainly due to the fact that the procedure is mostly performed blindly, relying solely on haptic cues and their interpretation by the anaesthetist. To perform such a gesture, the practitioner has to insert a Tuohy needle between two vertebrae while injecting a fluid using an epidural syringe. To perform the anaesthesia, the Tuohy needle has to go through several physical layers to finally reach the epidural space. Throughout the insertion, the anaesthetist experiences an increasing resistance coming both from the needle insertion and from fluid injection. This resistance reaches its maximum in the *ligamentum flavum* to then plummet creating what is commonly called the *loss of resistance* principle. It is this principle that allows physicists to know they reached the epidural space. Due to the relatively high forces exerted simultaneously on the needle and on the plunger of the syringe, the procedure is really demanding in term of precision as the epidural space is usually only about 4 mm wide. In consequence, this particular medical gesture has a quite steep learning curve and may require up to 90 attempts to be performed with only 80% efficiency [49], which cannot be enough for healthcare applications.

Some training support solutions have been developed in the form of manikins or complete robotic applications. Most of the proposed solutions have been listed and compared in [49], which considers 17 manikin based solutions and 14 computer based ones. According to this study, there existed no any ideal solution. Out of computer based solutions, we can highlight, for instance, the works of Manoharan et al. [29], Magill et al. [28], Dubey ct al. [13] or Thao et al. [45], who introduced haptic simulators for this particular procedure. These simulators however, provide an incomplete representation of epidural anaesthesia. Concerning needle insertion, these solutions often developed a custom haptic interface to generate the necessary haptic cues but the only included simplified representation of the procedure, limited, for example, in terms of degrees of freedom or ignoring potential bone contacts. Regarding the loss of resistance, the simulations are very simplified (if present), only using on/off solenoid valves.

Vaughan et al. [49] have drawn some outlines as to what would be the perfect epidural needle insertion simulator. It has to be customisable (to represent various patient types) and should provide a haptic interface for the syringe manipulation (with the LOR feel), paired with a 3D interface for the needle insertion. However, most of the current simulation options are only able to reproduce what is called 100% (binary) loss of resistance behavior, which is encountered only in the "average" and "easy" cases. The difficult cases require a continuous feedback with a specific force feedback pattern.

Therefore, we designed an haptic simulator able to reproduce the rendering of the loss of resistance experience with a generic haptic interface paired with a pneumatic cylinder. In this simulator, the electric haptic interface (a Virtuose™6D from Haption) is here to reproduce the haptic cues generated by the needle insertion only. The pneumatic cylinder (an Airpel Anti Stiction® double acting pneumatic cylinder), mounted on the effector of the Virtuose, is employed to emulate the syringe used in the real application and its purpose is to emulate the loss of resistance. The whole is visible in Fig. 6 (the Virtuose is partly hidden under the green cover, and the syringe simulation part is depicted in the right picture).

To tune accurately enough the prototype, we asked trained anaesthetists to try and assess it. It allowed us to apply some necessary adjustments to some simulation parameters, such as the values of the the resistant force created by the pneumatic cylinder. Once the parameters were set to obtain realistic simulations, we recorded complete procedures. The aim of the first experimental study was to know whether it would be possible to differentiate an experienced user from a novice through the use of this simulator.

We set up three kinds of patients for the first test sessions. The first patient type was the average patient, which represents an "easy" case and serves as a

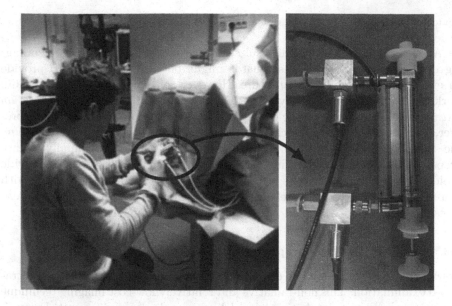

Fig. 6. Complete prototype setup (left) and Pneumatic cylinder setup (right)

basis from which we derived the others. The parameters changed for the other patient types were: the length of the derma layer, the length of the epidural space layer, and the LOR speed (i.e. the speed of transition between the highest resistant force in the ligamentum flavum to the lowest in the epidural space). The patient characteristics are available in [39].

The tests were performed with 2 experts and 6 novices. Up to 24 variables were simultaneously recorded. Details results are available in [39]. On initial examination, we analyzed the success rate per patient type, the velocity of the needle tip along the insertion path. Finally we proposed an *indicator* computed by dividing the proportion of emptied syringe by the distance d traveled by the haptic interface from the puncture of the derma to the end of the procedure along the elastic plane normal vector. Only the unskilled users failed some attempts. The overall results show that novices obtain a success rate of 60% (versus 100% for the experts). Their success rate is better for the calcified/average cases and worse for the overweight ones. This indicates that the LOR phenomenon reproduction played an important role in how the unskilled users detected the epidural space. This confirmed the need for a customizable LOR rendering instead of a binary LOR behavior. Also, regarding the aforementioned indicator, the experts tend to have a much higher score than the novices, showing that they usually use a greater portion of the cylinder length (d being constant regardless of the patient type). This might be a way for the skilled users to retrieve as much haptic information as possible. This may help them knowing their current position with much more precision than only detecting the LOR. Furthermore, concerning the novices, the standard deviation of the indicator is also quite high showing a wide variety of use.

As a conclusion, the overall results of this first study are quite encouraging and indicate that such a simulator might be realistic enough to provide an efficient training tool in the future. However some points might need more work such as introducing a very distinct feel of cutting through the *ligamentum flavum* which, according to our experts, feels like friction and cracklings. Moreover, according to their feedback, when the needle is in the *ligamentum flavum*, the syringe plunger should give the impression that it is locked. Finally, we are preparing an automatic classification of the users method in order to provide rapid, objective and automatic user assessment. This requires more trials with more users in order to generate a representative training data set. We will also assess the relevancy of such a tool in a teaching environment.

3.3   Remote Ultrasonography Haptic Master

Nowadays, more than one out of four emergency admissions requires an ultrasound examination. This non-radiative and relatively low-cost imaging technique is routinely used to help physicians to deliver a preliminary diagnosis. Depending on state health policies, an ultrasound imaging diagnosis is performed either by trained physicians or by specialized sonographers. In both cases, the physician/sonographer must be close to the patient to maintain and hold the ultrasound probe on the designated anatomic area to perform the examination. The sonographer integrates the position of the probe and the motion of his hand to analyze the resulting 2D ultrasound images. Since the late 1990s, in order to deliver equitable healthcare in medically isolated settings, several concepts of remote robotized ultrasonography have been developed, giving the sonographer the ability to move an ultrasound probe on a distant patient [10, 14]. TER [53] or Masuda [30] used fixed robots attached to a table. Current trends are light body-mounted robots [33, 34]: a paramedic holds the robot on the patient body while the distant sonographer controls the probe orientation using a dedicated input device, as in Fig. 7.

Ergonomics is a critical requirement as the sonographers should not be disturbed by the distance with their patient in order to only focus on the medical procedure. Hence, master devices have to provide sonographers with full transparency to perform a robotized remote ultrasound scanning as if they were next to the patient. The master device must be adapted to the sonographer's hand and to his/her expertise. It is possible to use off-the-shelf 3D haptic interfaces, but their kinematic chain is totally different from the one offered by the standalone ultrasound probes sonographers are used to. This means that the practitioner has to adapt his/her hand motions to the proposed input devices, which therefore disturbs the medical act. We proposed in [11] to provide the sonographer with a master ultrasound probe with no mechanical link with the environment, similar to a standard ultrasound probe. The tele-echography device was developed by Prisme laboratory (see Fig. 8) and industrialized under patent in the AdEchoTech company. The system is divided into two parts: a slave robot on

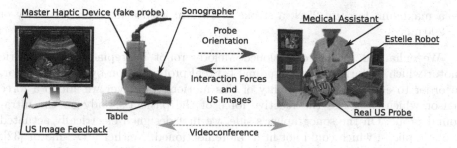

Fig. 7. Remote echography.

the patient side and a hand-free probe replica on the sonographer side without any mechanical connection between the two. A TCP/IP connection links the two parts.

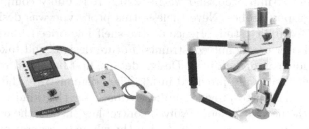

Fig. 8. Adechotech melody system.

With the haptic master probe concept, trained doctors or specialized sonographers should need less training to control the distant robot [50], as when performing an ultrasound examination, sonographers need to feel the interaction between the probe and the patient's body. Indeed, they need to feel when they touch hard body parts and when the body-probe interaction stiffness changes. This is even more true at distance as the practitioner does not have a direct view of the patient (only through a web camera). The master probe should thus be actuated to render the interaction forces (and stiffness) between the real remote ultrasound probe and the patient's body. Even if it concerns only one dimension, this force-feedback is a real ergonomic and technological challenge: it is important to preserve weight and dimensions comparable to standard ultrasound probes. During first experiments with sonographers, we identified the following requirements, which correspond to classical dimensions of ultrasonography probes and measures performed on our first prototype:

- a reversible mechanism with small dimensions (12 cm long, 6.5 cm wide and 3.5 cm thick at most),
- in the z direction (orthogonal to the patient's skin), a continuous force feedback level around 15 N,

– a maximum force of 25 N, a stroke of 50 mm, with a maximum velocity of 200 mm/s.

We enhanced the aforementioned Melody robot by replacing the electric motor which performs the longitudinal (z) real probe motion by a linear motor, in order to ensure the reversibility of this motion. Moreover we added a force sensor which measures the reactive force of the patient's body on the ultrasound probe. On the sonographer's side, we first designed electrically actuated probe replicas, which could not fit as aforementioned. Neither a DC-motor [32], nor a custom brushless motor [9], nor a linear motor [52] were able to meet the requirements: without any reduction, electric motors cannot deliver a such desired force-feedback –without motion– in so tiny a (hand held) volume. This is why we proposed to use a pneumatic cylinder as internal VSA for the fake probe, introduced in a bilateral control scheme (proposed earlier in [51]), to remotely control the robot which holds the real ultrasound probe. Figure 9 displays the pneumatic haptic probe we designed. The whole probe measures 120 mm high, 65 mm wide and 35 mm deep, and weighs 240 g. It is bulky compared to modern ultrasonography probes. Nevertheless, this prototype was designed for the purposes of a feasibility study. Hence off-the-shelf low-priced components have been preferred over ergonomic constraints. Future designs will take ergonomics into account more comprehensively. Design details, first basic control approaches and experiment results are provided in [2]. They demonstrate the feasibility of a pneumatic actuation to provide haptic feedback for this kind of application, and highlight the need for a more evolved control law such as the one exposed in Sect. 2.2. Future experimentations featuring the whole teleoperation loop will be held to evaluate the overall quality of the whole system. Also, these results will be accompanied by a psychometric study determining whether users are able to recognize common medical cases.

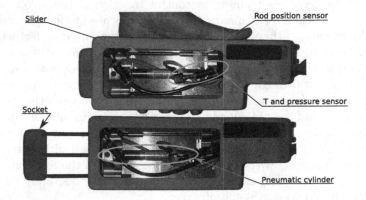

Fig. 9. Haptic pneumatic probe (with the rod completely in (top picture) and out (bottom picture).

4   Conclusion

In this paper, we illustrated the use of pneumatic actuators for medical applications. We recalled the properties of such actuators versus classical electric ones, and highlighted their interest for rendering haptic feedback, when controlled with advanced nonlinear control laws such as the one depicted in Sect. 2.2, in order to make them act as Variable Stiffness Actuators. We then presented several applications we have been working on, based on this approach: the childbirth simulator (birthSim), the epidural needle insertion simulator (periSim) and the haptic master probe for remote ultrasonography. The latter simulator is pneumatic based only, while the two others are hybrid: electric and pneumatic (birthSim and periSim). They all highlight the interest of such an approach, especially when the size and form constraints limit the feasibility of electric solutions.

Aside from Birthsim, these applications are in progress. The more advanced one is periSim as we are getting enough data to propose an effective and objective assessment method. The ultrasonography master probe requires further developments (such as the use of aforementioned nonlinear control law) and experiments in a teleoperation context to validate its effectiveness. We are also working on an intra-articular injection (it is one of the most used method by the rheumatologists to treat the shoulder pain) under ultrasonography simulator which will also use hybrid actuation. First results about the electric part are available in [5].

References

1. Abate, A.F., Acampora, G., Loia, V., Ricciardi, S., Vasilakos, A.V.: A pervasive visual-haptic framework for virtual delivery training. IEEE Trans. Inf. Technol. Biomed. **14**(2), 326–334 (2010)
2. Abdallah, I., Gatwaza, F., Morette, N., Lelevé, A., Novales, C., Nouaille, L., Brun, X., Vieyres, P.: A pneumatic haptic probe replica for tele-robotized ultrasonography. In: Basu, A., Berretti, S. (eds.) ICSM 2018. LNCS, vol. 11010, pp. 79–89. Springer, Cham (2018). https://doi.org/10.1007/978-3-030-04375-9_7
3. Abry, F., Brun, X., Sesmat, S., Bideaux, E., Ducat, C.: Electropneumatic cylinder backstepping position controller design with real-time closed-loop stiffness and damping tuning. IEEE Trans. Control. Syst. Technol. **24**(2), 541–552 (2016)
4. Abry, F., Brun, X., Sesmat, S., Bideaux, E.: Non-linear position control of a pneumatic actuator with closed-loop stiffness and damping tuning. In: Proceedings of the European Control Conference 2013 (2013)
5. Ma de los Angeles Alamilla, D., Moreau, R., Redarce, T.: A new method to render virtual walls for haptic systems: "tracking wall". In: Proceedings of the: 7th International Conference on Mechatronics and Control Engineering (ICMCE 2018), Netherlands, Amsterdam (2018)
6. Andersen, B.W.: The Analysis and Design of Pneumatic Systems. Wiley, Hoboken (1967)
7. Carneiro, J.F., de Almeida, F.G.: Using two servovalves to improve pneumatic force control in industrial cylinders. Int. J. Adv. Manuf. Technol. **66**(1–4), 283–301 (2013)

8. Cestari, M., Sanz-Merodio, D., Arevalo, J.C., Garcia, E.: Ares, a variable stiffness actuator with embedded force sensor for the atlas exoskeleton. Ind. Robot.: Int. J. **41**(6), 518–526 (2014)
9. Charron, G., et al.: Robotic platform for an interactive tele-echographic system: the PROSIT ANR-2008 project. In: Proceedings of Hamlyn Symposium on Medical Robotics, London, UK (2010)
10. Conti, F., Park, J., Khatib, O.: Interface design and control strategies for a robot assisted ultrasonic examination system. In: Khatib, O., Kumar, V., Sukhatme, G. (eds.) Experimental Robotics. Springer Tracts in Advanced Robotics, vol. 79. Springer, Heidelberg (2014). https://doi.org/10.1007/978-3-642-28572-1_7
11. Courreges, F., Novales, C., Poisson, G., Vieyres, P.: Modelisation, commande geometrique et utilisation d'un robot portable de tele-echographie: Teresa. J. Eur. Syst. Autom. JESA **43**(1), 165–196 (2009)
12. Deering, S., Brown, J., Hodor, J., Satin, A.J.: Simulation training and resident performance of singleton vaginal breech delivery. Obstet. Gynecol. **107**(1), 86–89 (2006)
13. Dubey, V., Vaughan, N., Wee, M.Y.K., Isaacs, R.: Biomedical engineering in epidural anaesthesia research. In: Adriano Andrade, editor, Practical Applications in Biomedical Engineering. InTech, January 2013
14. Gourdon, A., Poignet, P., Poisson, G., Vieyres, P., Marche, P.: A new robotic mechanism for medical application. In: Proceedings of the IEEE/ASME International Conference on Advanced Intelligent Mechatronics (AIM 1999), pp. 33–38 (1999)
15. Granry, J.-C., Moll, M.-C.: état de l'art (national et international) en matière de pratiques de simulation dans le domaine de la santé. Technical report, Haute Autorité de la Santé (HAS) (2012)
16. Groothuis, S.S., Rusticelli, G., Zucchelli, A., Stramigioli, S., Carloni, R.: The variable stiffness actuator vsaUT-II: mechanical design, modeling, and identification. IEEE/ASME Trans. Mechatron. **19**(2), 589–597 (2014)
17. Hayati S.: Hybrid position/force control of multi-arm cooperating robots. In: Proceedings of the 1986 IEEE International Conference on Robotics and Automation, vol. 3, pp. 82–89, April 1986
18. Herzig, N., Moreau, R., Leleve, A., Pham, M.T.: Stiffness control of pneumatic actuators to simulate human tissues behavior on medical haptic simulators. In: 2016 IEEE International Conference on Advanced Intelligent Mechatronics (AIM), pp. 1591–1597, July 2016
19. Herzig, N., Moreau, R., Redarce, T., Abry, F., Brun, X.: Nonlinear position and stiffness backstepping controller for a two degrees of freedom pneumatic robot. Control. Eng. Pract. **73**, 26–39 (2018)
20. Hogan, N.: Stable execution of contact tasks using impedance control. In: Proceedings of the 1987 IEEE International Conference on Robotics and Automation, vol. 4, pp. 1047–1054, March 1987
21. Huang, Y., Vanderborght, B., Van Ham, R., Wang, Q., Van Damme, M., Xie, G., Lefeber, D.: Step length and velocity control of a dynamic bipedal walking robot with adaptable compliant joints. IEEE/ASME Trans. Mechatron. **18**(2), 598–611 (2013)
22. Jafari, A., Tsagarakis, N.G., Sardellitti, I., Caldwell, D.G.: A new actuator with adjustable stiffness based on a variable ratio lever mechanism. IEEE/ASME Trans. Mechatron. **19**(1), 55–63 (2014)
23. Khalil, H.K.: Nonlinear Systems I. Prentice Hall, Upper Saddle River (2002)

24. Kheddar, A., Devine, C., Brunel, M., Duriez, C., Sibony, O.: Preliminary design of a childbirth simulator haptic feedback. In: Proceedings of the IEEE Intelligent Robots and Systems Conference, vol. 4, pp. 3270–3275. IEEE (2004)
25. Konrad, C., Schupfer, G., Wietlisbach, M., Gerber, H.: Learning manual skills in anesthesiology: is there a recommended number of cases for anesthetic procedures? Anesth. Analg. **86**(3), 635 (1998)
26. Lapeer, R.J., Chen, M.S., Villagrana, J.G.: Simulating obstetric forceps delivery in an augmented environment. In: Proceedings of AMI/ARCS Sattelite Workshop of MICCAI 2004, pp. 1–10 (2004)
27. Liu, H., Hirzinger, G.: Joint torque based Cartesian impedance control for the DLR hand. In: Proceedings of the IEEE/ASME International Conference on Advanced Intelligent Mechatronics, pp. 695–700 (1999)
28. Magill, J.C., Byl, M.F., Hinds, M.F., Agassounon, W., Pratt, S.D., Hess, P.E.: A novel actuator for simulation of epidural anesthesia and other needle insertion procedures. Simul. Healthc.: J. Soc. Simul. Healthc. **5**(3), 179–184 (2010)
29. Manoharan, V., van Gerwen, D., van den Dobbelsteen, J.J., Dankelman, J.: Design and validation of an epidural needle insertion simulator with haptic feedback for training resident anaesthesiologists. In: 2012 IEEE Haptics Symposium (HAPTICS), pp. 341–348. IEEE (2012)
30. Masuda, K., Kimura, E., Tateishi, N., Ishihara, K.: Three dimensional motion mechanism of ultrasound probe and its application for tele-echography system. In: Proceedings of the IEEE/RSJ International Conference on Intelligent Robots and Systems, vol. 2, pp. 1112–1116 (2001)
31. Moreau, R., Pham, M.T., Silveira, R., Redarce, T., Brun, X., Dupuis, O.: Design of a new instrumented forceps: application to safe obstetrical forceps blade placement. IEEE Trans. Biomed. Eng. **54**(7), 1280–1290 (2007)
32. Mourioux, G., Novales, C., Smith-Guerin, N., Vieyres, P., Poisson, G.: A free haptic device for tele-echography. In: Proceedings of International Workshop on Research and Education in Mechatronics (REM 2005), Annecy, June 2005
33. Najafi, F., Sepehri, N.: A novel hand-controller for remote ultrasound imaging. Mechatronics **18**(10), 578–590 (2008)
34. Nouaille, L., Vieyres, P., Poisson, G.: Process of optimisation for a 4 DOF tele-echography robot. Robotica **30**, 1131–1145 (2012)
35. Panait, L., Akkary, E., Bell, R.L., Roberts, K.E., Dudrick, S.J., Duffy, A.J.: The role of haptic feedback in laparoscopic simulation training. J. Surg. Res. **156**(2), 312–316 (2009)
36. Rahman, R.A., He, L., Sepehri, N.: Design and experimental study of a dynamical adaptive backstepping-sliding mode control scheme for position tracking and regulating of a low-cost pneumatic cylinder. Int. J. Robust Nonlinear Control **26**(4), 853–875 (2016)
37. Salisbury, J.K.: Active stiffness control of a manipulator in Cartesian coordinates. In: 1980 19th IEEE Conference on Decision and Control including the Symposium on Adaptive Processes, pp. 95–100, December 1980
38. Semini, C., Tsagarakis, N.G., Guglielmino, E., Focchi, M., Cannella, F., Caldwell, D.G.: Design of HyQ - a hydraulically and electrically actuated quadruped robot. Proc. Inst. Mech. Eng. Part I: J Syst. Control Eng. **225**, 831–849 (2011)
39. Senac, T., Lelevé, A., Moreau, R.: Control laws for pneumatic cylinder in order to emulate the loss of resistance principle. In: IFAC 2017 World Congress, Proceedings of the 20th World Congress of the International Federation of Automatic Control, Toulouse, France. IFAC, IFAC, July 2017

40. Shen, X.R., Goldfarb, M.: Simultaneous force and stiffness control of a pneumatic actuator. J. Dyn. Syst. Meas. Control.-Trans. **129**(4), 425–434 (2007)
41. Sielhorst, T., Blum, T., Navab, N.: Synchronizing 3D movements for quantitative comparison and simultaneous visualization of actions. In: Proceedings of the 4th IEEE/ACM International Symposium on Mixed and Augmented Reality, ISMAR 2005, pp. 38–47. IEEE Computer Society, Washington, DC (2005)
42. Silveira, R., Pham, M.T., Redarce, T., Betemps, M., Dupuis, O.: A new mechanical birth simulator: Birthsim. In: 2004 IEEE/RSJ International Conference on Intelligent Robots and Systems (IROS) (IEEE Cat. No. 04CH37566), vol. 4, pp. 3948–353, September 2004
43. Sutherland, C., Hashtrudi-Zaad, K., Sellens, R., Abolmaesumi, P., Mousavi, P.: An augmented reality haptic training simulator for spinal needle procedures. IEEE Trans. Biomed. Eng. **60**(11), 3009–3018 (2013)
44. Takaiwa, M., Noritsugu, T.: Development of pneumatic human interface and its application for compliance display. In: Proceedings of 26th Annual Conference of the IEEE Industrial Electronics Society (IECON 2000), vol. 2, pp. 806–811 (2000)
45. Dang, T., Annaswamy, T.M., Srinivasan, M.A.: Development and evaluation of an epidural injection simulator with force feedback for medical training. Stud. Health Technol. Inform. **81**, 97–102 (2001)
46. Toedtheide, A., Lilge, T., Haddadin, S.: Antagonistic impedance control for pneumatically actuated robot joints. IEEE Robot. Autom. Lett. **1**(1), 161–168 (2016)
47. Utkin, V., Guldner, J., Shi, J.: Sliding Mode Control in Electro-Mechanical Systems. Taylor and Francis, Milton Park (2009)
48. Van Ham, R., Sugar, T.G., Vanderborght, B., Hollander, K.W., Lefeber, D.: Compliant actuator designs. IEEE Robot. Autom. Mag. **16**(3), 81–94 (2009)
49. Vaughan, N., Dubey, N., Venketesh, M.Y., Wee, K., Isaacs, R.: A review of epidural simulators: where are we today ? Med. Eng. Phys. **35**(9), 1235–1250 (2013)
50. Vieyres, P., et al.: The next challenge for WOrld wide robotized tele-echography experiment (WORTEX 2012): From engineering success to healthcare delivery. In: Proceedings of TUMI II, Congreso Peruano de Ingeniera Biomedical Bioingeniera, Biotecnologica y Fisica Medica, Lima, Peru, May 2013
51. Vieyres, P., Poisson, G., Courreges, F., Smith-Guerin, N., Novales, C., Arbeille, P.: A tele-operated robotic system for mobile tele-echography: the otelo project. In: Istepanian, R.S.H., Laxminarayan, S., Pattichis, C.S. (eds.) M-Health, Topics in Biomedical Engineering, pp. 461–473. Springer, Heidelberg (2006). https://doi.org/10.1007/0-387-26559-7_35
52. Vieyres, P., et al.: An anticipative control approach and interactive GUI to enhance the rendering of the distal robot interaction with its environment during robotized tele-echography: Interactive platform for robotized tele-echography. Int. J. Monit. Surveill. Technol. Res. **1**(3), 1–19 (2013)
53. Vilchis Gonzales, A., et al.: TER: a system for robotic tele-echography. In: Niessen, W.J., Viergever, M.A. (eds.) MICCAI 2001. LNCS, vol. 2208, pp. 326–334. Springer, Heidelberg (2001). https://doi.org/10.1007/3-540-45468-3_39

ANA: A Natural Language System with Multimodal Interaction for People Who Have Tetraplegia

Maikon Soares[1]([⊠]), Lana Mesquita[2]([⊠]), Francisco Oliveira[2]([⊠]),
and Liliana Rodrigues[3]([⊠])

[1] Cear State University, Fortaleza, Cear, Brazil
maikon@dellead.com
[2] Federal University of Cear, Fortaleza, Cear, Brazil
{lana.beatriz,fran}@dellead.com
[3] Le@d Lab, Fortaleza, Brazil
liliana.rodrigues@dellead.com

Abstract. To interact with a computer, users with tetraplegia must to use special tools/devices that, in most cases, require a great effort. In online education, these tools normally become a distraction, which might hinder learning. Solutions like tongue mouses, smart glasses and computer vision systems, although promising, still face problems of use. This paper introduces ANA, a natural language system which can hear the student and see what is being presented on the interface. With new affordance, learning objects (LO)can have their own grammar, which allows a much more natural voice interaction. LOs respond either by audio or performing the requested action. Tests performed with people with tetraplegics show that the creation of such a shared workspace brings a statistically significant reduction in effort while taking on online lessons and their respective workshops.

Keywords: People with tetraplegia · Education · Smart agent

1 Introduction and Theoretical Background

Promoting accessibility and inclusion of people with disabilities (PwD) in the various teaching and learning environments, besides being an important step in the exercise of citizenship, brings dignity to the PwD. The popularization of the Internet allows educational contents to be distributed on large scale, which makes distance education (DE) a great ally on creating opportunities of skill development for that population. Tetraplegics need special devices like orthoses and rods in the mouth, normally mounted by a caregiver and installed in an environment adapted to their reality. In addition, this type of equipment can generate discomfort to the user, as well as making the interaction cumbersome, frustrating and effortful. Putting on Norman's terms, this is a problem of execution gulf: tetraplegic user must to over great lengths to have the computer understand their intentions.

© Springer Nature Switzerland AG 2019
M. Antona and C. Stephanidis (Eds.): HCII 2019, LNCS 11573, pp. 353–362, 2019.
https://doi.org/10.1007/978-3-030-23563-5_28

People Expect Other People to See What They See. When the locatability constraint is lifted, discourse production and comprehension effort is greatly reduced. We shall use Clark's candle example to illustrate the concept. Imagine two people sitting at a table, conversing. There is a candle on the table. Conversant A asks conversant B: "Would you please light the candle for me?". Since both are seeing the candle, conversant B needs no more extra detail to understand conversant A's intent. They are both sharing the same workspace. Communication effort is at its minimum, as they are physically co-present [1].

With ANA, we try to replicate the scenario above. The creation of a shared workspace between student and assistant allows discourse interaction with objects (either learning objects or just regular widgets). In this way, different from other common wizards or chatbots, ANA is able to see what the user sees. LOs can have their own grammar, which adds considerably more context, which contributes even more for lowering communicative effort. For instance, imagine a LO which implements a quizz is being displayed. The student, looking at the screen would say: "The answer is letter B", as a response to the object's onset.

Another way to show the benefits of our approach is that, in many circumstances, users have to become accustomed to the interaction, and it might take sometime. In this particular case, the navigation task becomes a second to the learning. Multiple tasks are not desirable, specially for online education, as they compete for the user's attentional resources. With time, the navigation becomes automatic and the "competition" lowers. We argue that, with ANA, because it is based on natural discourse interaction, time to reach maximum user performance, will much shorter than with other similar interactions.

This paper is organized as follows: after introduction and Theorical background, Sect. 2 introduces the scenarious of use and target people of the study and is followed by related Work in Sect. 3. Section 4 shows the learning environment and describes ANA. Section 5 describes how this study was conduced, the methodology used, subject recruitment and results. Section 6 shows the conclusion of the study.

2   Scenarious of Use

2.1   Target People

According to Torrecilha [2], spinal cord injury is defined as any impairment in the spinal cord that causes deficits in the motor, sensory, and visceral function. According to the American Spinal Injury Association [3], the total absence of sensory and/or motor and/or autonomic functions below the level of the lesion, including at the sacral levels, characterizes a complete lesion. While the preservation of some function sensory and/or motor and/or autonomic levels below the level of the lesion, including in the sacral segments, characterizes the incomplete lesion. Injuries to the C1, C2, C3, and C4 vertebrae are characterized as "high cervical lesion" and result in paralysis of the arms, hands, trunk, and legs; lack of control over spontaneous breathing; difficulties in speech. Lesions in vertebrae from C5 through C8 are characterized as "low cervical lesion" and result in paralysis in the hands, arms, trunk, intestine, and bladder, but the functions of cognition and speech remain.

Thus, people with injuries above C5 need the help of medical equipment, such as oxygen tubes to survive, and require the assistance of a third person for all other actions. Therefore, this research has targets students with low cervical lesion.

As previously mentioned, people with tetraplegia need special equipment to handle the computer. These devices can be corrective orthoses or even sticks that, placed in the mouth, allow the PwD to use the keyboard.

Fig. 1. Tetraplegic PwD using a computer

Figure 1 shows a quadriplegic PwD positioned in front of the computer with a stick in the mouth to compose a text in an electronic editor.

3   Related Work

Over the years, some research has been developed to improve somehow the way a person with quadriplegia use the computers. Steriadis [5] proposes an interface adapted for tetraplegics. His approach projects the interaction using widgets to highlight elements that can be clicked by the user using a single-switch input device. The intervention still builds a virtual keyboard with word prediction to lessen the user's effort. Although its results bring a decrease of clicks to type, the user still has to reach the keys and widgets using the single-switch input device.

Alqudah [4] researches on a mouse controller that, connected to the face of the PwD by electrodes, can capture muscular movements and translate to mouse/pointer actions. For that, a hardware component constructed with Arduino processes signals obtained from electrodes positioned on the face of the quadriplegic user. Although the results show that it is possible to interact with the electrodes, it was still necessary to couple them in the face of the user, and this can be annoying and reduce freedom of movement. Figure 2 displays Alqudah's system.

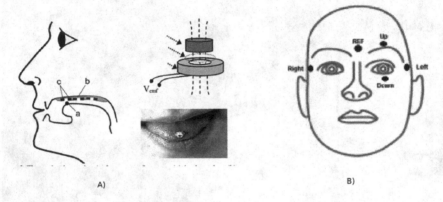

Fig. 2. (A) Tongwise mouse [5]. (B) face controlled mouse [4]

With ANA the student interacts with the computer through natural language, that is, no special equipment/gadget is required, unless, for reasons of comfort, one chooses to use a headphone.

Computer vision (CV) is a robust research topic within the scope of helping quadriplegic PwDs in the handling of the computer. Works such as Middendorp and others [6] investigate the use of visual tracking technology in online education systems for people with quadriplegia to navigate in classrooms. The study makes use of infrared cameras to capture the face image of the PwD. These images are processed and converted into commands. CV based systems either use computer's builtin camera or an external one (normally more expensive). They often need calibration, sometimes their accuracy are less than optimal and definitely constrain head movement, which is an issue for that population.

EID [7] proposes a visual tracking system for a quadriplegic PwD to send commands to a computer that controls a wheelchair. Soares and others [8], propose a gadget called TGlass. This equipment is a low budget smartglass designed to be distributed as material for online courses, respecting the anthropometry of users and providing them comfort and freedom of movement. This equipment has input and output peripherals and uses eye tracking to provide PwD and computer interaction. The works cited above use hardware resources in order to provide accessibility. This approach often encounters barriers to development and deployment cost; besides, physical devices need maintenance and may be defective and require replacement of parts or entire equipment during a course, which can bring about high costs of logistics as well as disrupt the progress of the student. Solutions that require to be coupled in some way to the PwD user's body may need medical-orthopedic follow-up so that they will not cause you physical harm.

4 ANA the Accessible Navigation Assistant

4.1 Learning Environment Description

Oliveira [9] proposed an accessible Virtual Learning Environment (LE). His lab offers, free of charge, professional training courses in information technology, especially computer programming. However, this LE offers, to date, courses for PwDs with deafness or blindness. Therefore, to increase the number of users who can use benefit from better professional training, it was decided to face the challenge of providing accessibility to PwD with tetraplegia.

Figure 3 shows the initial screen to the introduction to programming logic course. Each lesson contains several Learning Objects (LOs), which together offer all course content interactively. Some LO's are: Webaula, Forums, Workshops, Exercises, and evaluation. Of these, WebAula contains all the theoretical information and some typing exercises and is subdivided into topics with a few pages of content on each topic, so it is the main LO and gateway to all others.

ANA assists the user in various navigation and interaction functions. With ANA, one can navigate through pages and topics within a Webaula, respond to quizzes, click buttons, interact with videos, listen to text reading and image description, and click on hyperlinks.

To navigate to a page 3 of topic 2 in a Webaula, for example, one needs to click on the topic he/she wants and then navigates using the navigation arrows at the bottom of the screen. Using ANA, the user can go to the desired page just by giving a single command, "Open page 3 of topic 2".

Virtual assistants are a reality that we live with on a daily basis. We can find them in the various operating systems: Apple Siri, Amazon Alexa, and Google assistant, as well as some video game consoles. Although very "intelligent", unlike ANA, these assistants still do not implement speech-addressable objects and therefore can not see and interact with the screen elements next to the user.

4.2 ANA's Description

This work proposes the Accessible Navigation Assistant (ANA). ANA allows a student to send commands to the computer in the form of speech. This speech

Fig. 3. Screen of programming logic web class object tool

is processed in some command known by the agent and then the action is performed on the system that returns multimodal feedback to the student. Figure 4 illustrates this interaction process. To listen to the user, the system has a voice recognition agent, which listens to the microphone connected to the user's computer. Once the voice is picked up, the actuating agent is called to display what has been recognized on the screen. Parallel to this process, the speech recognition agent synthesizes the speech into text and then passes the text to the natural language processing agent (PLN) to search for a command pattern there. The PLN agent makes use of the dialog flow service [10]. This service provides a range of features for creating a chatbot. The web service receives a text corpus called "user says" and through machine learning translates it into an intent or returns an error if it can not find an intent that matches the input text.

In the web context, HTML structures do not necessarily represent the real role of the component it generates within the system, a list of links, for example, can represent from a menu up navigation tabs to a higher level of abstraction. For the end-user of the system, the abstraction can reach even higher levels, and navigation tabs can have other meanings, such as "Class topics" or other types of tools.

In the AVA to which ANA was inserted, the screen interaction objects were cataloged according to their roles. These objects can be: "icons", "topics",

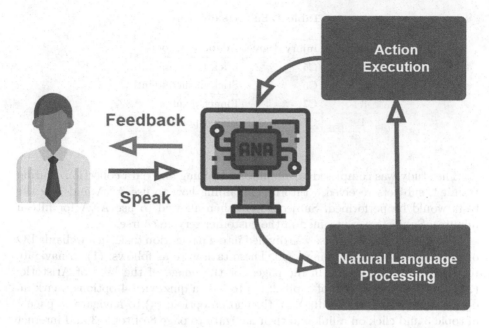

Fig. 4. ANA interaction flow

"pages", "quiz" or others according to the need for task interaction. Assigning roles to objects allows the creation of a shared workspace between student and assistant. Once the dialog flow response has been obtained, the assistant knows exactly where each interaction object is, i.e. ANA can see what the user sees and therefore can access each object directly with only one command, making LOs addressable by the speech.

5   The Study

5.1   Methodology

The study compares differences in the execution gulf [11] of a user with tetraplegia when navigating in a Webaula of course of programming logic using their standard adaptation and using ANA. Table 1 displays subject profile, type of injury and adaptation tool. All subjects were male and volunteers. No remuneration or reward was offered to participants. Each participant performed the tasks at the location and time of their choice and using their computers. Thus, a scenario of actual use of a distance learning course could be simulated more faithfully. Participants were free to leave the experiment at any time, before or during the study procedures.

The data collection took place in the within-subjects model, that is, each participant performed the task using their standard adaptation and also using ANA. All trials were recorded on video to later be counted the number of steps for the execution of each task (execution gulf).

Table 1. Subjects division

Subject	Injury degree	Adaptation tool
1	C5	Stick in the mounth
2	C5	Stick in the mounth
3	C7	Finger pointer
4	C7	Finger pointer

The study was comprised of two phases: training and data collection. During training, subjects received a tutorial explaining how to use ANA and how the tests would be performed. Subjects were then allowed to use ANA for fifteen minutes. Subjects were trained in the customer service course.

The data collection phase was divided into 4 navigation tasks in a webaula LO of the course of programming logic. These task were as follows: (1) to navigate to the topic 1 and search in the pages for the image of the bust of Aristotle; (2) to navigate to page 5 of topic 2; (3) to find a quiz with 4 options, mark an option and observe the feedback of the chosen option; (4) to navigate to page 5 of topic 3 and click on a link and then navigate to page 8 of topic 3 and interact with a video (Play, Pause, Forward and Back).

The Fig. 5 shows a screenshot of the webula LO used in experiment. In the figure one can see the topic navigation structure, a four option quiz and the page navigation structure.

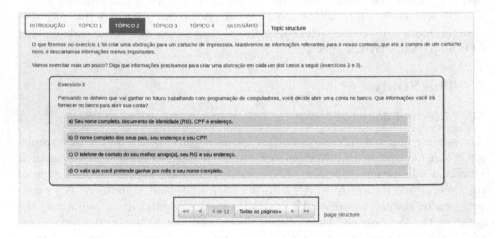

Fig. 5. Webaula LO

5.2   Results

Once data were collected, the number of steps performed by each participant took to perform the tasks using their standard means of accessibility and using ANA were counted. Tables 2 and 3 shows the number of steps of each subject in each task, without using ANA and using ANA, respectively.

Table 2. Step number of each subject to perform each task without ANA

Task	Subject 1	Subject 2	Subject 3	Subject 4
1	8	10	5	6
2	51	29	8	7
3	58	15	6	9
4	65	40	15	25

Table 3. Step number of each subject to perform each task with ANA

Task	Subject 1	Subject 2	Subject 3	Subject 4
1	4	6	2	6
2	5	5	4	5
3	4	6	3	3
4	15	10	10	7

Data do not conform to normal distribution (failed on Shapiro Wilk normality test [12]). Thus we ran Wilcoxon Signed-Ranked test at 95% confidence level. The results were: $z = 3.4078$ and p-value $= 0.00032$, quite significant.

The results show that using the ANA the student with tetraplegia can perform the proposed tasks with a smaller number of steps.

6   Conclusion

Provide accessibility in teaching people with tetraplegia is a major challenge for the academic community. Interaction must not get in the way of the user's goals. Making LOs "see" what the learner is seeing allowing a more natural and direct interaction can pave the way of online education for that population.

In this paper we introduce ANA, a conversational agent that enables the creation of LOs to whom PWD can talk while making discourse references to what is on the computer screen. We ran an initial study to compare the effort of tetraplegics while doing routine tasks on our accessible learning environment either with ANA or with their preferable method of interaction. Results show a great difference on the number of steps necessary to manifest their intentions to the computer.

This research is only in its infancy. We plan to develop a framework and specific grammars to enable content creators and curators ANA-like/google dialog interactions on their courses. Furthermore, more tests are needed to understand the impacts of ANA on learning outcomes.

References

1. Clark, H.H., Brennan, S.E.: Grounding in communication. Perspect. Soc. Shared Cogn. **13**(1991), 127–149 (1991)
2. Torrecilha, L.A., et al.: O perfil da sexualidade em homens com lesão medular. Fisioterapia em Movimento, pp. 39–48 (2014)
3. ASIA. American Spinal Injury Association. http://asia-spinalinjury.org/
4. Alqudah, A.M.: EOG-based mouse control for people with quadriplegia. In: Kyriacou, E., Christofides, S., Pattichis, C.S. (eds.) XIV Mediterranean Conference on Medical and Biological Engineering and Computing 2016. IP, vol. 57, pp. 145–150. Springer, Cham (2016). https://doi.org/10.1007/978-3-319-32703-7_30
5. Steriadis, C.E., Constantinou, P.: Designing human-computer interfaces for quadriplegic people. ACM Trans. Comput.-Hum. Interact. (TOCHI) **10**(2), 87–118 (2003)
6. Van Middendorp, J.J., et al.: Eye-tracking computer systems for inpatients with tetraplegia: findings from a feasibility study. Spinal Cord **53**(3), 221 (2015)
7. Eid, M.A., Giakoumidis, N., El-Saddik, A.: A novel eye-gaze-controlled wheelchair system for navigating unknown environments: case study with a person with ALS. IEEE Access **4**, 558–573 (2016)
8. Soares, M.I.D.S., et al.: VISUAL JO2: Um Objeto de Aprendizagem para o Ensino de Programação Java a Deficientes Físicos e Auditivos através do Estímulo Visual- Um Estudo de Caso. RENOTE **12**(2), 1–10 (2014)
9. Oliveira, F.C.D.M.B., et al.: IT education strategies for the deaf. In: Hammoudi, S., Maciaszek, L., Missikoff, M.M., Camp, O., Cordeiro, J. (eds.) Proceedings of the 18th International Conference on Enterprise Information Systems (ICEIS 2016), SCITEPRESS - Science and Technology Publications, Lda, Portugal, pp. 473–482. https://doi.org/10.5220/0005922204730482
10. DIALOGFLOW. DialogFlow https://dialogflow.com/
11. Norman, D.A.: Cognitive engineering. User Centered System Design, vol. 31, p. 61 (1986)
12. Shapiro, S.S., Wilk, M.B.: An analysis of variance test for normality. Biometrika **52**(3 and 2), 591 (1965)

An Investigation of Figure Recognition with Electrostatic Tactile Display

Hirobumi Tomita[1(✉)], Shotaro Agatsuma[1], Ruiyun Wang[1], Shin Takahashi[1], Satoshi Saga[2], and Hiroyuki Kajimoto[3]

[1] University of Tsukuba, 1-1-1 Tennoudai, Tsukuba, Ibaraki, Japan
{tomita,agatsuma,wang}@iplab.cs.tsukuba.ac.jp, shin@cs.tsukuba.ac.jp
[2] Kumamoto University, 2-39-1, Kurokami, Chuo-ku, Kumamoto, Japan
saga@saga-lab.org
[3] The University of Electro-Communications,
1-5-1 Choufugaoka, Choufu, Tokyo 182-8585, Japan
kajimoto@kaji-lab.jp

Abstract. The visually impaired must obtain shape information in a tactile manner. However, existing conventional graphics are static. We prepared a more useful, dynamic tactile display; we aimed to allow the visually impaired to recognize and draw figures via tactile feedback. We developed an electrostatic force-based tactile display and performed two preliminary evaluative experiments. We measured figure recognition rates and explored how users perceived figures that were displayed in a tactile manner. We describe the results and future planned improvements.

Keywords: Tactile feedback · Visually impaired · Tactile graphics

1 Introduction

The visually impaired find it very difficult to interact with computers, particularly to perceive figures. No good off-the-shelf tactile display presently allows the visually impaired to perceive and draw figures. Therefore, we developed an affordable dynamic device presenting spatial information in a tactile manner. The device features dynamic changes in tactile stimulation (electrostatic force based stimulation to a finger). The force magnitude changes by position; spatial information is thus imparted. Here, we describe our prototype and preliminary evaluation thereof. We explored whether figures were recognized when fingers were stimulated, and we discuss planned future improvements.

2 A Tactile Graphics Display

The visually impaired must recognize shapes in a tactile manner. However, conventional graphics are static. Several researchers have explored tactile displays

© Springer Nature Switzerland AG 2019
M. Antona and C. Stephanidis (Eds.): HCII 2019, LNCS 11573, pp. 363–372, 2019.
https://doi.org/10.1007/978-3-030-23563-5_29

of spatial information; for example, a pin array has been used to stimulate the fingers [1,2]. Ohka et al. developed a "tactile mouse"; this combined a computer mouse with a pin array to present graphical information to the fingers via simple rendering. For example, when the mouse moves inside a figure, the pins representing the figure protrude to stimulate the finger. Thus, the user captures the figure edges and can trace the outline. The pin array is a form of mechanical stimulation and is thus relatively easy to use. Another tactile feedback method features electro stimulation [3]. Uematsu et al. sought to display characters using this method. However, the recognition rate was only 76.2% for large characters, thus lower than that afforded by pin arrays. Also, in the cited system, the user must wear a finger pad to allow stimulation. We also developed an electrostatic-force-based, tactile stimulation apparatus featuring a high-voltage control circuit, an electrode, and an insulator, all made of easily available materials. Recently, lateral-force-based, tactile feedback devices employing static electric fields have been developed (Senseg Inc., Bau, et al. [4]), and many researchers have explored their parameters (input frequencies, waveforms, and amplitude modulations [4–8]).

Bateman et al. explored whether dots could be located using a tactile electrostatic device [9]; all subjects found the dots quickly. Xu et al. investigated whether simple shapes (a circle, square, and triangle) could be recognized using an electrostatic tactile display [10]. The average recognition rate was 56%. As only simple shapes were tested, it is unclear whether more complex shapes could be recognized. We added complex figures to basic shapes when deriving recognition rates and exploring the figures captured by users.

3   A Graphics System Featuring an Electrostatic Tactile Display

The device features a high-voltage generator, an electrode, and an insulator. The generator was developed by Kajimoto et al. The device includes an mbed LPC1768 microcontroller maintaining the output voltage at a maximum of 600 V by modifying the firmware. Various waveforms can be output to the electrode and used to impart different forms of electrostatic tactile feedback. The electrode is covered with an insulating plastic film 15 μm in thickness that the user touches.

An electrostatic force is generated only when the user slides his/her finger on the display. When a high voltage is applied to the electrode, dielectric polarization is generated in the finger. In this state, the electrode applies a static attractive force to the finger, but the force is too weak to feel. However, when s/he slides his/her finger on the display, s/he feels a tactile sensation (as if the surface texture has changed). Because our electrode is a single large sheet, the system delivers only one type of force at any time. Thus, when changing the force by finger position, the system should sense the finger position, and then changes the force accordingly. We incorporated an infrared-based touch sensor (zForce AirTM 295) to that end.

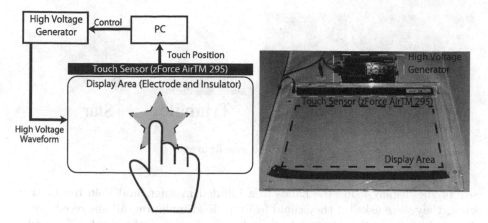

Fig. 1. A schematic of the tactile graphics system (left) and a photograph of the device (right).

The tactile display is enclosed in an acrylic frame; the user can easily grasp the display area. The width is 295 mm and the height 180 mm. A figure is represented as an area within the tactile region; the user can feel a tactile sensation within the figure but not outside. The shape presented can be switched using a PC (Fig. 1).

4   Experiment 1: Identification of Simple Figures

We explored whether it was possible to discern figures; we measured the identification rates of simple figures.

4.1   Outline of Experiment 1

We recruited three participants (all males; no handicapped subject). We prepared four figures: a circle, a square, a triangle, and a star (Fig. 2), all 8 cm high. The task was to choose the correct figure from among the four. The recognition rate was the proportion of correct answers. The input waveform for tactile feedback was a 100 Hz rectangular wave. The participants were asked to touch the tactile display using the right index finger, irrespective of the dominant hand. The tracing speed was limited to 30 cm/s to ensure that the sensor captured finger motion. We set no restriction on either finger pressure or the direction of movement.

Initially, the experimenter gave an overview of the task and asked all participants to complete a consent form. The device was covered during this introduction. At this point, participants were not given any information on the shapes to be displayed. Then, all participants were blindfolded while touching the display, to minimize visual information and reproduce the challenge faced by the visually impaired. Participants were then asked to review the location and the

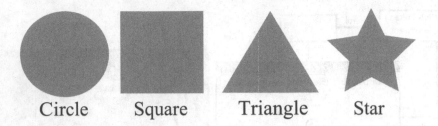

Circle   Square   Triangle   Star

Fig. 2. The four figures.

size of the display using the hands in a blinded manner, and then to practice. First, they were asked if they could feel tactile stimulation; all answered "yes". Then, they were allowed to slide their fingers freely on the display for 1 min. During the practice session, we used only the circle, and we did not tell the participants that this was displayed. During the experiment, we displayed the four shapes five times (20 trials in total). The shapes were described to the participants before the experiment commenced. When the participant gave the name of the shape s/he felt, the next figure was immediately presented; that participant was not told whether s/he was correct. Participants were allowed 5-min breaks after every five trials. Finally, they were asked to make short comments. The experiment took about 1 h.

4.2 Results and Discussion

The average correct identification rate was 68.3%. Table 1 shows the confusion matrix for each shape. The triangle and star were relatively easily identified; the circle and square identification rates were lower than those for the triangle and star, which were rarely mistaken for other shapes. However, the circle and square were often mutually mistaken.

Table 1. The correct answer rates for each figure.

Display \ Answer	Circle	Square	Triangle	Star
Circle	46.7	33.3	13.3	6.67
Square	26.7	60.0	13.3	0.00
Triangle	13.3	0.00	80.0	6.67
Star	13.3	0.00	0.00	86.7

The high recognition rates for the triangle and star may reflect their unique characteristics. Both have acute angles at the vertices. When tracing the top or bottom of the triangle or star, participants can feel these angles and identify the shapes correctly. However, tracing an edge with a finger is relatively difficult, which may explain the low recognition rates of the square and circle. When participants sought to identify shapes by tracing, they tried to move their fingers horizontally or vertically. However, because they were blindfolded, they often moved the fingers somewhat obliquely and misidentified the shape. For example, when a square is displayed, the finger expects a long edge, but if the participant traces the upper edge of the square at (even a slight) angle from the horizontal, stimulation is brief. Therefore, the participant thought that the line was curved and misidentified the square as the circle.

5 Experiment 2: Free Drawing

We next explored how accurately graphic information could be grasped in the absence of prior information, and also how shapes were captured in more realistic settings. Here, we asked participants to draw shapes felt on the device on paper.

5.1 Outline of Experiment 2

Five participants (no handicapped person; two females) were recruited. The participants were shown a shape in a tactile manner, and then asked to draw the exact shape sensed on the paper. Figure 3 shows the shapes presented by the display. By comparing the drawn to the original shapes, we explored how accurately a shape was recognized using our display.

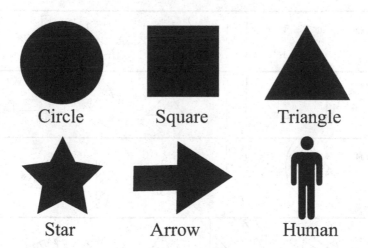

Fig. 3. The figures for experiment 2.

As for experiment 1, we first gave an overview of the test, blindfolded all participants, and had them operate the device. The input waveform was a rectangular wave of 100 Hz. After practice, each participant was presented with one of the shapes and asked to trace it. When each participant concluded that s/he "knew" the shape, the blindfold was removed and the shape drawn on paper. All participants were asked to preserve shape orientation when drawing. Furthermore, if they were not confident that they were correct, they were permitted to touch the tactile device again (as many times as they wished) wearing a blindfold. After drawing the figures, participants were asked to describe the figures orally, to explore how they recognized the shapes. After participants finished drawing, they took a 1-min break and then moved to the next figure. The figures were presented in random order. Finally, all participants were asked to give short comments. The experiment required about 1 h.

5.2   Results and Discussion

Figure 4 shows the figures drawn by the participants. In terms of simple shapes, the characteristics of the triangle and square were well captured. However, most

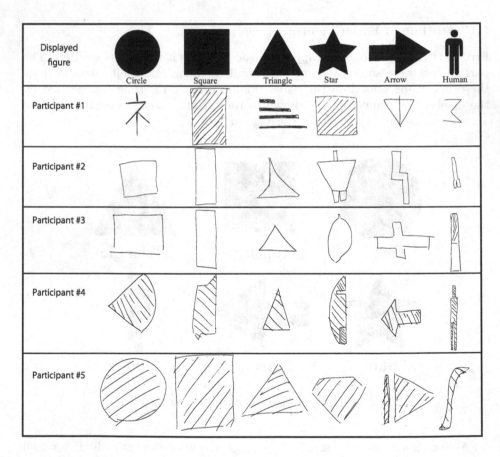

Fig. 4. The results of experiment 2.

participants were unable to capture the circle. As shown in Fig. 4, two participants thought that the circle was a square, as was also the case in Experiment 1. For the square, all participants captured the quadrilateral features well. However, several participants drew vertically elongated rectangles, although we displayed a square. In terms of the complex shapes (the star, arrow, and human), no participant fully captured the figures. However, parts thereof were captured by several. For the star, some participants captured protrusions of the upper or lower part. Similarly, for the arrow, the shape was well-grasped by several subjects. For the human, a foot or a bar-like structure was captured by several participants. However, none grasped the shape of the arm, and only some of the body was drawn.

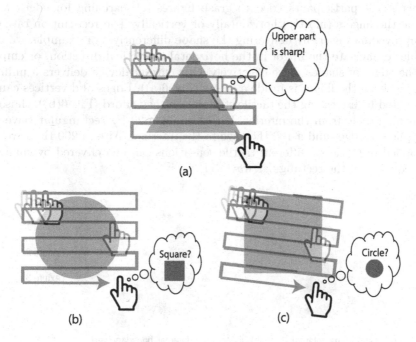

Fig. 5. Example of how to trace a finger.

Participants found it easy to capture figures (or parts thereof) featuring characteristic vertices and edges, thus triangles, squares, and arrows, but difficult to capture figures with curved edges such as circle and a human shape. Most of the participants sought to recognize figures by tracing horizontal or vertical lines (as in Fig. 5(a)). However, when the tracing interval is long (as in Fig. 5(b)) or the required tracing direction is not a horizontal/vertical line (as in Fig. 5(c)), mistakes were sometimes made. Particularly, for the equilateral triangle, it was simple to discern that the top was narrow and the bottom wide. In addition, as the width increases linearly from the top to the bottom, the triangle was relatively easy to identify.

It is possible that the rectangular shape of the display affects the finger sliding speed; participants thus tend to see the square as a rectangle. The width of the device is longer than the height. Therefore, when the participant is tracing, horizontal finger movement is faster than vertical movement; less time is required when tracing the square horizontally. Thus, the square tended to be recognized as a rectangle. Our method facilitates recognition of simple graphics such as triangles and squares, with acute or right angles, but it is difficult to capture the features of complex figures such as stars.

6   Future Work: Improvements

In this study, participants tried to grasp figures by searching for edges while moving the fingers (mostly) horizontally or vertically. The recognition rate can be improved, however, by delivering the shape differently. For example, we may be able to navigate the finger in the horizontal or vertical direction, or emphasize the edges of shapes more effectively. Our current device delivers a uniform stimulus when the finger is inside a shape (Fig. 6(a)). Edges and vertices can be highlighted by increasing the tactile stimulation they afford (Fig. 6(b)), thus distinguishing them from the interior. For example, a 20 Hz rectangular wave can be applied to edges and a 100 Hz wave to the interiors. Also, a 200 Hz wave can be applied to vertices. Different tactile sensations can be delivered by changing the frequency of the rectangular wave [11].

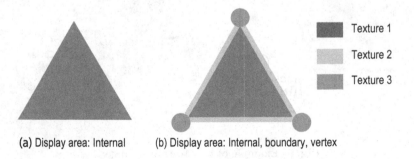

(a) Display area: Internal   (b) Display area: Internal, boundary, vertex

Fig. 6. The current display. The display area is wholly internal (left). We will aim to display three tactile sensations (right).

The shearing force applied to the finger also changes when the waveform is modulated. Saga et al. proposed that both large bumps and small textural changes could be simultaneously felt by changing the shearing force [12]. We will use this method to apply bumps to edges and vertices to emphasize the unique characteristics of different graphics.

7   Conclusion

We explored whether shapes were correctly identified using our electrostatic tactile display. We presented four shapes and asked volunteers to identify them. The shape recognition rate was 68%. Furthermore, we investigated to what extent a shape could be captured without prior information. We presented six types of shapes; participants were asked to draw them on paper. They found it easy to recognize graphics with characteristic vertices or edges, such as triangles and squares, but difficult to recognize curves. In the future, we will improve our method and enroll visually impaired subjects.

References

1. Ohka, M., Koga, H., Mouri, Y., Sugiura, T., Miyaoka, T., Mitsuya, Y.: Figure and texture presentation capabilities of a tactile mouse equipped with a display pad of stimulus pins. Robotica **25**(4), 451–460 (2007)
2. Mineta, T., Yanatori, H., Hiyoshi, K., Tsuji, K., Ono, Y., Abe, K.: Tactile display MEMS device with SU8 micro-pin and spring on SMA film actuator array. In: 2017 19th International Conference on Solid-State Sensors, Actuators and Microsystems (TRANSDUCERS), pp. 2031–2034. IEEE (2017)
3. Uematsu, H., Suzuki, M., Kanno, Y., Kajimoto, H.: Tactile vision substitution with tablet and electro-tactile display. In: Bello, F., Kajimoto, H., Visell, Y. (eds.) EuroHaptics 2016. LNCS, vol. 9774, pp. 503–511. Springer, Cham (2016). https://doi.org/10.1007/978-3-319-42321-0_47
4. Bau, O., Poupyrev, I., Israr, A., Harrison, C.: TeslaTouch: electrovibration for touch surfaces. In: Proceedings of the 23rd Annual ACM Symposium on User Interface Software and Technology, UIST 2010, pp. 283–292. ACM, New York (2010)
5. Mallinckrodt, E., Hughes, A., Sleator Jr., W.: Perception by the skin of electrically induced vibrations. Science **118**, 277–278 (1953)
6. Strong, R.M., Troxel, D.E.: An electrotactile display. IEEE Trans. Man-Mach. Syst. **11**(1), 72–79 (1970)
7. Meyer, D.J., Peshkin, M.A., Colgate, J.E.: Fingertip friction modulation due to electrostatic attraction. In: World Haptics Conference (WHC), pp. 43–48. IEEE (2013)
8. Vezzoli, E., Amberg, M., Giraud, F., Lemaire-Semail, B.: Electrovibration modeling analysis. In: Auvray, M., Duriez, C. (eds.) EUROHAPTICS 2014. LNCS, vol. 8619, pp. 369–376. Springer, Heidelberg (2014). https://doi.org/10.1007/978-3-662-44196-1_45
9. Bateman, A., et al.: A user-centered design and analysis of an electrostatic haptic touchscreen system for students with visual impairments. Int. J. Hum.-Comput. Stud. **109**, 102–111 (2018)
10. Xu, C., Israr, A., Poupyrev, I., Bau, O., Harrison, C.: Tactile display for the visually impaired using TeslaTouch. In: CHI 2011 Extended Abstracts on Human Factors in Computing Systems, pp. 317–322. ACM (2011)

11. Tomita, H., Saga, S., Kajimoto, H., Vasilache, S., Takahashi, S.: A study of tactile sensation and magnitude on electrostatic tactile display. In: 2018 IEEE Haptics Symposium (HAPTICS), pp. 158–162. IEEE (2018)
12. Saga, S., Raskar, R.: Simultaneous geometry and texture display based on lateral force for touchscreen. In: World Haptics Conference (WHC), pp. 437–442. IEEE (2013)

A Survey of the Constraints Encountered in Dynamic Vision-Based Sign Language Hand Gesture Recognition

Ruth Wario[✉] and Casam Nyaga

University of the Free State, Bloemfontein, South Africa
wariord@ufs.ac.za

Abstract. Vision-based hand gesture recognition has received attention in the recent past and much research is being conducted on the topic. However, achieving a robust real time vision-based sign language hand gesture recognition system is still a challenge, because of various limitations (The term limitation in this study is used interchangeably to mean constraint or challenge in respect to the problems that can or are encountered in the process of implementing a vision-based hand gesture recognition system.). These limitations include multiple context and interpretations of gestures as well as well as complex non-rigid characteristics of the hand. This paper exposes the constraints encountered in the image acquisition via camera, image segmentation and tacking, feature extraction and gesture classification phase of vision-based sign language hand gesture recognition. It also highlights the various algorithms that have been used to address the problems. This paper will be useful to new as well as experienced researchers in this field. The paper is envisaged to act as a reference point for new researchers in vision-based hand gesture recognition in the journey towards achieving a robust system that is able to recognize full sign language.

Keywords: Algorithms · Constraints · Gesture recognition · Sign language

1 Introduction

Gestures form an important aspect in human communication, to the point that people gesture even in telephone conversations. Gesture recognition can be viewed as the ability of a computer based system to decode the meaning of a gesture [1]. Hand gesture recognition has many application areas for instance sign language recognition, robotic arm control and Human Vehicle Interaction (HVI) [2].

In this study, the main application area of interest was sign language recognition. Hand gesture recognition has demonstrated to be more convenient over other conventional methods of human computer interactions like mouse and key board [3]. There are two approaches to hand gesture recognition, namely data glove and vision-based [1]. The vision-based approach can be categorized as appearance-based methods and 3D hand model-based methods. Appearance-based methods are preferred in real-time performance, because it is less complex to perform image processing on a 2D image.

M. Antona and C. Stephanidis (Eds.): HCII 2019, LNCS 11573, pp. 373–382, 2019.
https://doi.org/10.1007/978-3-030-23563-5_30

The 3D hand model-based method provides a better description of hand features. However, as the 3D hand models are articulated, deformable objects with many degrees of freedom require a very large image database to cover all the characteristic shapes under different views. Matching the query image frames from video input with all images in the database is time-consuming and computationally expensive [4].

The vision-based approach is considered to provide a more natural and intuitive human computer interface [3]. However, hand gesture recognition has proved to be quite challenging due to the multiple context and interpretations of gestures amid other challenges like the complex non-rigid characteristics of the hand [5]. Sign language (SL) is also primarily grounded on spatial characteristics and iconicity characteristics. Hand parameters like the shape, motion of the hand, position in space as well as lips movement, and facial expressions are used to decode meaning of a sign [6].

Past research indicates that most research in sign language recognition is confined to a small subset of the whole sign language due to the constraints associated with vision-based hand gesture recognition [7]. This paper outlines the constraints associated with vision-based sign language hand gesture recognition.

2   Objective

The objectives of this study are to:

- Analyze the constraints in the hand tracking and segmentation phase of a vision-based sign language hand gesture recognition system.
- Analyze the constraints in the feature extraction phase of a vision-based sign language hand gesture recognition system.
- Analyze the constraints in the classification phase of a vision-based sign language hand gesture recognition system.

3   Methodology

In this study, a qualitative research design was employed through desktop research. The research comprised document analysis, which can be defined as an orderly process for reviewing or assessing printed and electronic documents [8]. Document analysis has been applied in many research studies to triangulate other methods, but can also be used singly in research [9]. It has been argued by [10] to be less time consuming, because it involves data selection as opposed to data collection and hence suitable for repeated reviews [10].

Desktop research, as guided by [11], has been successfully employed by [2, 12] and many other authors to bring out important conclusions; hence this method of data collection was used in this study. Twelve papers were reviewed in this study. The papers were searched using the google scholar search engine using key words matching the objectives.

4 Technology Description

This paper was based on identification of the constraints associated with the implementation of a vision-based sign language hand gesture recognition system. Different authors have come up with different representations and terms of the phases that comprise a typical vision-based gesture recognition system. Below is Table 1, indicating some of the terms used.

Table 1. Vision-based hand gesture recognition system phases by different authors

Author	Phases				
[7]	Image acquisition from camera	Hand region segmentation	Hand detection and tracking	Hand posture recognition	Classified gesture (display as text or voice display as text or voice)
[13]	Capture image	Image preprocessing	Feature extraction	Gesture recognition system	Assign specific task
[14]	Image acquisition	Hand segmentation	Feature extraction	Gesture classification	Gesture recognition
[10]	Capture video	Hand tracking and segmentation	Feature extraction	Classification and recognition	Text application interface

As depicted in Table 1, the phases of a vision-based hand gesture recognition system are similar even though they represent different instances of different systems. The phase includes image acquisition, hand tracking and segmentation, feature extraction, classification and recognition. Below is a brief description of each phase and the constraints associated with the phase.

i. Image acquisition from camera

The first step in gesture recognition is to capture the gesture via a video camera, either attached to the computer or independent from the computer[1]. The constraints in this phase may be due to a number of factors. For instance, accuracy of gesture recognition may be affected by the following camera specifications: color range, resolution and accuracy, frame rate, lens characteristics and camera computer interface [5].

ii. Hand region segmentation

The main objective of the segmentation phase is to remove the background and noises, leaving only the Region of Interest (ROI), which is the only useful information

---

[1] Computer in this case refers to a desktop computer, tablet or even laptop computer that is used in the vision-based hand gesture recognition system.

in the image. This objective can be achieved in various ways like skin colour detection, hand shape features detection and background subtraction [3]. A Bayesian classifier, which a is supervised learning model, can be used for skin colour segmentation as well as an unsupervised model such as K-Mean clustering [3].

iii. Hand detection and tracking

Hand tracking is an important phase in gesture recognition and can be achieved through a number of algorithms. The algorithms return information such as the colour tracking, template matching, motion tracking and other cues, which can be returned in order to track the hand. These algorithms may include Kalman filtering, particle filtering, optical flow, camshaft, viola jones, and mean shift among others [3, 15].

In the tracking phase while using the skin color-based methods, the skin colour may vary from one person to another posing a major constraint. Hence the Hue Saturation and Value (HSV) and Yellow blue component and red component (YCBCr) colour models are used to give a better result than other models, because they separate luminance from chrominance components.

iv. Hand gesture classification and recognition

Classification of the gesture is also viewed as the point of recognition of the gesture, because it is the last step of a hand gesture recognition system. This phase involves matching the current gesture feature with stored features. The classification algorithms play an important role in the gesture recognition system as they determine the accuracy of the gesture. The speed of the classification algorithm is also important, especially for real time systems as speed is of the essence [9]. In this phase there are many algorithms, which can be applied. They can be categorized as mathematical model based algorithms such as Hidden Markov Model (HMM) and Finite State Machine (FSM), or as soft computing algorithms such as neural networks [3].

5   Result

Constraints as identified by different authors are summarized in Table 2.
Constraints arranged in the phase that they occur

(a) Constraints associated with image acquisition

Image acquisition is the first step in vision-based sign language hand gesture recognition. This is done via a camera attached to the system or attached on the system. Table 3 illustrates the constraints associated with image acquisition.

Table 2. Constraints as identified by different authors

Reference	Contributions (constraints identified)
[16]	• View point dependence constraint - vision-based hand gesture recognition systems require that the users have to position their hand facing the camera, which is a challenge to many and it compromises the naturalness of the system • Gesture intraclass variability constraint - a sign language gesture suffers from uniqueness of sign language dialect and it is also not possible to perform the same gesture the same way even by the same individual • Gesture start and stop detection constraint - most hand gesture recognition systems rely on classification based on frames, hence when multiple gestures are performed it may be challenging to detect when a gesture begins and ends, which may lead to inaccuracy of detection • Gesture context challenge - sign language hand gestures like any other language have a grammar context and in most cases, the hand gestures are performed with other cues like face expressions and lip reading. This makes it a very complex problem to solve • Use of both hands for hand gestures constraint - many sign languages use both hands in the process of performing hand gestures, which causes hand tracking and detection challenges
[5]	• There are many types of hand gestures, which have different meanings and are performed differently • The complexity of the hand and its ability to move in different directions according to its degrees of freedom makes recognizing it a challenge • The computer vision discipline is still not well understood by many people and the cameras used are also not up to the task, since they also suffer environmental challenges • The paper contributes to recognizing the challenges in the stages, which they are likely to manifest in a gesture recognition system
[18]	In this paper, the constraints that were identified include, occlusion, varying position of the person performing the gesture, loss of depth information by the camera, special temporal nature of the hand and difference in the signer's speed, and the co-articulation constraint not being able to know when the next gesture begins or when the previous one ends
[1]	In this paper, the constraints that were identified include, complex background, speed in real time tracking, feature selection and co-articulation

(b) Constraints in the hand tracking and segmentation phase of a vision-based sign language hand gesture recognition system

The main constraint in hand tracking is brought about by the ability of the hand to move in different directions depending on its 27 degrees of freedom. This constraint is referred to, by most researchers, as rotation. Other constraints in this phase include variation in the speed of hand gestures [3], variation in skin colour, illumination variation, background complexity, and occlusion. Table 4 below outlines the constraints associated with tracking and segmentation of hand gestures.

Table 3. Constraints associated with image acquisition [5]

Step	Constraint
Image acquisition	Camera specifications (colour range, resolution and accuracy, frame rate, lens characteristics, camera computer interface) and 3D image acquisition (depth accuracy, synchronization)

Table 4. Constraints associated with tracking and segmentation [5]

Step	Constraint
Segmentation	Illumination variation, complex background and dynamic background
Gesture detection	Hand articulation, occlusion

(c) Constraints in the feature extraction phase of a vision-based sign language hand gesture recognition system

The most notable constraints in this phase include rotation, scale and translation. Rotation constraintarises when the hand region is rotated in any direction in the scene. Scale constraint arises, because of the different sizes of people's hands making the gestures. The translation problem is the variation of hand positions in different images, which leads to erroneous representation of the features [19]. Table 5 indicates the constraints that can be encountered in the feature extraction phase of a vision-based sign language hand gesture recognition system.

Table 5. Constraints in the feature extraction phase [5]

Feature type	Examples	Constraint
Histogram-based	• Histogram of gradient (HoG) features	Complex background and image noise affect performance of the algorithm
Transform domain	•Fourier descriptor • Discrete Cosine Transform (DCT) descriptor • Wavelet descriptor	Challenge in differentiating gestures
Mixture of features	• Combined features	Compatibility may lower the recognition rate
Moments	• Geometric moments • Orthogonal moments	Moments cannot handle occlusion well
Curve fitting based	• Curvature scale space	Sensitive to distortion in the boundary

(d) constraints in the classification and recognition phase of a vision-based sign language hand gesture recognition system

An appropriate classifier identifies gesture features and categorizes them into either predefined classes (supervised) or by their similarity (unsupervised) [20]. Some of the limitations encountered in this phase include large data sets for classifier training in some algorithms, computational complexity, selection of optimum parameters and recognition of unknown gestures. Below is Table 6 outlining the constraints likely to be encountered in the classification phase.

Table 6. Constraints in the classification phase [5]

Classifier	Constraint
Dynamic time wrapping (DTW)	• Requires a huge training data set • Not computationally effective for large gesture vocabulary size
K nearest neighbour (k-XN)	• Computationally intensive for a large dataset • Performance degrades as the dimensionality of the feature space increases
Deep networks	• Difficult to get an optimized solution for non-convex and nonlinear systems • Needs a huge training data set
Finite state machine (FSM)	• Complex to manage large states, • The state transition conditions are rigid
Artificial neural networks (ANN)	• Difficult to set parameters (e.g. the optimal number of nodes, hidden layers, sigmoid functions) • Training is computationally intensive and requires a large set of training data for obtaining acceptable performance • It acts like a 'black box,' and hence, it is difficult to identify errors in a complex network
Conditional random fields (CRF)	• High computational complexity during training makes it difficult to re-train the model when new training gesture sequences become available • CRFs cannot recognize totally unknown gestures, i.e., gestures that are not present in the training dataset
k-means	• Dependent on initial cluster centre values • Sensitive to outliers • Does not perform well in the presence of non-globular clusters • Selecting an appropriate value of k is challenging
Support vector machines (SVM)	• Right kernel function selection is challenging • Computationally expensive • SVM is a binary classifier; and hence, multi-class classification requires multiple pairwise classifications • Multi-class SVMs based on single optimization are difficult to implement
Hidden Markov Model (HMM)	• The number of states and the structure of the HMM must be predefined • Statistical nature of an HMM precludes a rapid training phase. Well-aligned data segments are required to train an HMM • The stationarity assumption in HMM may not hold true for a complete gestural action

(continued)

Table 6. (*continued*)

Classifier	Constraint
Mean-shift	• The algorithm is computationally complex • Data needs to be sufficiently dense with a discernible gradient to locate the cluster centres • Susceptible to outliers or data points located between natural clusters

The constraints can also be categorized by the cause. The three causes include the hand itself, the system and equipment in use and environmental factors, as indicated in Fig. 1.

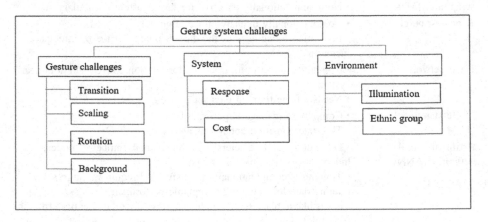

Fig. 1. Pictorial representation of gesture challenges or constraints [14]

6   Business Benefits

This study highlights the constraints encountered in vison-based system implementation in a logical way since the constraints are presented in the phases they are mostly likely to occur. This will help researchers and gesture recognition system developers to easily identify the constraints they want to address using new or a combination of existing algorithms. This can be beneficial in many hand gesture recognition application areas like robot control, game applications sign language recognition, amongst others.

The results of this study can provide a basis for a better sign language hand gesture recognition system capable of full sign language interpretation. Sign language interpretation systems are beneficial for communication, because they assist hearing impaired individuals to understand the non-hearing impaired and *vice versa*. Vision-based sign language interpretation systems enable communication in a natural way without the need for a human interpreter, hence they are likely to be more cost efficient. The vision-based gesture recognition interpretation systems can be deployed as software applications on mobile phones, computers, laptops and even tablets. This can facilitate communication for hearing impaired individuals in public facilities like banks, airports, churches and schools.

7 Conclusion

In this paper, the phases of a typical vison-based sign language hand gesture recognition system are identified. The constraints that can be encountered in each stage of a vision-based hand gesture recognition system are outlined. It is evident from the literature that the challenges begin right from the first phase, which is image acquisition where camera resolution and quality can affect the gesture recognition rate. Background noise and lighting also pose serious constraints.

These constraints coupled with many others as mentioned in this paper have resulted in development of many algorithms. Each of these algorithms has its strengths and weaknesses. Hence the choice of the algorithm to use for sign language application may vary from one researcher to another. Further work needs to be done in order to find better solutions to overcome the constraints.

References

1. Choudhury, A., Kumar, A. Kumar, K.: A review on vision-based hand gesture recognition and applications (2015)
2. Micheni, E., Murumba, J.: The role of ICT in electoral processes: case of Kenya (2018)
3. Zhu, Y., Yang, Z., Yuan, B.: Vision based hand gesture recognition (2013)
4. Chen, Q., Georganas, N., Petriu, E.: Real-time vision-based hand gesture recognition using haar-like features. In: IEEE Instrumentation and Measurement Technology Conference IMTC (2007)
5. Chakraborty, B., Sarma, D.. Bhuyan, M., Macdorman, K.: Review of constraints on vision-based gesture recognition for human – computer interaction (2018)
6. Braffort, A.: Research on computer science and sign language: ethical aspects. In: Wachsmuth, I., Sowa, T. (eds.) GW 2001. LNCS (LNAI), vol. 2298, pp. 1–8. Springer, Heidelberg (2002). https://doi.org/10.1007/3-540-47873-6_1
7. Bhuyan, P., Ghoah, D.: A framework for hand gesture recognition with application to sign language (2006)
8. Corbin, J., Strauss, A.: Basics of qualitative research: techniques and procedures for developing grounded theory (2008)
9. Bowen, G.: Document analysis as a qualitative research document analysis as a qualitative research method. Qual. Res. J. 9(2), 27–40 (2017)
10. Ghotkar, A.: Study of vision based hand gesture recognition using (2014)
11. McLeod, S.: Qualitative vs quantitative data simply psychology (2017)
12. Gamundani, A., Nekare, I.: A review of new trends in cyber attacks: a zoom into distributed database systems (2018)
13. Ahmed, T., Bernier, O., Viallet, J.: A neural network based real time hand gesture recognition system (2012)
14. Darwish, S., Madbouly, M., Khorsheed, M.: Hand gesture recognition for sign language: a new higher order fuzzy HMM approach. Hand 1, 18565 (2016)

15. Ghotkar, S., Kharate, G.: Study of vision based hand gesture recognition using indian sign language. Int. J. Smart Sens. Intell. Syst. **7**(1), 96–115 (2014)
16. Zabulis, X., Baltzakis, H., Argyros, A.: Vision-based hand gesture recognition for human-computer interaction. Gesture, 1–56 (2009)
17. Wachs, J., Kölsch, M., Stern, H., Edan, Y.: Vision-based hand-gesture applications. Commun. ACM **54**(2), 60 (2011)
18. Bauer, B., Karl-Friedrich, K.: Towards an automatic sign language recognition system using subunits. In: Wachsmuth, I., Sowa, T. (eds.) GW 2001. LNCS (LNAI), vol. 2298, pp. 64–75. Springer, Heidelberg (2002). https://doi.org/10.1007/3-540-47873-6_7
19. Simei, A., Wysoski, G., Marcus V., Susumu, K.: A rotation invariant approach on static-gesture recognition using boundary histograms and neural networks. In: IEEE 9th International Conference on Neural Information Processing (2002)
20. Mitra, S., Acharya, T.: Gesture recognition: a survey. IEEE Trans. Syst. Man Cybern. Part C Appl. Rev. **37**(3), 311–324 (2007)

Assistive Environments

Quantifying Differences Between Child and Adult Motion Based on Gait Features

Aishat Aloba, Annie Luc, Julia Woodward, Yuzhu Dong,
Rong Zhang, Eakta Jain, and Lisa Anthony[(✉)]

Dept of CISE, University of Florida, Gainesville, FL 32611, USA
{aoaloba, annieluc, julia.woodward, yuzhudolg}@ufl.edu,
{rzhang, ejain, lanthony}@cise.ufl.edu

Abstract. Previous work has shown that motion performed by children is perceivably different from that performed by adults. What exactly is being perceived has not been identified: what are the quantifiable differences between child and adult motion for different actions? In this paper, we used data captured with the Microsoft Kinect from 10 children (ages 5 to 9) and 10 adults performing four dynamic actions (walk in place, walk in place as fast as you can, run in place, run in place as fast as you can). We computed spatial and temporal features of these motions from gait analysis, and found that temporal features such as step time, cycle time, cycle frequency, and cadence are different in the motion of children compared to that of adults. Children moved faster and completed more steps in the same time as adults. We discuss implications of our results for improving whole-body interaction experiences for children.

Keywords: Whole-body interaction · Motion gestures · Children · Adults · Kinect · Gait · Spatial features · Temporal features

1 Introduction

Whole-body interaction requires robust and accurate recognition of the user's motions. Most whole-body interaction or recognition research has focused on adults [1, 2]. However, there is reason to believe that children's motions may differ significantly from those of adults based on differences in physiological factors such as body proportion [3] and development of the neuro-muscular control system [4]. In fact, previous work by Jain et al. [5] has shown that the motion of a child is perceivably different from the motion of an adult. The authors found that naïve viewers can perceive the difference between child and adult motion, when presented as *point-light displays* (points of lights representing each joint of the human body), at a rate significantly above chance, and with 70% accuracy for dynamic motions such as walking and running.

However, what exactly is being perceived has not been identified: what are the quantifiable differences between child and adult motion for dynamic actions, specifically, walking and running actions? If these quantifiable differences could be identified, we could improve recognition of child motion and enhance whole-body interaction experiences (e.g., exertion games [6]). To identify these differences, we considered features from literature on the analysis of gait. We chose to focus on these features

M. Antona and C. Stephanidis (Eds.): HCII 2019, LNCS 11573, pp. 385–402, 2019.
https://doi.org/10.1007/978-3-030-23563-5_31

because they have often been used to characterize walking and running motions from a physiological perspective [7–9].

This paper presents results on gait features that characterize the differences between adult and child motion. We chose a subset of walking and running motions from the Kinder-Gator dataset [10], collected by our research groups. This dataset includes child and adult motions from 10 children (ages 5 to 9) and 10 adults (ages 19 to 32) performing 58 actions forward-facing the Kinect [11]. Of the 58 actions in the dataset, we analyze four actions: walk in place (walk), walk in place as fast as you can (walk fast), run in place (run), and run in place as fast as you can (run fast). We grouped the gait features into two categories based on the gait literature: *spatial features*, dependent on distance, e.g., step width, step height, relative step height, and walk ratio; and *temporal features*, dependent on time, e.g., cadence, step time, cycle time, cycle frequency, and step speed. Each feature was computed on the motion data and results were analyzed statistically based on two factors: age group (child vs. adult) and action type.

We found a significant effect of age group across all temporal features except step speed. Hence, step time, cycle time, cycle frequency, and cadence are significantly different in the motion of children compared to that of adults. For spatial features, we found no significant effect of age group. There was a significant effect of action type in both temporal and spatial features; we found differences between pairs of actions across all features. The contributions of this paper are to (a) identify a set of nine features from gait analysis that can be used to (b) quantify the differences between child and adult motion. Our paper advances the understanding of child and adult motion in terms of measurable differences in their motion to inform the design and development of whole-body interactive systems (e.g., smart environments, exertion games). Our paper will also improve how motion recognition algorithms classify child and adult motion. These motion recognition algorithms can in turn help to improve how child actions are recognized in whole-body interaction applications.

2   Related Work

Prior work has studied recognition and perception of human motion, though most has focused on children or adults only, rather than comparing or contrasting the two.

2.1   Perception of Human Motion

Behavioral researchers [12–14] studied infants' preferences for point-light display representations of human motion compared to non-human motion, and asserted that humans develop the ability to perceive and detect biological motion during infancy. Point-light displays have also been used in various perception studies to understand what identifying information people can perceive. Cutting and Kowlozski [15] and Beardsworth and Buckner [16] found that adults could identify themselves and their friends from point-light display representations of walking motions. Furthermore, Wellerdiek et al. [17] studied a wider range of motions, such as spontaneous dancing and ping-pong, and showed that adults were able to detect their own motion from point-light displays. Golinkoff et al. [18] found that children could identify different motions such as walking and dancing from point-light display videos. Jain et al. [5] examined whether naïve viewers could perceive the difference between the motion of a child and an adult from point-light displays. They found that adults can perceive the difference between child and adult motion at levels significantly above chance, and with about 70% accuracy for dynamic actions such as "run" and "walk". However, what cues were helping participants identify whether the motion came from a child or an adult were not determined. In our study, we analyze child and adult motions to understand the features that differentiate child motion from adult motion.

2.2   Motion Recognition Using the Kinect

Human motion tracking has been studied extensively using image sequences [14, 19–23]. For example, Ceseracciu [20] analyzed people's gait by processing grayscale images from video recordings using multiple cameras. With the advent of low-cost commercial motion tracking devices such as the Microsoft Kinect [11], researchers shifted attention to recognizing actions using depth sensors [24, 25] and developing Kinect-based applications [26–29]. Nirjon et al. [29] used the 3D skeleton joint coordinates tracked by the Kinect to distinguish between aggressive actions, such as kicking from punching. The Kinect has also been applied to hand gesture recognition [26, 30, 31]. Oh et al. [26] created the Hands-Up system, which issues commands to a computer by tracking adults' hand gestures using a Kinect attached to the ceiling. Zafrulla et al. [30] and Simon Lang [31] both applied the Kinect to sign language recognition. Jun-Da Huang [27] created a Kinect-based system, KineRehab, to track the movements of young adults with muscle atrophy and cerebral palsy. KineRehab detects movements with 80% accuracy. Dimitrios et al. [28] used the Kinect to track adults' dance movements and to automatically align the movement to those of a dance expert in the real world. Few researchers have used the Kinect to successfully recognize children's motions. Zhang et al. [32] tracked children's movements in a classroom using multiple Kinect sensors, but did not attempt to classify specific motions. Connell et al. [33] used a Wizard-of-Oz approach to elicit gestures and body movements from children when interacting with a Kinect. Lee et al. [6] and Smith and Gilbert [34] have both created interactive games for children to solve math problems by performing gestures tracked by the Kinect, but did not include real-time recognition. In our study, we expand upon previous work by tracking and logging the motion of both children and adults with the Kinect in order to quantify the differences between these motions to inform better recognizers in the future.

3   Gait and Gait Analysis Background

Gait is defined as one's manner or style of walking [35]. The analysis of gait is defined as the systematic study of human locomotion [36, 37], using the cycles and steps in the motion (Fig. 1). A gait *cycle* (stride) is defined as the period between a foot contact on the ground to the next contact of the same foot on the ground again [7, 36]. A gait *step* is defined as the period between a foot contact on the ground to the next contact of the opposite foot, also known as half a gait cycle [7].

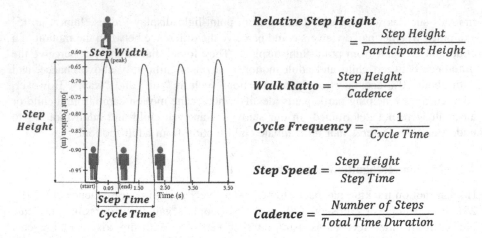

Fig. 1. Gait features we computed and how we extracted them from the motion capture data.

3.1 Gait Analysis

Wilheim and Eduard Weber [38] pioneered the study of spatial and temporal gait features by showing that human locomotion can be measured quantitatively. This finding led to the development of different quantitative methods for analyzing gait kinematics, of which the most commonly used is the placement of 3D markers along segments of the human body [39]. Since then, gait kinematics have been studied extensively. Researchers [7, 8, 36, 40–42] have placed reflective markers on adults walking at different speeds, and have analyzed distance and time features of their gait such as stride length, step length, walk ratio, stride time, cadence, and speed. Gait analysis has also been used to identify individuals from their gait. Gianaria et al. [43] achieved 96% accuracy on classifying adults by gait by extracting gait features from Kinect data and feeding the features into a support vector machine (SVM). Prior work has also analyzed features from children's gait [44–47]. Dusing and Thorpe [46] analyzed the cadence of children ages 1 to 10 walking at a self-selected pace, and found that cadence reduces as age increases. Barreira et al. [44] also studied the cadence of children walking freely in their environment. They found that children spent more time at lower cadences (0–79 steps/minute) compared to cadences signifying moderate or vigorous physical intensity (120 steps/min).

A limitation of the studies reviewed above is that they mainly focus on either children or adults. Some prior work has studied the comparison between child and adult motion, but they either focus on very young children [48] or older children [49] rather than a range of younger and older children. Davis [48] extracted and fed features collected from children's and adults' gait into a two-class linear perceptron to differentiate between the walking motion of young children (ages 3 to 5) and adults. He found that gait features can be used to differentiate between young children's and adults' walking patterns with about 93–95% accuracy. Oberg et al. [49] also compared gait features across ages from 10 to 79 years, and found that the speed of the gait and length of a step reduces with age. In our study, we extracted gait features such as cadence, step time, and step length from the motion of children ages 5 to 9 to quantify the differences between child and adult motion. We focus on ages within the gap of age ranges previously studied in the literature (ages 5 to 9) because of the rapid development of motor skills during this age [50, 51].

3.2   Gait Features

We surveyed the literature on gait analysis [7–9, 36, 40–42, 48, 49, 52], and identified ten features commonly used to characterize a person's gait. Gait analysis has been historically utilized to analyze walking or running motion that involves moving a distance away from the starting point. One feature that is commonly examined in gait analysis is the step length. The step length measures the distance between feet along the direction of motion, which for moving motions is parallel to the floor. However, the walking and running actions in the Kinder-Gator dataset [10] involve moving in place instead of moving away from a starting point. Therefore, the direction of motion is perpendicular to the floor instead of parallel to the floor, so we calculated the perpendicular distance (i.e., step height). This adaptation from step length to step height is valid because both measure the peak distance between feet.

Of the ten features we identified, we eliminated cycle length, which is the distance between successive placements of the same foot (measured as two step lengths). We eliminated this feature because, for in-place motions, the same foot returns to nearly the same location between steps. Measuring successive placements of the same foot using two step heights instead of two step lengths would imply the participant is moving continuously upward, e.g., climbing up a ladder, rather than the in-place actions in our dataset [10]. We categorized the remaining nine gait features into *spatial* and *temporal* feature groups (Fig. 1). Spatial features are distance-based; they include features which are dependent on the length (height) of a step in their computation. Temporal features are time-based. They include features which are dependent on time in their computation. We chose these nine features because prior research has shown that they are unique per person [41], that is, analogous to a fingerprint, and can be used as a biometric measure [53]. The nine features include:

Step Width (m). This is a spatial feature. The step width is the maximum lateral distance between feet during a step [9]. It is measured as the horizontal distance between the position of one foot and the other foot during a step. This feature evaluates how wide or narrow the step taken is.

Step Height (m). This is a spatial feature and is an adaptation of the step length. Step length is defined as the distance by which a foot moves in front of the opposite foot [7, 36, 41, 54]. Since the walking and running actions in the Kinder-Gator dataset [10] involve moving in place rather than forward over a distance, we define the step height as the distance a foot travels above the other foot during a step. It is measured as how high above the ground vertically a foot is during the highest part of a step.

Relative Step Height. This is a spatial feature, and is defined as the length of a step in relation to the height of the person [48]. It measures the ratio between the step length (we use step height because they are walking in place) and height of the person. The relative step height is an important feature to consider as it normalizes the step height by the person's height, hence eliminating differences in step heights due to variations in height across people (e.g., children and adults).

Walk Ratio (m/Steps/Minute). This is a spatial feature. It is an index used to characterize a person's walking pattern, and is measured as the ratio between the step length (we use step height) and the cadence (rate at which a person walks) [8, 40]. This feature is relevant to dynamic motions, as previous research notes that it reflects participants' balance and coordination when performing a motion [8, 40].

Step Time (s). This is a temporal feature which defines the time duration of a step [9]. It can further be defined as the time it takes a foot to complete one step. It is measured as the duration from when the foot leaves the ground to the time when the foot touches the ground again in completion of a step.

Cycle Time (s). This is a temporal feature and is also known as stride time. The cycle time can be defined as the time it takes a foot to complete one cycle (two steps). It can be measured as the time between two consecutive steps of the same foot along the horizontal (we use vertical) trajectory [48].

Cycle Frequency (1/s). This is a temporal feature and is also known as stride frequency. It is defined as the number of cycles per unit time and can be computed as the inverse of the cycle time [42, 48, 55]. Prior research shows that participants' preferred cycle frequency optimizes energy cost [42].

Step Speed (m/s). This is a temporal feature, and is defined as the ratio between the step length (we use step height) and the step time [9]. It defines how fast a step is completed and can help in understanding the pace of an action.

Cadence (steps/min). This is a temporal feature and is defined as the rate at which a person walks. It is measured as the number of steps taken per minute [36, 41, 42, 52], and reflects the level of energy being exerted.

4   Dataset Used

The data used in this study is from our publicly available dataset, Kinder-Gator [10]. Kinder-Gator is a dataset our research groups collected from 10 children, ages 5 to 9 (5 females), and 10 adults, ages 19 to 32 (5 females), performing 58 natural motions, in-place, forward-facing the Microsoft Kinect 1.0. The dataset includes 3D positions (x: horizontal, y: vertical, z: depth) of 20 joints in the body at 30 frames per second (fps). Since we are focusing on walking and running motions, we used the four walking/running motions in the dataset, namely: "walk in place" (walk), "walk in place as fast as you can" (walk fast), "run in place" (run), and "run in place as fast as you can" (run fast). Each participant-action pair contained at least 10 steps. We created point-light display videos [14], which present each joint as a white dot on a black background animating through the course of the motion, for each participant-action pair.

5   Analysis

To compute the gait features for each person-action pair, we depend on knowledge of the *stance phase* (when the foot is on the ground [41]) and *swing phase* (when the foot is away from the ground [41]). Hence, we needed to identify the frames of each motion that corresponded to the step boundaries. We manually extracted these frames from the point-light display videos using a video annotation toolkit called EASEL [56]. Two researchers annotated subsets of the point-light display videos of the Kinect data for all of the actions. To ensure balanced labeling, the videos were counterbalanced between each annotator by age group (child, adult) and action. Also, for similar actions (run & run fast, walk & walk fast), the same annotator annotated the same participant for both motions. For each video to be annotated, we created three tracks in EASEL. Frames for the start (foot is on the ground), peak (foot is at its maximum position), and end (foot is returned to the ground) were recorded on the first, second, and third track respectively. The frames start-peak-end are the frames within a step. Previous research [36] has suggested that the analysis of gait can be done with either the foot, knee, hip, or pelvis joint, so we used the left foot joint from the Kinect skeleton tracking data in our analysis. Once all the frames had been annotated, we exported the annotation session, which creates an output CSV file with all the frames and the corresponding tracks recorded. We used this file for feature computation based on the start-peak-end frames.

    We automated the feature computation process by extracting the corresponding foot positions and time stamps from the data for the frames we had manually extracted. The foot positions and time stamps were used to compute the gait features. For features involving computations per step or cycle, we averaged the values over the total number of steps or cycles in that motion. Therefore, each participant has one data point per action for each gait feature. A two-way repeated measures ANOVA was used to analyze the main effect of age group and action type and the interaction effect between them. Whenever we found no interaction effect between age group and action, we recomputed the two-way repeated measures ANOVA without the interaction effect in the model, and report that. For features where we found a significant effect of action, we conducted a Tukey post-hoc test to identify action pairs that are significantly different. We present results for all of the gait features we considered in our study. All means and standard deviations for features in the analysis can be found in Table 1, and they are expressed in units commonly used in the analysis of gait [57].

    Our results show that the following temporal features differ between child motion and adult motion: step time, cycle time, cycle frequency, and cadence.

5.1   Spatial Features

Distance-based features generally showed no significant effect of age group; hence, we conclude that they cannot be used to distinguish adult and child motion. However, these features show a significant effect of action type, which serves to validate our approach of using features from the analysis of gait, despite the differences in motion structure (i.e., in-place motions versus moving along a distance).

Table 1. Mean (and standard deviation: SD) of gait features by age group and actions. (*) denotes a significant effect at $p < 0.05$.

Age group

Gait feature	Child	Adult	p
Step width (m)	0.16 (0.04)	0.17 (0.06)	
Step height (m)	0.10 (0.07)	0.10 (0.11)	
Relative step height	0.09 (0.05)	0.06 (0.06)	
Walk ratio (m/steps/min)	0.00063 (0.00061)	0.00070 (0.00078)	
Step time (s)	0.33 (0.11)	0.43 (0.15)	*
Cycle time (s)	1.05 (0.47)	1.26 (0.49)	*
Cycle frequency (1/s)	1.15 (0.44)	0.90 (0.32)	*
Step speed (m/s)	0.32 (0.17)	0.25 (0.27)	
Cadence (steps/min)	203 (79)	166 (61)	*

Actions

Gait feature	Walk	Walk fast	Run	Run fast	p
Step width (m)	0.14 (0.05)	0.16 (0.05)	0.16 (0.05)	0.18 (0.04)	*
Step height (m)	0.10 (0.09)	0.08 (0.08)	0.09 (0.08)	0.13 (0.09)	*
Relative step height	0.07 (0.07)	0.06 (0.06)	0.07 (0.05)	0.10 (0.06)	*
Walk ratio (m/steps/min)	0.0010 (0.0010)	0.00051 (0.00057)	0.00048 (0.00044)	0.00061 (0.00051)	*
Step time (s)	0.53 (0.14)	0.37 (0.12)	0.32 (0.08)	0.29 (0.05)	*
Cycle time (s)	1.84 (0.30)	1.05 (0.34)	0.97 (0.25)	0.78 (0.14)	*
Cycle frequency (1/s)	0.56 (0.09)	1.07 (0.37)	1.14 (0.35)	1.34 (0.25)	*
Step speed (m/s)	0.17 (0.14)	0.24 (0.20)	0.28 (0.20)	0.44 (0.27)	*
Cadence (steps/min)	99 (16)	193 (66)	203 (56)	243 (48)	*

Step Width. Recall that the width of a step is computed as the horizontal distance between both feet during a step. A two-way repeated measures ANOVA on step width with a between-subjects factor of *age group* (child, adult) and a within-subjects factor of *action type* (walk, walk fast, run, run fast) found no significant effect of age group ($F_{1,18} = 0.18$, *n.s.*). The lateral placement of the feet for adults is roughly the same as that for children. This similarity may be because both adults and children are spending less effort to control the horizontal distance between their feet since in-place actions involve vertical movements. However, we found a significant main effect of action ($F_{3,57} = 5.27$, $p < 0.01$). Post-hoc tests identified a difference between walk and run fast ($p < 0.001$). People have wider step widths when running fast than when walking (see Table 1), irrespective of age group. Bauby and Kuo [57] asserted that wider steps are an advantage in stability; thus, participants widen their steps when running fast compared to walking to improve coordination. No difference in step width was found between other pairs of actions.

Step Height. The height of a step is computed as the vertical distance between when the foot is on the ground, and when the foot is at its maximum position (peak). We focused on the left foot step heights because we annotated the left foot joints. We computed the ground for each step as the minimum between the (start) and (end) position. A two-way repeated measures ANOVA on step height with a between-subjects factor of *age group* (child, adult) and a within-subjects factor of *action type* (walk, walk fast, run, run fast) found no significant effect of age group ($F_{1,18} = 0.002$, *n.s.*). Our analysis found that the average step height for children and adults is about the same (see Table 1). This finding is surprising given the typical difference in height between children and adults. The range helps to illuminate what is really happening: children: min: 0.01 m, max: 0.25 m, med: 0.11 m; adults: min: 0.008 m, max: 0.41 m, med: 0.06 m. Thus, adults can raise their feet higher than children, but children tend to raise them proportionally higher on average. We also found a significant main effect of action ($F_{3,57} = 5.03$, $p < 0.01$). Post-hoc tests showed the following action pairs differed: run/run fast ($p < 0.01$), walk/run fast ($p < 0.05$), and walk fast/run fast ($p < 0.01$). Participants have higher step heights when running fast compared to the other actions (see Table 1): this confirms their exertion level was higher when running fast.

Relative Step Height. The relative step height is computed as the ratio of the step height to the height of the person performing the action. We estimated the height of each Kinder-Gator participant using the difference between the head and the foot along the vertical dimension (*y*-axis). A two-way repeated measures ANOVA on relative step height with a between-subjects factor of *age group* (child, adult) and a within-subjects factor of *action type* (walk, walk fast, run, run fast) found no significant effect of age group ($F_{1,18} = 1.51$, *n.s.*). The large variance in the relative step height in children and adults may be the reason why we found no significant difference (Table 1). However, adults generally have a lower average relative step height compared to children. This finding follows, since their average step heights were the same, but adults are taller than children. We also found a significant effect of action ($F_{3,57} = 4.91$, $p < 0.01$). Post-hoc tests showed that the following action pairs differed: run/run fast ($p < 0.01$) and walk fast/run fast ($p < 0.01$). Like step height, children and adults have a higher relative step height when running fast compared to just running or walking fast (see Table 1).

Walk Ratio. The walk ratio, a measure of balance and coordination, is computed as the ratio of the step height and the cadence (a temporal feature). A two-way repeated measures ANOVA on walk ratio with a between-subjects factor of *age group* (child, adult) and a within-subjects factor of *action type* (walk, walk fast, run, run fast) found no significant effect of age group ($F_{1,18} = 0.07$, *n.s.*). The similarity in average walk ratio between children and adults (Table 1) may be because walking pattern is influenced by how high participants raise their foot during a motion, and there was no significant difference in the average step height between children and adults. The standard walk ratio for adults in the gait literature is 0.0065 m/steps/min [8, 40].

However, we found a lower walk ratio for adults (M = 0.00070 m/steps/min, SD = 0.00078). The lower walk ratio is because the maximum step *height* that has been achieved while moving in place in our dataset is much less than the average step *lengths* noted in the literature (M = 0.68 m) [40]. Children had an average walk ratio of 0.00062 m/steps/min (SD = 0.00061). We also found a significant effect of action ($F_{3,57}$ = 7.64, p < 0.001). Post-hoc tests showed that the following action pairs differed: walk/walk fast (p < 0.001), walk/run fast (p < 0.01), and walk/run (p < 0.001). Participant's walk ratios were higher when walking in place compared to any of the other actions (see Table 1). This result follows what we might expect: participants have more coordination and balance in the (slowest) walking action when performing the action in place.

5.2   Temporal Features

We discuss the computation of time-based gait features, and present our findings using the same statistical analysis we used for the spatial features. Time-based features (except step speed) show significant effects by both age group and action. Thus, these are promising features to use to differentiate between child and adult motion.

Step Time (Fig. 2a). The time for each step during an action is computed as the difference between the time stamp for the end frame (when the foot is back on the ground) and the timestamp for the start frame (when the foot first leaves the ground). A two-way repeated measures ANOVA on step time with a between-subjects factor of age group (child, adult) and a within-subjects factor of action type (walk, walk fast, run, run fast) found a significant main effect of age group ($F_{1,18}$ = 12.15, p < 0.01). Children moved faster compared to adults (Table 1). During collection of the Kinder-Gator dataset [10], we observed that, given the same prompts as adults (e.g., "run in place as fast as you can"), children were more energetic and enthusiastic when performing the actions. We also found a significant effect of action ($F_{3,54}$ = 55.86, p < 0.0001). Post-hoc tests showed that the following action pairs differed: walk/walk fast (p < 0.001), walk fast/run fast (p < 0.001), walk fast/run (p < 0.01), walk/run fast (p < 0.001), and walk/run (p < 0.001). As expected, people have a faster step time when running, and become slower when walking fast or walking (Table 1). We also found a significant interaction effect ($F_{3,54}$ = 4.74, p < 0.01). Children have a faster step time than adults when performing all of the actions except running fast (see Fig. 2a). It is possible that the prompt "run as fast as you can" encouraged adults to "pick up the pace" and exert themselves more than they did in the previous actions.

Cycle Time (Fig. 2b). The time for a cycle is computed as the time it takes to complete two consecutive steps of the same foot. A two-way repeated measures ANOVA on cycle time with a between-subjects factor of *age group* (child, adult) and a within-subjects factor of *action type* (walk, walk fast, run, run fast) found a significant main effect of age group ($F_{1,18}$ = 7.61, p < 0.05). Like step time, the children completed cycles with a shorter time duration compared to adults (Table 1). We also found a significant effect of action ($F_{3,57}$ = 102.13, p < 0.0001). Post-hoc tests showed that the following action pairs differed: walk/run (p < 0.001), walk/run fast (p < 0.001), walk/walk fast (p < 0.001), run/run fast (p < 0.05), and walk fast/run fast (p < 0.001).

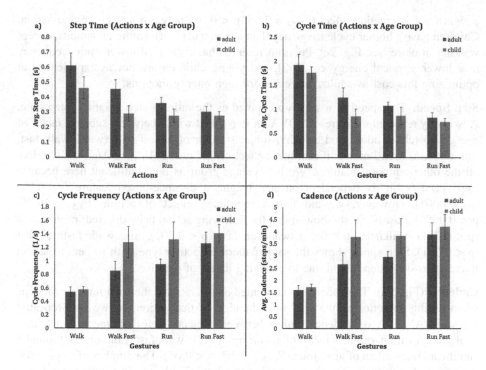

Fig. 2. Effect of action type and age group on selected gait features. Error bars indicate 95% confidence interval.

Intuitively, people exhibit the fastest cycle time when running fast, and cycle time is slowest for walking (Table 1). Compared to step time, participants completed individual steps faster when running than when walking fast, but the time to complete successive steps (in this case, two steps of the same foot) is roughly the same in both actions. On the other hand, unlike step time, we found no significant interaction between age group and action. Figure 2b shows a similar relationship in cycle time as Fig. 2a shows in step time except for the walk action. One possible explanation is that people may walk with each foot at a slightly different frequency, depending on their exertion level. Previous research [58] has shown people exhibit a strength imbalance on their non-dominant side, which could also lead to higher variability in the motion.

Cycle Frequency (Fig. 2c). The cycle frequency is computed as the inverse of the cycle time (1/cycle time). A two-way repeated measures ANOVA on cycle frequency with a between-subjects factor of *age group* (child, adult) and a within-subjects factor of *action type* (walk, walk fast, run, run fast) found a significant main effect of age group ($F_{1,18} = 10.53$, $p < 0.01$). Children have a higher cycle frequency compared to adults (Table 1). We also found a significant effect of action ($F_{3,54} = 46.53$, $p < 0.0001$). Like cycle time, post-hoc tests showed that the following action pairs differed: walk/run ($p < 0.001$), walk/run fast ($p < 0.001$) walk/walk fast ($p < 0.01$), run/run fast ($p < 0.01$), and walk fast/run fast ($p < 0.001$). Unlike cycle time, we found a significant interaction effect ($F_{3,54}$ = 3.55, p < 0.05) between age group and action. Children have a higher cycle frequency than adults when performing all actions, except walking in place (see Fig. 2c). Previous research has correlated lower cycle frequency to a lower physical energy cost [42], and young children are not as experienced at optimizing this cost as adults, especially for high energy motions.

Step Speed. The speed of a step is computed as the ratio of step height to step time. A two-way repeated measures ANOVA on step speed with a between-subjects factor of *age group* (child, adult) and a within-subjects factor of *action type* (walk, walk fast, run, run fast) found no significant main effect of age group ($F_{1,18}$ = 0.66, *n.s.*), unlike all the other temporal features. We believe age group is not significant here because step speed is highly dependent on step height, a distance-based feature (r = 0.90, p < 0.0001). However, we did find a significant effect of action ($F_{3,57}$ = 19.06, p < 0.0001). Post-hoc tests showed that the following action pairs differed: run/run fast (p < 0.001), walk/run (p < 0.05), walk/run fast (p < 0.001), and walk fast/run fast (p < 0.001). Step speed shows the same pattern by action type with respect to action intensity as does step height due to changing levels of exertion.

Cadence (Fig. 2d). The cadence is computed as the ratio of the total number of steps taken during an action to the total time duration of that action. A two-way repeated measures ANOVA on cadence with a between-subjects factor of *age group* (child, adult) and a within-subjects factor of *action type* (walk, walk fast, run, run fast) found a significant main effect of age group ($F_{1,18}$ = 5.87, p < 0.05). The number of steps taken per minute for children is higher compared to adults (Table 1). This higher number is expected because we know from step time that children move faster and are more enthusiastic in their motions than adults. Thus, it follows that children will complete more steps than adults in a similar time. Our analysis showed that adults in our dataset exhibited an average cadence of 96 ± 19 steps/min when walking in place, which is lower than the average walking cadence of 120 steps/min for adults observed in prior work [36]. This difference may be because adults did not walk long enough (<20 steps) during the collection of the Kinder-Gator dataset [10] compared to other gait studies, and may not have settled into a regular cadence. The finding could also be because the walking actions in the Kinder-Gator dataset [10] involved moving in place compared to moving forward along a distance. Similarly, children in our study had alower cadence (M = 103 steps/min, SD = 13) than the average cadence of 138 steps/min in D using and Thorpe's study for children ages 5 to 10 [46]. However, Barreira et al. [44] found that children spent more time in their open walking study walking at cadences of 100–119 steps/min compared to cadences of 120+ steps/min. We also found a significant effect of action ($F_{3,54}$ = 55.75, p < 0.0001). Post-hoc tests showed that the following action pairs differed: walk/run (p < 0.001), walk/run fast (p < 0.001) walk/walk fast (p < 0.001), run/run fast (p < 0.01), and walk fast/run fast (p < 0.001). Cadence exhibits a similar pattern by action type as cycle frequency and cycle time; we discuss the relationship between these features in the next section. We also found a significant interaction effect ($F_{3,54}$ = 2.99, p < 0.05). Children have a higher cadence than adults when performing all actions except walk (see Fig. 2d). Like cycle frequency, children display the same pattern of cadence by action type; they expend higher energy and exhibit lower coordination than adults.

6 Discussion

We analyzed the differences between child and adult motion based on spatial and temporal gait features from prior work in the analysis of gait [7–9, 36, 40–42, 52, 55]. Our results found significant differences in temporal features such as step time, cycle time, cycle frequency, and cadence. Children have a faster step time and cycle time and a higher cadence, but a lower cycle frequency compared to adults. Hence, these features are promising to use to differentiate between child and adult motion. We found no significant differences between child and adult motion for spatial features; thus, these features cannot be used to differentiate between the age groups.

6.1 Feature Dependency

We analyzed pairs of features for inter-dependency. Our results show a high correlation between cadence and cycle frequency ($r = 0.97$, $p < 0.001$). Although most papers we found define cadence as we have (number of steps taken per minutes [36, 41, 42, 52]), Cooper [41] acknowledged that cadence is highly related to step frequency (frequency for half a gait cycle). Our correlation analysis also showed that features that are an aggregation of two other features are more dependent on the feature in the numerator than the feature in the denominator. For example, walk ratio, which has step height in the numerator and cadence in the denominator, is more correlated with step height ($r = 0.86$, $p < 0.001$) than cadence ($r = -0.43$, $p < 0.001$). Similarly, step speed is more correlated with its numerator step height ($r = 0.90$, $p < 0.01$) than its denominator step time ($r = -0.16$, n.s.). As expected, we see a high negative correlation between cycle time and cycle frequency since cycle frequency is the inverse of cycle time ($r = -0.92$, $p < 0.001$). Furthermore, we also found a strong positive correlation between cadence and cycle time ($r = 0.91$, $p < 0.001$), and step time and cycle time ($r = 0.89$, $p < 0.001$). These results show that the correlations between features can be used to corroborate findings from different features. However, it is important to note that each individual feature is worth considering, because the information they provide about particular parts of a person's gait is unique.

6.2 Implications

Our findings have implications regarding improving whole-body recognizers for children and designing prompts for whole-body interaction experiences.

Whole-Body Recognition for Children. There are some proposed recognizers for whole-body interaction [32, 41, 59]. However, since most of these recognizers are not tailored for children, and our results show that children's movements are quantifiably different than adults' in ways that might affect tracking and recognition (e.g., speed and movement time), we hypothesize that children's motions will be poorly recognized. Thus, future work can examine tailoring whole-body recognizers to child motion qualities. For example, the coordination of a child's movement depends on the action being performed. In our study, children exhibited higher coordination during the walking in place action versus the running fast action. Recognizers will need higher tolerance for variance in the motion to be able to recognize less coordinated motion from children.

Whole-Body Interaction Prompts. Our post-hoc analysis found no significant difference between walk fast and run in any feature except step time, and conversely, we found a significant difference between run and run fast in all features except step time. Furthermore, our results for cycle frequency and cadence show that children exhibited more energy than adults when walking and running. Taken together, we can assert that children demonstrate higher exertion than adults for the same action prompts. This finding suggests that designers of motion applications should tailor interaction prompts given to users based on the age group and desired level of exertion. For example, for higher levels of exertion, designers need only prompt children to "run," but must prompt adults to "run fast."

7 Limitations and Future Work

We are limited by the small number of steps per action (<20 steps) in the Kinder-Gator dataset [10], compared to 100 or more steps commonly seen in the analysis of gait (e.g., [9, 57]). This disparity may have led to increased variability in some of the features, resulting in lack of sensitivity for significant differences. However, there are precedents for this approach when studying children's gait [46, 54, 60], likely due to their shorter attention spans [46]. Furthermore, we only focused on the left foot when computing the step features, and future work could consider both the left and right foot for the computation to investigate differences that may exist between the dominant and non-dominant foot. The increase in variability in some of the features may also be because people are less familiar with the experience of moving in-place, since the typical form of walking involves moving forward along a distance. However, it is worth noting that in-place motions are common in motion applications (e.g., exertion games [6]). Though the features we analyzed are representative of the most common features in the analysis of gait, they are not exhaustive, and future work could identify additional features which may be more discriminating for distinguishing child and adult motion. Lastly, future work can also classify children and adults based on their motions by training a binary classifier using these gait features.

8 Conclusion

In this paper, we aimed to answer the question: what are the quantifiable differences between child and adult motion for different whole-body motions? The main contributions of this paper are to (a) identify a set of nine features from the analysis of gait that can be used to (b) quantify the differences between child and adult motion. We analyzed data collected from 10 children (ages 5 to 9), and 10 adults performing four actions (walk in place, walk in place as fast as you can, run in place, run in place as fast as you) based on spatial and temporal gait features. We found that temporal features such as step time, cycle time, cycle frequency, and cadence are significantly different in the motion of children compared to that of adults. Children complete more steps in a faster time compared to adults. However, we found no significant difference in step speed, possibly due to its high dependence on step height, a spatial feature. Similarly, we found no significant difference in age group for all spatial features. Our work has implications for improving whole-body recognition algorithms for child motion and the design of whole-body interaction experiences.

Acknowledgements. This work is partially supported by National Science Foundation Grant Award #IIS-1552598. Opinions, findings, and conclusions or recommendations expressed in this paper are those of the authors and do not necessarily reflect these agencies' views.

References

1. Park, S., Aggarwal, J.K.: Recognition of two-person interactions using a hierarchical Bayesian network. In: ACM SIGMM International Workshop on Video Surveillance (IWVS 2003), pp. 65–76 (2003). http://dx.doi.org/10.1145/982452.982461
2. Yang, H.D., Park, A.Y., Lee, S.W.: Human-robot interaction by whole body gesture spotting and recognition. In: International Conference on Pattern Recognition, pp. 774–777 (2006). http://dx.doi.org/10.1109/ICPR.2006.642
3. Huelke, D.F.: An overview of anatomical considerations of infants and children in the adult world of automobile safety design. Assoc. Adv. Automot. Med. **42**, 93–113 (1998). https://doi.org/10.1145/982452.982461
4. Thelen, E.: Motor development: a new synthesis. Am. Psychol. **50**, 79–95 (1995). https://doi.org/10.1037/0003-066X.50.2.79
5. Jain, E., Anthony, L., Aloba, A., Castonguay, A., Cuba, I., Shaw, A., Woodward, J.: Is the motion of a child perceivably different from the motion of an adult? ACM Trans. Appl. Percept. **13**(2), Article no. 22 (2016). http://dx.doi.org/10.1145/2947616
6. Lee, E., Liu, X., Zhang, X.: Xdigit: An arithmetic kinect game to enhance math learning experiences. In: Fun and Games Conference, pp. 722–736 (2013)
7. Hunter, J.P., Marshall, R.N., McNair, P.J.: Interaction of step length and step rate during sprint running. Med. Sci. Sports Exerc. **36**, 261–271 (2004). https://doi.org/10.1249/01.MSS.0000113664.15777.53
8. Rota, V., Perucca, L., Simone, A., Tesio, L.: Walk ratio (step length/cadence) as a summary index of neuromotor control of gait: application to multiple sclerosis. Int. J. Rehabil. Res. **34**, 265–269 (2011). https://doi.org/10.1097/MRR.0b013e328347be02
9. Terrier, P.: Step-to-step variability in treadmill walking: influence of rhythmic auditory cueing. PLoS ONE **7**(10), e47171 (2012). https://doi.org/10.1371/journal.pone.0047171
10. Aloba, A., et al.: Kinder-Gator: the UF kinect database of child and adult motion. In: Diamanti, O., Vaxman, A. (eds.) Eurographics 2018 - Short Papers, pp. 13–16 (2018). http://dx.doi.org/10.2312/egs.20181033
11. Microsoft: Kinect for Windows. https://developer.microsoft.com/en-us/windows/kinect
12. Bertenthal, B.I., Proffitt, D.R., Kramer, S.J.: Perception of biomechanical motions by infants: implementation of various processing constraints. J. Exp. Psychol. Hum. Percept. Perform. **13**, 577–585 (1987). https://doi.org/10.1037/0096-1523.13.4.577
13. Fox, R., McDaniel, C.: The perception of biological motion by human infants. Science **218**, 486–487 (1982). https://doi.org/10.1126/science.7123249
14. Johansson, G.: Visual perception of biological motion and a model for its analysis. Percept. Psychophys. **14**, 201–211 (1973). https://doi.org/10.3758/BF03212378

15. Cutting, J., Kozlowski, L.: Recognizing friends by their walk: gait perception without familiarity cues. Bull. Psychon. Soc. **9**, 353–356 (1977). https://doi.org/10.3758/BF03337021

16. Beardsworth, T., Buckner, T.: The ability to recognize oneself from a video recording of one's movements without seeing one's body. Bull. Psychon. Soc. **18**, 19–22 (1981). https://doi.org/10.3758/BF03333558

17. Wellerdiek, A.C., Leyrer, M., Volkova, E., Chang, D.-S., Mohler, B.: Recognizing your own motions on virtual avatars. In: ACM Symposium on Applied Perception (SAP 2013), pp. 138–138 (2013). http://dx.doi.org/10.1145/2492494.2501895

18. Golinkoff, R.M., et al.: Young children can extend motion verbs to point-light displays. Dev. Psychol. **38**, 604–614 (2002). https://doi.org/10.1037/0012-1649.38.4.604

19. Wang, L., Hu, W., Tan, T.: Recent developments in human motion analysis. Pattern Recognit. **36**, 585–601 (2003). https://doi.org/10.1016/S0031-3203(02)00100-0

20. Ceseracciu, E., Sawacha, Z., Cobelli, C.: Comparison of markerless and marker-based motion capture technologies through simultaneous data collection during gait: Proof of concept. PLoS ONE **9**(3), e87640 (2014). https://doi.org/10.1371/journal.pone.0087640

21. Steele, K., Corazza, S., Scanlan, S., Sheets, A., Andriacchi, T.P.: Markerless vs. marker-based motion capture: a comparison of measured joint centers. In: North American Conference on Biomechanics Annual Meeting, pp. 5–9 (2009)

22. Nieto-Hidalgo, M., Ferrández-Pastor, F.J., Valdivieso-Sarabia, R.J., Mora-Pascual, J., García-Chamizo, J.M.: A vision based proposal for classification of normal and abnormal gait using RGB camera. J. Biomed. Inform., 82–89 (2016). doi:http://dx.doi.org/10.1016/j.jbi.2016.08.003

23. Bobick, A.F., Davis, J.W.: The recognition of human movement using temporal templates. IEEE Trans. Pattern Anal. Mach. Intell., 257–267 (2001). http://dx.doi.org/10.1109/34.910878

24. Asteriadis, S., Chatzitofis, A., Zarpalas, D., Alexiadis, D.S., Daras, P.: Estimating human motion from multiple kinect sensors. In: Conference on Computer Vision/Computer Graphics Collaboration Techniques and Applications (MIRAGE 2013), 6 p. (2013). http://dx.doi.org/10.1145/2466715.2466727

25. Berger, K., Ruhl, K., Schroeder, Y., Bruemmer, C., Scholz, A., Magnor, M.: Markerless motion capture using multiple color-depth sensors. Vision, Model. Vis., 317–324 (2011). http://dx.doi.org/10.2312/PE/VMV/VMV11/317-324

26. Oh, J., Jung, Y., Cho, Y., Hahm, C., Sin, H., Lee, J.: Hands-up: motion recognition using kinect and a ceiling to improve the convenience of human life. In: ACM SIGCHI Conference on Human Factors in Computing Extended Abstracts (CHI EA 2012), pp. 1655–1660 (2012). http://dx.doi.org/10.1145/2212776.2223688

27. Huang, J.-D.: Kinerehab: a kinect-based system for physical rehabilitation: a pilot study for young adults with motor disabilities. In: ACM SIGACCESS Conference on Computers and Accessibility (ASSETS 2011), pp. 319–320. (2011). http://dx.doi.org/10.1145/2049536.2049627

28. Alexiadis, D.S., Kelly, P., Daras, P., O'Connor, N.E., Boubekeur, T., Moussa, M.B.: Evaluating a dancer's performance using kinect-based skeleton tracking. In: ACM International Conference on Multimedia, pp. 659–662 (2011). http://dx.doi.org/10.1145/2072298.2072412

29. Nirjon, S., et al.: Kintense: a robust, accurate, real-time and evolving system for detecting aggressive actions from streaming 3D skeleton data. In: IEEE International Conference on Pervasive Computing and Communications (PerCom 2014), pp. 2–10 (2014). http://dx.doi.org/10.1109/PerCom.2014.6813937

30. Zafrulla, Z., Brashear, H., Hamilton, H.: American sign language recognition with the kinect. In: International Conference on Multimodal Interfaces, pp. 279–286 (2011). http://dx.doi.org/10.1145/2070481.2070532

31. Lang, S.: Sign language recognition using kinect. In: International Conference on Artificial Intelligence and Soft Computing, pp. 394–402 (2011)

32. Zhang, B., Nakamura, T., Ushiogi, R., Nagai, T., Abe, K., Omori, T., Oka, N., Kaneko, M.: Simultaneous children recognition and tracking for childcare assisting system by using kinect sensors. J. Sig. Inf. Process. **07**, 148–159 (2016). https://doi.org/10.4236/jsip.2016.73015

33. Connell, S., Kuo, P.-Y., Liu, L., Piper, A.M.: A Wizard-of-Oz elicitation study examining child-defined gestures with a whole-body interface. In: ACM Conference on Interaction Design and Children (IDC 2013), pp. 277–280 (2013). http://dx.doi.org/10.1145/2485760.2485823

34. Smith, T.R., Gilbert, J.E.: Dancing to design: a gesture elicitation study. In: ACM Conference on Interaction Design and Children (IDC 2018), pp. 638–643 (2018). https://dx.doi.org/10.1145/3202185.3210790

35. Fish, D., Nielsen, J.-P.: Clinical assessment of human gait. J. Prosthetics Orthot. **5**, 39–48 (1993). https://doi.org/10.1097/00008526-199304000-00005

36. Gage, J.R., Deluca, P.A., Renshaw, T.S.: Gait analysis: principles and applications. J. Bone Jt. Surg. **77**, 1607–1623 (1995)

37. Tao, W., Liu, T., Zheng, R., Feng, H.: Gait analysis using wearable sensors. Sensors **12**, 2255–2283 (2012). https://doi.org/10.3390/s120202255

38. Weber, W., Weber, E.: Mechanics of Human Walking Apparatus. Springer, Heidelberg (1992)

39. Benedetti, M., Cappozzo, A.: Anatomical landmark definition and identification in computer aided movement analysis in a rehabilitation context II (Internal Report). U Degli Stud. La Sapienza, 31 p. (1994)

40. Sekiya, N., Nagasaki, H.: Reproducibility of the walking patterns of normal young adults: test-retest reliability of the walk ratio (step-length/step-rate). Gait Posture **7**, 225–227 (1998). https://doi.org/10.1016/S0966-6362(98)00009-5

41. Cooper, R.A.: Rehabilitation Engineering Applied to Mobility and Manipulation. CRC Press, Boca Raton (1995)

42. Marais, G., Pelayo, P.: Cadence and exercise: physiological and biomechanical determinants of optimal cadences-Practical applications. Sports Biomech. **2**, 103–132 (2003). https://doi.org/10.1080/14763140308522811

43. Gianaria, E., Grangetto, M., Lucenteforte, M., Balossino, N.: Human classification using gait features. In: International Workshop on Biometric Authentication, pp. 16–27 (2014). http://dx.doi.org/10.1007/978-3-319-13386-7_2

44. Barreira, T.V., Katzmarzyk, P.T., Johnson, W.D., Tudor-Locke, C.: Cadence patterns and peak cadence in US children and adolescents: NHANES, 2005–2006. Med. Sci. Sports Exerc. **44**, 1721–1727 (2012). https://doi.org/10.1249/MSS.0b013e318254f2a3

45. Shultz, S.P., Hills, A.P., Sitler, M.R., Hillstrom, H.J.: Body size and walking cadence affect lower extremity joint power in children's gait. Gait Posture **32**, 248–252 (2010). https://doi.org/10.1016/j.gaitpost.2010.05.001

46. Dusing, S.C., Thorpe, D.E.: A normative sample of temporal and spatial gait parameters in children using the GAITRite(R) electronic walkway. Gait Posture **25**, 135–139 (2007)

47. Bjornson, K.F., Song, K., Zhou, C., Coleman, K., Myaing, M., Robinson, S.L.: Walking stride rate patterns in children and youth. Pediatr. Phys. Ther. **23**, 354–363 (2011). https://doi.org/10.1097/PEP.0b013e3182352201

48. Davis, J.W.: Visual categorization of children and adult walking styles. In: International Conference on Audio- and Video-Based Biometric Person Authentication (AVBPA 2001), pp. 295–300 (2001). http://dx.doi.org/1007/3-540-45344-X_43

49. Oberg, T., Karsznia, A., Oberg, K.: Basic gait parameters: reference data for normal subjects, 10–79 years of age. J. Rehabil. Res. Dev. **30**, 210–223 (1993)

50. Piaget, J.: Piaget's Theory. In: Mussen, P. (ed.) Handbook of Child Psychology. Wiley, New York (1983)

51. Thomas, J.R.: Acquisition of motor skills: information processing differences between children and adults. Res. Q. Exerc. Sport **51**, 158–173 (1980). https://doi.org/10.1080/02701367.1980.10609281

52. Moe-Nilssen, R., Helbostad, J.L.: Estimation of gait cycle characteristics by trunk accelerometry. J. Biomech. **37**, 121–126 (2004). https://doi.org/10.1016/S0021-9290(03)00233-1

53. Bhanu, B., Han, J.: Bayesian-based performance prediction for gait recognition. In: Workshop on Motion and Video Computing (MOTION 2002), pp. 145–150 (2002). http://dx.doi.org/10.1109/MOTION.2002.1182227

54. Rose Jacobs, R.: Development of gait at slow, free, and fast speeds in 3- and 5-year-old children. Phys. Ther. 1251–1259 (1983). http://dx.doi.org/10.1093/ptj/63.8.1251

55. Danion, F., Varraine, E., Bonnard, M., Pailhous, J.: Stride variability in human gait: the effect of stride frequency and stride length. Gait Posture **18**, 69–77 (2003). https://doi.org/10.1016/S0966-6362(03)00030-4

56. Wang, I., et al.: EGGNOG: a continuous, multi-modal data set of naturally occurring gestures with ground truth labels. In: IEEE International Conference on Automatic Face Gesture Recognition (FG 2017), pp. 414–421 (2017). http://dx.doi.org/10.1109/FG.2017.145

57. Bauby, C.E., Kuo, A.D.: Active control of lateral balance in human walking. J. Biomech. **33**, 1433–1440 (2000). https://doi.org/10.1016/S0021-9290(00)00101-9

58. Sadeghi, H., Allard, P., Prince, F., Labelle, H.: Symmetry and limb dominance in able-bodied gait: a review. Gait Posture **12**, 34–45 (2000). https://doi.org/10.1016/S0966-6362(00)00070-9

59. Yang, H.D., Park, A.Y., Lee, S.W.: Gesture spotting and recognition for human-robot interaction. IEEE Trans. Robot. **23**, 256–270 (2007). https://doi.org/10.1109/TRO.2006.889491

60. Lechner, D.E., McCarthy, C.F., Holden, M.K.: Gait deviations in patients with juvenile rheumatoid arthritis. Phys. Ther. **67**(9), 1335–1341 (1987). https://doi.org/10.1093/ptj/67.9.1335

Learning User Preferences via Reinforcement Learning with Spatial Interface Valuing

Miguel Alonso Jr.[✉]

School of Computing and Information Sciences,
Florida International University, Miami, FL 33199, USA
malonsoj@cs.fiu.edu
http://dsail.fiu.edu

Abstract. Interactive Machine Learning is concerned with creating systems that operate in environments alongside humans to achieve a task. A typical use is to extend or amplify the capabilities of a human in cognitive or physical ways, requiring the machine to adapt to the users' intentions and preferences. Often, this takes the form of a human operator providing some type of feedback to the user, which can be explicit feedback, implicit feedback, or a combination of both. Explicit feedback, such as through a mouse click, carries a high cognitive load. The focus of this study is to extend the current state of the art in interactive machine learning by demonstrating that agents can learn a human user's behavior and adapt to preferences with a reduced amount of explicit human feedback in a mixed feedback setting. The learning agent perceives a value of its own behavior from hand gestures given via a spatial interface. This feedback mechanism is termed Spatial Interface Valuing. This method is evaluated experimentally in a simulated environment for a grasping task using a robotic arm with variable grip settings. Preliminary results indicate that learning agents using spatial interface valuing can learn a value function mapping spatial gestures to expected future rewards much more quickly as compared to those same agents just receiving explicit feedback, demonstrating that an agent perceiving feedback from a human user via a spatial interface can serve as an effective complement to existing approaches.

Keywords: Human computer interaction ·
Interactive machine learning · Reinforcement learning ·
Artificial intelligence

1 Introduction

Reinforcement Learning (RL) is an area of machine learning that seeks to create agents that learn what to do in uncertain, stochastic environments. That is,

Supported in part by Kynetic AI, LLC https://kynetic.ai.

M. Antona and C. Stephanidis (Eds.): HCII 2019, LNCS 11573, pp. 403–418, 2019.
https://doi.org/10.1007/978-3-030-23563-5_32

RL agents learn how to map situations to actions with the goal of maximizing some type of reward [24]. A core tenant of RL that is touted as a key benefit is that agents learn to do this autonomously, without human intervention, in an unsupervised fashion. The typical engineering workflow in the design of RL agents is (1) a reward function is designed, (2) input and output channels are selected, and (3) an RL learning algorithm is designed [2]. And once the RL pipeline has been designed and implemented, little if any human intervention is needed during the learning process.

However, for real-world situations, where the high complexity of the world cannot easily be modeled in simulation environments, autonomous learning is often not feasible, primarily because of three key issues: the difficulty in specifying reward functions [20], the constraints on exploration due to the presence of potentially catastrophic outcomes for agents and humans alike [9], and difficulty in communicating goals to RL agent so as to avoid being misinterpreted [3]. As a result, many researchers have proposed introducing a human in the loop of training RL agents, to varying degrees of success [1,2,12,13,17,18].

These examples not only have elements of machine learning, but also elements of human-computer interaction (HCI). First coined in 1983 by Card et al. [5,6], HCI as a discipline is concerned with the design, evaluation and implementation of interactive computing systems for human use and with the study of major phenomena surrounding them [10]. Similarly, there has been some work related to having RL agents learn user preferences [7,26]. For example, Veeriah et al. [26] demonstrated that an RL agent can learn a human user's intent and preferences via ongoing interactions where the human user provides feedback through facial expressions, as well as, through explicit negative rewards. Collectively, the interface between machine learning and human computer interaction is referred to as Interactive Machine Learning (iML). In this paper, I explore combining Reinforcement Learning techniques with spatial interfaces to create agents that learn a value function that relates a user's body language, specifically a thumbs up or thumbs down gesture, to expectations of future rewards.

1.1 Interactive Machine Learning

Interactive Machine Learning (iML) is a subfield of Machine Learning that creates systems that operate in environments alongside human operators, where the human and the machine collaborate to achieve a task and both the human and ML system can be agents in the environment [11]. Many times, iML is used to extend or amplify the capabilities of a human in cognitive or physical ways. For this to be successful, the machine must adapt to the users' intentions and preferences, learning about their human counterparts' behavior. In much of the current research in iML, a user must convey feedback in some way to the iML agent.

One popular method is directly, such as through the click of a mouse, or through spoken word. This is known as explicit human feedback. Implicit human feedback on the other hand, is a mechanism through which a human can guide an iML agent's learning process through subtle cues, such as body language.

And yet a third type of feedback is known as mixed human feedback which combines explicit and implicit feedback [4]. Although all three forms of feedback impose some cognitive load on the human, explicit feedback carries the heaviest cognitive burden, especially in real-world settings where humans have additional cognitive loads due to the environment. The main objective of this work is to extend the current state of the art in iML by demonstrating that a learning agent can learn a human user's behavior and adapt to the human's preferences with a reduced amount of explicit human feedback in a mixed feedback setting.

1.2 Related Work

With the growth of the machine learning field over the last decade, there has been much effort in the community to create successful interactions between humans and machine learning systems with the goal of increase performance across a myriad of tasks. Many approaches to human-in-the-loop machine learning have focused primarily on agents learning from humans via explicit rewards. For example, Thomaz and Breazel [25] used a simulated RL robot and had a human teach the robot in real-time to perform a new task. They presented three interesting findings: (1) humans use the reward channel for feedback, as well as for future-directed guidance; (2) humans exhibited a positive bias to their feedback using the signal as a motivational channel; and (3) humans change their behavior as they develop a mental model of the robotic learner.

More recently, Knox & Stone [14,15] introduced the TAMER framework (Training an Agent Manually via Evaluative Reinforcement). TAMER's system for learning from human reward is novel in three key ways:

- TAMER addresses delays in human evaluation through credit assignment
- TAMER learns a predictive model of human reward
- and at each time step, TAMER chooses the action that is predicted to directly elicit the most reward, disregarding consideration of the action's effect on future state (i.e., in reinforcement learning terms, TAMER myopically values state-action pairs using a discount factor of 0).

According to Knox & Stone, "TAMER is built from the intuition that human trainers can give feedback that constitutes a complete judgement on the long-term desirability of recent behavior." One drawback of this method, however, is that when the user needs to modify the agent's behavior, the model would need to be changed, for example, via additional rewards from the user.

Another interesting approach is by Christiano et al. [7]. Here, the agent is trained from a neural network known as the 'reward predictor', instead of the classical RL approach of using the rewards it collects as it explores an environment. There are three processes running in parallel:

- A RL agent explores and interacts with its environment
- Periodically, a pair of 1–2 s clips of its behavior is sent to a human who is then asked to select the best one indicating steps toward fulfilling the desired goal

– The human's choice is used to train a reward predictor, which is then used to train the agent

They showed that overtime, the agent learns to maximize the reward from the predictor and improve its behavior according to the human's preferences. However, this methodology introduces substantial lag between the human feedback and the agent's learning.

Another approach by Veeriah et al. [26] uses body language as one of the drivers of learning. In contrast to the other approaches mentioned, this approach is concerned with designing a general, scalable agent that would allow a human user to change the agent's behavior according to preferences with minimal human feedback. In this work, they used facial expressions to provide the feedback, not as a channel for control, but rather as a means of valuing the agent's actions.

This work is an extension of the work by Veeriah et al. [26] to hand gestures captured by a 3D spatial interface.

1.3   Problem Statement

There are many domains for which human and computing systems, enabled by some form of machine learning, work collaboratively to achieve various tasks. Once such domain of human computer interaction is in the realm of prostheses, whereby a human operator controls an electronic prosthetic limb [8,16]. Grip selection, for example, is one of the main tasks a human operating a prosthetic limb performs. Performing grasping of common objects with the prosthesis requires choosing the correct grip from an array of various configurations. Most systems cycle through grips via some feedback mechanism, typically a roll or pronation/suprenation [21].

Thus, a grip selection task environment was created in order to evaluate the machine learning system. This problem is taken from the real-world task of selecting an appropriate grip pattern for grasping a given object by a user that is operating a prosthetic arm, as mentioned above. Typically for most prosthetic arms, there are a set of n discrete grips and depending on the type of object or situation, the correct grip is defined according to the user's preference. These preferences are normally setup by a clinician and must be periodically revisited as the user gains experience with the prosthesis [21]. For this experiment, the agent must select the correct grip for a random object presented, move forward, and grasp the object. The episode ends when the agent successfully grasps the object. In this way, the agent learns the user's gripping preferences in an ongoing and online fashion, reducing the need for multiple clinical visits.

2   Reinforcement Learning Algorithm Background

Reinforcement Learning is a branch of Machine Learning that allows agents to decide what to do, that is, how to map situations to actions with the goal of maximizing a numerical reward signal. RL agents typically interact with an environment, either real-world or simulated, gaining experience with each inter-action, and improving their performance, as measured by the reward signal. Mathematically, Markov Decision Processes (MDPs) are used to describe RL problems. There are several algorithms that solve MDPs, many in optimal ways. The RL learning algorithm that was investigated for the grip selection task was SARSA(λ), which is an on-policy temporal difference (TD) control algorithm for MDPs [24].

2.1 Markov Decision Processes

A Markov Decision Process is a tuple $(S, A, P_s a, \gamma, R)$, where:

- S is a set of *states*, $S \in \{s_0, s_1, \ldots s_m\}$
- A is a set of *actions*, $A \in \{a_0, a_1, \ldots a_m\}$
- P_{sa} are the state *transition probabilities*
- $\gamma \in [0, 1)$ is the *discount factor*, and controls the influence of future rewards in the present
- $R : S \times A \mapsto \mathbb{R}$ is the *reward function*, which returns a real value every time the agent moves from one state to the other due to an action

For each state $s \in S$ and action $a \in A$, P_{sa} is a probability distribution over the state space and gives the distribution over what states the system will transition to if action a is taken when the system is in state s. Starting from an initial state, s, the agent can choose to take an action $a \in A$. The state of the MDP then randomly transitions to a successor state s'. The successor state is drawn according to $s' \sim P_{sa}$. In MDPs, the transition model depends on the current state, the next state, and the action of the agent. This process happens sequentially over time, with each new action, generating a new successive state, which leads to further actions.

The reward function, $R(s, a)$, is the primer driver of RL algorithms. Having the correct $R(s, a)$ can make or break RL algorithms in that the reward function is essentially the "teacher" in the learning algorithm. It indicates to the RL algorithm which state-action pairs are more desirable than others by assigning values. For example, one methodology is to assign positive value to state-action pairs that are desirable and a zero or negative values to those that don't matter or are not as desirable.

Thus, the goal in reinforcement learning is to maximize the reward over time, choosing states via actions that increase the return over time, and avoiding states that decrease the return over time. If we start at time t in state s_t and choose an action a_t, where s_i and a_i are states and actions in a sequence, and $t = 0, 1, 2, \ldots$, we can represent the the MDP as follows:

$$s_t \xrightarrow{a_t} s_{t+1} \xrightarrow{a_{t+1}} s_{t+2} \xrightarrow{a_{t+2}} s_{t+3} \cdots \tag{1}$$

Thus, the total reward at time t if we take $a_0, a_1, \ldots$ actions and visit $s_0, s_1, \ldots$ states over time $t+1, t+2, t+3, \ldots$ is given by:

$$R_t(s,a) = R_{t+1}(s_t, a_t) + \gamma R_{t+2}(s_{t+1}, a_{t+1}) + \gamma^2 R_{t+3}(s_{t+2}, a_{t+2}) + \cdots$$

$$R_t(s,a) = \sum_{i=1}^{\infty} \gamma^{i-1} R_{t+i}(s_{i-1}, a_{i-1}) \tag{2}$$

Thus, the goal would be to maximize the expected value of the total reward:

$$E[R_t(s,a)] \tag{3}$$

The solution to this problem is to find a *policy*, π, which returns the action that will yield the highest reward for each state. A policy is any function that maps states to actions: $\pi : S \mapsto A$. When we *execute* a policy π when in a state s, the agent takes action $a' = \pi(s,a)$. There can potentially be many policies to choose from, but only one can be considered an optimal policy, which is denoted by π^* and yields the highest expected reward over time, which is call the *action value function*. Thus, an action value function for a policy π is:

$$q^\pi(s,a) = E[R_t | S_t = s, A_t = a, \pi] \tag{4}$$

where $q^\pi(s,a)$ is the expected sum of discounted rewards when starting in state s and taking action a according to policy π.

The goal is to find an *optimal policy*, π^* that maximizes the action value function:

$$\pi^*_{(s,a)} = \underset{\pi}{\operatorname{argmax}} \, q^\pi(s,a) \tag{5}$$

For most practical applications of RL, $q^\pi(s,a)$ is not known and must be learned by interacting with the environment. This type of RL agent, one that learns the action-value function not from a transition model for the environment, but rather from direct interaction with the environment, is called a model-free RL agent.

2.2 SARSA(λ)

In RL, rewards are viewed as short-term signals of the quality of an action, where as the action value function, $Q^\pi(s,a)$ represents the long-term value of a state-action pair. Temporal difference (TD) learning is a class of model-free methods that estimates Q^π as the agent interacts with the environment. The agent samples transitions and then updates the estimate of Q^π using observed and the estimate of the values of the next action. Typically, the agent makes these observations and updates the action value function at every time step, according to the following update rule:

$$Q(S_t, A_t) \leftarrow Q(S_t, A_t) + \alpha \delta_t \tag{6}$$

where Q is an estimate of q^π, α is the step size, and δ_t is the TD error. SARSA is a TD learning algorithm that samples states and actions using an ϵ-greedy policy and then updates the Q values using Eq. 6 and a δ_t as follows:

$$\delta_t = R_{t+1} + \gamma Q(S_{t+1}, A_{t+1}) - Q(S_t, A_t) \tag{7}$$

The term $R_{t+1} + \gamma Q(S_{t+1}, A_{t+1})$ is called the target. It consists of the reward plus the discounted value of the next state and next action.

SARSA is known as an on-policy method which arises from the fact that the behavior policy u is the same as the target policy π. That is, the TD target in SARSA consists of $Q(S_{t+1}, A_{t+1})$, where A_{t+1} is sampled using μ. The target policy π, is used to compute the TD target. Although On-policy methods may result sub-optimal policies in certain instances, it has been shown that policies learned in on-policy methods tend to be safer when the risks are greater because SARSA takes the action selection into account [24].

Eligibility traces, a method of including information about not just the current time step, but information from multiple time steps, are a key mechanisms in reinforcement learning. For example, in TD(λ) algorithm, the λ refers to the use of an eligibility trace. Almost any TD method, such as SARSA, can be combined with eligibility traces to obtain a more general method that may learn more efficiently [24]. In this study, SARSA(λ) is used to improve efficiency of the learning agent.

2.3 Function Approximation Using Tile Coding

In order to implement Eqs. 6 and 7, an estimate of Q must be maintained and updated. Since there is no analytical way of expressing Q as a function, a method of funciton approximation is often used. Formally, function approximation is a technique for "representing the value function concisely at infinitely many points and generalizing value estimates to unseen regions of the state- action space" [23]. *Tile coding* is a linear function approximation method that is flexible and computationally efficient. In tile coding, the variable space, which is typically composed of states and actions, is partitioned into tiles, with each partition called a tiling. The method uses several overlapping tilings and for each tiling, maintains the weights of its tiles. The approximate value of an input is found by summing the weights of the tiles, one per tiling, in which it is contained. Given a training example, the method adjusts the weights of the involved tiles by the same amount to reduce the error on the example [23].

2.4 Action Selection Policy

As the action-value function is being learned via a function approximation method such as Tile Coding, the agent needs a method to select an action from the current set of actions whose action-value is known. Equation 5 always chooses the action that yields the largest action value. This is known as *greedy* action selection. But since the agent learns this function through experience, there may be other actions that the agent may never see that will yield better values. In order to remedy this, three popular action selection strategies are often used: ϵ-greedy, ϵ-soft, and soft-max [24].

ϵ-greedy. In ϵ-greedy, the action with the highest value is chosen most of the time. However, every so often, an action is selected at random with ϵ probability. The action is selected uniformly, independent of the action values. This guarantees that every action will be explored sufficiently to find the optimal policy. Larger values of ϵ favor more exploration of actions, while smaller values favor exploitation of the greediest action. This is known as the exploration vs exploitation trade-off.

ϵ-soft. ϵ-soft action selection is similar to ϵ-greedy, except that the best action is selected with probability $1 - \epsilon$ and a random action is selected the rest of the time.

Softmax. A limitation of ϵ-greedy and ϵ-soft is when selecting random actions, they are selected uniformly. The worst action to take is selected with equal probability to the best action to take. Enter softmax. In softmax action selection, first, a rank or weight is assigned to each of the actions, according to their action-value estimate. Then, a random action is selected with regards to the weight associated with each action. This means that as the action-value function is learned, poor actions would be less likely to be chosen. This is useful, particularly the worst actions have unfavorable consequences.

For this study, an ϵ-greedy action selection strategy was used for its simplicity.

3   System Design and Experimental Setup

While having a working prosthetic to develop a solution for learning grip preferences using reinforcement is desirable, prosthetic devices are both expensive and challenging to configure and interface with. Thus, in order to carry out the agent design and experimentation in a low-cost way that would be accessible by other researchers or iML practioners, a simulation environment was created. The simulation consisted of a grip selection task with object grasping and was created using two popular and well supported open source projects, Pybullet and OpenAI gym.

3.1   Simulation Environment

The grip selection task was simulated using the Bullet physics simulator[1] and the OpenAI gym interface[2]. A robotic arm with three degrees of freedom was modeled and imported into the Bullet environment. The robotic arm consists of two grippers, each with one degree of freedom, and the arm portion with one degree of freedom. The arm is attached to a table on which objects of variable

---

[1] https://pybullet.org/wordpress/.
[2] https://gym.openai.com/.

size are placed, one at a time. Figure 1 shows different scenarios that are possible in the simulation.

Although the Bullet environment offers an API that can be used to control the environment to do things such as restart an episode, control the arm, or add an object of random size, the OpenAI gym interface provides a consistent, common API that can be used to develop agents that can operate in multiple environments, as opposed to developing environment specific agents. Thus, the OpenAI gym interface was used to abstract the specific environment details of the grasping environment away from the agent. This allows for de-coupling of the agent development from the environment development and allows other researchers to use the grasping environment with their own custom agents. By using the Bullet physics engine along with the OpenAI gym environment, several experiments were carried out with little effort.

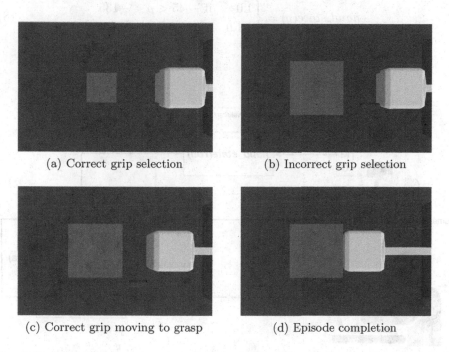

(a) Correct grip selection   (b) Incorrect grip selection

(c) Correct grip moving to grasp   (d) Episode completion

Fig. 1. Simulation environment in pybullet using the OpenAI gym interface. In (a), the arm is in the starting position called the "grip changing station". It has selected the correct grip for the object size. In (b), the incorrect grip has been selected, although the arm still remains in the grip chaning station. In (c), the correct grip for the object size as been enabled and the arm is moving forward to grasp the object. Finally, (d) illustrates the completion of an episode.

3.2   Spatial Interface Valuing

In order to have an agent learn from a human, some form of feedback is required.
In this work, a learning agent perceives a value of its own behavior from human
hand gestures given via a spatial interface, which I term Spatial Interface Valuing
(SIV). In order to capture hand gestures, a Leap Motion Controller[3] (LMC) is
used as the spatial interface device. The LMC provides tracking information for
the left and right hands, as well as simple gesture recognition. The gestures of
interest for this study are a thumbs up or thumbs down. Unfortunately, that is
not one of the stock gestures implemented. Instead, the roll (ρ) of the right hand
was used to determine whether the user was indicating a thumbs up or thumbs
down. A simple piece-wise function was used to determine the state of the hand:
thumbs up (1.0), thumbs down (−1.0).

$$hand\_state(\rho) = \begin{cases} 1.0 & \text{if} - 45 < \rho < -135 \\ -1.0 & \text{otherwise} \end{cases} \tag{8}$$

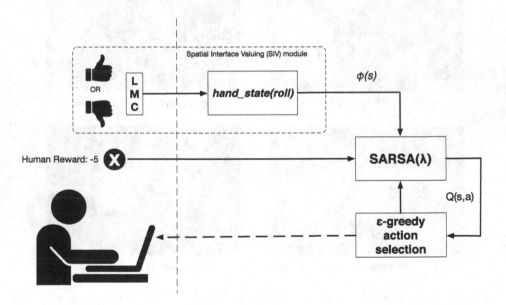

Fig. 2. Overview of the SIV agent setup

3.3   Experimental Setup

The experimental setup consists of a user observing the simulated grip-selection
task and assisting the agent during the training process by (1) signaling a thumbs

---
[3] https://www.leapmotion.com/.

up or thumbs down via the spatial interface to signal approval or disapproval of the agents behavior and (2) pushing the space bar on the keyboard when the agent was not behaving as expected, giving the agent a negative reward. A SARSA(λ) agent both with and without SIV, as well as baseline agent was evaluated, each for 3 runs with 15 episodes per run. All of the experiments were carried out without the user knowing which of the three agents was currently being evaluated, i.e. in a blind setting. Additionally, the order in which the experiments were carried out was randomly selected over the combination of runs, episodes, grips, object sizes, and agent. Figure 2 shows an overview of the experimental setup, with the only difference between agent the SIV/No SIV agents is the feedback from the hand state.

Note: Three distinct simulation environments were created for this study, a baseline environment for comparison, one environment with SIV feedback, and one environment without SIV. The experimental setup (excluding the SIV module) is the same, with the only difference being the inclusion (or exclusion) of the SIV module. For both simulation environments, the agent makes observations of the state space every *one-tenth of a second* and must take an action at every time step.

Table 1. State spaces for each agent under test: SIV vs no SIV

Agent type	State space variables
Baseline	current grip size, current object size, bias
With SIV	current grip size, hand state, bias
Without SIV	current grip size, bias

State Space. As mentioned above, three SARSA(λ) [22,24] agents, a baseline, one with SIV and one without SIV, were implemented to determine how well the agents learned a user's preferences for the grip selection task. Table 1 describes the state space design for all three agents. The SARSA(λ) baseline agent uses the size of the object and the current grip, along with a bias term as the state space vector $\phi(s)$:

$$\phi(s) = [current\_grip\_size, current\_object\_size, bias] \tag{9}$$

This state space was chosen in order to allow the agent to learn the best grip for each object, since both values are known form within the simulation, they can be observed and are used to establish a baseline level of performance with which to compare the SIV an No SIV enabled agents. The maximum grip size, maximum object size, number of grips, and number of objects are all parameters that can be configured prior to running each episode.

For the agent enabled with SIV, which is ideally for deployment in a real prosthetic device, the grip size is known to the agent, but for real-world grasping tasks, object size is not known. This is where SIV steps in. Having knowledge of the user's gesture takes the place of exact knowledge of the object size (as in the baseline agent). Thus, the state vector for the SIV agent is:

$$\phi_{SIV}(s) = [current\_grip\_size, hand\_state, bias] \tag{10}$$

And lastly, the state space for the agent that does not use SIV as a feedback channel, only has knowledge of the current grip.

$$\phi_{NoSIV}(s) = [current\_grip\_size, bias] \tag{11}$$

Action Space. Modeled after the grip selection task in [26], the complete action space for the agents consists of the following:

$$\mathcal{A}_{complete}(s) = \{grip_1, grip_2, \dots grip_n, \leftarrow, \rightarrow\} \tag{12}$$

where the first $1 \dots n$ actions are grip selections amongst the n grips. The remaining actions $\{\leftarrow, \rightarrow\}$, when taken, move the arm one step closer to the object ($\leftarrow$) or one step closer to the grip changing station ($\rightarrow$). However, the actions that are available to the agent depending on the position of the arm relative to the grip changing station, the object, and the reward. When the arm is in the grip changing station, the actions available are:

$$\mathcal{A}_{grip\_change}(s) = \{grip_1, grip_2, \dots grip_n, \leftarrow\} \tag{13}$$

Once the grip has been selected and the agent moves forward one step towards the object, the actions available are:

$$\mathcal{A}_{move}(s) = \{\leftarrow, \rightarrow\} \tag{14}$$

And lastly, if the agent receives an explicit negative reward from the human user, the only available action is to return to the grip changing station:

$$\mathcal{A}_{return}(s) = \{\rightarrow\} \tag{15}$$

Table 2. Hyperparameters for SARSA(λ) agents

Parameter	Value
λ	0.0
step size	0.5
γ	1.0
ϵ	0.1

4 Experiments

4.1 Two Objects with Four Grips

In order to validate SIV as an RL method that can effectively learn a user's preference during grip selection, several trials of a grip selection task were carried out with a user in a blind setting, as mentioned in Sect. 3.3. At the beginning of each episode, the environments were setup to randomly select one of two object sizes: a small object, much smaller than the largest grip size, and a large object, approximately the size of the largest grip. The number of grips for the grip selection task was set to four for all episodes, requiring that the agent learn the users grip preference in two distinct grasping situations. The experiments were formulated as an episodic MDP with 0 discount ($\gamma = 1.0$), 0 reward at every time step and a 0 reward for completing the episode. The hyperameters for SARSA(λ) agents are show in Table 2.

4.2 Experimental Results

The results for the experiment are shown in Fig. 3. Overall, the baseline agent performs the best when considering the average steps per episode (Fig. 3(a)). This is to be expected, because once the baseline agent learns the user preference, it can execute those preferences directly because the baseline agent has knowledge of both the object size and current grip and has learned the action value function and thus, the optimal policy. Once trained, it does not need to rely on the human for any form of input. However, observing the performance of the agent using SIV, although on average it took more steps to complete an episode than the baseline, it out performed the agent without SIV on two of the three runs. This may be due to the human user themselves getting accustomed to providing the feedback to all of the agents.

Another interesting finding is that both the baseline agent and the agent with SIV learned with approximately the same number of average reward button pushes, where as the agent learning the grip selection task without SIV needed many more button pushes to successfully complete the runs. This indicates that the agent with SIV learns as quickly as the baseline agent, while the agent without SIV takes longer to learn the user's preference. This is likely because the agent without SIV only has the reward channel to try to learn the human users preference. This is shown in Fig. 3(b) and (e). Similarly, the amount of total reward needed and total number of button pushes needed (Fig. 3(d) and (e)) for the baseline and SIV agents is approximately half what is needed for the agent without SIV.

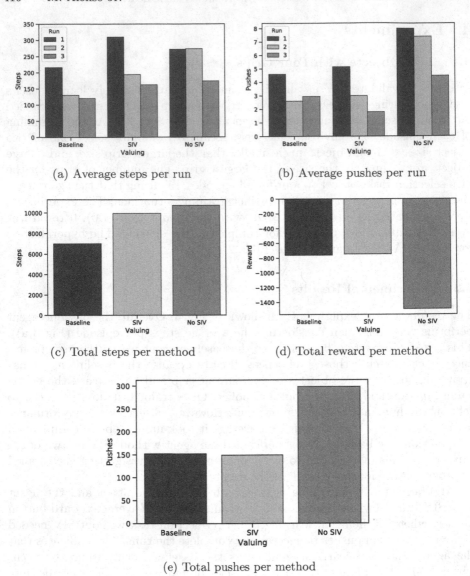

(a) Average steps per run

(b) Average pushes per run

(c) Total steps per method

(d) Total reward per method

(e) Total pushes per method

Fig. 3. Results from evaluating agents using SIV against a Baseline agent with full knowledge of the current grip and object size, and an agent not using SIV with knowledge of only the current grip.

5   Conclusions and Discussion

This study introduces a new approach called *Spatial Interface Valuing* that uses a 3D spatial interface device to adapt an agent to a user's preferences. I showed that SIV can produce substantial performance improvements over an agent that does not use SIV. Through SIV, the agent learns to adapt its own performance and learn user preferences through gestures, reducing the amount of explicit feedback required. SIV delivers implicit feedback from a user's hand gestures to an agent, allowing the agent to learn much more quickly and require less explicit human generated reward. The SIV agent learned to map hand gestures to user satisfaction, codifying satisfaction as a action value function using temporal methods for Reinforcement Learning. This technique is task agnostic and I believe it will easily extend to other settings, tasks, and forms of body language.

6 Future Work

The current results are encouraging, however, more work can be done on several fronts. This study was limited to studying the SARSA(λ) RL algorithm, an on-policy agent that uses tile coding, a function approximation technique, to estimate the action value function and learn a policy that encodes user preferences. It may be useful to study other types of agents such as off-policy agents using Q-learning, for example. Additionally, deep learning techniques such as Deep Q Networks [19] and DeepSARSA [27] may perform better. Additionally, simply extending the size of the experiment by conducting more runs with more users, different hyper-parameters and random seeds, and different object size/grip number combinations in a randomized control experiment will allow for a more robust statistical analysis of the performance.

References

1. Abbeel, P., Ng, A.Y.: Apprenticeship learning via inverse reinforcement learning. In: Proceedings of the Twenty-First International Conference on Machine Learning. ICML 2004. ACM, New York (2004)
2. Abel, D., Salvatier, J., Stuhlmüller, A., Evans, O.: Agent-agnostic human-in-the-loop reinforcement learning. In: NIPS Workshop on the Future of Interactive Learning Machines 2016, January 2017
3. Amodei, D., Olah, C., Steinhardt, J., Christiano, P., Schulman, J., Mané, D.: Concrete problems in AI safety. arXiv:1606.06565 [cs], June 2016
4. Boukhelifa, N., Bezerianos, A., Lutton, E.: Evaluation of interactive machine learning systems. arXiv:1801.07964 [cs], January 2018
5. Card, S.K., Moran, T.P., Newell, A.: The keystroke-level model for user performance time with interactive systems. Commun. ACM 23(7), 396–410 (1980)
6. Card, S.K., Newell, A., Moran, T.P.: The Psychology of Human-Computer Interaction. L. Erlbaum Associates Inc., Hillsdale (1983)
7. Christiano, P., Leike, J., Brown, T.B., Martic, M., Legg, S., Amodei, D.: Deep reinforcement learning from human preferences. arXiv:1706.03741 [cs, stat], June 2017
8. Edwards, A.L., et al.: Application of real-time machine learning to myoelectric prosthesis control: a case series in adaptive switching. Prosthet. Orthot. Int. 40(5), 573–581 (2016)
9. García, J., Fernández, F.: A comprehensive survey on safe reinforcement learning. J. Mach. Learn. Res. 16, 1437–1480 (2015)

10. Hewett, T.T., et al.: ACM SIGCHI curricula for human-computer interaction. Technical report. ACM, New York (1992)
11. Holzinger, A.: Interactive machine learning for health informatics: when do we need the human-in-the-loop? Brain Inform. **3**(2), 119–131 (2016)
12. Knox, W.B., Setapen, A., Stone, P.: Reinforcement learning with human feedback in mountain car. In: AAAI Spring Symposium: Help Me Help You: Bridging the Gaps in Human-Agent Collaboration (2011)
13. Knox, W.B., Stone, P.: Augmenting reinforcement learning with human feedback. In: ICML Workshop on New Developments in Imitation Learning, p. 8 (2011)
14. Knox, W.B., Stone, P.: Framing reinforcement learning from human reward. Artif. Intell. **225**(C), 24–50 (2015)
15. Knox, W.B., Stone, P., Breazeal, C.: Teaching agents with human feedback: a demonstration of the TAMER framework. In: Proceedings of the Companion Publication of the 2013 International Conference on Intelligent User Interfaces Companion, pp. 65–66 Companion. ACM, New York (2013)
16. Li, C., Ren, J., Huang, H., Wang, B., Zhu, Y., Hu, H.: PCA and deep learning based myoelectric grasping control of a prosthetic hand. BioMedical Eng. OnLine **17**, 107 (2018)
17. Liu, F., Su, J.B.: Reinforcement learning based on human-computer interaction. In: Proceedings. International Conference on Machine Learning and Cybernetics, vol. 2, pp. 623–627, November 2002
18. Mathewson, K., Pilarski, P.M.: Simultaneous control and human feedback in the training of a robotic agent with actor-critic reinforcement learning. In: Interactive Machine Learning Workshop at IJCAI 2016, p. 7 (2016)
19. Mnih, V., et al.: Human-level control through deep reinforcement learning. Nature **518**(7540), 529–533 (2015)
20. Ng, A.Y., Russell, S.J.: algorithms for inverse reinforcement learning. In: Proceedings of the Seventeenth International Conference on Machine Learning, ICML 2000, pp. 663–670. Morgan Kaufmann Publishers Inc., San Francisco (2000)
21. Resnik, L., Meucci, M.R., Lieberman-Klinger, S., Fantini, C., Kelty, D.L., Disla, R., Sasson, N.: Advanced upper limb prosthetic devices: implications for upper limb prosthetic rehabilitation. Arch. Phys. Med. Rehabil. **93**(4), 710–717 (2012)
22. Rummery, G.A., Niranjan, M.: On-line Q-learning using connectionist systems. Cambridge University Engineering Department, Technical report (1994)
23. Sherstov, A.A., Stone, P.: Function approximation via tile coding: automating parameter choice. In: Zucker, J.-D., Saitta, L. (eds.) SARA 2005. LNCS (LNAI), vol. 3607, pp. 194–205. Springer, Heidelberg (2005). https://doi.org/10.1007/11527862_14
24. Sutton, R.S., Barto, A.G., Bach, F.: Reinforcement Learning: An Introduction, 2nd edn. A Bradford Book, Cambridge (2018)
25. Thomaz, A.L., Breazeal, C.: Teachable robots: understanding human teaching behavior to build more effective robot learners. Artif. Intell. **172**(6–7), 716–737 (2008)
26. Veeriah, V., Pilarski, P.M., Sutton, R.S.: Face valuing: training user interfaces with facial expressions and reinforcement learning. CoRR abs/1606.02807 (2016)
27. Zhao, D., Wang, H., Shao, K., Zhu, Y.: Deep reinforcement learning with experience replay based on SARSA. In: 2016 IEEE Symposium Series on Computational Intelligence (SSCI), pp. 1–6, Decemner 2016

Adaptive Status Arrivals Policy (ASAP) Delivering Fresh Information (Minimise Peak Age) in Real World Scenarios

Basel Barakat[1]($\boxtimes$), Simeon Keates[2]($\boxtimes$), Ian Wassell[3]($\boxtimes$),
and Kamran Arshad[4]($\boxtimes$)

[1] Faculty of Engineering and Science, University of Greenwich, Kent ME4 4TB, UK
bb141@gre.ac.uk
[2] School of Engineering and the Built Environment, Edinburgh Napier University,
Edinburgh EH10 5DT, UK
[3] Computer Laboratory, University of Cambridge, Cambridge CB2 1TN, UK
[4] Ajman University, Ajman, UAE

Abstract. Real-time systems make their decisions based on information communicated from sensors. Consequently, delivering information in a timely manner is critical to such systems. In this paper, a policy for delivering fresh information (or minimising the Peak Age of the information) is proposed. The proposed policy, i.e., the Adaptive Status Arrivals Policy (ASAP), adaptively controls the timing between updates to enhance the Peak Age (PA) performance of real-time systems. Firstly, an optimal value for the inter-arrival rate is derived. Afterwards, we implemented the policy in three scenarios and measured the ASAP PA performance. The experiments showed that ASAP is able to approach the theoretical optimal PA performance. Moreover, it can deliver fresh information in scenarios where the server is located in the cloud.

Keywords: Real time systems · Cloud · Adaptive systems ·
Information freshness · Peak Age of Information

1 Introduction

Recently, several applications rely on real-time communications had been investigated such as autonomous cars, tactile internet and telehealth. Delivering the information in a timely manner is critical for these applications. For instance, in a robotic surgery, excessive delays might be life-threatening. Hence, several researchers proposed polices for delivering information with as low latency as possible. However, most of the work done is based on assumptions that might be oversimplifying. These assumptions and the whole latency current paradigm should be questioned.

Consider a vegetable market as an analogy, an insightful question would be what do we mostly care about: (i) the speed of the vegetable carrier, (ii)

© Springer Nature Switzerland AG 2019
M. Antona and C. Stephanidis (Eds.): HCII 2019, LNCS 11573, pp. 419–430, 2019.
https://doi.org/10.1007/978-3-030-23563-5_33

the time the vegetables took to arrive at the market or (iii) the freshness of the vegetables? Usually, one only care about the freshness of the vegetables. It is obvious that freshness is affected by both the speed of transporting the vegetables and the time it took for them to arrive at the market. Moreover, the freshness is also affected by when the vegetables were grown. This analogy can represent several systems where the information has a window of time in which it is useful and after that window, the information loses its usefulness, for example, an adaptive control system or systems with online machine learning algorithms. In queuing theory/data networks terms, the speed of transmitting information in a network is called the data throughput, the time taken to communicate a piece of information is the delay or latency and the time to grow the vegetables is the time to generate the package/information. However, until recently no metric has focused on information freshness.

To evaluate the freshness of an update, let us assume that we have a stopwatch that starts counting as soon as an update was generated. The stopwatch stops immediately after receiving the next update at the destination. The time elapsed until the stopwatch is stopped is called Age of the Information (AoI) [1]. The final value, the stopwatch shows is the Peak Age of the Information (PA) [2]. The PA is defined as the maximum value of AoI [3]. It can be noticed that the PA consists of the inter-arrival time and the delay time. Unlike the conventional queuing theory delay metric, PA evaluates the freshness from the destination's perspective. In other words, PA reflects the information's freshness, not the time it took to communicate it.

Several policies have been proposed to minimise PA in the literature and can be classified into two main categories [1]. The first category controls the buffer size, which regulates the maximum number of updates waiting in the buffer, e.g., in [3]. The second category attempts to minimise the updates' waiting time and hence reduce the PA. For instance, the Zero-Wait (ZW) policy minimises the waiting time by only permitting the updates' source to communicate a new update after it has ensured that the destination is idle, by waiting for an Acknowledgement (ACK) from the destination after every transmitted update [4]. The state-of-art policies reduce PA, however, their performance does not always approach the minimum PA, as shown in Sect. 2.

In this paper, we propose a policy that is able to reach a near-optimal PA performance. The proposed policy, i.e., Adaptive Status Arrivals Policy (ASAP), adapts the status updating inter-arrivals rate to reach the optimal PA performance. The ASAP was tested in three scenarios, as detailed in Sect. 2. The first scenario is a single queue with single service rate (μ). The second scenario also has a single queue, however, the mean service time can take four values, that emulates a system where the load on the server changes between four service durations or a wireless channel with adaptive coding and modulation [6,7]. In the third scenario, the updates were transmitted through the internet to a destination located in the cloud. The ASAP continuously changes the inter-arrivals rate (λ) to reach the optimal value, which is derived in Sect. 3. The PA performance of the ASAP presented in Sect. 4 shows that it can deliver the information with a freshness that is close to the theoretical optimal freshness. This paper is concluded in Sect. 5.

2   Problem Statement and System Model

2.1   Peak Age Metric

The PA metric was proposed as a policy for calculating the peaks of a typical AoI sawtooth profile [1]. The AoI is defined as the time elapsed from the generation of the last successfully received message [1]. In Fig. 1, an illustration of AoI evolution is presented. When an update is generated, the AoI is zero and as time passes, the AoI grows linearly until the receipt of the next update. The highest value of AoI, which forms the peak of the AoI pattern, is PA, which represents the highest value of AoI.

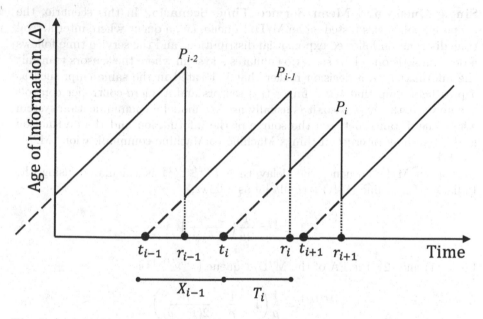

Fig. 1. Age of information (AoI.pdf) illustration, the AoI for update (i) starts when an update is generated (t_i) and keeps counting until the server receives the next update (R_{i+1}). The maximum value of AoI is called Peak Age (P_i), which is equal to inter-arrival time (X) plus the system time (T).

As shown in Fig. 1, the PA for a general queue with a single server (G/G/1) can be obtained as follows [3],

$$P \triangleq \mathbb{E}[X] + \mathbb{E}[T] = \mathbb{E}[X] + \mathbb{E}[S] + \mathbb{E}[W]. \qquad (1)$$

where $\mathbb{E}[.]$ is the expectation operator, X is the inter-arrival time, i.e., $1/\lambda$ and T is the delay time, i.e., the service time (S) plus the queuing time (W).

The service time and the waiting time affect the delay time and hence the PA. The service time is either deterministic or follows a random distribution, depending on the time the server takes to process the information. The waiting time depending on the service time, inter-arrival time, queue service discipline, number of servers and the maximum queue length.

2.2   Tested Scenarios

In this paper, the proposed policy is tested on three main scenarios. The first scenario is a single First in First Out (FIFO) queue and the service time follows a single distribution. The second scenario is similar to the first scenario, however, the mean service time follows several values. The final scenario is that the server is located in the cloud.

Single Queue and Mean Service Time Scenario. In this scenario, the proposed policy was tested on an M/D/1 queue, i.e., a queue where inter-arrival time distribution followed exponential distributions and the service time follows a deterministic one. This scenario emulates a system where the sensors transmit the information to a decision maker that is located in the same chip, such as a prosthetic limp that has a finger tips sensors and a micro-controller controls the power of the grip. Also, it is usually used to model a communication system where the distance between the source of the information and the destination is short such as Internet of things Machine to Machine communication (M2M) [10,11].

For an M/D/1 queue, the delay time $(T^{M/D/1})$ is calculated using the Pollaczek-Khinchine (P-K) formula [5] as follows,

$$T^{M/D/1} = \frac{1}{2\mu}\left(\frac{2 - \rho^2}{1 - \rho}\right).\tag{2}$$

From (1) and (2), the PA of the M/D/1 queue $(P^{m/D/1})$ is,

$$P^{M/D/1} = \frac{1}{\mu}\left(1 + \frac{1}{\rho} + \frac{\rho}{2(1 - \rho)}\right).\tag{3}$$

Single Queue with Several Mean Service Time Values. In this scenario, the proposed policy was tested on a server with a mean service rate that can take several values. This scenario emulates a server that has several probable data processing speeds e.g. a data communication channel with Adaptive Modulation and Coding (AMC) such as Long Term Evolution (LTE) which has 15 Channel Quality Indicator (CQI) ranges [6,7]. In this scenario, we have tested the ASAP for a server has four mean service rates as shown in Fig. 2.

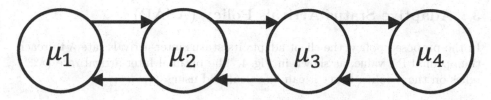

Fig. 2. Second Tested Scenario. In this scenario, the server mean service can take one of four mean values.

Status Updating Through the Internet. Several major cloud services offer Infrastructure as a service for Internet of Things (IoT) applications, e.g., Amazon and Google. However, delivering information that is as fresh as possible to the cloud might be challenging. Hence, the models proposed in the literature, only consider a single queue. In this scenario, the source located at University of Greenwich, Medway Campus, will be transmitting information through the internet to a server located in a cloud service is shown in Fig. 3. The inter-arrival time follows an exponential distribution with mean $1/\lambda$, where λ is the inter-arrival rate. The sever service time mean is deterministic and is equal to $1/\mu$, where μ is the service rate.

Fig. 3. Third tested scenario. In this scenario, the server is located in a cloud services provider.

Delivering fresh information in the third scenario, is more challenging than the previous two scenarios. On the other hand, it might be critical for the next generations of communication networks. Since several new applications and systems rely on the remote control of other machines that might be very distant from the controller. For example, the control of the *Cambridge Minicar*, which the control of a fleet of *Minicars* [8]. Also, the control of Base Stations antenna tilting in a cooperative self organisation network [9].

3   Adaptive Status Arrivals Policy (ASAP)

In the proposed policy, the client adapts its status inter-arrivals rate λ to reach the optimal PA value, as shown in Fig. 4. The optimal inter-arrival rate (λ^{opt}) relies on the μ value, hence it can be calculated using [5],

$$\lambda^{\text{opt}} = \rho^{\text{opt}} \times \mu, \tag{4}$$

where, ρ^{opt} is the optimal server utilisation value. Now ρ^{opt} depends on the scenario, hence in the next section, ρ^{opt} is derived for the tested scenarios.

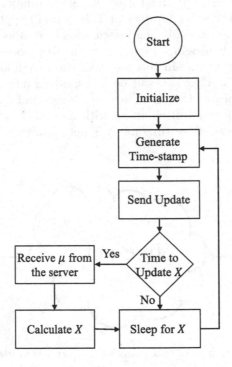

Fig. 4. ASAP client flow chart. The client generates a time-stamp, sends the update and sleeps for the duration of the inter-arrival time (X). When its time to update the (X) it would receive the server service time μ and adapt its X accordingly.

3.1   Optimal Server Utilisation for Minimising the Peak Age

An optimisation problem with its objective to minimise the PA with respect to ρ was formulated, as follows:

$$
\begin{aligned}
P^{\text{opt}} \triangleq \min_{\rho} \quad & P(\rho) \\
\text{subject to:} \quad & \rho < 1 \\
& \lambda \leq \lambda^{\text{max}}.
\end{aligned} \tag{5}
$$

The P^{opt} refers to the optimal PA value, the first constraint is to ensure that the queue is stable, since if $\rho \to 1$ then the delay time $T \to \infty$ [5]. The second constraint ensures that λ does not exceed the λ^{max}, which is determined by the device capabilities (for instance the sensor clock cycle).

For an M/D/1 queue, the ρ^{opt} can be obtained as follows,

$$\frac{d}{d\rho} P^{M/D/1}(\rho) = 0. \tag{6}$$

From (3) and (6), ρ^{opt} can be derived by,

$$\frac{d}{d\rho}\left[\frac{1}{\lambda} + \frac{1}{2\mu}\left(\frac{2-\rho^2}{1-\rho}\right)\right] = 0 \tag{7}$$

$$\frac{d}{d\rho} P^{M/D/1}(\rho) = \frac{1}{2\rho - 2} - \frac{2(\rho - 2)}{(2\rho - 2)^2} - \frac{1}{\rho^2} = 0. \tag{8}$$

The optimal server utilisation for the M/D/1 queue is,

$$\rho^{opt} \approx 0.5858. \tag{9}$$

The PA value for M/M/1 and M/D/1 queues are plotted in Fig. 5. It can observed that optimal ρ derived in (9) achieve the minimum PA value.

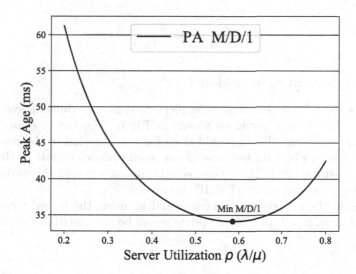

Fig. 5. Peak Age versus server utilisation, showing minimum value for M/D/1 queue with $\mu = 100$. The optimal Server Utilisation value for M/D/1 queues is equal to 0.5858.

The derived optimal values for ρ, are only applicable to a single queue. However, the PA in the third scenario (where the server is located on the cloud) the

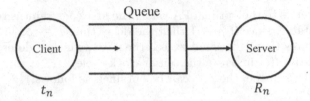

Fig. 6. Client-Server network model. The updates are generated in the Client and sent to the Server. The time of generating the update is t_n and the time of receiving the update is r_n.

internet load would affect the inter-arrival time, hence it must be considered. The Inter-arrival time in this scenario (X), as observed from the server, is

$$X(n) = X^* + \epsilon(n). \tag{10}$$

Where X^* refers to the inter-arrival time and $\epsilon(n)$ is a random value representing the time it takes the information (n) to be transmitted through the internet.

From (10), the optimal inter-arrival rate in the third scenario is,

$$\lambda(n)^* = \frac{1}{X^*} = \frac{1}{X(n) - \epsilon(n)}. \tag{11}$$

Hence,

$$\rho(n)^* = \frac{1}{(X(n) - \epsilon(n)) \times \mu}. \tag{12}$$

3.2   Experimental System Model

To evaluate the PA performance, we implemented an experimental system, consisting of a Client and Server as shown in Fig. 6. The Client sends statutes updates, to the server, then it would sleep for the duration of the inter-arrival time, as shown in Fig. 7. In the experiment, each update consists of the instantaneous time-stamp (t_n). The server records time it received the updates (r_n). The updates were sent using TCP/IP protocol.

The PA in the experiment was calculated by using the logged time-stamps. As shown in Fig. 8 [12], the PA of update n can be calculated as follows,

$$P_n = r_{n+1} - t_n. \tag{13}$$

The experimental PA was obtained by taken the median value of the PA of all the updates sent,

$$P = \tilde{P}_{(1,2,...,N)} \tag{14}$$

where, $\tilde{P}$ refers to the median value and N refers to the total number of transmitted updates.

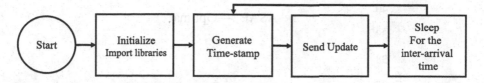

Fig. 7. Client flow chart. The Client initially import the socket libraries, then generate the instant time-stamp using the Time module, after sending the update to the Server it would sleep for the inter-arrival duration.

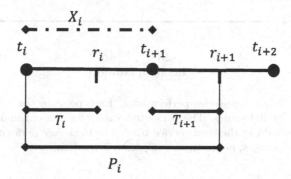

Fig. 8. Peak age illustration. For the experiment the Peak Age value (P) is equal to the inter-arrival time (X) plus the update system time (T). Consequently, in the experiment, the Peak Age value of an update (i) can be obtained by subtracting the value of generating an update t_i from the time of receiving the next update (r_{i+1}).

4 ASAP Peak Age Performance

The ASAP PA performance is presented in the three scenarios. In Fig. (9) the time series PA performance for an M/D/1 for the first scenario is shown. It can be observed that the ASAP PA performance varies with time as it keeps changing its inter-arrival time, i.e., λ, according to the server mean service time, μ. In our experiments, the server sends its service rate after receiving 100 updates and the sent value is the median service rate $(\tilde{\mu})$. Figure 10, presents the mean PA value of the updates shown in Fig. 9. It is observed that the PA approaches the theoretical optimal PA value.

The ASAP PA performance for the second scenario is presented in Fig. 11, where the service rate can take four possible mean values, the ASAP managed to deliver the updates with almost optimal freshness. It can be noted that the achieved performance is very close to the theoretical optimal performance. It is worth mentioning the achieved instantaneous PA performance might outperform the theoretical optimal PA, as the optimal represent the mean (average) performance.

In the third scenario, the ASAP managed to handle the internet load fluctuations as shown in Fig. 12. It is worth mentioning that the presented optimal performance in Fig. 12 represents the optimal PA for a single queue with the

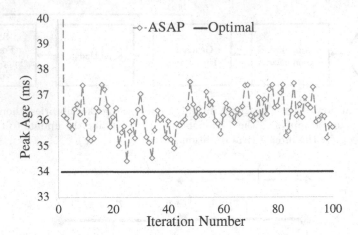

Fig. 9. ASAP peak age time-series performance. Each point in the ASAP represents the median value of 100 values. The presented values for the optimal, represent the theoretical value for PA at the used service time. The Peak Age performance of ASAP policy continually changes, hence the service time is random.

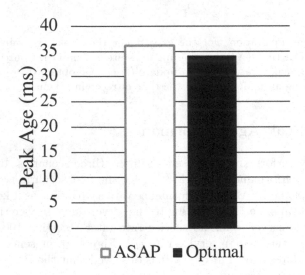

Fig. 10. ASAP mean peak age performance. The values presented are the mean value of the results presented in Fig. 9.

service time equal to the service time of the server plus the approximated value internet delay, hence the internet delays are random and hard to predicate.

Changing the inter-arrival rate makes ASAP a dynamic policy that is able to change its sampling rate to best fit the server. This feature can be critical in real-world applications where the server might have several background processes running on it. Using ASAP, instead of the client imparing an extra load on the server, it can reduce its transmissions but maintain a near optimal freshness performance.

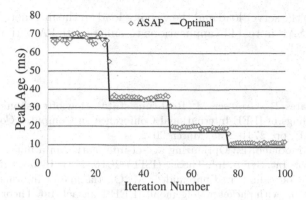

Fig. 11. ASAP peak age performance in the second scenario. Each point in the ASAP represents the median value of 1000 values. The presented values for the Optimal, represent the theoretical value for PA at the used service time. The peak age performance of ASAP policy continually changes, hence the service time is random. ASAP can outperform the Optimal value if the majority of the updates service time is less than the mean service time. Consequently, in the experiment the ASAP can outperform the Optimal value.

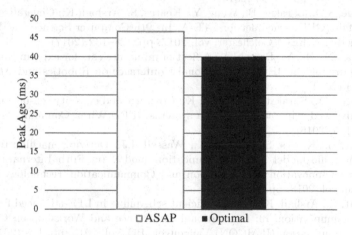

Fig. 12. ASAP peak age performance in the third scenario. Each bar in the ASAP represents the median value of 1000 values. The presented values for the optimal, represent the theoretical value for PA at the used service time for a single queue.

5   Conclusions

In this paper, the ASAP policy was proposed for minimising the Peak Age of Information. The policy regulates the inter-arrival time of status updates. The performance was measured by conducting experiments on three scenarios. The ASAP PA performance in the tested scenarios approaches the optimal value. Moreover, it can adapt to the server load and the varying load on the internet. The next step is to use the ASAP in real life applications to enhance its PA performance.

References

1. Kaul, S., Yates, R., Gruteser, M.: Real-time status: How often should one update?. In: Proceedings of IEEE International Conference on Computer Communications (INFOCOM), pp. 2731–2735, March 2012
2. Huang, L., Modiano, E.: Optimizing age-of-information in a multi-class queueing system. IEEE Int. Symp. Inf. Theory (ISIT) **2015**, 1681–1685 (2015)
3. Costa, M., Codreanu, M., Ephremides, A.: On the age of information in status update systems with packet management. IEEE Transact. Inf. Theory **62**(4), 1897–1910 (2016). https://doi.org/10.1109/TIT.2016.2533395
4. Sun, Y., Uysal-Biyikoglu, E., Yates, R.D., Koksal, C.E., Shroff, N.B.: Update or wait: how to keep your data fresh. IEEE Transact. Inf. Theory **63**(11), 7492–7508 (2017)
5. Bertsekas, D., Gallager, R.: Data Networks. Prentice-Hall, Englewood Cliffs (1987)
6. Barakat, B., Arshad, K.: An adaptive hybrid scheduling algorithm for LTE-Advanced. In: 2015 22nd International Conference on Telecommunications (ICT), Sydney, NSW, pp. 91–95 (2015)
7. Aramide, S.O., Barakat, B., Wang, Y., Keates, S., Arshad, K.: Generalized proportional fair (GPF) scheduler for LTE-A. In: 9th Computer Science and Electronic Engineering (CEEC) Colchester, vol. 2017, pp. 128–132 (2017)
8. Hyldmar, N., He, Y., Prorok, A.: A fleet of miniature cars for experiments in cooperative driving. In: IEEE International Conference on Robotics and Automation (ICRA) (2019)
9. Sharsheer, M., Barakat, B., Arshad, K.: Coverage and capacity self-optimisation in LTE-advanced using active antenna systems. IEEE Wirel. Commun. Netw. Conf. **2016**, 1–5 (2016)
10. Barakat, B., Keates, S., Arshad, K., Wassell, I.J.: Deriving machine to machine (M2M) traffic model from communication model. In: Fifth International Symposium on Innovation in Information and Communication Technology (ISIICT). Amman, vol. 2018, pp. 1–5 (2018)
11. Barakat, B., Arshad, K.: Energy efficient scheduling in LTE-advanced for machine type communication. In: International Conference and Workshop on Computing and Communication (IEMCON) Vancouver, BC, vol. 2015, pp. 1–5 (2015)
12. Barakat, B., Yassine, H., Keates, S., Wassell, I., Arshad, K.: How to measure the average and peak age of information in real networks? In: 25th European Wireless Conference on European Wireless 2019, Aarhus, Denmark (2019)

A Feasibility Study of Designing a Family-Caregiver-Centred Dementia Care Handbook

Ting-Ya Chang[1,2] and Kevin C. Tseng[1,2(✉)]

[1] Product Design and Development Laboratory, Taoyuan, Taiwan
ktseng@pddlab.org
[2] Department of Design, National Taiwan Normal University, Taipei, Taiwan

Abstract. This study aims to explore the feasibility of designing a family-caregiver-centred dementia care handbook. The objectives of this study were to (1) test the readability and understandability of existing written health education materials (WHEMs), (2) identify barriers to meeting dementia family caregivers' (DFCs') information needs in the context in which caregiving occurs, and (3) propose best-practice strategies and recommendations for redesigning WHEMs for DFCs with diverse health literacy skills. An innovative product design and development (IPDD) approach was implemented for the design and development process of the proposed WHEM for DFCs named IDEA. In-depth interviews with healthcare experts were conducted and analysed to determine their limitations and actionable recommendations for possible changes. Based on the research findings, we clarified current barriers to information experienced by DFCs and designated nine prominent themes and three essential elements to be included in thedesignprotocol. Finally, best-practice strategies and recommendations were proposed for redesigning a family-caregiver-centred dementia care handbook that may help to enhance the role of DFCs as active caregivers.

Keywords: Dementia family caregivers · Health literacy ·
Innovative product design and development · Written health education material

1 Introduction

Providing requisite information regarding dementia care benefits informal caregivers, particularly the spouses and daughters of people with dementia (PWD). To realize proper home care and help to resolve conflicts between dementia family caregivers (DFCs) and PWD, disseminating accurate and supportive written health educational materials (WHEMs) is of great importance. However, most of the dementia care-related WHEMs have not been written based on DFCs' information needs [1, 2]. Without obtaining specific knowledge and training regarding the progressive and complex nature of dementia, most DFCs are constantly confused and frustrated, especially in their early years of caregiving [3, 4]. They often experience caregiver burden (CB) from the extensive physical and emotional effort involved in caregiving, resulting in high levels of depression, social isolation and physical health problems. Moreover, along with the disease progression, more and more behavioural and psychological symptoms of dementia (BPSDs) occur; given the disease's complex nature, both PWD and DFCs are

© Springer Nature Switzerland AG 2019
M. Antona and C. Stephanidis (Eds.): HCII 2019, LNCS 11573, pp. 431–444, 2019.
https://doi.org/10.1007/978-3-030-23563-5_34

left on their own to cope with negative psychological, physical and economic impacts over an extended period of time [4–6].

Well-written WHEMs for DFCs not only guide them through the lengthy caregiving path but also enhance their level of engagement with the caregiving role [7]. In 1994, the National Institute of Health (NIH) in the U.S. laid out clear guidelines for developing print materials for low-literacy populations, emphasizing their effectiveness from both the content and visual representation perspectives. However, the latest research has revealed that current dementia healthcare handbooks have not yet met DFCs' information needs. Low literacy and inadequate education have become significant barriers to quality dementia healthcare [3, 8, 9]. A qualitative study conducted in 2017 using the sociotechnical systems approach identified nine barriers that DFCs confront in obtaining needed information about behavioural symptoms, which are a key clinical feature of dementia of any aetiology [5]. Based on these categories of barriers and their interactions within the sociotechnical system, three critical information needs were identified as follows: (1) timely access to information, (2) access to information that is tailored or specific to DFC needs and contexts, and (3) usable information that can directly inform how DFCs manage behaviours.

Previous studies have showed that the design of quality education and support materials could be greatly improved by utilizing an iterative participatory design methodology, rooted in the action research approach and philosophy of person-centred care [3, 5, 10, 11]. In 2018, Papachristou, Hickeys and Iliffe conducted a Think-Aloud method to gather information about DFCs' perspectives of booklets on food-related activities for PWD [12]. Their findings indicated that printed materials were considered economical, beneficial, and effective in assisting DFCs in coping with and adapting to changes in their lives, and that well-tailored information may enhance the confidence of DFCs and support them in making decisions during the caregiving process. The results also suggested that a handbook that raised awareness, illustrated coping skills, offered practical tips, and included quotes from other DFCs helped users engage with the provided information and take on a more positive role as primary caregivers.

To assist DFCs in comprehending textual information, the application of infographics as an innovative and engaging method of visual communication has piqued significant interest in recent years, especially within the healthcare sector [13–15]. Infographics have various strengths that are relevant to the current digital age. For the general public who are often unwilling and unable to spend time understanding lengthy and complicated information, infographics become a quick and efficient way to assimilate information into their frenetic lives. Furthermore, as the "baby boomer" generation has reached the age at which they are at greater risk of illness, there emerges an increasing demand for high-quality healthcare information in the form of infographics. A recent article by Harvard Medical School Professor David P Steensma evaluated the potential role of infographics in counselling patients affected by myelodysplastic syndromes [16]. Steensma concluded that graphical representation might be useful for some of his patients in aiding counselling about treatment complications and could also contribute to treatment and related counselling of other similarly complex medical conditions.

To advance knowledge regarding designing and developing a dementia WHEM from a user-centred perspective, the objectives of this article are to (1) test the readability of the original WHEM texts using Chinese Readability Index Explore (CRIE2.3), (2) identify barriers to meeting DFCs' information needs in the context in which caregiving occurs, and (3) propose best-practice strategies and recommendations for redesigning a WHEM for DFCs with diverse health literacy skills.

2 Methods

An innovative product design and development (IPDD) approach was used to systematically guide and ensure that the proposed WHEM (named as IDEA) would meet the needs and preferences of the target population [17]. In this study, the development process consisted of three phases: (1) opportunity identification, (2) opportunity understanding, and (3) opportunity conceptualization (see Fig. 1).

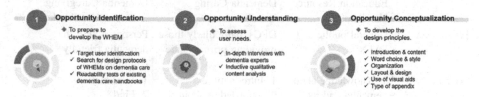

Fig. 1. A flowchart depicting the research framework implemented with IPDD approach

The design and development process involved several iterative steps to determine the target users and further identify their information needs regarding each DFC's distinct caregiving experience.

2.1 Opportunity Identification (Phase 1)

The opportunity identification phase was intended to explore usage problems of the existing dementia WHEM and inform the process of revision. Given the nature of disease and the current caregiving environment in Taiwan, we determined that our target users for the dementia care handbook were the informal caregivers of PWD, particularly spouses and daughters.

After the selection of our target users, we searched specified databases (e.g., PubMed and Science Direct) for clinical practice guidelines in articles that had been published in the past five years, in English or Chinese and that presented clinical evidence on dementia caregiving in homecare settings. We discovered that previous studies mainly focused on developing physical interventions and educational programmes, with very limited research on the improvement ofWHEMs for dementia caregiving, let alone research considering DFCs' health literacy during the design process.

Furthermore, existing dementia WHEMs often failed to satisfy DFCs' information needs. We selected three existing WHEMs on dementia care that were designed by representative government entities in Taiwan. Table 1 compares the information contained in these three handbooks; the appraising items were slightly modified according to the rating topics in understandability section of the patient education materials assessment tool (PEMAT) [18]. The readability score of each handbook was measured with the Chinese Readability Index Explore (CRIE2.3) based on multilevel linguistic features (difficult words, average sentence length, sentences with complex semantic categories, and conjunctions) [19].

Table 1. Comparison of existing WHEMs on dementia care

Items	Existing WHEMI	Existing WHEM II	Existing WHEM III
Publisher (year)	Ministry of Health and Welfare (2016)	New Taipei City Government (2016)	Taipei City Government (2018)
Handbook title	Dementia Health and Education Resource Handbook	Eliminate Loss: Dementia Caring Handbook	Memorize Together: Dementia Caregiving Resource Handbook
Target user	General public and DFCs	DFCs, particularly those providing primary care	Persons within the dementia friendly community
Content	1. Knowledge about dementia and its prevention 2. Caregiving skills 3. Care resources 4. Support for caregivers 5. Appendix	1. Introductions 2. Knowledge about dementia 3. Dementia progression and care 4. Basic dementia care principles 5. Management of special situations 6. Appendix	1. Introductions 2. Find 3. Care 4. Help 5. Community 6. Appendix
Use of visual aids	1. Run-in headings in separate colours 2. Charts for lists of available resource	1. Illustrations of complex medical terms 2. Small icons before run-in headings in separate colours and font size 3. Information boxes to indicate tips for DFCs 4. Charts for chronological information 5. Half-page illustrations in the hand-drawing style 6. QA format	1. Whole page illustration of geometric shapes as section breaks 2. Separate colour schemes for each chapter 3. Half-page illustrations in the comic style 4. Charts for comparison of complex ideas and chronological information 5. Run-in headings in separate colours and font size
Length (without appendix/ total)	38/60	37/54	104/132

(continued)

Table 1. (*continued*)

Items	Existing WHEMI	Existing WHEM II	Existing WHEM III
Type of appendix	1. Lists of community service spots awarded by the MOHW 2. Lists of long-term care management centres in Taiwan 3. Lists of day-care centres in Taiwan 4. Lists of group homes for PWD 5. Lists of dementia care areas in facilities	1. Procedures of how to make "life story books" for PWD 2. A wealth of resources 3. Standard health and education courses 4. Health diary	1. Lists of health service centres in the 12 districts at Taipei city 2. Lists of centres for integrated dementia care and community service spots 3. Lists of homecare services 4. Lists of day-care services 5. Taipei city long-term care management centre 6. Registered homecare and long-term care facilities 7. References
Readability (difficult words)	Between G4 and G5-7 level	At G5-7 level	At G4 level
Readability (average sentence length)	At G7 level	Slightly above G8 level	Way above G8 level
Readability (sentences with complex semantic categories)	Between G3 and G4 level	At G3 level	Between G2 and G3
Readability (conjunctions)	Between G7 and G8	Slightly above G5-7 level	At G5-7 level

2.2 Opportunity Understanding (Phase 2)

Having demonstrated that existing WHEMs show poor readability that overwhelms most target users' (DFCs') health literacy skills, it was necessary to identify their information needs. However, most DFCs were uncomfortable with expressing their opinions to strangers or sharing their actual experiences in academic settings. To successfully assess what they truly wanted, needed to know from WHEMs, we conducted two in-depth interviews with healthcare experts. The study involved healthcare specialists from Taipei and its vicinity who represented the overall dementia caregiving contexts in the region. To be included in the expert interviews, all participants needed to be (1) at least twenty years of age, (2) unrelated to the researchers, and (3) have a minimum of three months of previous work experiences with DFCs and (4) were preferred to have participated in the writing, editing, and/or assessment of WHEMs.

The in-depth interviews were conducted in a semi-structured style with open-ended questions in two pilot sessions. The interviews explored experts' observations of and experience with DFCs regarding the (1) contexts for receiving and applying WHEMs, (2) required information and practical skills for dementia care, and (3) preferred displaying format and style for DFCs. The interviews were recorded in audio form and subsequently transcribed verbatim and analysed using an inductive qualitative content analysis.

2.3   Opportunity Conceptualization (Phase 3)

Based on the insights of the previous phases, we integrated the global design strategy [11], art and story implementation [20], and application of health infographics [21] into the IPDD approach. In this way, we generated practical recommendations and specific design protocols regarding the (1) introduction and content, (2) word choice and style, (3) organization, (4) layout and design, (5) use of visual aids, and (6) type of appendix. Best-practice strategies and recommendations for redesigning WHEMs for DFCs regardless of their health literacy skills are described in the following section.

3   Results

3.1   Participants and Recruitment

Ethical approval was obtained from the National Taiwan Normal University Research Ethics Committee (Ref No. 201810HS001). All of the participants were adults over the age of twenty. Participants were assured that information collected in this study would be kept confidential and coded, that their participation would be voluntary, that there would be no penalty for refusing participation and that they could refuse to participate or withdraw participation at any stage of the data collection period. The informed consent for both the expert interviews and focus group validation experiments was sought for each stage. Coded information is being used in data management. Data are stored on a password-protected computer. The consent forms, data collection forms and verbatim transcripts will be kept for a ten-year period.

Two participants were recruited for this feasibility study. They were recruited through local dementia associations and among government employees within the public health sectors, using the e-form introduction article and emails. Specialists who showed interest in the study were provided with an email attachment containing the research proposal, informed consent form, and the semi-structured interview outline. These healthcare specialists were from distinct professional fields; one is experienced in the clinical diagnosis and medication of PWD and has conducted numerous in-depth interviews with DFCs, and the other has become well acquainted with DFCs through lecturing and providing them with psychological support (see Table 2).

Table 2. Details of the participants

Code #	Code #A1	Code #b1
Profession	Nurse and employee within dementia care associations	Doctor specialized in Neurology
Age	58	41
Gender	Female	Male
Dementia healthcare related experience (yrs)	Organization management, dementia care, family support, education programmes, academic research, service for young-onset dementia (10)	Clinical diagnosis, dementia care and treatment, homecare, palliative care (8)
Date of interview	2018/10/25	2018/12/27
Time (mins)	39	73

3.2 Data Analysis

The transcripts from the interviews were organized into separate data sets for each interviewee using a Microsoft Excel spreadsheet. The data sets were manually coded separately in the first instance. The researcher used an inductive approach to code and analyse each data set [22]. The data were coded and then the codes were grouped into themes. The analysis was conducted by assessing each transcript for information on the contexts for use of WHEMs, required content in WHEMs, and the preferred format and style of the WHEM for DFCs. A second researcher then checked the codes and themes against the data to ensure credibility and trustworthiness. An iterative approach was used as the two researchers discussed any variations in the coding and gradually came to a consensus.

Table 3. Main themes from transcripts

Category	#A1	#B1
Contexts for use	1. Physical access to WHEMs (1) 2. The relation between PWD and DFCs (1) 3. The impact factors of caregiving conflicts (2) 4. Considerations for providing WHEMs (2)	Physical access to WHEMs (1)
	6/22 (27.4%)	1/23 (4.4%)
Required contents	1. Considerations for designing the contents of WHEMs (1) 2. Critical themes of dementia care (1) 3. Mnemonic phrases of dementia care (1) 4. Caregiving tips for each disease progression (5)	1. Considerations for designing the contents of WHEMs (6) 2. Critical themes of dementia care (6) 3. Tips for designing mnemonic phrase (1)
	8/22 (36.3%)	13/23 (56.5%)

(continued)

Table 3. (*continued*)

Category	#A1	#B1
Preferred format and style	1. Preferred layout (2) 2. Preferred format (2) 3. Preferred style (3) 4. The application of portable cards (1)	1. Preferred layout (2) 2. Preferred format (1) 3. Preferred style (1) 4. Considerations for the applications of portable cards (5)
	8/22 (36.3%)	9/23 (39.1%)

Table 3 shows the themes that were identified from the thematic analysis. While interviewee #A1 discussed the three categories themes relatively equally, interviewee #B1 focused more on the "required content" theme and showed very little concern for the "contexts for use". The findings below reflect the experts' views on WHEMs, as informed by their own interactions and caregiving experiences with DFCs. The transcripts were carefully analysed in terms of how they fit with the aims of the study, which were to obtain the perspectives of experts on the contexts, contents, and format of the best-practice dementia handbook for DFCs in Taiwan.

Contexts for Receiving and Applying WHEMs. Both experts were in agreement that each DFC should receive an effective WHEM at the moment when their relative is diagnosed with dementia, though different amounts of information may be needed based on each individual's distinct preference.

> *A well-functioned integrated dementia care centre should provide the DFCs with related health and education instructions at the time their relative was diagnosed with dementia. Right at the instance of diagnosis! ...But the provider should consider each recipient's extents of acceptance of the disease. If the DFC shows low acceptance towards the disease, you can't give too much information... The service provider should prepare various education materials, such as courses, supportive groups, handbooks, and videos... I have numerous mediums of educational materials for the DFCs to choose from based on, in which would best satisfy with their information needs. (expert interview #A1)*

Some of the doctors directly provided DFCs with WHEMs at the clinics or instructed them to obtain one from the integrated dementia care centre in the hospital. Unfortunately, most DFCs were passive in obtaining this needed information on dementia care.

Required Information and Practical Skills on Dementia Care. The basic caregiving principles for each phase of disease progression were summarized by the interviewed experts, as is shown Table 4.

As for suggestions for enhancing the value of the WHEMs, the designer should add content that previous materials have not yet included. This additional content could include the latest healthcare issues, either those disseminated only in Taiwan or innovative concepts already adopted abroad, among other choices.

> *I intend to develop a WHEM, that is... that describes the basic concepts (of dementia care) ... so that the general public would find it easy to use... people can learn about most of the latest issues in the currently available WHEMs... but we should also add something surprising... this is where the value of the proposed WHEM comes from, in my opinion... even if there are only one or two additional chapters, you will find it very valuable. (expert interview #B1)*

Table 4. Basic caregiving principles at different disease progression stage

Disease progression stage	Basic caregiving principles
Before diagnosis	To provide checklists for dementia warning signs and educate the DFCs on how to confirm a solid diagnosis as soon as possible
Mild dementia	To assist the DFCs in overcoming negative emotions from caregiving by listing possible interventions for delaying the disease progression
Moderate dementia	To teach the DFCs practical coping skills of BPSDs
Severe dementia	To illustrate specific techniques about physical care so that the PWD could still retain considerable scores on activities of daily scale (ADLS) and cognitive functions
End of life/ palliative care	To prepare the DFCs for the significant decision making, such as whether to accept emergency medical services (EMS) or to go through a peaceful natural death

Preferred Displaying Format and Style for DFCs. Given the older ages of the DFCs in Taiwan (approximately mid to late 50s), printed educational materials are preferred to e-forms (websites and apps). Among the different options for display formats, integrated infographics provide the best understandability for a range of DFC health literacy skills. However, the experts expressed contradictory opinions regarding the application of portable cards as an appendix of the WHEM. One expert confessed that she could not imagine that portable cards regarding dementia care would be feasible:

The application of portable cards would be hard... very hard. For instance, when the BPSDs occur today, you could not assist my understanding of the situation by giving me these cards... Dementia is unlike diabetes. Diabetes caregivers may need these cards to provide tips regarding which food the patient can and cannot eat. But for dementia caregiving, I can't think of many things that I need to take out and memorize at this moment. (expert interview #A1)

Meanwhile, the other expert showed a very positive attitude toward the application of portable cards:

...Yes, definitely yes. The portable cards could be useful for summarizing tips for the DFCs... You could also design ID cards for Alzheimer's and dementia, which would assist the DFCs in discreetly describing the reasons for the many weird behaviours of PWD without telling others aloud and in turn hurting the feelings of PWD... These portable cards could be designed to be the same size as the old version of iPhones (expert interview #B1)

Therefore, recruiting focus groups comprised of DFCs using the Patient Education Materials Assessment Tool (PEMAT) to verify the understandability and feasibility of the redesigned WHEMs is necessary for the next phase of the project.

4   Discussion

The main objectives of this article were to (1) test the readability and understandability of existing WHEM on dementia care, (2) identify barriers to meeting DFC's information needs when caregiving occurs, and (3) propose best-practice strategies and recommendations for redesigning a WHEM for DFCs from a family-caregiver-centred perspective.

Three phases of the IPDD design process (opportunity identification, opportunity understanding, and opportunity conceptualization) were implemented and executed collaboratively with healthcare experts and the researcher (designer). Three different methods (global design strategy, art and story implementation, and health infographics application) and two verification tools (PEMAT and CRIE 2.3) were used which generated three essential elements to inform the design protocol for an effective WHEM. A qualitative study was conducted to focus on DFCs' information needs and preferences regarding WHEMs from the perspective of dementia care experts. The IPDD approach proved to be an engaging and cost-efficient means of exposing product designers to feedback, which included user difficulties, ideas, and opinions. Best-practice design and development protocols for WHEMs are as follows:

Show Tailored Information in Integrated Infographics. DFCs have very limited time and effort to devote to researching, comprehending, and obtaining needed information in real caregiving contexts. Therefore, integrated critical information in the form of info-graphics received the most positive feedback from DFCs (see the examples in Fig. 2).

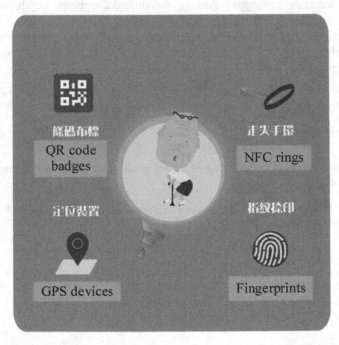

Fig. 2. Proposed infographic depicting strategies to prevent PWD from wandering and getting lost.

Organize Content by Thematic Topic. Although dementia is progressive and can be classified into five stages (1) before diagnosis, (2) mild dementia (3) moderate dementia, (4) severe dementia, and (5) end of life or palliative care, many BPSDs overlap occur across several stages and vary greatly for each individual. Therefore, a structure based on the general disease progression may cause confusion among DFCs. An information structure that categorizes critical topics in dementia care (see Table 4) is thus more intuitive and accessible to DFCs (Table 5).

Table 5. Critical thematic topics for dementia care.

#	Topic	Disease progression
1	Advancedcare planning (ACP)	End of life/ palliative care
2	Building a safe environment for PWD	Mild dementia Moderate dementia
3	Communication issues and the transforming roles	Mild dementia Moderate dementia
4	Coping with BPSDs	Mild dementia Moderate dementia
5	Getting a diagnosis and knowledge about the disease progression	Before diagnosis
6	Getting help with caregiving	Before diagnosis Mild dementia Moderate dementia Severe dementia
7	How to take care yourself	Mild dementia Moderate dementia End of life/ palliative care
8	Providing everyday care for PWD	Mild dementia Moderate dementia
9	Remote family care promotion	Before diagnosis Mild dementia Moderate dementia Severe dementia End of life/ palliative care

Only Select Content that is Highly Relevant to Real Caregivers. DFCs have shown very little interest in disease research findings such as the aetiology of or medical treatment for dementia. Since their loved ones have already been diagnosed with dementia and cannot be cured, DFCs only care about practical tips for caregiving. Sharing actual caregiving experiences and feelings in WHEMs tend to attract the interest of DFCs (see Fig. 3).

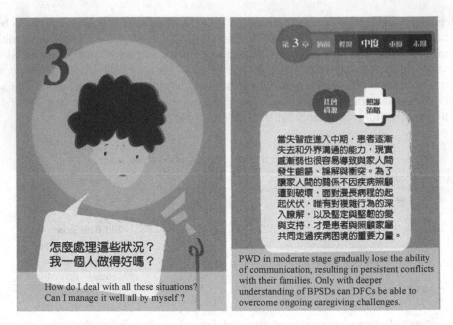

Fig. 3. Visual design of the proposed WHEM (IDEA) handbook, embedding caregiving feelings from various roles within the family.

5 Conclusion

This study explores the feasibility of designing a family-caregiver-centred dementia care handbook and identifies the IPDD process proven to be useful in developingbest-practice strategies and recommendations for meeting DFC information needs and enhancing the activeness of their roles as caregivers. It enables us to overcome communication barriers with DFCs in the early design stage, and ensure a more insightful exploration of their needs from expert perspectives. A utility test using PEMAT in focus groups of randomly selected target users is however neeeded to verify the understandability and feasibility of the proposed WHEM (named IDEA). Further exploration of particular tools that appeal to the participants in the form of appendix represents an exciting area for researchers and healthcare providers. Finally, the study participants were recruited only in Taipei and its vicinity. Thus, the proposed IDEA could be further developed with the IPDD approach, ant its impact on caregivers thoroughly evaluated in a larger study.

Acknowledgments. This work was supported in part by the Ministry of Science and Technology of Taiwan, ROC under Contracts MOST 106-2628-H-182-002-MY3 and MOST 108-2622-8-003-001-TM1. The funders had no role in the study design, data collection and analysis, decision to publish, or preparation of the manuscript.

References

1. Steiner, V., Pierce, L.L., Salvador, D.: Information needs of family caregivers of people with dementia. Rehabil. Nurs. **41**, 162–169 (2016)
2. Peterson, K., Hahn, H., Lee, A.J., Madison, C.A., Atri, A.: In the Information Age, do dementia caregivers get the information they need? Semi-structured interviews to determine informal caregivers' education needs, barriers, and preferences. BMC Geriatr. **16**, 164 (2016)
3. Whitlatch, C.J., Orsulic-Jeras, S.: Meeting the informational, educational, and psychosocial support needs of persons living with dementia and their family caregivers. Gerontologist **58**, S58–S73 (2018)
4. Bull, M.J.: Strategies for sustaining self used by family caregivers for older adults with dementia. J. Holist. Nurs. **32**, 127–135 (2014)
5. Werner, N.E., et al.: Getting what they need when they need it. Identifying barriers to information needs of family caregivers to manage dementia-related behavioral symptoms. Appl. Clin. Inform. **8**, 191–205 (2017)
6. Thompson, G.N., Roger, K.: Understanding the needs of family caregivers of older adults dying with dementia. Palliat. Support. Care **12**, 223–231 (2014)
7. Rathnayake, S., Moyle, W., Jones, C.J., Calleja, P.: Development of an mHealth application for family carers of people with dementia: a study protocol. Collegian **26**(2), 295–301 (2018)
8. Haralambous, B., Mackell, P., Lin, X., Fearn, M., Dow, B.: Improving health literacy about dementia among older Chinese and Vietnamese Australians. Aust. Health Rev. **42**, 5–9 (2018)
9. Sand-Jecklin, K.: The impact of medical terminology on readability of patient education materials. J. Commun. Health Nurs. **24**, 119–129 (2007)
10. Smith, F., Wallengren, C., Öhlén, J.: Participatory design in education materials in a health care context. Action Res. **15**, 310–336 (2017)
11. Cardenas, C., et al.: Global design strategy for cancer patient education materials: Haiti pilot case study. Des. Manag. J. **11**, 15–31 (2016)
12. Papachristou, I., Hickeys, G., Iliffe, S.: Involving caregivers of people with dementia to validate booklets on food-related activities: a qualitative think-aloud study. J. Appl. Gerontol. **37**, 644–664 (2018)
13. McCrorie, A., Donnelly, C., McGlade, K.: Infographics: healthcare communication for the digital age. Ulster Med. J. **85**, 71 (2016)
14. Barros, I.M., Alcântara, T.S., Mesquita, A.R., Santos, A.C.O., Paixão, F.P., Lyra, D.P.: The use of pictograms in the health care: a literature review. Res. Soc. Adm. Pharm. **10**, 704–719 (2014)
15. Locoro, A., Cabitza, F., Actis-Grosso, R., Batini, C.: Static and interactive infographics in daily tasks: a value-in-use and quality of interaction user study. Comput. Hum. Behav. **71**, 240–257 (2017)
16. Steensma, D.P.: Graphical representation of clinical outcomes for patients with myelodysplastic syndromes. Leuk. Lymphoma **57**, 17–20 (2016)
17. Tseng, K.C.: An IPDD approach for systematic innovation of products, processes, and services: a case study on the development of a healthcare management system. In: The 5th IASDR World Conference on Design Research. Tokyo (2013)
18. Shoemaker, S.J., Wolf, M.S., Brach, C.: Development of the patient education materials assessment tool (PEMAT): a new measure of understandability and actionability for print and audiovisual patient information. Patient Educ. Couns. **96**, 395–403 (2014)

19. Sung, Y.-T., Chang, T.-H., Lin, W.-C., Hsieh, K.-S., Chang, K.-E.: CRIE: An automated analyzer for Chinese texts. Behav. Res. Methods **48**, 1238–1251 (2016)
20. Kagan, S.H.: Using story and art to improve education for older patients and their caregivers. Geriatr. Nurs. **39**, 119–121 (2018)
21. Lo, C.-W.J., Yien, H.-W., Chen, I.-P.: How universal are universal symbols? An estimation of cross-cultural adoption of universal healthcare symbols. HERD: Health Environ. Res. Des. J. **9**, 116–134 (2016)
22. Braun, V., Clarke, V.: Using thematic analysis in psychology. Qual. Res. Psychol. **3**, 77–101 (2006)

Occupational and Nonwork Stressors Among Female Physicians in Taiwan: A Single Case Study

Kuang-Ting Cheng[1,2] and Kevin C. Tseng[1,2(✉)]

[1] Product Design and Development Laboratory, Taoyuan, Taiwan
ktseng@pddlab.org
[2] National Taiwan Normal University, Taipei, Taiwan

Abstract. The high suicide rate among doctors is a significant issue in many countries, especially among female doctors, for whom the rate is more than two times that of the general population. Compared to many countries, Taiwan has a much lower proportion of female physicians relative to male physicians, which has been suggested as a negative factor in affecting the suicide rate. Previous studies of female physician stressors are few and focus mainly on occupational stress. Nonwork stress has not been well-researched. This study aims to explore the feasibility of providing a comprehensive evaluation of all stressors in female doctors' daily lives by examining a cohort of Taiwanese female doctors. Maslach burnout inventory (MBI) and the Brief Symptom Rating Scale (BSRS-5) are used to screen participants for occupational stress and depressive attributes respectively. In this study, an interview is conducted with a participant, and factors contributing to lifestyle and occupational stress are identified. The study results indicate that family issues, primarily child-rearing, acts as the largest stressor in the participant's life, outweighing even traditionally studied occupational stressors for female physicians.

Keywords: Depression · Female physician ·
Occupational and nonwork stressor · Suicide · Qualitative analysis

1 Introduction

Studies show that physicians have a higher risk of suicide, which is about 1 to 3 times for male physicians and 2 to 6 times for female physicians when compared to the normal population [1]. There is only few research about suicide in Taiwan, one study shows the risk is 2 times for physicians aged 25–44 years compared with the general population [2]. From previous research, we found that the most important evidence-based risk factors for suicide are psychiatric disorders (mostly depression and schizophrenia), past or recent social stressors, suicide in the family or among friends or peers, low access to psychological help, and access to methods for committing suicide [3]. Several studies have emphasized the importance of depression on suicide risk, particularly for female physicians [4, 5]. In a study of more than 1300 male medical graduates from Johns Hopkins University, the lifetime prevalence of depression was found to be 12.00% [6]. An analogous study of 4500 female physician manifests the

© Springer Nature Switzerland AG 2019
M. Antona and C. Stephanidis (Eds.): HCII 2019, LNCS 11573, pp. 445–454, 2019.
https://doi.org/10.1007/978-3-030-23563-5_35

prevalence is 19.50% [4]. The above statistics imply the gender difference regarding the issue of physicians' stress. Although the etiologies of depression haven't totally been understood yet, stress has been found to be an impactful factor on depression which changes many mechanisms in the human body [7].

One might expect physicians to have excellent access to mental healthcare services and medication. However, postmortem toxicology testing of physicians who committed suicide shows low rates of antidepressant use. They are more likely to use self-prescribed medication like antipsychotics, benzodiazepines and barbiturates. Inadequate medical treatment and increased problems related to job stress are potentially modifiable risk factors that can reduce suicidal death among physicians [8]. Nevertheless, there are barriers which prevent physicians from seeking help, including confidentiality, discrimination in medical licensing, hospital privileges, concerns about the impact on professional advancement, and the risk of being stigmatized [5, 9].

Numerous studies describe occupational stress as one of the foremost causes of physician mental illness and depress [10–12]. Imbalance among job, family, and personal growth is mentioned but few studies have been done to understand physicians' nonwork stress. Existing studies also tend to focus on male physicians, who have traditionally been much more prevalent in the healthcare system. Social and occupational roles for male and female physicians can be very different. There are also significant differences between male and female physicians in the statistics of suicide and depression prevalence. These factors, in addition to the dearth of previous research on female physicians, lead us to specifically study both occupational and nonwork stress of female physicians.

We first discuss occupational stress. Although physicians with different genders have similar missions in their jobs, careers and opportunity structures, power, benefits, and occupational networks can be significantly different, so gender differences in physician stress are expected. Due to female physicians' minority status/prejudice, they may experience a lack of role models/mentors/sponsors and role strain. Sixty-sixpercent of female physicians indicated at least one very difficult period in their lives [13]. According to OECD Health Statistics in 2016 [14], only 21.02% of doctors are women in Japan. Most countries have a proportion of female physicians that is less than half. In Taiwan, the proportion of female positions is much less, only 19.80% in 2018 [15]. The differences diminished when the members of minority become more, so we can guess the problem is more serious in Taiwan. Ann et al. [16] identified that work stressors may increase the risk of suicide for female physicians. Taiwan has a much lower proportion of female physicians relative to male physicians, which has been suggested as a negative factor in affecting the suicide rate. Previous studies of female physician stressors are few and focus mainly on occupational stress.

There are significant gender-related differences in nonwork stressors as well. Grace et al. [17] mentioned that among women, family role stressors are more strongly correlated to mental illness, especially depression, when compared to work-related stressors. For men, work-related stressors have a more significant impact on mental health than family role stressors, indicating that overall, women are more vulnerable to the negativity of family stress. Studies [17] also found that women show a higher level of depression than men due to marital dissatisfaction. Also, women's primary responsibility for the wellbeing of others, especially children, causes poorer mental health of women [17]. The multitude of roles that women must fulfil has traditionally also been a source of stress for women. The environment of high demands and low control by the mother, in contrast to one of the low demands and high control by the father, is one of the reasons that women can be more stressed than men [18]. Therefore, to research female physicians' stressors, nonwork stressors cannot be ignored.

Previous studies about female physicians are few and mainly focus on clinical job settings. However, female physicians face more challenges than male physicians regarding work-family balance [19]. This research aims to understand the stressors thoroughly in female physicians' daily lives, including occupational stressors, nonwork stressors, stress relievers, and the interrelation between all of them. By understanding the stressors in the whole picture, we can prevent physician suicide in the very early beginning.

2 Materials and Method

There are three steps in our research method. First, we screened participants via a questionnaire. Second, Day Reconstruction Method (DRM) [20] was used, followed by interviewing participants. Finally, we conducted qualitative analysis by ATLAS.ti CLOUD [21] to code the interview data [22].

In the questionnaire, several types of demographic data of female physicians were collected, including residency, age, salary, department, working hours, working organization, working years, and family members, etc. The questionnaire was only open to licensed female physicians from Taiwan, who were trained in western medicine, thus excluding interns, eastern medicine doctors, and dentists. The Chinese version of the Brief Symptom Rating Scale (BSRS-5) [23, 24] and Maslach burnout inventory (MBI) [25] were used to evaluate physicians' general mental health and occupational stress.

BSRS-5 is designed to capture general mental health status in order to prevent suicide. Based on the results of a study that shows the high overlap between burnout and depression [26], we chose MBI to record participants' occupational stress. MBI measures three dimensions of burnout: emotional exhaustion, depersonalization, and personal accomplishment. High scores indicate a greater extent in each dimension, greater burnout in emotional exhaustion and depersonalization, and less burnout in personal accomplishment. For this case study, we selected one physician to be interviewed based on the results of a screening questionnaire.

In the second step, we conducted Daily Reconstruction Method by sending out a format of multiple diary tables, in which participants jot down, for one week, every event and the concurrent emotion and stress level. The interview will be conducted based on those events, and the diary table will be reviewed by participants as a reference. Participants can choose to keep the diary for privacy or have it returned to researchers. The set of interview questions is as follows:

- Please describe your life this week.
- Why do you have that emotion in this event?
- What's the main stressor in your life?

- What do you like to do to relieve yourself of stress?
- If you had the ability to change anything in your current life, what would you change?

The interview was also voice recorded. We typed the whole conversation word for word after the interview.

In the third step, we do qualitative analysis by coding to indexing what the participants said in order to establish a framework for this female physician's stressors.

3 Result

The participant is an attending physician working in a physical medicine and rehabilitation clinic for about 25 h per week. She lives with her physician husband and her 4 years and 8 months old child. More demographic details are shown in Table 1. Her results on the Brief Symptom Rating Scale and Maslach burnout inventory are as Table 2.

Table 1. The participant's characteristics information

Characteristics	Description
Gender	Female
Age	36 years and 3 months old
Habitat	Hsinchu, Taiwan
Occupation	Attending physician, the 3rd year
Specialty	Physical medicine and rehabilitation
Work place	Local clinic
Marital status	Married
Working hours/week	25 h/week
Number of family members	4 (Husband, one 4-year-old child, and one 8-year-old child)
Household income	1,500,000 to 2,000,000 NTD

Table 2. The participant's brief symptom rating scale score

Questions	Score
Question 1–5	2
Question 6: having suicidal thoughts	0

Note. Each question is evaluated by participant with a scale from 0 to 4 (not at all to extremely). A total score on the BSRS-5 above 14, or a score of more than 1 on the suicide survey item, may indicate a severe mood disorder. Total score of the participant: <6: normal; no suicidal thoughts

Table 3. The participant's Maslach burnout inventory score

Dimension	score/total
Emotional exhaustion	12/54
Depersonalization	3/30
Personal accomplishment	35/48

Note. High scores indicate greater burnout in emotional exhaustion and depersonalization, and less burnout in personal accomplishment.

Overall, the participant emphasized that children are the main stressor in her current life as an attending physician. She spent more than half of the interview duration speaking about things related to her children.

"Most of my stressors come from children."
"I like to have shifts because it's possible to have overnight sleep, which is impossible when I stay home because I have to wake up at night and breastfeed my baby."
"The most frustrating thing that happened last week is that my child cheated, which conflicts my upbringing principle."
"I feel that I was being misunderstood when my brother suggested another way to educate my child."

When asked about the childcare related issues shared by the physician couple, the participant professed to take more responsibilities for their children.

"I have to wake up once every night for breastfeeding. My husband has tried that before, but he needs much sleep and can't wake up. I've tried to wake my husband up, but he is too slow and the child burst into tears. It's better and more efficient that I do it alone."
"Children wake up early and my husband cannot, so I have to wake up even when I've been interrupt when sleeping for breastfeeding."
"When the baby was sick and couldn't breathe smoothly, I'm afraid that the sputum blocked his airway so I would wake up to check my baby if his breathing sound was abnormal. My husband sleeps very deeply so I'm the one who wakes up."

As shown in the MBI score (Table 3), the participant scores high in personal achievement, which agrees with the interview results. She is in general satisfied with her current job. She only feels frustrated when her patients' treatment didn't meet expectation.

"If I can find what I want to learn in my job and apply it, I'll be happy, and this is what I like to do. I would feel satisfied if my current job continues, because it's hard to find a boss who has the same values as you and is willing to support you. Many bosses here don't want to buy an ultrasound, but my boss did. Ultrasound is my key technique, so I can do what I want to do here."
"There was a patient on Friday who complained and questioned my decision. I had explained things before treatment, but she still felt the result didn't meet her need. It's hard to communicate."

When asked what most stressful aspects of being a doctor are, she mentioned two of her biggest stressors: night shifts in internal medicine department during postgraduate year-1(PGY), and having a child in residency.

> *"The most unforgettable stressors in my life would be two things; one is the night shifts in the internal medicine department in PGY. It was very terrible... because I was the front-line resident. The biggest stressor is the fear of being sued. Senior residents were often doing CPR in intensive care centre, so the backup support is not enough. I was very afraid of malpractice and being sued."*
>
> *"Another one is having children in residency. There are so many things that have to be done as a resident. If your child has issues that you need to spend additional time on, such as refusing to go to bed on time or getting sick. You don't know when you can start your own work, which makes me anxious. That's so terrible"*

According to the interview, her medical career thus far can be divided into three periods: PGY, residency, and attending physician. Her occupational stress went down after being an attending physician. However, her primary source of nonwork stress, her children, has remained a constant source of stress. (Figure 1) The stress in residency comes not only from the responsibilities of childcare but also from her job, so the occupational and nonwork stress are both high in her residency.

> *"When you are a resident, you need to react based on the head of the department. At times, he would say something threatening. For example, he would say 'when I was a resident, a young couple wouldn't dare become pregnant during residency. You may need to stay one more year if you delay too much.'"*
>
> *"I was lucky because senior residents in my department supported me a lot. They were really kind and didn't badmouth me. I would do things they didn't want to do like paperwork as a return."*

The Participant's values and personality are also shown in the interview.

> *"My principle is seeking the truth from facts. When my son has questions, I'll try my best to answer him right. I won't say it's ok if we don't know the answer. Instead, I'll say let's try to find out the answer together."*
>
> *"When there are patients that could be treated by new skills therapists don't want to learn, I feel this attitude conflicts with my principles, seeking the truth from facts. Why don't we do it if it's doable with more effort?"*
>
> *"By my educational values, girls are the same as boys. I would not educate my children differently based on their gender."*

For stress relief, the participant likes to read books about patient treatment or upbringing in her free time. She also mentioned that she feels happy sometimes when she can leave children for a while, such as when she has shifts, which means her husband will take care of the children. Also, the support from senior residents helped release her stress in residency.

Figure 2 shows the relationship between stress and stressors for this participant. During her medical career, the occupational stressors are incongruent of treatment expectation between her and the patients, the lack of skill in clinical practice, lack of support for clinical work, and lack of support when pregnant; the nonwork stressor are children related issues, including her children's education, health, concern for other issues, such as when her children are negatively affecting others.

Her primary forms of occupational stress relief come from the support of her colleagues and from breaks from the working space. Her relief from nonwork, mostly childcare stress, comes from the support of her husband, and from opportunities to take

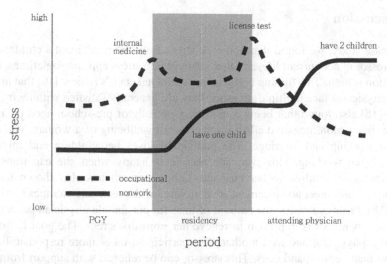

Fig. 1. Participant's stress in different medical career path. The degree of stress (low to high) is based on the participant's own drawing.

temporary breaks from taking care of her children. Other forms of stress relief, such as her hobby, reading the books on medical treatment and education, relieve both of her occupational and nonwork stress. The participant also mentioned that nonwork stress affected her job more than occupational stress affected her family.

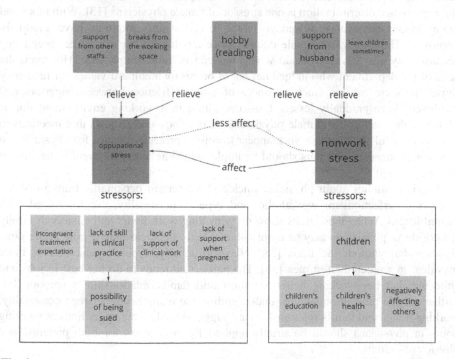

Fig. 2. The relationship between occupational stressor, nonwork stressor, and stress reliever

4 Discussion

In this case study, we found that nonwork stress from the participant's children is the main stressor in her current life, and that although job stress appears sometimes, her job satisfaction is high. This finding is consistent with Gail et al.'s review [19] that although female physicians face multiple stressors, they are generally satisfied with their careers. Barnett [18] also found that being a mother, especially of pre-school-aged or multiple children, has a more negative effect on the overall wellbeing of a woman, when compared to working and marriage. The participant loves her children and misses her children when working. However, she also feels happy when she can transfer the responsibility of children to her husband. Other research has also shown that both employment and career advancement are correlated with positives outcomes in women's health [18]. From our study, the amount of time for our female physician participant to pursue her own goal is important to relieve her nonwork stress. The goal is from two roles, as a physician, and as a mother. The participant takes more responsibilities for childcare than her husband does. This stressor can be relieved with support from others and from the enjoyment of her own hobbies. Thus the support of the husband in child-rearing is highly important for the mental health of employed women [27]. For mothers, learning to manage one's expectations for children and expressing these feelings in a healthy and productive way can also lower childcare related stress.

Lack of support in clinical practice is the participant's main occupational stressor. The types of support she seeks can be categorized as support for pregnancy-related needs and support for her clinical work. Marjorie et al. mentioned that minority status/prejudice/discrimination is one stressor of female physicians [13]. With more and more female physicians nowadays, some parts of the situation have gradually improved. The participants' male doctor coworkers helped her during her pregnancy because they also have physician wives and can be more empathetic. However, the head of the department who judged her based on his own cultural values for residency caused pressure on her. Such instances of cultural friction between superiors and employees have gradually lessened, also resulting in a working environment that is friendlier than before for female physicians. A previous study shows that meetings to discuss stressful situations at work appear to offer a protective factor for physicians in different residencies, and thus should be implemented as a regular event in healthcare institutions [16].

Previous studies about physician suicide show certain personality traits in physicians, i.e. perfectionism, workaholic, and type A personality, are associated with mental illness. When these traits show up with long work hours and unsmooth family relationships, physicians may be more vulnerable to life stress than people in the same situation but without the traits [28, 29]. Additionally, type A behaviour is more prevalent in women than in men [30]. The participant tends to do her best in jobs and childcare, but her working hours are short and family relationship is smooth. For further suicide prevention studies, understanding the relationship between personality, working hours, and family relationships are suggested. Also, the upper limit of working hours in physicians should be strictly applied by institutions for both patients' and physicians' safety.

5   Conclusion

To improve the suicide issue for physicians, female physicians should be discussed separately from male physicians due to differences in the family and professionally in the healthcare system. In the professional setting, female physicians are a minority group and thus also face unique workplace issues related to this. The Taiwanese healthcare system is a useful setting for this investigation, particularly because it has a much smaller proportion of female physicians when compared to other countries. From this case study, the occupational stressors are lack of support for clinical work, lack support for pregnancy-related issues, and the incongruent treatment expectations between doctor and patient. The nonwork stressor, which is this female attending physician's current main stressor, mainly come from her child-rearing responsibilities. Both occupational and nonwork stressors can be relieved by support from others and by the enjoyment of her own hobbies. Other suggestions for stress relief are provided based on previous studies. Further research with more participants is needed to get a comprehensive picture of female physicians' stress model.

Acknowledgments. This work was supported in part by the Ministry of Science and Technology of Taiwan, ROC under Contracts MOST 106-2628-H-182-002-MY3 and MOST 108-2622-8-003-001-TM1. The funders had no role in the study design, data collection and analysis, decision to publish, or preparation of the manuscript.

References

1. Schernhammer, E.S., Colditz, G.A.: Suicide rates among physicians: a quantitative and gender assessment (meta-analysis). Am. J. Psychiatry **161**, 2295–2302 (2004)
2. Chen, I.M., Liao, S.C., Lee, M.B.: Physician suicide in Taiwan: a nationwide retrospective study from 2000–2013. Aust. J. Psychiatry Behav. Sci. **1**, 1008 (2014)
3. World Health Organization: Prevention of mental disorders: effective interventions and policy options. World Health Organization (2004)
4. Frank, E., Dingle, A.D.: Self-reported depression and suicide attempts among U.S. women physicians. Am. J. Psychiatry **156**, 1887–1894 (1999)
5. Center, C., et al.: Confronting depression and suicide in physicians: a consensus statement. JAMA **289**, 3161–3166 (2003)
6. Ford, D.E., Mead, L.A., Chang, P.P., Cooper-Patrick, L., Wang, N., Klag, M.J.: Depression is a risk factor for coronary artery disease in men: The precursors study. Arch. Int. Med. **158**, 1422–1426 (1998)
7. Yang, L., Zhao, Y., Wang, Y., Liu, L., Zhang, X., Li, B., Cui, R.: The effects of psychological stress on depression. Curr. Neuropharmacol. **13**, 494–504 (2015)
8. Gold, K.J., Sen, A., Schwenk, T.L.: Details on suicide among US physicians: data from the national violent death reporting system. Gen. Hosp. Psychiatry **35**, 45–49 (2013)
9. Schwenk, T.L., Gorenflo, D.W., Leja, L.M.: A survey on the impact of being depressed on the professional status and mental health care of physicians. J. Clin. Psychiatry **69**, 617–620 (2008)
10. Wang, L.-J., Chen, C.-K., Hsu, S.-C., Lee, S.-Y., Wang, C.-S., Yeh, W.-Y.: Active job, healthy job? Occupational stress and depression among hospital physicians in Taiwan. Ind. Health **49**, 173–184 (2011)

11. May, H.J., Revicki, D.A., Jones, J.G.: Professional stress and the practicing family physician. South. Med. J. **76**, 1273–1276 (1983)
12. Balch, C.M., Freischlag, J.A., Shanafelt, T.D.: Stress and burnout among surgeons: understanding and managing the syndrome and avoiding the adverse consequences. Arch. Surg. **144**, 371–376 (2009)
13. Bowman, M.A., Allen, D.I.: Female physician stress. In: Bowman, M.A., Allen, D.I. (eds.) Stress and Women Physicians, pp. 129–141. Springer, New York (1990). https://doi.org/10. 1007/978-1-4684-0267-4_9
14. OECD: Health Care Resources: physicians by age and gender. OECD (2016)
15. The Gender Equality Committee: Gender statistics of various medical personnel in Taiwan. Ministry of Health and Welfare, Taipei (2019)
16. Fridner, A., Belkic, K., Marini, M., Minucci, D., Pavan, L., Schenck-Gustafsson, K.: Survey on recent suicidal ideation among female university hospital physicians in Sweden and Italy (the HOUPE study): cross-sectional associations with work stressors. Gend. Med. **6**, 314–328 (2009)
17. Baruch, G.K., Biener, L., Barnett, R.C.: Women and gender in research on work and family stress. Am. Psychol. **42**, 130 (1987)
18. Barnett, R.C., Biener, L., Baruch, G.K.: Gender and Stress. Free Press, New York (1987)
19. Robinson, G.E.: Stresses on women physicians: consequences and coping techniques. Depress. Anxiety **17**, 180–189 (2003)
20. Kahneman, D., Krueger, A.B., Schkade, D.A., Schwarz, N., Stone, A.A.: A survey method for characterizing daily life experience: the day reconstruction method. Science **306**, 1776–1780 (2004)
21. ATLAS.ti Scientific Software Development GmbH: ATLAS.ti CLOUD. vol. BETA. ATLAS.ti Scientific Software Development GmbH (2018)
22. Hwang, S.: Utilizing qualitative data analysis software: a review of Atlas.ti. Soc. Sci. Comput. Rev. **26**, 519–527 (2008)
23. Lee, M.-B., et al.: Development and verification of validity and reliability of a short screening instrument to identify psychiatric morbidity. J. Formos. Med. Assoc. **102**, 687–694 (2003)
24. Lu, I.C., Yen Jean, M.-C., Lei, S.-M., Cheng, H.-H., Wang, J.-D.: BSRS-5 (5-item Brief Symptom Rating Scale) scores affect every aspect of quality of life measured by WHOQOL-BREF in healthy workers. Q. Life Res. **20**, 1469–1475 (2011)
25. Maslach, C., Jackson, S.E., Leiter, M.P., Schaufeli, W.B., Schwab, R.L.: Maslach Burnout Inventory. Consulting Psychologists Press, Palo Alto (1986)
26. Bianchi, R., Schonfeld, I.S., Laurent, E.: Burnout–depression overlap: a review. Clin. Psychol. Rev. **36**, 28–41 (2015)
27. Belle, D.: Gender differences in the social moderators of stress. In: Monat, A., Lazarus, R.S. (eds.) Stress and Coping: An Anthology, pp. 257–277. Columbia University Press, New York (1991)
28. Wallace, J.E., Lemaire, J.B., Ghali, W.A.: Physician wellness: a missing quality indicator. Lancet **374**, 1714–1721 (2009)
29. Gunter, T.: Physician death by suicide: problems seeking stakeholder solutions (2016)
30. Sorensen, G., Jacobs, D.R., Pirie, P., Folsom, A., Luepker, R., Gillum, R.: Relationships among type a behavior, employment experiences, and gender: The Minnesota heart survey. J. Behav. Med. **10**, 323–336 (1987)

Classification of Physical Exercise Intensity Based on Facial Expression Using Deep Neural Network

Salik Ram Khanal[1](✉) , Jaime Sampaio[1,2] , Joao Barroso[1,3] ,
and Vitor Filipe[1,3]

[1] Universidade de Trás-os-Montes e Alto Douro, Vila Real, Portugal
{salik,ajaime,jbarroso,vfilipe}@utad.pt
[2] Institute for Systems and Computer Engineering, Technology and Science
(INESC TEC), Porto, Portugal
[3] Research Center in Sports Sciences,Health Sciences and Human Development
(CIDESD), Vila Real, Portugal

Abstract. If done properly, physical exercise can help maintain fitness and health. The benefits of physical exercise could be increased with real time monitoring by measuring physical exercise intensity, which refers to how hard it is for a person to perform a specific task. This parameter can be estimated using various sensors, including contactless technology. Physical exercise intensity is usually synchronous to heart rate; therefore, if we measure heart rate, we can define a particular level of physical exercise. In this paper, we proposed a Convolutional Neural Network (CNN) to classify physical exercise intensity based on the analysis of facial images extracted from a video collected during sub-maximal exercises in a stationary bicycle, according to standard protocol. The time slots of the video used to extract the frames were determined by heart rate. We tested different CNN models using as input parameters the individual color components and grayscale images. The experiments were carried out separately with various numbers of classes. The ground truth level for each class was defined by the heart rate. The dataset was prepared to classify the physical exercise intensity into two, three, and four classes. For each color model a CNN was trained and tested. The model performance was presented using confusion matrix as metrics for each case. The most significant color channel in terms of accuracy was Green. The average model accuracy was 100%, 99% and 96%, for two, three and four classes classification, respectively.

Keywords: Physical exercise intensity · Convolutional neural network · Heart rate

1 Introduction

Physical exercise reduces the risk of developing and/or dying from cardiovascular disease by maintaining various types of physiological parameters (Heart rate, blood pressure etc.) and blood components (Blood sugar, cholesterol, triglycerides, etc.). It enhances and maintains physical fitness, increases muscle strength, reduces

M. Antona and C. Stephanidis (Eds.): HCII 2019, LNCS 11573, pp. 455–467, 2019.
https://doi.org/10.1007/978-3-030-23563-5_36

depression and anxiety, and reduces various types of diseases [1, 2]. The benefits from physical exercise could be increased by proper monitoring in real time [3]. Exercises for the elderly and for rehabilitation may lead to many accidents, which may be caused by lack of proper monitoring of the exercise in real time [4]. Measurement of intensity (how hard it feels for the person to perform exercise) can be performed in subjective or objective ways. Basically, there are three ways of monitoring physical exercise intensity: by extracting or monitoring physiological parameters, such as heart rate (HR), respiratory rate (RR) etc., the rated perceived exertion scale, and the talk test (how hard it is for a subject to talk). Usually, physical exercise intensity is synchronous to heart rate; therefore, if we measure heart rate, we can define a level of physical exercise.

In general, physical exercise intensity is considered "very light" at the beginning and "very hard" at the end of the physical exercise in a submaximal graded exercise. The intensity of the exertion depends on many overall body responses, including heart rate (HR), respiratory rate (RR), blood lactate, physical status, mood state, etc. Therefore, a proper measurement or estimation of these parameters during exercise helps to monitor the physical exercise.

Borg scale is a common way to classify the feeling of hardness during physical exercise [5]. He proposed a subjective technique to classify the perceived exertion during exercise which is called rate of perceived exertion (RPE). The measurement of these parameters during exercise is quite challenging because the feeling is based on the individual, and the subject must be familiar with this scale, making the measurement of these levels quite difficult for people who do not have enough knowledge of the Borg scale. Nowadays, monitoring of physical exercise can be carried out by extracting physiological features, using invasive or non-invasive techniques. The subjective way of defining the level of exercise intensity has been used for a long time and has considerable validity [5]. There are several instruments using invasive techniques or contact-sensor technology to measure physiological signals, including heart rate, respiratory rate, and blood lactate etc. [6, 7]. By measuring these parameters, we can correlate them with physical exercise intensity level. The non-invasive technique to identify/classify exercise intensity can consist of measuring physiological data and converting it into exercise intensity level or classes, or directly recognizing exercise intensity using computer vision technique. It is commonly observed that if the person gets tired his/her facial expression and facial color changes, which could be an important cue for the classification of level of intensity.

Most of the recent research trends for measurement of physical exercise intensity include facial image analysis using feature points analysis [8, 9], facial color analysis, mouth and eye blink analysis [10], body movement tracking [11], etc. In the literature, we can find various ways to measure physical exercise intensity using non-invasive methods. Fatigue can be detected by analyzing the pattern of movement of the muscles [12]. The head motion pose analysis can be measured using feature points tracking, which can be analyzed using statistical and machine learning algorithms [8]. Haque [12] presented an efficient non-contact system for detecting non-localized physical fatigue from maximal muscle activity using facial videos, where the video was taken in a realistic environment. Salik [9] proposed exercise intensity classification using facial feature point analysis.

In computer vision, deep learning is an emerging area to classify images. The classification of physical exercise intensity is more likely to use facial expression analysis since the facial expression changes when a person feels a higher intensity in the exercise [9]. Nowadays, facial emotion analysis using deep learning techniques is also very common and has achieved better results than traditional machine learning techniques [13–16]. Deep learning can also be implemented to analyze or monitor physical exercise by analyzing body parameters. Gordienko [17] proposed a multi-modal approach to estimate the fatigue using deep learning techniques, where the input parameters were extracted using wearable sensors.

The exercise intensity level has been classified using several subjective techniques for a long time, but using objective techniques to achieve this task is still challenging. In this paper, an objective or quantitative technique is proposed to classify exercise intensity using computer vision technique. The ground truth class/level of exercise intensity was defined according to the incremental HR. The intensity level of (class) exercise at the beginning or minimum HR is the initial class 'light' and at the end of exercise, the maximum HR is the final class or 'hard' level. The other classes or levels are also defined by the HR. The deep learning approach using convolutional neural network was applied to classify the facial images, where images were extracted from a video collected during submaximal exercises.

2 Methods

2.1 Dataset Description

Twenty university students (mean age = 26.88 ± 6.01 years, mean weight = 72.56 ± 14.27 kg, mean height = 172.88 ± 12.04 cm, 14 males and six females, and all white Caucasian) participated in the study. An informed consent form was signed by each participant prior to data collection and they were informed of the study protocol before the recordings. The test consisted of a submaximal ramp exercise protocol in a Wattbike Cycloergometer (Wattbike Ltd, Nottingham, UK), after a 5-min warmup with a constant power output of 60 W. The initial power output was 75 W, which was increased by 15 W min-1 until participants reached 85% of their maximal heart rate (calculated as 208 – (0.7*age) or until they were unable to maintain cadence to generate the required power output throughout this stage. Heart rate data was collected at 100 Hz using the Polar T31 cardiofrequencimeter, (Polar Electro, Kempele, Finland) synchronized to the Wattbike load cell for power output measures, sampled at 100 Hz. For the facial tracking, facial video (25 Hz with spatial resolution of 1080 × 1920 pixels) was recorded during the test using a video camera placed in (90° angle with face and camera) the frontal plane view to capture the participants' face while performing the exercise. The participants were not allowed to talk during the test but could express their feelings freely with facial expression throughout.

For the purpose of this study, a dataset containing various classes of images with different levels of tiredness was prepared. The image frames were manually assigned the categories, accordingly to the heart rate. Considering two classes, the initial 500 frames of each video were considered as class one (not tired faces) and the last 500 frames were considered as class two (tired faces). Since there were 20 subjects, the total number of images for a class were 10,000 and the total number of images in the dataset was 10,000 times the number of classes. The dataset was prepared for two, three, and four classes separately (see Fig. 1).

The allocation of time slots in a video was based on the incremental HR value. In the case of more than two class classification, the middle classes are considered according to the Heart rate value. For instance, if the minimum heart rate is 80 and the final heart rate is 180, then the image frames at the time of 130 bpm are considered as second class or middle class. Likewise, the time slots for more classes are considered by synchronizing the heart rate with the frame number.

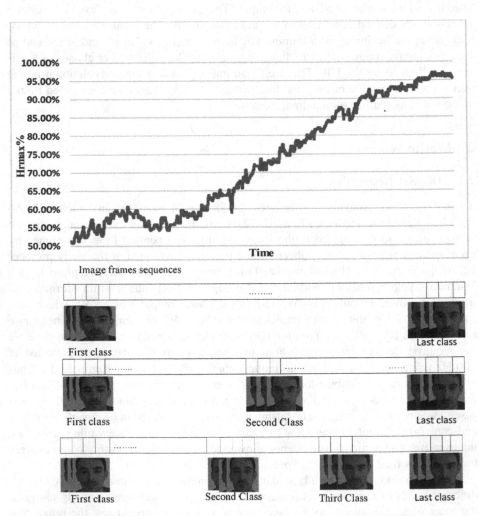

Fig. 1. Allocation of time slots to extract images with the initial class (Minimum exercise intensity), the intermediate classes, and the final class (Maximum exercise intensity).

2.2   Pre-processing

Before feeding the neural network with inputs, various image pre-processing techniques were applied. All the pre-processing before the neural network is shown in Fig. 2. Since the images were extracted from the video recorded with a moving object (head movement), the frames extracted may have some blurred effects. Therefore, the first pre-processing consists of detecting and removing any blurring effect on the image frame. In the second step, the face was detected in the frame so that we could specifically analyze the face, not the whole frame. The well-known Viola Jones algorithm [18] was applied to detect the face. After detecting the face, we cropped it and down-sampled it into 96 × 96, which is the size of the input layer. One of the basic purposes of this research is to find out the best color channel to classify the physical exercise; therefore, the experiments were performed with separate raw 2D image representing each color channel (Red, Green and Blue) and Grayscale. So, after cropping the face and resizing, RGB frames were split into R, G, B, and converted to Grayscale image.

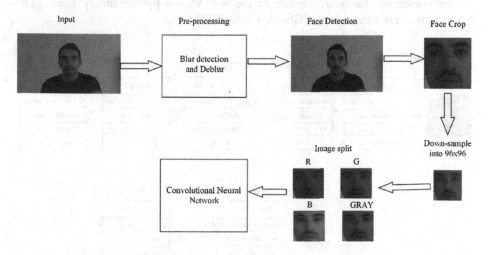

Fig. 2. Block diagram with detail preprocessing with output images.

2.3   Proposed CNN Architecture

A deep neural network based on the Convolutional Neural Network (CNN) or ConvNet was designed with five hidden layers and two fully connected layers, as shown in Fig. 3. Three main types of layers are used to build ConvNet architectures: Convolutional Layer, Pooling Layer, andFully-Connected Layer (exactly as seen in regular Neural Networks). These layers were stacked to form a full ConvNetarchitecture:

- Input Layer [96 × 96]: holds the raw pixel values of 2D the image of the faces.
- CONV layer: computes the output of neurons that are connected to local regions in the input, each computing a convolution of their weights, as well as a small region they are connected to in the input volume.
- Fully-connected layer: computes the class scores, resulting in volume of size [1 × 1 × n], where n is the number of classes. As with ordinary Neural Networks, and as the name implies, each neuron in this layer will be connected to all the numbers in the previous layer.
- The activation function chosen was ReLU.
- Maxpooling with Pool size (2, 2).
- 25% dropout results in the maximum amount of regularization.

The first part of each layer consists of a convolutional layer (Conv2d) which can have spatial batch normalization, Maxpooling, dropout and ReLU activation. Each layer consists of these five tasks. After 5 convolutional layers, the network is led to 2 fully connected layers that always have Affine operation and ReLU activation.

We implemented this architecture in the well-known python library Keras. The experiments were carried out in Google Colab GPU.

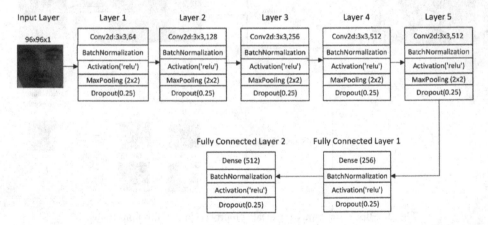

Fig. 3. Proposed convolutional neural network architecture.

2.4 Experiments

The first convolutional layer consists of 64 3 × 3 filters; the second one had 128 3 × 3 filters; the third one had 256 3 × 3; the forth one had 512 3 × 3 filters; and the last one also had 512 3 × 3 filters. In all the hidden layers a stride size of 1, batch normalization, max-pooling of size 2 × 2, dropout of 0.25 and ReLU as the activation function. These five hidden layers are followed by two fully connected layers with 256 neurons and 512 neurons respectively. Both the fully connected layers had batch normalization, dropout and ReLU with the same parameters. SoftMax is also used as an out-loss function. Figure 3 shows our deep neural network architecture.

The training was performed in 75 epochs with the batch size of 64. From the dataset of 10,000 images of each class, the dataset was randomly split into training, validation and testing set in the ratio of 80:10:10. For two classes (tired and not tired) the total number of images was 20,000, where 16,000 were for training, 2,000 for validation, and 2,000 for testing. Experiments with two, three, and four classes were also performed. To reduce overfitting, we used dropout and batch normalization in addition to L2 regularization.

3 Experimental Results and Discussion

Separate experiments were done in order to determine the classification into two, three, and four classes and the accuracy of each case was analyzed. In the experiments, the color images were split into Red, Green and Blue components and the original RGB images were converted into Grayscale. The green color component provides the best accuracy of classification. The confusion matrix was drawn in each case. Most of the cases, the accuracy of classification using a two-class classification is more than 99% (See Table 1). From the overall results, classification into two and three classes was accurate and resulted in very high classification accuracy, whereas the classification with four classes had a lower classification performance in each algorithm.

Table 1. The average accuracy of classification into two, three, and four classes, using red, green, blue and gray channels.

Color component	Two classes	Three classes	Four classes
Red	100	99.86	95.60
Green	100	99.86	99.75
Blue	100	99.93	97.47
GRAY	100	99.76	99.20

Based on the result presented on the table, the classification into two classes has 100% of accuracy in all the cases. It also shows that the best raw color channel is Green which obtained the average accuracy of 100%, 99.86%, and 99.75% in two, three, and four class classification, respectively. From these results it is concluded that the level of tiredness or physical exercise intensity is better reflected by the Green color channel. Therefore, in the remaining part of this article, all the experimental results and slots will only be based on the Green color channel.

The accuracy and loss history during training 75 epoch is shown in Fig. 4(a) and (b) respectively. Only the plot of the Green channel is shown, since Green channel resulted the best average prediction accuracy among all other color channels (Tables 2, 3 and 4).

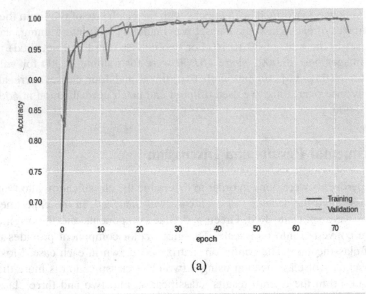

(a)

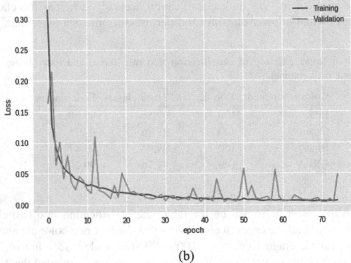

(b)

Fig. 4. Training and validation accuracy and loss vs. epoch of green color for four class classification. (Color figure online)

Table 2. Classification accuracy of each class in the classification of physical exercise intensity into two classes.

Color component	Classes	Accuracy	Average
R	C1	100	100
	C2	100	
G	C1	100	100
	C2	100	

(*continued*)

Table 2. (*continued*)

Color component	Classes	Accuracy	Average
B	C1	100	100
	C2	100	
GRAY	C1	100	100
	C2	100	

Table 3. Classification accuracy of each class in the classification of physical exercise intensity into three classes.

Color component	Classes	Accuracy	Average
R	C1	99.7	99.86
	C2	100	
	C3	99.9	
G	C1	99.9	99.86
	C2	99.7	
	C3	100	
B	C1	99.8	99.93
	C2	100	
	C3	100	
GRAY	C1	99.4	99.76
	C2	99.9	
	C3	100	

Table 4. Classification accuracy of each class in the classification of physical exercise intensity into four classes.

Color component	Classes	Accuracy (%)	Average
R	C1	89.6	95.6
	C2	98.9	
	C3	95.2	
	C4	99.7	
G	C1	99.9	99.75
	C2	99.2	
	C3	99.9	
	C4	100	
B	C1	93.9	97.47
	C2	99.5	
	C3	97.8	
	C4	98.9	
GRAY	C1	98.7	99.20
	C2	98.5	
	C3	99.6	
	C4	100	

The confusion matrixes of the green color channel in all the class classifications are presented as shown in Figs. 5, 6 and 7. The accuracy for the two-class classification is 100% and it shows that when using convolutional neural network, it is very easy to classify normal and fully tired face. The test set contains randomly selected 2000 images, where 1037 images were normal faces and 963 images were tired faces.

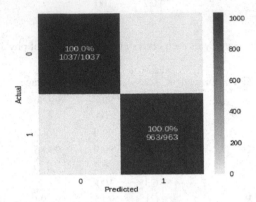

Fig. 5. Confusion matrix of the two-class classification of physical exercise intensity.

Similarly, the confusion matrix of the three-class classification is shown in Fig. 6. In this case, the misclassification is only for the first and second classes. The last class is 100% accurate. None of the other classes are classified into this class, nor this class is classified into other class. In the case of class one, among 1025 images, only one class

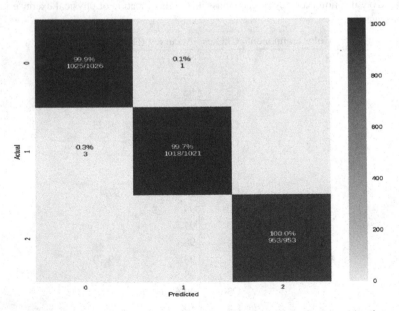

Fig. 6. Confusion matrix of the three-class classification of physical exercise intensity.

was misclassified as class two. Likewise, in the case of class two, three images out of 1018 were misclassified as class one.

Likewise, the recognition of fully tired faces was easier when compared to the others. The misclassification rate was always greater for the nearest class. For example, the first class, or normal faces (not tired), are mostly misclassified into second class, second class is misclassified into first class and third class, and so on.

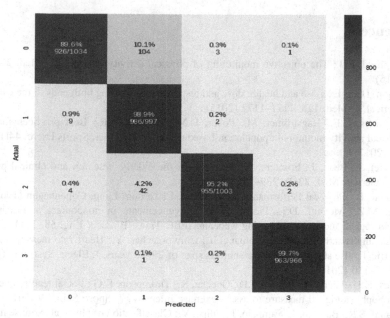

Fig. 7. Confusion matrix of the four-class classification of physical exercise intensity.

4 Conclusion

Based on various experiments with various types of image datasets, the deep learning approach for exercise intensity classification based on facial expression can be a potential method to classify exercise intensity in two, three, four or more levels. In case of a two-class classification, the accuracy rate is 100% and even for the three-class classification it is also around 99%. From all the experiments, it can be concluded that the best color channel for the raw input image is Green, in terms of its classification accuracy. The training and testing dataset were randomly prepared from the same subjects; therefore, this approach is more appropriate for personalized physical exercise monitoring.

Future work can be extended two classify images into more than four classes. The experiments were done with only 20 subjects with little diversity in age and origin. To generalize this model, we can train the model with more diversity and a greater number of subjects, in order to improve the test accuracy. Considering that the training and testing were performed from the same subjects, this approach might be more appropriate for personalized exercise monitoring systems, where the system can be trained from the same subject with image datasets taken in various exercise sessions.

Acknowledgement. This article is a result of the project INOV@UTAD, NORTE-01-0246-FEDER-000039, supported by Norte Portugal Regional Operational Programme (NORTE 2020), under the PORTUGAL 2020 Partnership Agreement, through the European Regional Development Fund (ERDF).

References

1. Shephard, R.J.: The objective monitoring of physical activity. Prog. Prev. Med. **2**(4), 1–7 (2015)
2. Simon, H.B.: Exercise and health: dose and response, considering both ends of the curve, the americal. J. Med. **128**, 1171–1177 (2015)
3. Matthews, C.E., Hagströmer, M., Pober, D.M., Bowles, H.R.: Best practices for using physical activity monitors in population-based research. Med. Sci. Sports Exerc. **44**(1), s68–s76 (2012)
4. Frankel, J., Bean, J., Frontera, W.: Exercise in the elderly: research and clinical practice. Clin. Geriatr. Med. **22**(2), 239–256 (2005)
5. Gunnar, B.: Physical Performance and Perceived Exertion. Lund, Copenhagan (1962)
6. Poh, M.Z., McDuff, D.J., Picard, R.W.: Advancements in noncontact, multiparameter physiological measurements using webcam. IEEE Trans. Biomed. Eng. **58**, 7–11 (2011)
7. Chirakanphaisarn, N., Thongkanluang, T., Chiwpreechar, Y.: Heart rate measurement and electrical pulse signal analysis for subjects span of 20–80 years. J. Electr. Syst. Inf. Technol. **5**, 112–120 (2016)
8. Miles, K.H., Clark, B., Périard, J.D., Goecke, R., Thompson, K.G.: Facial feature tracking: a psychophysiological measure to assess exercise intensity? J. Sport. Sci. 1–9 (2017)
9. Khanal, S.R., Barroso, J., Sampaio, J., Filipe, V.: Classification of physical exercise intensity by using facial expression analysis. In: 2018 Second International Conference on Computing Methodologies and Communication (ICCMC), Erode, India (2018). https://doi.org/10.1109/ICCMC.2018.8488080
10. Khanal, S.R., Fonseca, A., Marques, A., Barroso, J., Filipe, V.: Physical exercise intensity monitoring through eye-blink and mouth's shape analysis. In: 2nd International Conference on Technology and Innovation in Sports, Health and Wellbeing (TISHW), Thessaloniki, Greece (2018). https://doi.org/10.1109/TISHW.2018.8559556
11. Schmal, H., Holsgaard-Larsen, A., Izadpanah, K., Brønd, J.C., Madsen C.F., Lauritsen, J.: Validation of activity tracking procedures in elderly patients after operative treatment of proximal femur fractures. Rehabil. Res. Pract., 1–9 (2018). Article ID 3521271
12. Haque, M.A., Irani, R., Nasrollahi, K., Moeslund, T.B.: Facial video based detection of physical fatigue for maximal muscle activity. IET Comput. Vis. **10**, 323–330 (2016)
13. Alizadeh, S., Fazel, A.: Convolutional Neural Networks for Facial Expression Recognition, CoRR, vol. abs/1704.06756, pp. 1–8 (2017)
14. Burkert, P., Trier, F., Afzal, M.Z. Dengel, A., Liwicki, M.: DeXpression: Deep Convolutional Neural Network for Expression Recognition, CoRR, vol. abs/1509.05371 (2015
15. TeixeiraLopes, A., Aguiar, E.D., Souza, A.: Facial expression recognition with convolutional neural networks. Pattern Recognit. **61**, 610–628 (2017)

16. Sang, D.V., Dat, N.V., Thuan, D.P.: Facial expression recognition using deep convolutional neural networks. In: 9th International Conference on Knowledge and Systems Engineering (KSE), Hue, Vietnam (2017)
17. Gordienko, Y., Stirenko, S., Kochura, Y., Alienin, O., Novotarskiy, M., Gordienko, N.: Deep Learning for Fatigue Estimation on the Basis of Multimodal Human-Machine Interactions, CoRR, vol. abs/1801.06048, pp. 1–12 (2017)
18. Viola, P., Jones, M.: Robust real-time object detection. Int. J. Comput. Vis., 1–25 (2001)

Effect of Differences in the Meal Ingestion Amount on the Electrogastrogram Using Non-linear Analysis

Fumiya Kinoshita[✉], Kazuya Miyanaga, Kosuke Fujita, and Hideaki Touyama

Toyama Prefectural University,
5180 Kurokawa, Imizu-shi, Toyama 939-0398, Japan
f.kinoshita@pu-toyama.ac.jp

Abstract. This paper reports a study of the impact of differences in meal ingestion amount on electrogastrograms. The study was performed by recording an electrogastrogram and an electrocardiogram of eight young men for 60 min once before and once after each test subject ingested meals with ingestion amounts of 800 kcal and 400 kcal. The results showed that meal ingestion affected the power spectral density of the tachygastria range (3.7–5.0 cpm) by significantly increasing its value after the meal. The differences in meal ingestion amount are expressed in the power spectral density of the colon range (6.0–8.0 cpm) by significantly increasing its value after the meal, but only when the subject ingested an 800-kcal meal.

Keywords: Electrogastrogram (EGG) · Maximum entropy method (MEM) · Non-linear analysis · Wayland algorithm

1 Introduction

The aging of the global population is accompanied by a steady increase in expectations regarding the health care field. The rate of aging in Japan is among the highest in the world, attracting companies' interest in medical treatment and health care from various foreign countries. A problem facing aging societies is a rise in the number of elderly people requiring nursing care, accompanied by the rising cost of medical treatment and burden of nursing care. Thus, the problem of how to reduce the number of elderly people who require nursing care is one challenge that must be overcome quickly. In addition, abnormal peristaltic movement of the gastrointestinal tract is related to a decline in physical and psychological quality of life, so early discovery and prompt treatment of abnormalities in peristaltic movement is an effective way to overcome this challenge [1].

One type of alimentary canal obstruction movement function investigation is an electrogastrogram (EGG), which is a noninvasive low-constraint measurement of the electrical activity of the gastrointestinal tract performed from the body surface [2, 3]. Regular electrical activities that repeatedly electrically depolarize and repolarize are seen in the stomach and intestines, just as in the heart. Pacemaker cells that control the

© Springer Nature Switzerland AG 2019
M. Antona and C. Stephanidis (Eds.): HCII 2019, LNCS 11573, pp. 468–476, 2019.
https://doi.org/10.1007/978-3-030-23563-5_37

electrical activity of the stomach are in the top 1/3 of the greater curvature of the stomach, and from there, electrical activity in the body is transmitted toward the pyloric region at a rate of three waves per minute (three cycles per minute (cpm)). The pacemaker cells are governed by parasympathetic nervous system activities but cause spontaneous cyclical electrical activities. This is a result of a network of islands called interstitial cells of Cajal (ICCS) [4–7]. This electrical activity is categorized as either electrical response activity (ERA), which is accompanied by peristaltic movement, or electrical control activity (ECA), which is not accompanied by peristaltic movement [8], but an electrogastrogram cannot distinguish between these, so peristaltic movement is not directly recorded [9]. However, an electrogastrogram definitely records the electrical activity of the stomach [10], and it is thought to be possible to discover abnormal peristaltic movement based on a response confirmation test.

In the case where it is difficult to visually assess the waveform shown on an electrogastrogram, it is necessary to perform spectral analysis of the waveforms shown on the electrogastrogram. The dominant frequency of the electrogastrogram obtained by spectral analysis is considered to be 2.4–3.7 cpm, with those smaller than 2.4 cpm considered to be bradygastria, and those larger than 3.7 cpm considered to be tachygastria.

Response confirmation of an electrogastrogram during meal ingestion often uses enteral nutrition or solid nutritional supplements, etc., and from 5 to 15 min after meal ingestion by a healthy person, its period falls to between 0.2 and 0.5 and its amplitude increases by 1 to 3 times. This response is called "postprandial dip" [11, 12]. Regarding postprandial dip, there is no reference to its impact in frequency ranges outside of the dominant frequency, and the specific ingestion amount that causes postprandial dip is not clear. This research studied the impact of differences in meal ingestion amount on electrogastrograms. A study comparing a measured electrogastrogram with the numerical solution obtained from a stochastic differential equation the authors propose was also performed to determine the impact of differences in meal ingestion amount on the mathematical model of an electrogastrogram.

2 Experimental Method

2.1 Subjects and Materials

The subjects were eight young men between 21 and 28 years of age (average ± standard differential: 22.8 ± 1.4) who had no previous history nor presented symptoms of digestive disease. The experiment was first fully explained to the subjects, and their agreement to participate in the experiment was obtained by having them enter answers on a form that presented the purpose and significance of the research, guarantee of privacy, handling of data. The electronic data obtained in the experiment were recorded using aliases that cannot be linked to the test subjects, and the experiment obtained the approval of the Ethics Committee of Toyama Prefectural University.

2.2   Protocol

The experiments were performed by recording electrogastrograms and electrocardiograms for 60 min with the subjects resting in a supine position before and after meal ingestion. The meal load used solid portable foods, and the test subjects ingested 800-kcal meals and 400-kcal meals. The measurements of each test subject were performed on different days, and considering the effects of sequence, the sequences were set randomly.

The electrogastrograms were acquired using disposable electrodes used to perform electrocardiograms (blue sensors, METs), applied as shown in Fig. 1. The electrodes were applied after skin resistance was adequately lowered using disinfection-use ethanol. Each electrogastrogram was recorded using bipolar leads, amplification was performed by a biological amplifier (Biotop mini, East Medic Corporation), and the data were recorded on a PC using an analog input/output card (ADA16-32/2(CB)F, CONTEC). The biological amplifier measurement conditions were a low pass filter of 0.02 Hz and a high pass filter of 0.5 Hz. To unify the time food was retained in the stomachs of all the subjects, they were ordered not to eat for two hours prior to the start of the experiment.

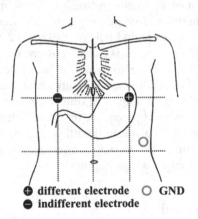

⊕ different electrode   ○ GND
⊖ indifferent electrode

Fig. 1. Electrode positions.

2.3   Analytical Indices

In this study, time series data were obtained by A/D conversion of the electrogastrograms and electrocardiograms, which were recorded at 1 kHz. In the electrogastrogram time series, to remove electrical noise caused by the foreign electromyogram or electronic equipment, a band pass filter with a cutoff frequency from 0.015–0.15 Hz was applied to the time series data that were obtained. In addition, because the normal frequency of an electrogastrogram has a relatively slow fluctuation of approximately 3 cpm, the 1-kHz electrogastrogram time series was resampled at 1 kHz. This research applied running spectral analysis to analyze the time series that was obtained. The electrogastrogram time series was shifted and divided between 1,024 points (approx.

17 min) time windows at 300 point intervals, and each was analyzed. In this paper, among analysis intervals, divided start times are entered as the representative value of each analysis interval, as shown below.

The divided electrogastrogram time series used a translational error estimated statistically according to a Wayland algorithm in addition to frequency analysis [13, 14]. The translational error (Etrans) estimated according to the Wayland algorithm is an index that quantitatively evaluates the smoothness of the track of an attractor embedded in topological space. If the track of an attractor reconstituted in embedded space is smooth, the time series has determinism. If the translational error is a positive value and is close to zero, and the numerical model that forms the time series is deterministically large, it can be considered probabilistic. When the object has Brownian motion, in particular, the value of the translational error is estimated to be 1.

The frequency analysis of each electrogastrogram time series was accomplished by performing the maximum entropy method (MEM) for each of the divided time histories and employing frequency analysis using a stochastic method [15, 16]. As the spectrum calculation method for use in the MEM, algorithms based on the Yule-Walker method and Burg method have been proposed, but for this study, the Burg method was adopted because it permits stable spectrum estimation with high resolution, even using sparse data. In this paper, the smallest value for which the autocorrelation function is $1/e$ or lower (correlation is considered to be zero) was defined as the lag value of the auto-correlation function. From the power spectrum that was calculated, the power spectral density (PSD) in these frequency bands was calculated focusing on bradygastria (1.1–2.4 cpm) and tachygastria (3.7–5.0 cpm) [17]. It has been reported that fluctuation of approximately 7 cpm in an electrogastrogram reflects the electrical activity of the colon [18], so PSD was also calculated for the 6.0–8.0 cpm range.

The electrocardiograms recorded at the same time as the electrogastrograms were analyzed based on heart rate variability (HRV). HRV can quantify indices of the sympathetic nervous system and the parasympathetic nervous system by analyzing the RR interval, which is the stroke of the heart from the time range and frequency range. The RR interval time series that was abstracted was divided by shifting time windows at 512 points at 300-s intervals, and each was analyzed, corresponding to the divided times of the electrogastrogram time series. In this paper, with the low-frequency component LF of PSD as 0.04–0.15 Hz, and the high-frequency component HF of PSD as 0.15–0.4 Hz, HF, which represents the activity index of the parasympathetic nerves, and LF/HF, which represent the activity index of the sympathetic nerves, were separately calculated.

For the calculated analysis indices, the average values recorded for each time before meal ingestion and for each time after meal ingestion were compared using the Wilcoxon signed-rank test (this study used a significance level of .05).

3   Results

Figure 2 shows typical examples of electrogastrogram waveforms from 10 min to 20 min after start of measurement for the same test subject. Regardless of the meal ingestion amount, the measured waveform after meal ingestion has a higher amplitude than that before meal ingestion. In addition, the measured waveform when 400 kcal was ingested was compared with the measured waveform when 800 kcal was ingested, showing a clear change of 3 cpm. In the measured waveform when 800 kcal was ingested, on the contrary, the high-frequency component is irregularly superimposed by a change of approximately 3 cpm.

PSD and translational error in each range—bradygastria (1.1–2.4 cpm), tachygastria (3.7–5.0 cpm), and colon (6.0–8.0 cpm)—were calculated for divided electrogastrogram time series. The results for 800-kcal ingestion are shown in Fig. 3, and the results for 400-kcal ingestion are shown in Fig. 4. In the bradygastria range, consistent trends before and after meal ingestion, regardless of meal ingestion amount, were not seen (Figs. 3A and 4a), but in the tachygastria range, the value after meal ingestion was significantly higher than it was before meal ingestion (Figs. 3b, and 4b). This trend appeared beginning 5 min after meal ingestion and was sustained until 35 min after ingestion. In the colon range, when 400 kcal was ingested, no significant change was seen after meal ingestion; when 800 kcal was ingested, the value after meal ingestion was significantly higher than it was before meal ingestion (Figs. 3c and 4c). In the results for translational error, on the contrary, no consistent trend, regardless of the meal ingestion amount, was seen before or after meal ingestion (Figs. 3d and 4d).

Next, the electrocardiograms obtained at the same time as the electrogastrograms were analyzed. When a 400-kcal meal was ingested, both HF and LF/HF showed no significant difference before or after meal ingestion (Figs. 5b and 5d). When an 800-kcal meal was ingested, on the contrary, the HF was significantly reduced, and LF/HF tended to increase significantly. While HF showed this trend at all times after meal ingestion (Fig. 5a), LF/HF increased significantly from 0–20 min. Beginning at 25 min, it disappeared (Fig. 5c).

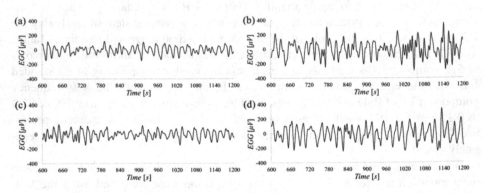

Fig. 2. Typical EGGs for one participant: (a) before 800 kcal meal, (b) after 800 kcal meal, (c) before 400 kcal meal, and (d) after 800 kcal meal.

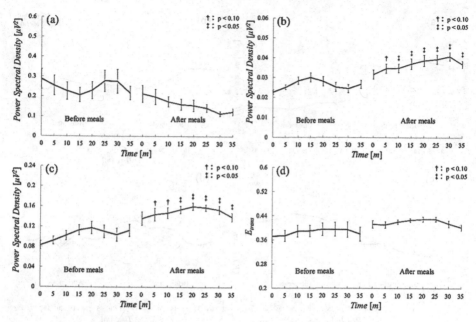

Fig. 3. Average PSD and E_{trans} (mean ± SE) for an 800-kcal meal: (a) 1.1–2.4 cpm, (b) 3.7–5.0 cpm, (c) 6.0–8.0 cpm, and (d) E_{trans}.

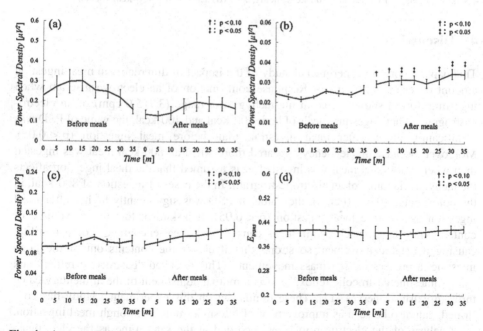

Fig. 4. Average PSD and E_{trans} (mean ± SE) for a 400-kcal meal: (a) 1.1–2.4 cpm, (b) 3.7–5.0 cpm, (c) 6.0–8.0 cpm, and (d) E_{trans}.

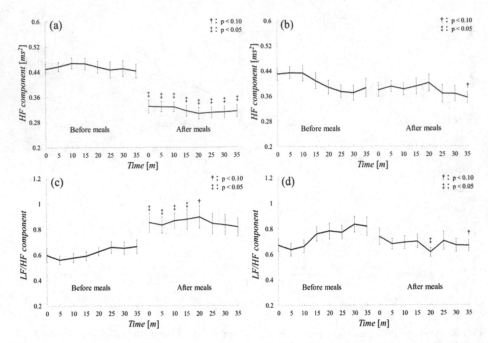

Fig. 5. AverageHF and LF/HF components (mean ± SE): (a) HF for 800-kcal meals, (b) HF for 400-kcal meals, (c) LF/HF for 800-kcal meals, and (d) LF/HF for 400-kcal meals.

4   Discussion

This paper reports an experimental study of the impact of differences in meal ingestion amount on electrogastrograms. Response confirmation of an electrogastrogram when ingesting a meal showed that, in the tachygastria range (3.7–5.0 cpm) of an electro-gastrogram, when ingesting meals of both 800 kcal and 400 kcal, the value of PSD was significantly higher after meal ingestion than before meal ingestion ($p < 0.05$). Moreover, because this tendency appeared more when an 800-kcal meal was ingested than when a 400-kcal meal was ingested, it is assumed that the meal ingestion affects the tachygastria range of an electrogastrogram. In the cases of ingestion of 800 kcal, in the colon range (6.0–8.0 cpm), the value of PSD was significantly higher after meal ingestion than before meal ingestion ($p < 0.05$). It is known that the movement of contents into the stomach or intestines caused by a meal encourages colon motility, causing peristaltic movement so strong that it cleans the contents out of the large intestine, a process called "mass movement." This is called "gastrocolic reflex" and "small intestine gastrocolic reflex," and is a mutual adjustment of the alimentary canal through extrinsic nerves [17]. These trends, which this experimental study has con-firmed, can also be seen as improvement of intestinal motility through meal ingestion.

Analysis of the electrocardiograms recorded at the same time as the electrogas-trograms confirmed that after ingestion of 800-kcal meals, HF declines significantly and LF/HF increases significantly. When ingesting a meal, the flow of blood into the abdomen increases and alimentary canal hormones expand blood vessels to improve gastrointestinal tract activity. The tendency shown by heart rate fluctuation analysis in 800-kcal ingestion cases can be seen in the restriction of improvement of sympathetic nerve activity and pneumogastric nerve activity by the actions that maintain apparent blood pressure as a result of the decline of peripheral vascular resistance throughout the body caused by the expansion of blood vessels [19, 20]. This is assumed to ensure improvement of activity in the colon range on the electrogastrograms after meal ingestion.

5 Conclusion

This paper reports a study of the impact of differences in meal ingestion amount on electrogastrograms. The results show that the meal ingestion affects the power spectral density in the tachygastria range of an electrogastrogram, significantly increasing its value after meal ingestion above its value before meal ingestion. Differences in meal ingestion amount also affect the power spectral density in the colon range, increasing its value significantly after meal ingestion above its level before meal ingestion only when an 800-kcal meal is ingested. These trends, which were confirmed by the electrogastrograms, can be seen as an improvement in intestinal activity by meal ingestion. In the future, we will aim to construct evidence that will be of use in early discovery of abnormal peristaltic movement of the gastrointestinal tract based on a multifaceted discussion of the complexity in the generators of an electrogastrogram using analysis by a mathematical model.

References

1. Aro, P., Talley, N.J., Agréus, L., et al.: Functional dyspepsia impairs quality of life in the adult population. Aliment. Pharmacol. Ther. 33(11), 1215–1224 (2011)
2. Alvarez, W.C.: The electrogastrogram and what is shows. J. Am. Med. Assoc. 78, 1116–1119 (1922)
3. Kenneth, L.K., Robert, M.: Handbook of Electrogastrography. Oxford University Press, Oxford (2004)
4. Nakamura, E., et al.: Cellular mechanism of spontaneous activity of the stomach smooth muscle. Nihon Yaku-rigaku Zasshi 123(3), 141–148 (2002)
5. Torihashi, S.: Structure and functions of the Cajal cells. Pediatr. Surg. 37(4), 467–472 (2005)
6. Takayama, I., Horiguchi, K., Daigo, Y., Mine, T., Fujino, M.A., Ohno, S.: The interstitial cells of Cajal and a gastroenteric pacemaker system. Arch. Histol. Cytol. 65(1), 1–26 (2002)
7. Thomsen, L., et al.: Interstitial cells of Cajal generate a rhythmic pacemaker current. Nature Med. 4, 848–851 (1998)
8. Smout, A.J.P.M., Van Der Schee, E.J., Grashuis, J.L.: What is measured in Electrogastrography? Dig. Dis. Sci. 25(3), 179–187 (1980)
9. Chen, J.Z., McCallum, R.W.: Electrogastrography: Principles and Applications. Raven Press, Ely (1994)
10. Pezzolla, F., Riezzo, G., Maselli, M.A.: Electrical activity recorded from abdominal surface after gastrectomy or colectomy in humans. Gastroenterology 97(2), 313–320 (1989)

11. Sakakibara, Y., Asahina, M., Suzuki, A., et al.: Gastric myoelectrical differences between Parkinson's disease and multiple system atrophy. Mov. Disord. **24**, 1579–1586 (2009)

12. Seligman, W.H., Low, D.A., Asahina, M., Mathias, C.J.: Abnormal gastric myoelectrical activity in postural tachycardia syndrome. Clin. Auton. Res. **23**(2), 73–80 (2013)

13. Wayland, R., Bromley, D., Pickett, D., Passamante, A.: Recognizing determinism in a time series. Phys. Rev. Lett. **70**(5), 580–582 (1993)

14. Takada, H., Simizu, Y., Hoshita, H., Shiozawa, T.: Wayland tests for differenced time series could evaluate degrees of visible determinism. Bull. Soc. Sci. **19**(3), 301–310 (2005)

15. Hino, M.: Spectral Analysis. Asakura Shoten, Tokyo (1977)

16. Minami, S.: Handling Wave Form Data for Scientific Measurements. CQ Shuppansha, Tokyo (1986)

17. Japan Society of Neurovegetative Research: Autonomic nerve function examination. vol. 5. Bunkodo Co., Ltd. (2007)

18. Homma, S.: Isopower mapping of the electrogastrogram (EGG). J. Auton. Nerv. Sys. **62**(3), 163–166 (1997)

19. Koike, Y.: Differences between postprandial hypotension and orthostatic hypotension. Did you know? Postprandial hypotension – New blood pressure abnormality clinic, Nanzando, pp. 80–83 (2004)

20. Hirayama, M.: Expression mechanisms from the perspective of hemodynamics. Did you know? Postprandial hypotension – New blood pressure abnormality clinic, Nanzando, pp. 93–99 (2004)

MilkyWay: A Toolbox for Prototyping Collaborative Mobile-Based Interaction Techniques

Mandy Korzetz[1(✉)], Romina Kühn[1], Karl Kegel[1], Leon Georgi[1], Franz-Wilhelm Schumann[1], and Thomas Schlegel[2]

[1] Software Technology Group, TU Dresden, Dresden, Germany
{mandy.korzetz,romina.kuehn,karl.kegel,leon.georgi,
franz-wilhelm.schumann}@tu-dresden.de
[2] Institute of Ubiquitous Mobility Systems, Karlsruhe University of Applied Science,
Karlsruhe, Germany
thomas.schlegel@hs-karlsruhe.de

Abstract. Beside traditional multitouch input, mobile devices provide various possibilities to interact in a physical, device-based manner due to their built-in hardware. Applying such interaction techniques allows for sharing content easily, e.g. by literally pouring content from one device into another, or accessing device functions quickly, e.g. by facing down the device to mute incoming calls. So-called mobile-based interaction techniques are characterized by movements and concrete positions in real spaces. Even though such interactions may provide many advantages in everyday life, they have limited visibility in interaction design due to the complexity of sensor processing. Hence, mobile-based interactions are often integrated, if any, at late design stages. To support testing interactive ideas in early design stages, we propose MilkyWay, a toolbox for prototyping collocated collaborative mobile-based interaction techniques. MilkyWay includes an API and a mobile application. It enables easily building up mobile interactive spaces between multiple collocated devices as well as prototyping interactions based on device sensors by a programming-by-demonstration approach. Appropriate sensors are selected and combined automatically to increase tool support. We demonstrate our approach using a proof of concept implementation of a collaborative Business Model Canvas (BMC) application.

Keywords: Prototyping · Mobile devices · Device-based interaction · Collocated interaction · Collaboration · Interaction design

1 Introduction

Smartphones as mobile devices have become a companion of our everyday life. Current devices innately provide numerous built-in sensors such as accelerometers and gyroscopes for sensing motions or orientation sensors and magnetometers for determining positions. Thus, mobile devices cover a wide range

© Springer Nature Switzerland AG 2019
M. Antona and C. Stephanidis (Eds.): HCII 2019, LNCS 11573, pp. 477–490, 2019.
https://doi.org/10.1007/978-3-030-23563-5_38

of interaction techniques and can support users in various situations. For example, performing a short shaking gesture facilitates to provide feedback to map applications. Furthermore, turning a mobile device enables muting the device instead of pushing hardware buttons or navigating through complex menus. In addition, working and learning situations often involve collaborative activities which are increasingly supported by mobile device usage to avoid digital disruptions [19]. Collaborative activities comprise joining and leaving a group, creating and editing content, presenting and comparing results as well as sharing content.

In recent years, research has investigated various specific mobile-based interaction techniques to support these activities, for example, connecting devices by touching another device to add it to the group [9], comparing by ordering devices in a row to rank displayed content [17], or sharing by figuratively pouring content from one device to another [15]. Integrating mobile phones in a physical manner aims at fostering face-to-face collaboration because users are enabled to pay more attention to group activities instead of looking at screens. Although, there are numerous examples from research, mobile device-based interaction techniques still have limited visibility in interaction design [16].

As mobile-based interaction techniques base on sensors for connecting and building groups of collocated devices and also for detecting physical and motion-based interactions, designers and developers need advanced technical knowledge about hardware. Hence, mobile-based interactions are often integrated, if any, at late design stages. But to develop highly usable and useful systems along with addressing the user needs as a whole, it is particularly necessary to test and evaluate ideas in early design stages with interactive prototypes. Existing prototyping tools focus on single-device interactions with mobile phones [2] or specific device sensors [11]. Nevertheless research shows that mobile-based interaction techniques are more versatile. Leigh et al. [21] generally describe that smartphones are used as a tangible interfaces. Rico and Brewster [26] illustrate interactions that involve touching and moving mobile devices directly regardless of specific sensors. Furthermore, Lucero et al. [22–24] demonstrate various interactions with multiple devices where users share their mobile phones in collocated situations. As these examples show the versatility of these interaction techniques, prototyping tools are currently limited because they only address specific aspects of mobile-based interaction possibilities.

The *Milky Way* toolbox is intended to support prototyping interactions that are generally invoked by deliberate device movements. In contrast to other tools, *Milky Way* tries to minimize the required technical know-how, which means designers of mobile-based interactions can create and adjust prototyped interactions without any knowledge of sensor characteristics. Additionally, our toolbox supports developers with an API for easily building collaborative multi-device applications without further equipment, so-called mobile co-spaces [16]. Mobile co-spaces are formed by connecting multiple mobile devices that allow collaboration of multiple collocated users.

To describe our approach, we structured the remainder of the paper as follows: First, we give an overview of related work focusing on several kinds of prototyping tools. Then, we define design goals that consider the user requirements for implementing this kind of interaction techniques. Taking the design goals into account, we present our *MilkyWay* toolbox describing the including tools that target the development support. We illustrate and discuss the usage of the toolbox by describing the mobile application *MobiLean*. Concluding, we give an outlook on next steps.

2 Related Work

As stated by Ledo et al. [20], toolkit research plays an important role in the HCI community. In their work, they collected several representative papers that focus on toolkits or toolkit research. Thereby, they give a broad overview of related research but mainly investigate how these papers address evaluation aspects. For our work, it is important how toolkits facilitate easy prototyping of movement-based interactions for mobile phones without further equipment and devices, e.g. desktop PCs, but also allow prototyping of mobile co-spaces.

Interactions in ubiquitous environments, mid-air gestures and also tangible interfaces are related to mobile-based interaction techniques because they also deal with human movements. In the following, we therefore describe approaches that address prototyping in these research areas. One of the early works is the iStuff approach [4]. This extensive toolkit supports interaction designer to quickly create events in ubiquitous environments. Using additional equipment (so-called PatchPanels), they can complement everyday objects physically and utilize them to execute different actions, e.g. turning on and off the light by adding PatchPanels wherever wanted within a room. Keller et al. [10] extend Ballagas' et al. approach by providing virtual interactive surfaces to interact with everyday objects in a ubiquitous way. On top of the iStuff approach, Ballagas et al. [3] built iStuff mobile that is a visual programming environment, which enables the low-fidelity prototyping of mobile interactions. The toolkit provides several predefined sensors and interactions that can be combined individually. However, this limits the way the devices can be used because new sensors have to be added before they can be used. This leads to high maintenance effort due to the need of constantly updating the tool.

The requirements for authoring mid-air gestures are similar to the prototyping of mobile device-based interactions, because they also deal with human movements. Baytaş et al. [5] give an overview of tools for authoring mid-air gestures. They distinguish between tools that use graphs of the data from the sensors that detect movement, tools that use an own visual markup language, and tools that provide a timeline of frames. MAGIC [2] captures mid-air gestures of on-body devices such as wristwatches and visualize the data in form of sensor graphs within a desktop application. Developing motion-based gestures is done by demonstrating but allows only single device and capturing of a wristwatch device. EventHurdle [12] is a visual tool for authoring gesture events. The gestures are drawn in a separate application, code is generated and integrated into the prototype to trigger the interaction. EventHurdle is specialized on simple two-dimensional gestures, e.g. for touch pads, but lacks more complex three-dimensional gestures. In contrast, M.Gesture [11] allows authoring directly on a mobile device by using a visual metaphor of a mass-spring, but concentrates on gestures based solely on accelerometers. As Baytaş et al. propose, we combine the two programming approaches demonstration and declaration to prototype mobile-based interactions.

Hartmann et al. [8] describe their approach on authoring sensor-based interactions by demonstrating using direct manipulation and pattern recognition. Their tool *Exemplar* visualizes data streams of connected sensors and enables a direct manipulation of the data streams using the PC's mouse. Although their tool support is extensive, recording and editing the interactions is done on another device, a desktop PC. This leads to a device-switch with a higher learning and execution effort than authoring directly with the sensor device. Klompmaker et al. [14] present the INDiE approach that consists of a network protocol, a device abstraction and a software development kit (SDK). Whereas the network protocol facilitates connection establishment and data exchange, the device abstraction approach aims at supporting rapid prototyping of multimodal interactions with focus on virtual reality (VR). Similar to Hartmann et al. [8], this approach needs a server component to manipulate and aggregate the received data from the sensors. This leads to a higher dependency on a fast WiFi connection to communicate with the server component as well as the server itself. Ajaj et al. [1] describe their Real/Virtual-Device/Task (RVDT) model that aims at supporting the design of multimodal/multi-view interfaces. The presented design space binds input devices to spatial tasks. Although the approach is extensive in terms of the degrees of freedom, it only addresses graphical output modalities. Klemmer et al. [13] present a Wizard of Oz (WOz) approach for tangible user interfaces to support designers by providing WOz generation and removal of input. This approach is useful for simulating hardware that is not (yet) available.

Although, there are already various approaches that present toolkits and frameworks to support developers, current work lacks approaches that allow for prototyping mobile-device interactions which on the one hand are based on motions and on the other hand are often combined with other mobile devices to allow collaboration. Moreover, most approaches require additional applications and PCs, so that learning is more difficult and end-users can not be involved easily.

3   Design Goals

Klemmer et al. [13] derived functionality that should be provided for tangible user interfaces. We extended their work and tailored the functional requirements to collaborative mobile-based interaction techniques. Geiger et al. [6] also describe requirements for an easy-to-use framework for prototyping hybrid user interfaces that combines 2D, 3D and haptic interfaces. Based on these works, we derived the following requirements for the design of the *Milky Way* toolbox.

Collaborative Mobile-Based Interaction Techniques. For a better understanding of the domain of mobile-based interaction techniques, we investigated several

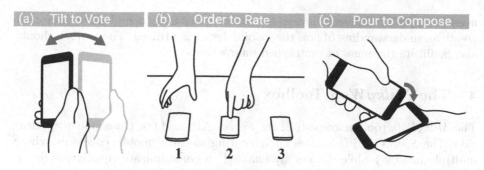

Fig. 1. Examples for mobile-based interaction techniques: (a) Tilt to Vote [18], (b) Order To Rate [17], and (c) Pour to Compose [15].

interactions from literature (see Fig. 1). Interaction characteristics can be distinguished between interactions that are for individual use, e.g. tilting the device to accept or reject a solution (Fig. 1a) or facing down the device to mute incoming calls, as well as for collaborative use, e.g. ordering devices on a table to rate the displayed content (Fig. 1b) or merging content by literally pouring content from one device to another (Fig. 1c). Prototyping tools should support easily building up such mobile interactive spaces between multiple collocated devices. The examples also show the important role of *spatiality* and *motion*. The intended interaction either can depend on relative positions between or a specific arrangement of devices that trigger system functions as shown in Fig. 1b, or is caused by a movement (Fig. 1a).

User-Centered Design of Interactions. Nielsen et al. [25] stated that gestures and thus mobile-based interactions should not be motivated by the easiness of implementation (technology-based approach), but instead should be developed together with end-users to ensure an intuitive and ergonomic interface. Our tools aim at providing user involvement at different prototyping stages. This enables collecting interaction ideas to specific system functions with participants of a user group, improving interim results, for example, by gathering different variants of how to perform interactions, and evaluating interactions to either test the interaction itself or to test the interplay with other interactions and system functions.

Facilitating the Processing and Use of Device Sensors. Especially designers but also developers often have an extensive knowledge about graphical aspects of creating mobile user interfaces, but lack advanced technical knowledge in terms of sensor hardware. Prototyping of mobile-based interactions which are based on device sensors therefore should reduce the programming effort. To tackle this issue, we propose an easy programming-by-demonstration approach where relevant sensors are selected and combined automatically. By using the respective device for prototyping, there is no need for additional equipment such as a PC for editing the prototype or an additional server for collecting sensor data. Furthermore, the usage of the respective devices reduces effort both regarding the need of applying different devices for prototyping and to find metaphors that create an understanding of how the mobile device is utilized. The toolbox should also facilitate the reuse of prototyped interactions.

4   The *Milky Way* Toolbox

The *Milky Way* toolbox consists of the *Spaces* API and the *Gaia* mobile application. The *Spaces* API facilitates implementing so-called mobile co-spaces, where multiple nearby mobile devices are enabled to communicate equally privileged with the objective to collaborate. This aim includes, for example, sharing created content within a group or voting on several working results. The *Gaia* application aims at mainly supporting designers of mobile device-based interaction techniques by enabling an easy recording by demonstration of interactions. The tools can be used in parallel by designers and developers of mobile-based interaction techniques. The following sections describe the tools in detail as well as how they can be utilized for prototyping.

4.1   The *Spaces* API

The application programming interface (API) *Spaces* supports developers in prototyping group communication between multiple mobile devices. *Spaces* provides developers with methods to easily built-up and maintain stable connection within mobile co-spaces. Further, the API automates handling connection-based functionality such as sending and receiving data or joining and leaving co-spaces. The *Spaces* API is based on the Nearby Connections 2.0 API[1], an API that is originally intended to enable position-based information to users, e.g. for advertising. For convenience, connections between mobile devices are fully-offline by combining Bluetooth, Bluetooth Low Energy (BLE), and WiFi and therefore enable fast, secure data transfers. Moreover, users are not prompted to turn on Bluetooth or WiFi, because the API enables these features as they are required. The Nearby Connections 2.0 API provides very extensive functions specialized for position-based advertising, but for prototyping and implementing mobile co-spaces developers have to write a lot of boilerplate code[2] and the effort for configuring peer-to-peer networks which is needed for the communication within a mobile co-space is very high. *Spaces* combines the API functionality to multi-device communication so that configuration effort is low and using the API becomes more simple and intuitive. As collaborative scenarios require a stable connection between mobile devices, the *Spaces* API enables a fully automated reconnection management in case of connection failures.

---

[1] https://developers.google.com/nearby/connections/.

[2] Boilerplate code is code that has to be included in many places with little or no alteration.

4.2 The *Gaia* Mobile Application

Current mobile devices include numerous built-in sensors to measure motion, orientation and other environmental conditions[3]. To support designers and developers with little technical knowledge about sensor characteristics, the *Gaia*[4] application facilitates the development of mobile-based interactions. Designers, developers, and also end-users can easily record interactions with the mobile application. For this purpose, app users have to demonstrate the respective movement or device constellation they want to achieve by directly manipulating the device(s). To ensure a higher recognition rate, the interaction has to be repeated a couple of times, approximately 5 to 10 repetitions. *Gaia* estimates dynamically which sensors are appropriate to recognize the demonstrated interaction by an algorithm which compares absolute sensor values. Advanced *Gaia* users can adjust the suggested sensors post hoc. Interrelated interactions can be grouped, e.g. simple moving left, right, up and down on a table can be grouped as "sliding interactions". To compare a stored recorded interaction with a new record, *Gaia* uses an optimized dynamic time warping algorithm for each relevant sensor type. Advanced users can have a look at the sensor graphs to understand single recordings.

For prototyping collaborative interactions that consist of multiple parts distributed on multiple devices, e.g. pouring content from one device to another device [15], *Gaia* allows composing interaction parts. Users of the *Gaia* application can choose from a list which interaction parts belong together. Composing interaction parts is implemented as simple event processing architecture, which means if two or more interaction events occur, the composed interaction is recognized. For this purpose, *Gaia* itself uses the *Spaces* API to easily exchange event messages.

4.3 Prototyping with the *MilkyWay* Toolbox

The prototyping process by using *MilkyWay* is structured as shown in Fig. 2. Whereas developers are especially supported by using the *Spaces* API to prototype system functionality for collaborative interactions, designers can work with the *Gaia* mobile application to prototype the mobile-based interactions related to the system functions. Designers typically collect different ideas for interacting mobile-based with the system. The ideas can be demonstrated with *Gaia* once to capture it. To refine first ideas and capture performing variants, interaction ideas have to be recorded multiple times. In case the interaction involves multiple devices, each part has to be repeated about 5 to 10 times. The last step includes composing single interaction parts to the collaborative mobile-based interaction. Designers and developers of mobile device-based interactions can work efficiently in parallel. At any stage they can exchange interim results, e.g. integrating first

---

[3] As example the sensor types of typical mobile devices with Android operating system: https://developer.android.com/guide/topics/sensors/sensors_overview.

[4] Gaia is a spacecraft of the European Space Agency (ESA) for astrometry and measures the positions, distances and motions of stars with unprecedented precision.

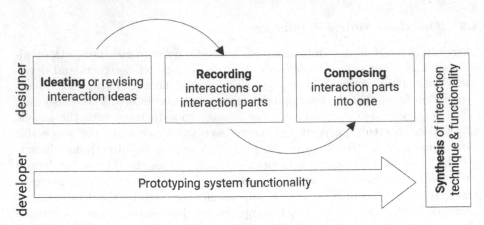

Fig. 2. Prototyping process with the *Milky Way* toolbox. Designers and developers of mobile device-based interactions can work efficiently in parallel and exchange interim results. Users can be involved at any prototyping stage.

ideas for a mobile-based interaction, or vice versa, working results of test functionality can be used to generate mobile-based interactions with users itself. To enable a user-centered design approach, users can be involved at any prototyping stage to develop easy to perform and remember interactions. In the first stage, users can directly demonstrate their interaction ideas by using the *Gaia* app either by simulating system functionality or using prototyped functionality from the developer. In the next stage, the recording stage, users can be involved to record variants of concrete interaction ideas. The main goal of integrating users at this stage is to ensure that different performances can be recognized and concrete interactions and interaction parts can be refined. At the third stage, users should be involved to test the prototyped interactions and give feedback for improvements. The last step of prototyping comprises combining prototyped system functionality and revised mobile-based interaction techniques to an extensive interactive mobile app prototype.

5   Applying *Milky Way* to a Mobile System for Creating BMCs Collaboratively

Collaboration usually includes working together as well as working individually, so-called mixed-focus collaboration [7]. Kühn and Schlegel [19] identified several activities that are typically performed in mixed-focus collaboration, e.g. creating and sharing content. Furthermore, they presented an example scenario for the key activities addressing the creation process of a business model canvas[5] (BMC) with several building blocks. In their example, a group of students performs this collaboration process using mobile phones. The activities involve initiating collaboration and assigning blocks to group members, creating text entries within

---

[5] The traditional paper tool for BMC can be found on https://strategyzer.com/.

(a) Initiating collaboration.

(b) Creating and presenting results.

(c) Selecting content.

(d) Sharing content.

Fig. 3. Initiating collaboration by detecting blocks using image recognition (Fig. 3a). Creating text entries collaboratively by adding, editing or removing content and presenting results by highlighting important content (Fig. 3b). Selecting relevant content for sharing (Fig. 3c). Sharing content by merging (preselected) entries from different blocks after working individually (Fig. 3d).

the blocks, presenting and discussing their interim results, sharing and merging individual results to receive an overall solution and finishing collaboration. We use this example scenario to illustrate how the *MilkyWay* toolbox supports several collaboration activities and respective prototyping of mobile-based interaction techniques.

We implemented *MobiLean* (Fig. 3), a mobile application for Android devices that enables the creation of BMCs collaboratively. The usage of *MilkyWay* supports developers in implementing collaborative interactions and designers in trying several interaction techniques for different activities. The application has the following functionalities according to the *collaboration activities* [19]:

- Initiating collaboration (Fig. 3a): connecting devices; joining sessions; inviting to sessions.
- Creating content (Fig. 3b): adding, editing, and removing text entries within BMC blocks; creating new projects and sessions.
- Presenting content (Fig. 3b, right): highlighting text entries.
- Comparing results: voting on results.

- Sharing content (Fig. 3d): selecting content; sharing complete entry sets; merging entry sets with other entries.
- Finishing collaboration: disconnecting devices; leaving sessions.

While we implemented core functionality that is specific for BMCs, e.g. building different blocks and editing block entries, we used the *MilkyWay* toolbox to create functionalities for collaborating and to establish mobile interactive spaces (mobile co-spaces) between several devices. *MobiLean* uses the *Spaces* API to connect the devices easily. A device can act as host and can invite further devices (guests) to join a session. Each guest can easily join such hosted sessions. As it is common in mixed-focus collaboration to work individually, too, re-joining a session can be performed automatically. Especially for designers and developers, the *Spaces* API is an easy tool to establish multi-device connections.

The *Gaia* app is used to teach the interaction techniques by applying the programming-by-demonstration approach. In the current version of *MobiLean*, direct voting interactions [17] and merging interactions [15] (see Fig. 3d) are implemented. Applying the *Gaia* app, these interaction techniques could already be recorded to be used during the BMC creation. As described in Sect. 4.2, each interaction was performed several times while *Gaia* estimated the utilized sensors. Especially for developers, there is an advanced setting to adjust the suggested sensors post hoc. For example, for *Shift to Vote*, one of Kühn's et al. voting interaction technique [17], several single up and down movements of the device on a table are grouped as the mentioned interaction. Furthermore, designers can group interactions that belong to the category of "voting". Designers can use this functionality to create and refine interaction techniques. Furthermore, end-users can record different interactions to extend the variety of the way how interactions can be performed. The *MilkyWay* toolbox simplifies the development of several interaction techniques tremendously, especially for those, who are not familiar with programming. We aim at extending *MobiLean* with additional interaction techniques to further investigate our toolbox and its application. While implementing *MobiLean* we could observe, so far, that prototyping collaborative mobile-based interaction techniques can be speed up with *MilkyWay*. We plan to investigate the exact speed up rate in future work.

6   Discussion

We implemented *MobiLean* as proof of concept application to investigate the usage of our *MilkyWay* toolbox. Although, the application is still in development, we got a better understanding of how the toolbox can support designers, developers, and end-users of mobile-based interaction techniques. Based on our design goals for *MilkyWay* in Sect. 3 and the *MobiLean* implementation (Sect. 5), we discuss the following aspects.

6.1   Coverage of Collaborative Mobile-Based Interaction Techniques

The *MilkyWay* toolbox allows prototyping of interaction techniques that base on movements of one or more devices, but also on more static constellations of one or multiple devices if built-in sensors are involved. The toolbox therefore covers a wide range of possible mobile-based interactions that are appropriate for individual but also for collaborative use. As we build on all available sensors of a device and automatically select and combine them, designers and developers are not restricted concerning the used sensor data.

6.2 Involving Users to Interaction Design

To ensure intuitive and ergonomic interactions with the system, that has to be developed, *MilkyWay* enables to involve users at any stage of interaction design. They can be involved (1) to collect a set of interaction ideas, (2) to record variants of interactions, and (3) to conclusively evaluate interactions as described in Sect. 4.3. As interactions are captured by directly manipulating the mobile devices themselves, end-users of mobile-based interactions can easily be involved.

6.3 Facilitating the Use of Device Sensors

Understanding sensor characteristics needs advanced technical knowledge, which is seldom pronounced within the group of designers and developers of mobile applications. *MilkyWay* simplifies the usage of sensor data by allowing a fully automated selection and combination of sensors. Users only have to demonstrate their interaction idea with the respective device and *Gaia* handles the translation into code. In doing so, no additional equipment is needed such as a PC/server for, e.g. adjusting or processing raw sensor data. Moreover, implementing multi-device communication is facilitated by the *Spaces* API by providing and enclosing methods that are needed for collaborative interactions.

6.4 Limitations and Future Work

There remains some future work to improve the usability of static mobile-based interactions (e.g. *Order to Vote* [17]). As interactions have to be demonstrated for prototyping, each possible arrangement has to be performed. So, prototyping of static interactions is possible but cumbersome. We would like to find strategies to simplify prototyping of interactions such as *Order to Vote* with an improved version of *MilkyWay*. Combining parts of interactions with *Gaia* is currently realized by a simple "if this than that" rule. We could imagine that there are cases that require more flexibility. So, we plan to implement algorithms to allow further combining strategies. We also plan to improve reusing of prototyped mobile-based interactions. Currently, recognition of interactions works best on the device type where they were recorded because sensor data of each device type slightly differs. In next steps we will test how this can be improved, e.g. with mixing training data of different devices to enhance accuracy to enable reusing of interaction data for further development stages.

7   Conclusion

We presented *MilkyWay*, a toolbox for prototyping mobile-based interaction techniques already in early design stages. The toolbox aims at supporting designers and developers to prototype complex mobile-based interaction techniques based on movements and/or relative positions detected by the devices' built-in sensors. Therefore, users of *MilkyWay* need no or little knowledge of sensor characteristics. And as interactions are prototyped with the respective device, end-users can be easily involved. Developers are supported by an API that facilitates and aims at speeding up the implementation of collaborative communication between devices. As a proof of concept, we implemented *MobiLean*, a collaborative BMC application that applies the *MilkyWay* toolbox.

As described in Sect. 6 we plan to facilitate prototyping of static mobile-based interactions, expand combination methods of interaction parts, and improve reusing interaction data. We would also like to use parts of *MilkyWay* to enable individualizing interaction techniques for end-users in mobile applications that are production-ready or already on the market.

We currently use *MilkyWay* to prototype collaborative mobile-based interactions for creating BMCs. We intend to implement further techniques, e.g. anonymous voting interactions [18]. Our next steps will also include an evaluation with mobile designers and developers to get additional impressions and suggestions for improvements of the *MilkyWay* toolbox, and further ideas for potential additional tools to support prototyping of mobile-based interaction techniques.

Acknowledgments. The European Social Fund (ESF) and the German Federal State of Saxony have funded this work within the project CyPhyMan (100268299). We would also like to thank Lukas Büschel for his work and support.

References

1. Ajaj, R., Jacquemin, C., Vernier, F.: RVDT: a design space for multiple input devices, multiple views and multiple display surfaces combination. In: Proceedings of the International Conference on Multimodal Interfaces (ICMI-MLMI 2009), pp. 269–276 (2009). https://doi.org/10.1145/1647314.1647372
2. Ashbrook, D., Starner, T.: MAGIC: a motion gesture design tool. In: Proceedings of the 28th International Conference on Human Factors in Computing Systems (CHI 2010), pp. 2159–2168. ACM (2010). https://doi.org/10.1145/1753326.1753653
3. Ballagas, R., Memon, F., Reiners, R., Borchers, J.: iStuff mobile: rapidly prototyping new mobile phone interfaces for ubiquitous computing. In: Proceedings of the SIGCHI Conference on Human Factors in Computing Systems (CHI 2007), pp. 1107–1116 (2007). https://doi.org/10.1145/1240624.1240793
4. Ballagas, R., Ringel, M., Stone, M., Borchers, J.: iStuff: a physical user interface toolkit for ubiquitous computing environments. In: Proceedings of the SIGCHI Conference on Human Factors in Computing Systems (CHI 2003), pp. 537–544 (2003). https://doi.org/10.1039/F29767200988

5. Baytaş, M.A., Yemez, Y., Özcan, O.: User interface paradigms for visually author-ing mid-air gestures: a survey and a provocation. In: 1st International Workshop on Engineering Gestures for Multimodal Interfaces (EGMI 2014), pp. 8–14 (2014). http://ceur-ws.org/Vol-1190/

6. Geiger, C., Fritze, R., Lehmann, A., Stöcklein, J.: HYUI: a visual framework for prototyping hybrid user interfaces. In: Proceedings of the 2nd International Confer-ence on Tangible and Embedded Interaction (TEI 2008), pp. 63–70 (2008). https://doi.org/10.1145/1347390.1347406

7. Gutwin, C., Greenberg, S.: Design for individuals, design for groups: tradeoffs between power and workspace awareness. In: Proceedings of the 1998 ACM Con-ference on Computer Supported Cooperative Work (CSCW 1998), pp. 207–216 (1998). https://doi.org/10.1145/289444.289495

8. Hartmann, B., Abdulla, L., Mittal, M., Klemmer, S.R.: Authoring sensor-based interactions by demonstration with direct manipulation and pattern recognition. In: Proceedings of the SIGCHI Conference on Human Factors in Computing Sys-tems (CHI 2007), pp. 145–154 (2007). https://doi.org/10.1145/1240624.1240646

9. Jokela, T., Chong, M.K., Lucero, A., Gellersen, H.W.: Connecting devices for col-laborative interactions. Interactions **22**(4), 1–5 (2015). https://doi.org/10.1145/2776887

10. Keller, C., Kühn, R., Engelbrecht, A., Korzetz, M., Schlegel, T.: A prototyping and evaluation framework for interactive ubiquitous systems. In: Proceedings of the First International Conference on Distributed, Ambient, and Pervasive Inter-actions, pp. 215–224 (2013). https://doi.org/10.1007/978-3-642-39351-8_24

11. Kim, J.W., Kim, H.J., Nam, T.J.: M.gesture: an acceleration-based gesture author-ing system on multiple handheld and wearable devices. In: Proceedings of the 2016 CHI Conference on Human Factors in Computing Systems (CHI 2016), pp. 2307–2318. ACM (2016). https://doi.org/10.1145/2858036.2858358

12. Kim, J.W., Nam, T.J.: EventHurdle: supporting designers' exploratory interaction prototyping with gesture-based sensors. In: Proceedings of the SIGCHI Conference on Human Factors in Computing Systems (CHI 2013), pp. 267–276 (2013). https://doi.org/10.1145/2470654.2470691

13. Klemmer, S.R., Li, J., Lin, J., Landay, J.A.: Papier-Mâché: toolkit support for tangible input. In: Proceedings of the SIGCHI Conference on Human Factors in Computing Systems (CHI 2004), Vienna, pp. 399–406 (2004). https://doi.org/10.1016/j.jjcc.2014.08.007

14. Klompmaker, F., Schrage, K., Reimann, C.: INDiE: a framework for human com-puter interaction in distributed environments. In: Proceedings of the 6th Interna-tional Conference on Mobile Technology, Application & Systems (Mobility 2009), pp. 26–29 (2009). https://doi.org/10.1145/1710035.1710061

15. Korzetz, M., Kühn, R., Heisig, P., Schlegel, T.: Natural collocated interactions for merging results with mobile devices. In: Proceedings of the 18th International Conference on Human-Computer Interaction with Mobile Devices and Services Adjunct (MobileHCI 2016), pp. 746–752. ACM (2016). https://doi.org/10.1145/2957265.2961839

16. Korzetz, M., Kühn, R., Schlegel, T.: Turn it, pour it, twist it: a model for designing mobile device-based interactions. In: Proceedings of the 5th International Confer-ence on Human-Computer Interaction and User Experience in Indonesia (CHIuXiD 2019). ACM (2019). https://doi.org/10.1145/3328243.3328246

17. Kühn, R., Korzetz, M., Büschel, L., Korger, C., Manja, P., Schlegel, T.: Natural voting interactions for collaborative work with mobile devices. In: Proceedings of the 2016 CHI Conference Extended Abstracts on Human Factors in Computing Systems (CHI EA 2016), pp. 2570–2575. ACM (2016). https://doi.org/10.1145/2851581.2892300

18. Kühn, R., Korzetz, M., Büschel, L., Schumann, F.W., Schlegel, T.: Device-based interactions for anonymous voting and rating with mobile devices in collaborative scenarios. In: Proceedings of the 15th International Conference on Mobile and Ubiquitous Multimedia (MUM 2016), pp. 315–317. ACM (2016). https://doi.org/10.1145/3012709.3016067

19. Kühn, R., Schlegel, T.: Mixed-focus collaboration activities for designing mobile interactions. In: Proceedings of the 20th International Conference on Human-Computer Interaction with Mobile Devices and Services (MobileHCI 2018), pp. 71–77. ACM (2018). https://doi.org/10.1145/3236112.3236122

20. Ledo, D., Houben, S., Vermeulen, J., Marquardt, N., Oehlberg, L., Greenberg, S.: Evaluation strategies for HCI toolkit research. In: Proceedings of the 2018 Conference on Human Factors in Computing Systems (CHI 2018), p. 17 (2018). https://doi.org/10.1145/3173574.3173610

21. Leigh, S.W., Schoessler, P., Heibeck, F., Maes, P., Ishii, H.: THAW: Tangible interaction with see-through augmentation for Smartphones on computer screens. In: Proceedings of the Ninth International Conference on Tangible, Embedded, and Embodied Interaction (TEI 2014), pp. 89–96. ACM (2015). https://doi.org/10.1145/2677199.2680584

22. Lucero, A., Holopainen, J., Jokela, T.: Pass-them-around: collaborative use of mobile phones for photo sharing. In: Proceedings of the 2011 Annual Conference on Human Factors in Computing Systems (CHI 2011), pp. 1787–1796 (2011). https://doi.org/10.1145/1978942.1979201

23. Lucero, A., Keränen, J., Jokela, T.: Social and spatial interactions: shared co-located mobile phone use. In: CHI 2010 Extended Abstracts on Human Factors in Computing Systems (CHI EA 2010), pp. 3223–3228. ACM (2010)

24. Lucero, A., Keränen, J., Korhonen, H.: Collaborative use of mobile phones for brainstorming. In: Proceedings of the 12th International Conference on Human Computer Interaction with Mobile Devices and Services (MobileHCI 2010), pp. 337–340. ACM (2010). https://doi.org/10.1145/1851600.1851659

25. Nielsen, M., Störring, M., Moeslund, T.B., Granum, E.: A procedure for developing intuitive and ergonomic gesture interfaces for HCI. In: Camurri, A., Volpe, G. (eds.) GW 2003. LNCS (LNAI), vol. 2915, pp. 409–420. Springer, Heidelberg (2004). https://doi.org/10.1007/978-3-540-24598-8_38

26. Rico, J., Brewster, S.: Usable gestures for mobile interfaces: evaluating social acceptability. In: Proceedings of the 28th International Conference on Human Factors in Computing Systems (CHI 2010), p. 887. ACM (2010). https://doi.org/10.1145/1753326.1753458

@HOME: Exploring the Role of Ambient Computing for Older Adults

Daria Loi[✉]

Intel Labs, Hillsboro, OR 97124, USA
daria.a.lois@intel.com

Abstract. Building on results of a recent global study as well as additional exploratory research focused on *Aging in Place*, this paper reflects on the role that intelligent systems and ambient computing may play in future homes and cities, with a specific emphasis on populations aged 65 and beyond. This paper is divided into five sections. The first section provides an introductory background, which outlines context, vision, and implications around the development of ambient computing and smart home technologies for the 65 + population. The second part of the paper overviews the methodological approaches adopted during the research activity at the center of this paper. The third section summarizes pertinent findings and a discussion on the opportunities offered by intelligent, ambient systems for the 65+ population follows. While this fourth section will specifically focus on the smart home, it will also provide reflections on opportunities and applications in the context of autonomous vehicles and smart cities. The fifth and last section offers conclusive remarks, including implications for developers and designers that are shaping ambient computing usages and technologies for the 65+ population. The paper ultimately advocates for adopting Participatory Design [1] approaches, to ensure that intelligent and ambient technologies are developed **with** (instead of **for**) end users.

Keywords: Artificial intelligence · Ambient computing · User experience

1 Background

The research here discussed focuses on enabling and grounding the development of ambient computing and smart technologies for the 65+ population. An increasing number of older adults live in isolated conditions, without the opportunity of aging in place and in emotionally stable conditions. Grounded in this knowledge, the project here featured acknowledges the high economic, psychological and social burden caused by such a reality. *Aging in Place* means making a conscious decision to live in the residence of one's choice for as long as one can with comforts that the individual sees as important and with the ability to leverage supplementary services that facilitate living conditions and maintain quality of life. This section outlines context, vision, and implications around the development of ambient computing and smart home technologies for older adults.

M. Antona and C. Stephanidis (Eds.): HCII 2019, LNCS 11573, pp. 491–505, 2019.
https://doi.org/10.1007/978-3-030-23563-5_39

1.1   Context

The so-called *population ageing*, a phenomenon related to fertility decline and life expectancy rising, is occurring globally. The number of people aged 60 years and over is projected to more than double by 2050 and more than triple by 2100, rising from 962 m globally in 2017 to 2.1b in 2050 and 3.1b in 2100 – this population segment is growing at a faster rate than all younger age groups [2]. By 2050 hyper-ageing societies will represent a large part of the global population [3]. This data highlights an urgent need to focus on diverse technological means to cater to this fast growing segment.

In the US, AARP [4] reports that while 90% of seniors want to live at home as they age, many cannot do so suitably because homes and communities cannot accommodate their particular needs. In 2011, 5.6% of the US elderly, community-dwelling Medicare population was for instance completely or mostly homebound [5], with 75 as average age and 30% living alone. Semi-homebound would be an additional $\sim$ 20%. Being homebound and living alone implies reduced opportunities for social interaction and research shows that mortality is higher among more socially isolated, lonely individuals and that social isolation had the most significant association with mortality [6]. It is urgent we address this segment's diverse needs and complexities.

Additionally, shifts in the *dependency ratio* (estimate of the pressure on productive population) and projections of further shifts [7], alongside a *Bean Pole effect* (family trees get taller, thin, with few people per generation, due to children decrease and life span increase) are impacting society's caregiving capabilities. Several societal changes are aggravating this situation: increase of divorce, re-partnering, and more complex family ties [8–12]; welfare state provision expansion (Europe) and decreased need for family support [13]; women's higher labor-force participation (Europe) and challenges for family caring [14–16]; and processes of individualization, secularization and emancipation, alongside greater emphasis on individual needs and personal happiness [17, 18].

The nature of caregiving has therefore changed and long-distance caregivers are emerging, with $\sim$ 5–7 m long-distance caregivers ($\sim$ 15% of total) in the U.S., with numbers projected to double by 2020 [19]. Long-distance caregivers, however, represent higher annual expenses (compared to co-resident caregivers or those who care for a loved one nearby) [20], their distance from clients is 450 miles on average [21], and are more likely to report emotional distress [22]. Regardless of who provides caregiving support, research also shows that emotional support is a key role they typically fulfil [23]. These caregiving shifts imply a need to provide economically, practically and emotionally sustainable support structures and tools: smart technologies offer the opportunity to address some of these needs and urgencies.

Another important point to consider is that, while better health and quality of life often lowers societal burden and costs, technology can lower the cost of maintaining wellness. In the U.S. for instance, functional limitations such as difficulty to bathe, dress or walk are often reason behind older adults' institutionalization, and 1 in 3 older adults report having trouble using some feature of their home [24]. Long-term care costs keep increasing [25] and nursing homes and assisted living care costs are growing at rates higher than overall inflation [26]. Yet, home care technology can bring substantial cost savings over using human provided care [27]. There is an opportunity to favor, where applicable, home-based care and, since tech-enabled home care could help address many medical conditions [28], there is an opportunity to create a high-tech home health care market that leverages the use of sensor technology to lower elder care costs [29]. Additionally, technology is increasingly accepted and familiar to older adults and less of a barrier to a high-tech home health care market – internet use by 65 + individual for instance increased from 14% in 2001 to 66% in 2018 and during the same timeframe social media use increased from 14% to 37% [30].

Finally, given the substantial shift in educational attainment over the past several decades [31], there is an opportunity to introduce high-tech care tools that, by utilizing and nurturing older adults' existing competencies, provide a platform to share their skills, knowledge, experience and wisdom with their communities. This thinking is central to recent initiatives, such as *The Amazings* [32] or initiatives where students and older adults share facilities in retirement homes [33].

The above discussion and details not only point out a global urgency: they point out a need for products, services, infrastructure and systems with a clear focus on and understanding of aging populations. As cities worldwide devise their smart city plans, it is clear that not addressing the needs and realities of this growing segment would have fatal consequences, with cross-segment repercussions. Because of that, many cities are designing or adjusting their comprehensive plans to address their aging populations.

1.2   Vision and Implications

Given the context outlined in the previous sections, my research endeavors in this space advocate for the use of unobtrusive home-based sensors technologies to (1) support older adults' emotional, intellectual, social wellness, (2) enabling them to live safely in their homes for as long as possible and (3) retaining their sense of independence and self-confidence longer. My hypothesis is that intelligent systems and technologies can help creating social connections and communities that leverage the skills and intellectual capital of homebound seniors, hence promoting their emotional, intellectual, social wellness (Table 1), and therefore addressing their overall wellness while lowering burden and costs.

Table 1. Dimensions of wellness [34, adapted from 35].

Dimensions of wellness	Definition
Occupational	Ability to contribute unique skills to personally meaningful and rewarding paid or unpaid work
Social	Ability to form and maintain positive personal and community relationships
Intellectual	Commitment to lifelong leaning through continual acquisition of skills and knowledge
Physical	Commitment to self-care through regular participation in physical activity, healthy eating, and appropriate health care utilization
Emotional	Ability to acknowledge personal responsibility for life decisions and their outcomes with emotional stability and positivity
Spiritual	Acquiring purpose in life and a value system

Existing literature grounded the decision to focus specifically on emotional, intellectual, and social wellness. Firstly, research not only shows a connection between physical activity and quality of life in older adults [36], but also that physical and emotional wellness are intertwined [37–39] and that positive emotions initiate upward spirals toward enhanced emotional wellbeing [40]. Secondly, while there are correlations between social support and physical health [41], social support, companionship, and control/regulation [42] impact health and can be provided by diverse social partners. Thirdly, cognitive decline may be prevented, slowed or reversed when engaged in creative, challenging, stimulating activities [43]. While having few social ties, poor integration and social disengagement are risk factors for cognitive decline [44, 45], those receiving more emotional support have better baseline cognitive performance [45]. Finally, there are correlations between emotional wellness and social behavior [46] as well as between social support and emotional wellbeing [47, 48].

In light of the above, investigations and developments I here report focused on using intelligent systems and technologies to provide to older adults the opportunity to: be - and feel - connected on their own terms; feel independent, enabled and valued; and be intellectually active. Specifically, experimentations focused on leveraging sensing technology, sensor fusion, emotion understanding and activity recognition to:

- Facilitate connections benefiting older adults by leveraging their own knowledge, past history, preferences, emotional state or patterns, needs, and capabilities;
- Act as emotional and intellectual companions (when others are not around, needed or wanted) by tracking and building on vocal and behavioral cues; and
- Recommend activities or trigger context-centric actions that do not burden to overcome social isolation or downward emotional spirals.

While the vision outlined in this section represents my ultimate goal of the project, the research effort is still in progress. The next sections discuss progress to date.

2 Methodological Approach

The effort at the center of this paper focuses on the use of unobtrusive sensor technologies in domestic environments, to detect older adults' behavioral patterns and then automate voice-based and screen-based interventions when pattern changes are detected. To achieve this, my work incorporates diverse techniques:

- Secondary research, global surveys, interviews and participatory workshops;
- Development of ad-hoc unobtrusive sensor-based systems and technologies and identification of off-the-shelf options;
- Analysis of existing datasets, to identify patterns and correlations, utilizing literature-derived inferences;
- In-home data collection with unobtrusive sensors and technologies, alongside user experience research (e.g. interviews and surveys) and telemetry from users' PCs;
- In-lab data collection with unobtrusive sensors and technologies while users engage in scripted or unscripted tasks and/or to ensure prototypes' usability, functionality, durability and overall value proposition.

This paper overviews data from secondary research (market analysis), global surveys (in US, PRC and Germany) preliminary interviews and workshops (in US), which focused on: perspectives of intelligent systems, with emphasis on smart home, autonomous vehicles (AV) and smart workspaces; the role of Ambient Computing [49] and Affective Computing [50] in everyday contexts; and participants' routines.

The market analysis, which grounded the decision to focus subsequent phases on smart home, AV and smart workspaces, looked at on AI through the lens of diverse verticals (home, office, factory, retail, entertainment, public transport, automotive, classroom, learning) and vectors (players, products, academic research, investments, partnerships, associations, mergers and acquisitions, policies, events).

Survey (~600 participants, of which ~200 were 65+) and 18 initial in-home interviews focused on two key areas: perceptions, attitudes, thresholds and expectations of intelligent systems; and perspectives toward specific applications in home, AV, and workspace contexts. It should be noted that this first part of the research was used to develop design criteria for those that develop intelligent systems [51]. The screener used to recruit participants focused on diverse criteria (e.g. age; gender; device ownership; purchase intention) and had soft quotas (e.g. family composition; income) as well as a natural fallout for intelligent system knowledge and Intel's segmentation. Through jargon-free descriptions and specific usage examples, I engaged participants in a series of activities: general discussion on intelligent systems and assessment of their comfort level with AI in four contexts (i.e. home; car; workspace; classroom); clustering exercise of AI usages in four areas (i.e. must have; nice to have; do not want; not sure); and assessment of their comfort level with and comparative evaluations of specific home, AV and workspace usages.

For each tested item, I collected a series of metrics to facilitate comparative analysis:

- One to five ratings to identify comfort levels or assess concepts on seven parameters (relevance, uniqueness, appeal, quality, comfort, excitement, trustworthiness);
- Word-based criteria, to gather associative feedback by selecting three items from a list of value-centric adjectives (e.g. exciting; creepy); and
- Emotion-based criteria, to gather emotional feedback by selecting three items from a list of emotions (e.g. love/desire; worried/fearful).

In-home interviews (two hours/participant) I mixed observational techniques (e.g. home tour) with a semi-scripted interview approach that mirrored the survey's protocol and criteria. After survey and in-home interviews, I invited some interviewees to a workshop, to explore key themes and co-create a manifesto to regulate intelligent systems futures. I then co-conducted similar workshops in conference settings, to gather expert input [52, 53].

Subsequent exploratory research on *Aging in Place* leveraged data from additional interviews (still in progress) and in-lab data collection using distributed sensors. This second round of interviews focuses on gaining a high-level understanding of participants' preferences, routines and attitudes toward smart home systems as well as an in-depth knowledge of each participants' daily routines, including discussions on home care (e.g. ironing; cleaning; storing; etc.), pets or garden care, sleeping habits, hygiene, cooking and eating habits, entertainment activities and hobbies, and social habits. In additional ethnographic sessions, I will also include a home tour and routine simulations within participants' everyday environments. In-lab data collection focused on a series of scripted usages and leveraged a number of distributed technologies installed in a laboratory setup to simulate kitchen and living room areas. The setup included point cloud cameras, RGB HD resolution cameras, low resolution and ultra-low resolution thermal cameras, microphones, pressure array mat, and several networking and storage devices. Additionally, during some interviews, we experimented with live or post-production sketching as additional research tool. This last technique will be discussed in future publications.

3 Key Findings to Date

While key findings from all users engaged in survey and initial home interviews are included in a previous publication [51], this section provides a summary of finding from all to-date activities that relate to older adults.

At a high level, 65+ participants expressed concerns with the potential for AI-based systems to impact their privacy (*Not opposed, but cautious about how information that is collected will be used.* Jim 65), security (*These things can be hacked so I expect them to be designed so they are safe.* Sonia 66) and sense of autonomy (*If it's the way it works, fine. However, I still want to make my own choices.* Carla 66). When reviewing usages that leverage Affective Computing [50], they saw benefits yet expressed great skepticism (*The issue is not with discomfort with the action but doubts that it can do it properly and reliably.* Sheila 69) as well as irritation (*Too personal, too close. Here is a (new) device to get mad at for checking out my emotions.* Carla, 66). To the notion of systems with their own personality and autonomy, participants provided negative (*Too much control. It's trying to act like a person and I do not want a machine to do that.* Jane, 65), annoyed (*Maybe it would assume things that are not true... maybe I am not that predictable.* Jim, 65) and entertaining commentary (*Have enough personalities in my life, thank you very much!* Monica 79). Again, these usages greatly challenged their sense of autonomy (*I do not want this as then it is no longer a helper.* Sonia 66) and self-worth (*I do not want a machine to do what I am capable of doing.* Sheila, 69).

When surveying older adults' comfort levels with the presence of AI systems in different locations (home, workspace, classroom and car) on a 1 to 5 scale, about a third expressed very high comfort with home and car-based applications (32% and 31% respectively). In response to specific (yet location-agnostic) usages, participants had very clear preferences (Table 2).

As depicted in Table 2, participants for instance negatively reacted to usages implying machine-autonomy in contexts where the end user may not have self-determination (e.g. provide childcare/babysitting) and to usages that referred to emotion (e.g. detect/react to emotions to personalize experiences) or that alluded to machines with their own agency (e.g. have its own personality and perspectives). On the other hand, they responded positively to usages implying a clear power relationship structure, where machines are subordinates (e.g. remind me of tasks and meetings) and where the human is in full charge (e.g. ask before acting or automating). At the same time, participants were clearly willing and open to embrace and leverage an intelligent machine's recommendations and abilities, to prevent issues (e.g. know what I do to prevent mistakes) or even be challenged to avoid them (e.g. challenge wrong decisions).

Table 2. Usage clustering (N = 200; ages 60 + ; usage descriptions shortened).

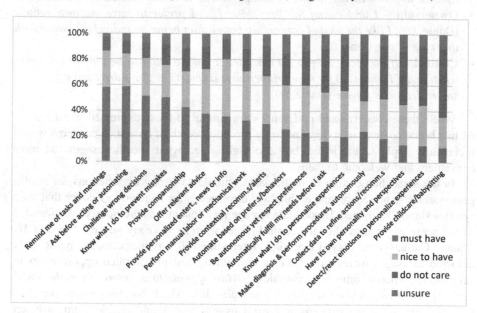

Interestingly, a high number of 60 + participants responded very positively to the notion of leveraging artificially intelligent technologies and services to provide companionship, specifically in the context of elder care and care of other adults in need. In fact, 74% of respondents aged 60–69 (n = 100) labelled the companionship usage as "must have" or "nice to have" and 67.5% of respondents aged 70+ (n = 77) labelled the same usage as "must have" or "nice to have". Interviews clearly confirmed this specific data point, because, as Jane (65), who takes care of her aging mother, pointed out: *There are so many people that do not have ability or family to stay with or have around… people get ill faster because of that.*

Additionally, older adults that participated in the research had much to say in relation to specific usages that leveraged AI in home and autonomous vehicle (AV) contexts. From an AV context, I identified for instance a series of concerns and preferences:

- **Reliability** – *What about reliability? How does it keep up to date?* (Carla 66); *I need it to be as bulletproof as possible in a car that drives itself* (Dante, 69)
- **Societal impact** – *What is this contributing to with regard to society? what happens to drivers? will they be displaced?* (Carla, 66); *I think it's going to be hard… in the US people equate themselves to their cars* (Jim, 65)

- **Practicalities** – *How can it take into account all passengers preferences? I do not have something to hide but it could take away my individuality. Is the system transferred to the new car or you keep it in the car?* (Nadia, 65)
- **Purpose** – *That seems unnecessary with all the things that need to be done in the world* (Monica, 79); *I can see you enjoying more what is in and out of the car* (Sheila, 69); *Not driving to do something else... valuable time saving* (Dante, 69)
- **Ownership** – *I understand the fleet thing (...) I prefer to have my own vehicle* (Dante, 69); *I like the idea of carpooling better than being an individual owner with this type of car* (Jane, 65)
- **Control** – *If the tech is right, I'd consider it but I'd want to be able to take over when and if I want (…) for now I rather have a smart car that I drive but that makes smart decisions for me* (Dante, 69).

While these concerns and preferences are similar to those expressed in relation to AI in a home context, interviews not only provided depth of how such concerns would specifically apply to the home, but also highlighted what specific usages felt most convenient and welcome to them.

In the smart home context, participants articulated strong expectations for intelligent systems to be secure, and highlighted clear needs for hands-on evidence that such systems should be trusted. Brand, data storage location and, as previously articulated, a strong sense of predictability and control (*I still want to make my own choices. It's My Choice.* Carla, 66) all played key roles. Interesting, 60 + participants felt particularly attracted to usages focused on distance monitoring and that implied opportunities to save money, energy, time and frustrations (*What appeals to me more about the smart home is that it can help me do things from afar.* Jim, 65). From this perspective, they provided openness and excitement toward usages that would enable them to better maintain, upgrade, and protect their homes.

More in relation to their specific age bracket, older participants positively remarked on usages with focus on use of audio, light switches or projections to help them find items; activity monitoring to prevent or address accidents; tracking of valuable data with automatic inclusion to their health chart; and tracking of physical activity to detect early sign of disease. On the other hand, older participants provided rather negative responses to home usages focused on monitoring or facilitating kids' play or school activities without parental supervision.

When comparing the global survey's pre- and post-ratings for AV and smart home on a scale 1 to 5, data shows interesting differences. Table 3 outlines ratings shifts, by age groups, when we asked the same question on purchase likelihood (*how likely are you to consider purchasing something like that in future?*) twice: at start (based on existing knowledge) and end of the survey (based on knowledge accumulated through the research).

While we anticipated a decline in ratings in relation to age brackets, we somehow expected higher margins. Besides interesting granularity seen by comparing the 60–69 versus the 70+ bracket, there are interesting differences in how older adults responded to AV versus smart home and in their higher openness to embrace smart home systems, which have arguably a higher potential to be perceived as intrusive.

Table 3. Pre and Post ratings for AV and smart home (<50 Yrs. n = 328; 50–59 Yrs. n = 102; 60–69 Yrs. n = 100; >70 Yrs. n = 77)

	<50		50-59		60-69		>70	
AV Likely to Purchase (Would consider definitely/probably)	CD Pre: (A) 78%	Post: CD 74%	CD Pre: (B) 75%	Post: D 68%	Pre: (C) 53%	Post: 57%	Pre: (D) 60%	Post: 45%
Smart Home Likely to Purchase (Would consider definitely/probably)	BCD Pre: (A) 81%	Post: D 77%	D Pre: (B) 71%	Post: D 73%	Pre: (C) 61%	Post: D 69%	Pre: (D) 55%	Post: 53%

4 Opportunities

As stated earlier, the hypothesis at the center of my research is that intelligent systems and technologies can help creating social connections and communities that leverage the skills and intellectual capital of homebound seniors, hence promoting their emotional, intellectual, social wellness (Table 3), and therefore addressing their overall wellness while lowering burden and costs. Insights outlined in Sect. 3, alongside ongoing research activities, are not only deepening and reiterating the reasons why (and how) intelligent systems should and could bring value to older adults, but also appear to support the offered hypothesis.

Given identified insights and existing literature, I propose that ambient computing opportunities for this population segment have ramifications in three entwined areas, all fundamental in the contest of future urban life and smart cities.

The first area, which I call *Community and Companionship*, is about leveraging distributed sensing, contextual understanding and Affective Computing [50] to equip older adults with opportunities to reach out, participate, belong, contribute – ultimately feeling connected, empowered and supported. This approach would help break down social isolation and increase older adults' sense of agencyand self-worth, while enabling them to contribute to society and their communities and promoting the sentiment that one's skills are recognized, cherished and valuable. Imagine an ambient system that:

- Enables sharing and co-management of resources with neighbors, through crowd-sourced local sensor data;
- Detects social isolation to provide recommendations and enable social and community interactions by integrating real-time emotion status, contextual details, local services, and knowledge of one's history and preferences;
- Detects emotional or social distress to prompt communication with loved ones;
- Uses humor and multi-modal techniques and tools to deflate argument or aid distressing social situations;
- Auto-records family memories when user-defined trigger words are used;
- Provides transparent user data in community living spaces; or
- Connectsone's skills and superpowers with local needs (for instance, an ex-mathematician may become the perfect tutor for a struggling teen that lives nearby).

The second area, *Wellness and Care*, focuses on leveraging ambient intelligence, context analytics and Affective Computing [50] to advance older adults' ability to self-care and be cared for, on their terms. This means using distributed intelligence to scaffold (vs direct) their ability to nurture their physical as well as emotional, social and intellectual wellness and independence. Imagine an ambient system that:

- Tracks everyday activity to detect early signs and symptoms of cognitive decline;
- Tracks diverse sets of valuable data and automatically adds it to one's health chart;
- Tracks activities and provides practical support such as health prompts, assistance calls, incident alerts or even service ordering;
- Leverages vision and voice activation to leverage and access private (e.g. in home or care facility) and public (e.g. transportation, shared facilities) services;
- Detects anomalies in everyday patterns to automate behaviors, recommendations and interventions that focus on nurturing and enabling personal wellness;
- Leverages multi-modal means (e.g. audio, visual, haptic, projected) to coach and provide instructions to self-care, address concerns or cope with distress; or
- Leverages voice, text, or visuals to promote cognitive wellness, delay cognitive deterioration or complement post-trauma therapy.

The third area, *Independency and Management*, focuses on leveraging ambient intelligence and natural user interfaces (UIs) to address key areas such as mobility, and life management. Here voice and other natural modalities can facilitate hands-free usages that are engaging and do not overwhelm less tech savvy users. Imagine, for instance, an ambient system that:

- Keeps track of and finds common objects such as keys, glasses or remote controls;
- Controls and manages one's property, providing alerts and instructions when maintenance or repairs are required, should be considered or are imminent;
- Understands who is giving commands and context to personalize actions;
- Adapts to feedback to refine future behaviors;
- Supports coordination and sense of control in diverse contexts and life stages;
- Offsets tasks through home automation and smart devices based on past behaviors or preferences;
- Provides context-appropriate reminders leveraging rich dialog capabilities;
- Contextually adjusts interaction and dialogue styles (e.g. adjusts speech calibration) to facilitate desired outcomes, based on real-time feedback, preference settings or tracking of behavioral patterns;
- Automatically provides to drivers or smart vehicles (including autonomous and shared) key details to facilitate or customize a transportation event.

5 Conclusive Remarks

In this paper, I shared insights from an ongoing research endeavor focused on the opportunities offered by distributed sensing and ambient computing in support of growing aging populations. Reflecting on data gathered to date, I proposed that ambient computing opportunities for this population segment have ramifications in three entwined areas, all fundamental in the contest of future urban life and smart cities: Community and Companionship; Wellness and Care; and Independency and Management.

Given the discussed context, it is clear that there is a global urgency to design and develop products, services, infrastructure and systems with a clear focus on and understanding of aging populations. As cities worldwide devise their smart city plans, many are focusing on addressing the needs and realities of this growing segment.

In the US for instance, following guidance from AARP's Network of Age-Friendly Communities [54], many cities launched comprehensive plans in 2017 with focus on *aging in place*. Equipping one to successfully age in place is not an easy task, as it requires focus on multiple, diverse needs (e.g. home, community or city supporting mobility) and barriers (e.g. reduced mobility). Equipping one to successfully age in place through ambient technologies comes with additional complications. If we just focus on health-related technologies, for instance, a number of challenges arise. Dishman, Matthews and Dunbar-Jacob [55] articulated this very well when they pointed at six key challenges: going beyond contemporary clinical and computing models (imagination); finding and prioritizing problems to pursue (identification); concept testing and refinement (iteration); deep dives on enabling technologies (infrastructure); exploration of human-machine interaction (interfaces); and testing whole systems in situ (integration).

Putting challenges aside and inspired by existing literature (e.g. [28, 29, 55–57]), this paper argues that ambient computing could play a key role in scaffolding *aging in place* in multiple contexts, from the home to the city. Moreover, insights from current research analysis show that older adults are open to embrace ambient computing in diverse contexts provided that ambient intelligent systems:

- Have clear purpose and no societal impact;
- Respect older adults' sense of and need for autonomy and equip them with a strong sense of predictability and control;
- Are designed to be and remain subordinates that ask before acting;
- Do not have autonomy, especially when the user does not have self-determination;
- Do not have their own personality, as this implies that they may have autonomy;
- Focus on assisting (versus controlling, deciding, or dictating) and providing recommendations to enable or prevent issues;
- Fully meet their stringent expectations in terms of privacy, security and reliability;
- Provide *helper* usages to save money, energy, time and frustrations, to maintain, upgrade, protect their assets, and to provide companionship to those in need; and
- Are clear on how emotion recognition is utilized, and why.

I argue that the need for intelligent products, services, infrastructure and systems for aging populations goes alongside the need for developers and designers that have the ability to shape ambient computing usages and technologies in ways that are respectful of and grounded in an understanding of older adults' everyday life – their practices, desires, expectations, and thresholds.

In a previous publication [51], I discussed ten guidelines for intelligent futures and offered them to designers and developers as practical people-centric recommendations to "spark a healthy debate on the processes used to develop intelligent systems and the agency that designers and developers have and should have in such processes". While I do not go in details on such guidelines in this paper, at a high level they are as follows:

1. Take a firm, unambiguous ethic stand – be a trusted brand;
2. Adopt the minimize intrusion mantra and a less-is-more approach;
3. Design socially trusted & trustworthy platforms;
4. Do not make systems human, but capable of helping humans;
5. Prioritize usages that matter – helper usages;
6. Design systems with consistent behaviors, yet design for serendipity;
7. Make people feel unique and empower their unique goals;
8. Create multiple and diverse educating tools;
9. Design on-boarding mechanisms that grow and evolve; and
10. Create families of products.

While these ten guidelines apply to any intelligent system for any end user type, I propose that, in light of research insights discussed earlier, some guidelines may be particularly critical when designing ambient systems for aging populations. Specifically, I suggest that five of the ten guidelines are key. While open to ambient computing opportunities, older adults shared stringent expectations in terms of privacy, security and reliability – because of this, I suggest great emphasis on guidelines 2 (*Adopt the minimize intrusion mantra and a less-is-more approach*) and 3 (*Design socially trusted & trustworthy platforms*). Secondly, this population segment clearly expressed a need for intelligent systems to respect their autonomy and to remain subordinates without personality, while equipping them with a strong sense of predictability and control – due to this, guidelines 4 (*Do not make systems human, but capable of helping human*) and 6 (*Design systems with consistent behaviors, yet design for serendipity*) are of key interest. Finally, specific discussions of what value and scenarios older adults wish to see, it is clear that they see value in practical directions, to help them minimize (cost, energy consumption, time loss or frustrations) and protect (as well as maintain and upgrade). Guideline 5 (*Prioritize usages that matter – helper usages*) is consequently another key parameter to consider. As mentioned, details and a discussion on each of the 5 recommended guidelines are available in a previous publication [51].

Regardless of guidelines and because of the specific target segment, designers and developers tackling *Aging in Place* contexts not only have the moral and ethical responsibility to engage with how intelligent systems and ambient computing futures are being (and will be) shaped: they must equip themselves with a deep understanding of older adults and with appreciation for the many nuances that categorize their realities. Only by deepening one's knowledge on older adults' everyday life – practices, desires, expectations, and thresholds – and by grounding design and development in such an understanding, truly meaningful and life-changing futures for this growing population segment will be achievable. Given this, I strongly advocate for adopting Participatory Design [1] approaches as a mean to ensure that intelligent and ambient technologies are developed **with** (instead of **for**) end users. A participatory approach will provide unbeatable opportunities to cater to the diverse, complex, nuanced realities of older adults' everyday life.

References

1. Schuler, D., Namioka, A.: Participatory Design: Principles and Practices. Erlbaum, Hillsdale (1993)
2. United Nations, Department of Economic and Social Affairs (UNDESA), Population Division: World Population Prospects: The 2017 Revision, Key Findings and Advance Tables. Working Paper No. ESA/P/WP/248. United Nations, New York (2017)
3. UNDESA Population Division: World Population Prospects: The 2015 Revision, DVD Edition. United Nations, New York (2015)
4. Farber, N., Shinkle, D., Lynott, J., Harrell, R.: AARP public policy institute and national conference of state legislatures research report. 2011–13. Washington, DC (2011)
5. Ornstein, K., et al.: Epidemiology of the Homebound Population in the United States. JAMA Int. Med. **175**(7), 1180–1186 (2015)
6. Steptoe, A., Shankar, A., Demakakos, P., Wardle, J.: Social isolation, loneliness, and all-cause mortality in older men and women. PNAS **110**(15), 5797–5801 (2013)
7. U.S. Census Bureau: 2010 Census Summary, Census of Population and Housing (2012)
8. Dykstra, P., Fokkema. T.: Relationships between parents and their adult children: a West European typology of late-life families. In: Ageing and Society, pp. 1–25. Cambridge University Press (2010)
9. Bengtson, V.L.: Beyond the nuclear family: the increasing importance of multigenerational bonds. J. Marriage Family **63**(1), 1–16 (2001)
10. Hagestad, G.: The aging society as a context for family life. Daedalus **115**(1), 119–139 (1998)
11. Matthews, S., Sun, R.: Incidence of four-generation family lineages: is timing of fertility or mortality a better explanation? J. Gerontol. Soc. Sci. **61B**(2), S99–S106 (2006)
12. Seltzer, J., et al.: Explaining family change and variation: challenges for family demographers. J. Marriage Family **67**(4), 908–925 (2005)
13. Esping-Andersen, G.: Social Foundations of Postindustrial Economies. Oxford University Press, Oxford (1999)
14. Blossfeld, H.: The New Role of Women: Family Formation in Modern Societies. Westview, Boulder (1995)
15. Blossfeld, H., Huinink, J.: Human capital investments or norms of role transition? How women's schooling and career affect the process of family formation. Am. J. Sociol. **97**(1), 143–168 (1991)
16. Hakim, C.: Work-Lifestyle Choices in the 21st Century: Preference Theory. Oxford University Press, Oxford (2000)
17. Hareven, T.: Historical perspectives on the family and aging. In Blieszner, R., Bedford, V. (eds.), Handbook of Aging and the Family, pp. 13–31, Greenwood, Westport CN, (1995)
18. Lewis, J.: The End of Marriage?. Individualism and Intimate Relations. Edward Elgar, Cheltenham (2001)
19. National Council on Aging: Nearly 7 Million Long-Distance Caregivers Make Work and Personal Sacrifices (2006)
20. AARP Public Policy Institute: Valuing the Invaluable: The Economic Value of Family Caregiving (2008)
21. National Alliance for Caregiving and the MetLife Mature Market Institute: Miles Away: The MetLife Study of Long-Distance Caregiving (2004)
22. National Alliance for Caregiving and AARP: Caregiving in the U.S. (2004)
23. Taylor, P., Parker, K., Patten, E., Motel, S.: The Sandwich Generation: Rising Financial Burdens for Middle-Aged Americans. Pew Res. Center, Washington DC (2013)

24. U.S. Census Bureau: American Community Survey (2011–2015) and American Housing Survey (2011)
25. Gurnon, E.: The staggering prices of long-term care (2017). https://www.forbes.com/sites/nextavenue/2017/09/26/the-staggering-prices-of-long-term-care-2017/#294d318f2ee2. Accessed 30 Jan 2019
26. Moeller, P.: Long-term care costs favor home-based treatment. https://money.usnews.com/money/blogs/the-best-life/2013/04/09/long-term-care-costs-favor-home-based-treatment. Accessed 30 Jan 2019
27. Caring, LLC: Technologies to reduce care costs and allow safe aging at home (2016)
28. Kayyali, B., Kimmel, Z., Van Kuiken, S.: Spurring the market for high-tech home health care. McKinsey (2011)
29. Deloitte: Using sensors technology to lower elder care costs. Wall Street Journal (2014)
30. Per Research Center: Internet and social media use by age. http://www.pewinternet.org/
31. U.S. Census Bureau and Census and Community Survey 1-year estimates: Educational Attainment by age and sex: 1970, 2010, and 2040 (1973, 2012, 2010)
32. Cargo Collective: The Amazings. https://cargocollective.com/cookie/The-Amazings. Accessed 30 Jan 2019
33. Thielking, M.: Baked fish, chair yoga, and life lessons: to learn to care for elderly, students move into retirement home. Stat News, 28 April (2017)
34. National Wellness Institute: Six dimension of wellness, https://www.nationalwellness.org/page/Six_Dimensions. Accessed 30 Jan 2019
35. Hettler, W.: The six dimension of wellness (1976), http://www.hettler.com/sixdimen.htm. Accessed 30 Jan 2019
36. Acree, L., et al.: Physical activity is related to quality of life in older adults. Health Q. Life Outcomes 4(37) (2006)
37. Fredrickson, B.L.: What good are positive emotions? Review of general psychology. J. Div. 1 Am. Psychol. Assoc. 2(3), 300–319 (1998)
38. Fredrickson, B.L.: Cultivating Positive Emotions to Optimize Health and Well-Being. Prev. Treat. 3(1), 1a (2000)
39. Stegeman, M.: The relations between health and wellbeing. Thesis, Twente University (2014)
40. Fredrickson, B., Joiner, T.: Positive emotions trigger upward spirals toward emotional well-being. Psychol. Sci. 13(2), 172–175 (2002)
41. Clark, C.: Relations Between Social Support and Physical Health. http://www.personalityresearch.org/papers/clark.html. Accessed 30 Jan 2019
42. Rook, K., August, K., Sorkin, D.: Social network functions and health. In: Contrada, R., Baum, A. (eds.) The Handbook of Stress Science: Biology, Psychology, and Health. Springer, New York (2011)
43. Strout, K., Howard, E.: Five dimensions of wellness and predictors of cognitive health protection in community-dwelling older adults. J. Holist. Nurs. 33(1), 6–18 (2015)
44. Zunzunegui, M., Alvarado, B., Del Ser, T., Otero, A.: Social networks, social integration, and social engagement determine cognitive decline in community-dwelling spanish older adults. J. Gerontol. Ser. B 58(2), S93–S100 (2003)
45. Seeman, T., Lusignolo, T., Albert, M., Berkman, L.: Social relationships, social support, and patterns of cognitive aging in healthy, high-functioning older adults: MacArthur studies of successful aging. Health Psychol. 20(4), 243–255 (2001)
46. Isen, A.: Positive affect, cognitive processes, and social behavior. Adv. Exp. Soc. Psychol. 20, 203–253 (1987)
47. Abbey, A., Abramis, D., Caplan, R.: Effects of different sources of social support and social conflict on emotional well-being. Basic Appl. Soc. Psychol. 6(2), 111–129 (1985)

48. Blazer, D.: Social support and mortality in an elderly population. Am. J. Epidemiol. **115**, 684–694 (1982)
49. Gunnarsdottir, K., Arribas-Ayllon, M.: Ambient intelligence: a narrative in search of users. Discussion Paper. Lancaster University (2011)
50. Picard, R.: Affective Computing. MIT Press, Cambridge (1997)
51. Loi, D.: Ten Guidelines for intelligent systems futures. In: FTC2018 Future Technologies Conference 2018, Vancouver BC, Canada (2018)
52. Loi, D., Raffa, G., Arslan Esme, A.: Design for affective intelligence. In: 7th Affective Computing and Intelligent Interaction conference, San Antonio, TX (2017)
53. Loi, D., Lodato, T., Wolf, C., Arar, R., Blomberg, J.: PD manifesto for AI futures. In: Proceedings of the 15th Participatory Design Conference2, Belgium (2018)
54. AARP: The AARP Network of Age-Friendly Communities. https://www.aarp.org/livable-communities/
55. Dishman, E., Matthews, J., Dunbar-Jacob, J.: Technologies for Adaptive Aging. National Academies Press, Washington (2004)
56. Kaye, J.: Making pervasive computing technology pervasive for health & wellness in aging. Pub. Policy Aging Report **27**(2), 53–61 (2017)
57. Kaye, J., et al.: Methodology for establishing a community-wide life laboratory for capturing unobtrusive and continuous remote activity and health data. J. Vis. Exp. **137**, e56942 (2018)

Designing and Evaluating Technology for the Dependent Elderly in Their Homes

Maria João Monteiro[1] , Isabel Barroso[1] , Vitor Rodrigues[1] ,
Salviano Soares[2] , João Barroso[2] , and Arsénio Reis[2(✉)]

[1] University of Trás-os-Montes and Alto Douro, Vila Real, Portugal
{mjmonteiro, imbarroso, vmcpr}@utad.pt
[2] INESC TEC, University of Trás-os-Montes and Alto Douro,
Vila Real, Portugal
{salblues, jbarroso, ars}@utad.pt

Abstract. The ageing population and the increasing longevity of individuals is a challenging reality for healthcaretoday. Longevity often leads to increased dependence and the need for continued care, whichis often left to informal caregivers given the inability of the elderly care network to provide. The informal care provided to the dependent elderly occurs either at the caregiver's or the elderly person's home. Technology application in healthcare has been attracting the attention of engineers for a long time, especially in providing support for health recovery and maintaining therapy practices. Due to major advances in technology, particularly in movement capture optic systems and information extraction through digital image analysis, support systems are being created to monitorhow therapeutic plans are carried out, as well as to evaluate people's physical recovery or to assist healthcare professionals and informal caregiver to provide care for dependent elders.

This paper reports on the accomplishments of a project with the general objective of using information and communication technologies to develop a prototype for a system focused on monitoring and assisting the execution of a therapeutic plan integrating physical mobilization and medication.

Keywords: Technology · The elderly · Surveillance · Caregivers

1 Introduction

The ageing population and the increasing longevity of individuals is a challenging reality forhealthcaretoday. Longevity often leads to increased dependence and the need for continued care, whichis often left to informal caregivers, given the inability of the elderly care network to provide The informal care provided to the dependent elder occurs either at the caregiver's or the elderly person's home. This caregiver may be a family member, a friend or a neighbor, as they are someone who is part of the elderly person's informal social support network, who assumed the responsibility for the care, either voluntarily or as an "obligation". Care provision requires continuous organization and effort from the elderly person and from the caregiver. Assuming responsibility for the care commonly occurs when the caregiver experiences difficulties in carrying

© Springer Nature Switzerland AG 2019
M. Antona and C. Stephanidis (Eds.): HCII 2019, LNCS 11573, pp. 506–510, 2019.
https://doi.org/10.1007/978-3-030-23563-5_40

out the therapeutic plans established by the health team. Technology application in healthcare has been attracting the attention of engineers for a long time, especially in providing support for health recovery and maintaining therapy. Due to major advances in technology, particularly in movement capture optic systems and information extraction through digital image analysis, support systems are being created to monitor how therapeutic plans are carried out, as well as to evaluate people's physical recovery or to assist healthcare professionals and informal caregiver to provide care for dependent elders. In other projects related to accessibility and elderly support, technology has already been successfully adopted [1, 2].

This paper reports on a project proposal with the general objective of using information and communication technologies to develop a prototype for a system focused on monitoring and assisting the execution of a therapeutic plan integrating physical mobilization and medication.

As part of the project, an aid and surveillance program was designed, which was backed up by a sensor network. In addition, a methodology was created to assess its impact on promoting the health care of the dependent elderlyin their homes.

The assessment was designed to include six stages:

- Evaluating the dependent elderly person's health statusat home, under the care of an informal caregiver.
- Identifying the established therapeutic plan and the major difficulties found by the informal caregiver while executing this plan.
- Monitoring the user/caregiver's adherence to the therapeutic plan.
- Implementing a therapeutic plan assistance and vigilance program through a sensor network.
- Evaluating the impact of the therapeutic plan assistance and vigilance program through a sensor network on promoting the elderly person's health.
- Evaluating the impact of the therapeutic plan assistance and vigilance program through a sensor network on assisting the informal caregiver.

The project is halfway through the implementation phase, therefore this paper focuses on the challenges and options of the activities and prototype design, as well as on the adoption of the assessment methodology.

2 The Portuguese Population

According to the PORDATA data from the Francisco Manuel dos Santos Foundation [3], the ageing index of the Portuguese population, defined as the ratio of the number of elderly people of an age when they are generally economically inactive (65 and over) to the number of young people (from 0 to 14), is dramatically increasing. In 2017 the ageing index was 153, 2%, combined with the increase of longevity and the increase of dependency.

The elderly above 60 years old are in the age range that will increase the most. In the next few years it is estimated that the number of people over 65 years old will surpass the number of children under 5 years old [4].

Considering the ageing and dependency statistics and also the fact that, according to the Portuguese Health Regulator Agency, Portugal has the highest rate of informal home caring by a resident in the same home, it is essential that action is taken to provide support to the informal care givers in their daily care activities. Many of them dedicate most, if not all, their working day to providing care, consequently there is a need to empower them with tools and to improve their quality of life.

3 Ageing and Comorbidity

Some of the elderly assisted by informal care givers in home caring environments have mobility issues and disabilities, meaning they are often bedridden. The lack of movement can cause musculoskeletal pathophysiology changes that promote deformities and postural alterations, leading to the on-set of pressure injuries. In the absence of regular care and proper interventions, these changes can affect the quality of life of the elderly and predispose them to the appearance of diseases [5].

A Pressure Injury (LPP) is defined as localized damage to the underlying skin and/or soft tissue, striking regions of bone saliences or in regions of continued contact with equipment or devices that cause continued or intense pressure, combined or not, with friction and/or shear. The pressure on bone saliences affects the blood circulation promoting cell death and the consequent appearance of these lesions in places of greater risk, such as the occipital, scapular, sacral, ischial, trochanter, iliac crest, knee, malleolus, and calcaneus [6].

Providing care to an elderly person may require the caregiver to restructure their lives, depending on the care necessity, its complexity or duration. The care giver is often required to change routines and behaviors, which is not always easy, leading to feelings of tension, anguish, and in some cases, even overload. For this reason, the caregiver may either choose to ignore their own needs orneglect the care of those for whom they are responsible r [7].

An important issue in providing care to the elderly is the necessity to comply with their doctor's instructions and treatment prescriptions, in particular regarding medicine administration and schedules. Most of the elderly suffer from several comorbidities and take several medicines throughout the day, often in a complex system, which can be prone to error. Their non-compliance increases the risks of incorrect medicine usage [8].

Benefits of installing a network of sensors:

- To programalerts for positioning warningsat predetermined hours;
- To define alarms in case of forgetfulness or delay in positioning the elderly on the bed;
- To establish alerts for medication schedules.

These alerts/alarms can take the form of audible sounds or written messages and should be available to the caregiver and to other health professionals. To interact with the elderly population, itis very important that the user interface establishes a common context and provides an effective communication as experienced in previous projects [9, 10].

4   The Project

To develop the prototype, a project was designed and implemented with an intervention plan to select and install the sensors network. The project is the SAICT-POL/23428/2016 – "IPAVPSI – Impact of an aid program and surveillance of the therapeutic plan supported by a sensor network, in promoting the health of the dependent elderly in their homes", referenced as: NORTE- 01-0145-FEDER-023428, financed by the Foundation for Science and Technology and co-financed by the Regional Development European Fund (FEDER), through the North Regional Operational Program(NORTE2020).

Two main criteria were used to select the homes where the project would be developed:

- The geographical proximity of the homesto the University of Trás-os-Montes and Alto Douro, where the project is based;
- The availability and willingness of the caregivers to work together with the project.

The initial plan is to include two groups, a control group and an experimental group, with 15 elderly people and 15 caregivers in each group, sharing similar characteristics, such as age, gender and level of dependency.

Subsequently, the research team will apply the ESC scale which evaluates the objective and subjective overload of the informal caregiver, the Barthel Scale to assess the degree of dependence of the person in the accomplishment of their daily life activities and Braden scale to assess the patient's risk of developing pressure injuries.

Acknowledgements. This work was supported by the project "IPAVPSI-The impact of an aid and surveillance programme backed up by a sensor network in the health care promotion of the dependent elder at their homes", (NORTE-01-0145-FEDER-023428), funded byFundação para a Ciência e Tecnologia e co-financed by Fundo Europeu de Desenvolvimento Regional (FEDER), through the Programa Operacional Regional do Norte (NORTE2020).

References

1. Reis, A., et al.: Designing autonomous systems interactions with elderly people. In: Antona, M., Stephanidis, C. (eds.) UAHCI 2017. LNCS, vol. 10279, pp. 603–611. Springer, Cham (2017). https://doi.org/10.1007/978-3-319-58700-4_49
2. Reis, A., Martins, P., Borges, J., Sousa, A., Rocha, T., Barroso, J.: Supporting accessibility in higher education information systems: a 2016 update. In: Antona, M., Stephanidis, C. (eds.) UAHCI 2017. LNCS, vol. 10277, pp. 227–237. Springer, Cham (2017). https://doi.org/10.1007/978-3-319-58706-6_19
3. PORDATA: The Portuguese resident population (2019). https://www.pordata.pt/en/Europe/Ageing+index-1609. Accessed 1 Mar 2019
4. Le Deist, F., Latouille, M.: Acceptability Conditions for telemonitoring gerontechnology in the elderly optimising the development and use of this new technology. IRBM **37**, 284–288 (2016). https://doi.org/10.1016/j.irbm.2015.12.002

5. Assis, V., Vidal, A., Dias, F.: Avaliação postural e de deformidades em idosos acamados de uma instituição …: Sistema de descoberta para FCCN. Revista Brasileira de Ciências Do Envelhecimento Humano, **12**(2), 123–133 (2015). https://eds.a.ebscohost.com/eds/pdfviewer/pdfviewer?vid=3&sid=493c3d95-295d-4be0-b306-898e676a2ccf%40sdc-v-sessmgr05

6. Silva, J., Santos, C., Zoche, D., Argenta, C., Ascari, R.: DIAGNÓSTICOS E CUIDADOS DE ENFERMAGEM PARA PACIENTES COM RISCO DE LESÃO POR…: Sistema de descoberta para FCCN. Braz. J. Surg. Clin. Res., **20**(1), 2317–4404 (2017). https://eds.b.ebscohost.com/eds/pdfviewer/pdfviewer?vid=2&sid=48726cf9-4529-42c2-af71-a434c040c2b2%40pdc-v-sessmgr03

7. Nunes, D., Brito, T., Corona, L., Alexandre, T., Duarte, Y.: Idoso e demanda de cuidador: proposta de classificação da necessidade de cu..: Sistema de descoberta para FCCN. Revista Brasileira de Enfermagem, **71**(2), 897–904 (2018). https://eds.a.ebscohost.com/eds/pdfviewer/pdfviewer?vid=6&sid=083d717c-1501-4979-919e-665971f75c22%40sdc-v-sessmgr02

8. Marin, M., Rodrigues, L., Druzian, S., Cecílio, L.: Nursing diagnoses of elderly patients using multiple drugs. Revista Da Escola de Enfermagem Da USP **44**(1), 47–52 (2010). https://doi.org/10.1590/S0080-62342010000100007

9. Khanal, S., Reis, A., Barroso, J., Filipe, V.: Using emotion recognition in intelligent interface design for elderly care. In: Rocha, Á., Adeli, H., Reis, L.P., Costanzo, S. (eds.) WorldCIST'18 2018. AISC, vol. 746, pp. 240–247. Springer, Cham (2018). https://doi.org/10.1007/978-3-319-77712-2_23

10. Reis, A., da Guia, E.B., Sousa, A., Silva, A., Rocha, T., Barroso, J.: An information system to remotely monitor oncological palliative care patients. In: Rocha, Á., Adeli, H., Reis, L.P., Costanzo, S. (eds.) WorldCIST'18 2018. AISC, vol. 746, pp. 388–396. Springer, Cham (2018). https://doi.org/10.1007/978-3-319-77712-2_37

Applying Universal Design Principles in Emergency Situations

An Exploratory Analysis on the Need for Change in Emergency Management

Cristina Paupini[✉] and George A. Giannoumis

Department of Computer Science, Oslo Metropolitan University, Oslo, Norway
paupini.cristina@gmail.com, gagian@oslomet.no

Abstract. The United Nations Convention on the Rights of Persons with Disabilities (CRPD) obligates States' to take all necessary measures to ensure the protection and safety of persons with disabilities in emergency situations. While these requirements represent one aspect of this article's aims, it also focuses on how another paradigm, universal design, can and should offer a useful approach to emergency situations and management. Referring once more to the CRPD, universal design is defined as the "design of products, environments, programs and services to be usable by all people, to the greatest extent possible, without the need for adaptation or specialized design". Consequently, universal design should provide a valid framework for identifying and removing usability and accessibility barriers in emergency situations. Using a heuristic analysis, this article intends to offer a preliminary reflection on the following question: "To what extent can universal design principles be applied to emergency management situations?".

Keywords: Emergency management · Universal design · Accessibility · Disability · Emergency situations · Situational disability · Disability rights

1 Introduction

According to the United Nations Convention on the Rights of Persons with Disabilities (CRPD) and under international law, international humanitarian law, and international human rights law, States Parties have an obligation to take all necessary measures to ensure the protection and safety of persons with disabilities in emergency situations. Emergency situations can refer to a range of events that pose immediate threats to an individual's life, health, or property (Fitzpatrick 1994 in Kutty 2007) and may include, among others, armed conflicts, natural disasters, epidemics, or famine. From a disability rights perspective, emergency situations pose unique barriers both for persons with physical, sensory, cognitive or psychosocial disabilities as well as those with temporary or situational disabilities. Scholarship has conceptualized disability as the interaction between an individual with impairments, the activities with which they want to engage, and the environmental and attitudinal barriers that limit or prevent them from performing those activities (Shakespeare et al. 2006). Situational disabilities

© Springer Nature Switzerland AG 2019
M. Antona and C. Stephanidis (Eds.): HCII 2019, LNCS 11573, pp. 511–522, 2019.
https://doi.org/10.1007/978-3-030-23563-5_41

describe the experience of persons with temporary forms of impairments. For example, a person experiencing sleep deprivation, due to jetlag or other causes, may experience a temporary cognitive impairment and as a result is situationally disabled. Other forms of situational disability, such as hearing impairments, may occur as a result of overexposure to sounds at loud volumes or visual impairments that occur when a person is exposed to smoke during a fire.

Accessibility is one of the primary aims of the disability rights movement, and has been enshrined in national and international legislation including the CRPD under Article 9 (Lid 2010). Under the CRPD, States Parties have an obligation to ensure access to, among other things, the physical environment, to information and communications technology (ICT), and to facilities and services open to the public. In other words, national governments have an obligation to ensure the accessibility of all relevant environments, systems, and services used in emergency situations for persons with disabilities. Article 9 of the CRPD goes on to state, in section 2(b) that accessibility includes the elimination of barriers that persons with disabilities may experience accessing "information, communications and other services, including electronic services and emergency services". While these obligations for accessibility represent one facet of this article's ambitions, another paradigm, universal design, provides a useful basis for considering situational disabilities. According to the CRPD, universal design refers to the "design of products, environments, programs and services to be usable by all people, to the greatest extent possible, without the need for adaptation or specialized design". As such, universal design provides a useful framework for considering the design of emergency management systems and may provide a basis for identifying and removing usability and accessibility barriers in emergency situations.

However, research has, only to a limited extent, examined the usability and accessibility barriers that persons with disabilities and everyone experiences in emergency situations (Christensen et al. 2007; Rowland et al. 2007; Kett and Van Ommeren 2009; Alexander et al. 2012) and has yet to fully examine the accessibility and universal design of emergency management systems and services (Rauschert et al. 2002; Kailes and Enders 2007; Malizia et al. 2008; Nick et al. 2009) This article aims to report on research-in-progress and to provide a preliminary and exploratory analysis of the application of universal design principles to emergency situations. The question this article asks is: "To what extent can universal design principles be applied to emergency management situations?". Using a modified form of a heuristic analysis, this article poses an initial consideration of universal design in emergency situations, and concludes that, among other things, future practitioners could usefully integrate universal design principles in emergency management systems and services to ensure that persons with disabilities and everyone do not experience usability or accessibility barriers in emergency situations. In addition, this article recommends that future scholarship use a universal design framework for empirically examining the universal design of emergency situations from a social equality, human diversity, usability and accessibility, and participatory process perspectives.

2 Analytic Framework

2.1 Human Rights and Emergency Situations

Notwithstanding all the improvements in international rights protection since the Universal Declaration of Human Rights in 1948, national and international governments are not close to achieving an acceptable standard of conduct by States during internal conflict (Fitzpatrick 1994 in Kutty 2007). Research suggests that public emergencies pose a heightened threat of serious and systematic human rights abuse when States employ extraordinary powers to address threats to public order (Oraá 1992). In 1966, the UN adopted the International Covenant on Civil and Political Rights, which obligates States to notify the international community promptly when they suspend their human rights obligations during national crises (art. 41). Two criteria are central to the derogation articles in international human rights treaties: the presence of a public emergency threatening the life of the State, and the necessity to adopt emergency measures due to the exigencies of the situation (Fitzpatrick 1994 in Kutty 2007). Nevertheless, governments all over the world have declared states of emergency in response to a wide variety of crises, including political unrest, general civil unrest, criminal or terrorist violence, labor strikes, economic emergencies, the collapse of public institutions, the spread of infectious diseases, and natural disasters (U.N. Treaty Collection Database). The "Lawless" criteria for declaring a state of emergency, on the other hand, affirms that the threat must be present or imminent, exceptional, and a "threat to the organized life of the community". As a consequence, even the threat posed by terrorist groups such as Al Qaeda seldom justify a declaration of emergency (Criddle and Fox-Decent 2010).

Scholars have suggested an agreement on four non-derogable human rights: (1) the right to life; (2) prohibition of torture; (3) prohibition of slavery; (4) prohibition of retroactive penalties for crimes (Fitzpatrick 1994 in Kutty 2007). To this list, the International Covenant on Civil and Political Rights adds three more non-derogable rights: the prohibition of imprisonment for breach of contract, the right to recognition as a person before the law and the right to freedom of thought, conscience and religion. Left apart the theoretical aspect, there appears to be no universal acceptance of any of these (Fitzpatrick 1994 in Kutty 2007). In the context of the aims of this article, research on human rights and emergency situations suggests that although States may suspend their human rights obligations, it is only under a narrow criteria of what constitutes a state of emergency. It is unclear, however, whether and to what extent obligations for accessibility constitute a non-derogable right for persons with disabilities.

2.2 State of the Art in Emergency Management

Decision Support Systems in Emergency Management

Technology has permeated every aspect of our lives, intertwining itself with the very fabric of our society (Schwab 2016) and bringing important changes and innovations, including widespread and broadly accessible internet, smaller, cheaper and more powerful sensors, artificial intelligence and machine learning (Kuruczleki et al. 2016). This dramatic development naturally brings extensive efforts to apply these technologies to the design of systems for support of systematic risk analysis, decision support systems for operating crews during plant disturbances and accident control, and for support of the general emergency management organization (Andersen and Rasmussen 1988). Numerous analyses have demonstrated how investing in information technologies to a regional, national or international level may vitally improve the capacity to respond to disasters (Hamit 1997; Li and Yang 1997; Xu et al. 1996; Zhu and Stillman 1995).

In recent years emergency preparedness has become a prominent component in national contingency programs in order to improve public health and safety, and emphasis has been placed upon building computer-based decision support systems (DSS) (Chang et al. 1997). Especially now, emergency response organizations face complex and unpredictable events with high risk of catastrophic losses. To assist emergency response organizations in responding to these events, new models must be developed, and the traditional command and control structure of decision-making must be revised to accommodate greater flexibility and creativity by teams (Mendonça et al. 2001). A good decision support system for managing emergencies must be an integrated system, using most advanced computer and communication technologies and should contain strategic components such as a state-of-the-art Geographic Information Systems (GIS), sophisticated models for damage projections and appropriate models for the estimation of the evacuating population and their behavior (Tufekci 1995). Designing emergency response strategies is often complicated by the necessity of simultaneously considering large amounts of relevant data, applicable simulation models, computational speed, display resolution, and required spatial analysis (Chang et al. 1997), therefore in Tufekci's view a valid decision support system would not be merely a collection of processors. On the contrary, it would be a distributed computers and communication network interconnected with intelligence and in constant interaction with one another, like a multiagent architecture (Tufekci 1995). From a universal design perspective, DSS may provide new opportunities for securing the safety of persons with disabilities including persons with situational disabilities in emergency situations.

ICT in Emergency Management

Communication in emergencies, especially among partner organizations, is crucial in order to make informed decisions under uncertain conditions and to engage individuals and communities in collaborative efforts to prepare for, respond to, and recover from disasters (Fleischer 2013). For these purposes in the last years, several information communication technologies (ICTs) have been introduced in emergency management at various levels (Shneiderman and Preece 2007; Jaeger et al. 2007a, b). The still emerging field of ICT in emergency management suggests that service providers continue to face challenges integrating ICTs with contemporary emergency management practices to improve performance and to strategically align ICT use with the needs and requirements of emergency management (Vogt et al. 2011). Adopting ICT in emergency management networks could lead organizations to a better understanding of their potential and to enhance communication and coordination between organizations, strengthening their performance (Hu and Kapucu 2016). Nevertheless, few systematic empirical studies have investigated ICT utilization patterns in emergency management organizations and their potential for easing coordination and communication among those organizations (Bunker and Smith 2009).

Various ICT tools are currently utilized in emergency management both from organizations and local residents, such as geographic information systems (GIS) and global positioning systems (GPS) (Cutter et al. 2007; Vogt et al. 2011) and information systems that contribute to emergency managers' decision-making gathering and processing information (Carver and Turoff 2007). In times of emergency, GIS and GPS can allow organizations to receive satellite information and produce accurate location information about the affected areas and during evacuation procedures locate members of the community, in particular persons with disabilities, to better serve them and, potentially, save their lives (Hu and Kapucu 2016). ICTs can help improve information dissemination and reduce communication costs. When ICTs first started to boom in the 1990s, scholars called attention to integrating information technology into decision-making in emergency scenarios to help mitigate the high levels of complexity and uncertainty associated with such emergencies (Comfort 1993). While traditional radio and TV news remain crucial channels for emergency broadcasting and updates to the general public, the broadly accessible Internet and wireless technologies allow for new channels and methods of communication (Cutter et al. 2007). From a universal design point of view, ICT tools in emergency management may allow organizations to consider more accurately the complex variety of needs that arises in emergency situations.

Social Media in Emergency Management

Disaster situations are non-routine events that result in non-routine behaviors. In times of disaster, people and organizations adapt and improvise (Wachtendorf 2004) to suit the conditions. While traditionally citizens affected by disaster situations relied on emergency officials and news media to provide them with information, in recent years social media has expanded access to information and increased the speed at which information can be distributed (Hughes and Palen 2012).

With growing access to the Internet, the pervasive adoption of mobile technology, and an explosion of social networking services, exponential amounts of socially-generated data are publicly available. For instance, users can share disaster-related information in real-time and seek support through their networks of users, which challenges the belief that emergency officials are the only legitimate source of information (Perng et al. 2012). Diverse social media platforms such as Twitter and Facebook have also been used to support grassroots participation by citizens in emergency management (Hughes and Palen 2012; Veil et al. 2011). Even emergency response organizations, which are strongly organized around locally- and federally-mandated protocols, adapt to accommodate the situation in terms of warning, rescue, and recovery (Sutton et al. 2008), especially since they face new demands by members of the communities to provide information through social media services (American Red Cross 2011). As a consequence of the pressure to incorporate information from the public into emergency response effort, the formal institutions of emergency management are shifting (Palen and Liu 2007). In the United States, the National Incident Management System (NIMS), specifies the structure and procedures for these formal response organizations. Though questions of how and when to participate in the social media arena remain, many PIOs incorporate information and communication technology (ICT) into the public communication aspects of their work (Hughes and Palen 2012). Taken from a universal design perspective, including Social Media in emergency management procedures may crucially improve users' participation in dealing with extraordinary situations, while taking into account the variety of human necessities.

Disability in Emergency Management

According to the U.S. Census Bureau, disadvantaged populations include persons with disabilities, elderly, indigent and illiterate, and cover more than 50% of the U.S. population. Given this demographic data, continuing to use the term "special needs" does a disservice to every group included and greatly weakens the chances of planning for specific needs and providing an effective, comprehensive response (Kailes and Enders 2007). Notably, disasters, terrorism and other emergency situations instantly increase the number of people with new disabilities and functional limitations, both as a physical or psychological consequence of the traumatic event and due to the environment's conditions. In addition, emergencies can intensify an individual's vulnerabilities and fears (Nick et al. 2009).

Disaster preparation and emergency response processes, procedures, and systems can be improved for the population as a whole by eliminating the use of a "special needs" category and integrating consideration of human diversity into the fabric and the culture of emergency management and disaster planning. As long as disability and other socially disadvantaged groups are viewed as unique or special, the system's existing inefficiencies and inefficacies will continue (Kailes and Enders 2007). Adopting the universal design perspective is therefore a necessity in order to overcome the term "special needs" and start considering population as a totality of different individualities, all deserving consideration.

3   Method

Using a heuristic analysis, this article explores the relevant accessibility and universal design factors in emergency situations. Heuristic analyses are typically used to examine the usability of ICT user interfaces, especially regarding websites (Youngblood and Youngblood 2013; Bonastre and Granollers 2014) and this article aims to adapt the heuristic analysis approach to apply it as a framework for an exploratory analysis of human rights in emergency situations. In particular, the heuristic analysis will help frame the necessity for adopting universal design in emergency situations.

A heuristic evaluation or expert review of a web or mobile site, is based on a set of predetermined heuristics or qualitative guidelines often grounded on a set of ten criteria elaborated by Nielsen (1994):

1. Visibility of system status - The system should always keep users informed about what is going on, through appropriate feedback within reasonable time.
2. Match between system and the real world - The system should speak the users' language, follow real-world conventions, making information appear in a natural and logical order.

3. User control and freedom - Users often choose system functions by mistake and the system should support undo and redo.
4. Consistency and standards - Users should not have to wonder whether different words, situations, or actions mean the same thing. Follow platform conventions.
5. Error prevention - a careful design which prevents a problem from occurring in the first place.
6. Recognition rather than recall - Minimize the user's memory load by making objects, actions, and options visible.
7. Flexibility and efficiency of use - Allow users to tailor frequent actions.
8. Aesthetic and minimalist design - Dialogues should not contain information which is irrelevant or rarely needed.
9. Help users recognize, diagnose, and recover from errors - Error messages should be expressed in plain language, precisely indicate the problem, and constructively suggest a solution.
10. Help and documentation - Any information should be easy to search, focused on the user's task, list concrete steps to be carried out, and not be too large.

Using a heuristic analysis, this article explores the application of three sets of universal design conceptualizations and sets of principles. The first set of principles comes from the originally definition of universal design set down by the North Carolina State University in the United States. The second set of principles comes a new definition of universal design posed by the WBDG Accessibility Committee, and the third set of principles comes from in a forthcoming article on universal design from a human rights lens (Giannoumis and Stein 2019).

4 Universal Design in Emergency Situations

4.1 Traditional Conceptualizations of Universal Design

As initially conceived, universal design focused on usability issues. "The design of products and environments to be usable by all people, to the greatest extent possible, without the need for adaptation or specialized design" (Mace 1985). A working group of architects, product designers, engineers and environmental design researchers from the North Carolina State University established seven principles of universal design, with the aim to guide a wide range of design applications and disciplines and for educational purposes. The first principle is the equitable use, and states that the design must be useful and marketable to people with diverse abilities. The second principle affirms that the design should accommodate a wide range of individual preferences and abilities, while the third adds that it also must be simple and intuitive to use. The fourth principle asserts the importance of perceptible information: the design must communicate necessary information effectively to the user, no matter ambient conditions or the user's sensory abilities. To align itself to the fifth principle, the design should have a valid tolerance for error and minimize hazards and negative consequences of unintended actions. The sixth principle establishes the need of low physical effort, allowing the users to handle it comfortably and efficiently and with minimum fatigue. The seventh and last principle states the necessity for appropriate space and size for approach, reach, manipulation and use of the design, regardless of users' body size, posture, or mobility (The Center for Universal Design 1997).

To adopt these principles in emergency management implies that the design of emergency procedures itself has to take into consideration equitable use and contemplate as many human characteristics as possible, in the smoothest way possible. One area to consider in an emergency situation, is the provision of clear information, which is exactly the focus of the fourth principle. Emergency response organizations are faced with complex, unpredictable events with the risk of catastrophic losses (Mendonça et al. 2001) and, therefore, applying the tolerance of error principle would accommodate greater flexibility and creativity by teams. The last two principles highlight the necessity to optimize the action for the lowest physical effort and the most appropriate size and space.

4.2   Universal Design as a Means for Social Inclusion

In the last ten years, the community of universal design practitioners have increased their attention on social inclusion issues and a new definition of universal design was posed by the WBDG Accessible Committee, Steinfeld and Maisel. It defined universal design as "a process that enables and empowers a diverse population by improving human performance, health and wellness, and social participation" (Steinfeld and Maisel 2012), making life easier, healthier and friendlier for all. In order to achieve said definition, Steinfeld and Maisel also indicated eight criteria for universal designing:

- *Body fit*. Accommodating a wide a range of body sizes and abilities
- *Comfort*. Keeping demands within desirable limits of body function
- *Awareness*. Insuring that critical information for use is easily perceived
- *Understanding*. Making methods of operation and use intuitive, clear, and unambiguous
- *Wellness*. Contributing to health promotion, avoidance of disease, and prevention of injury
- *Social integration*. Treating all groups with dignity and respect
- *Personalization*. Incorporating opportunities for choice and the expression of individual preferences
- *Cultural appropriateness*. Respecting and reinforcing cultural values and the social, economic and environmental context of any design project.

In this perspective, universal design provides a tool to develop a better quality of life for a wide range of individuals and to reduce the economic burden of special programs and services designed to assist individual citizens, clients, or customers while supporting people to be self-reliant and socially engaged. It also reduces stigma by putting people with disabilities on an equal playing field. While it would not substitute assistive technology, universal design would definitely benefit people with functional limitations and society as a whole. In the same way, tailoring emergency responses to the widest number of possibilities will reduce the need for specific, "special needs", intervention (Kailes and Enders 2007).

4.3   Universal Design from a Human Rights Lens

The last set of criteria this article considers is elaborated by Giannoumis and Stein (2019) and, while recognizing universal design as crucial in promoting accessibility for persons with disabilities, reframes universal design as a means to promote substantive equality for everyone. The article acknowledges that universal design arises from the complex relationship between human rights, disability rights, and access to and use of technology and, consequently, the article argues that universal design can promote equality through four principles that shift the focus from universal design as an outcome to universal design as a process:

- *Social Equality:* a more structured approach for understanding universal design should use equality and non-discrimination as a reference point for implementing universal design in policy and practice. Such an approach would position universal design, similar to accessibility, as a mechanism for promoting equality.
- *Human Diversity:* universal design can ensure a truly universal experience by considering the barriers that people experience across all forms of disadvantage, as well as the complex, overlapping, and multidimensional barriers that exist at the intersection of multiple forms of disadvantage.
- *Usability and Accessibility:* taking into account access as an interdependent component of use extends universal design considerations from how the design is used, to include considerations on whether and to what extent it can be accessed.
- *Participatory Processes:* a set of principles for universal design should take into account user participation as an integral element in the design and development of ICT. (Giannoumis and Stein 2019)

Applying this set of criteria to emergency management implies a shift in perspective to truly consider universal design as a process instead of an outcome. Using this set of criteria universal design in emergency situations can move beyond the accessibility of specific user interfaces and environments to considering the equality of people across the diversity of the human experience when it comes to accessing and using emergency management systems and services. One of the key features of Giannoumis and Stein's (2019) approach, in terms of emergency situations, is the active and substantive involvement of all relevant stakeholders and their representative organizations in the design and development of these systems and services. This could take into account the experiences of, for example, persons with physical or psychological disabilities, women, students and youth, people from minorities, elderly, families and more. Each individual has different experiences and different needs regarding every aspect of life and approaching to ICT is no exception, especially extra-ordinary situations.

5   Conclusions and Future Work

According to the United Nations CRPD, States Parties have an obligation, to take all necessary measures to ensure the protection and safety of persons with disabilities in emergency situations. While these obligations for accessibility represent one facet of this article's ambitions, another paradigm, universal design, provides a useful basis for considering situational disabilities that often occur in emergency situations. Universal design is defined as the "design of products, environments, programs and services to be usable by all people, to the greatest extent possible, without the need for adaptation or specialized design". Hence, universal design offers a valuable framework for considering the design of emergency management systems and for identifying and removing usability and accessibility barriers in emergency situations. By means of a heuristic analysis, this article poses an initial consideration of universal design in emergency situations, and concludes that universal design principles could be usefully integrated in emergency management practices, to ensure that persons with disabilities, and everyone, do not experience usability or accessibility barriers in emergency situations. Furthermore, this article recommends a universal design framework for empirically examining the universal design of emergency situations from a social equality, human diversity, usability and accessibility, and participatory processes perspectives.

References

Alexander, D., Gaillard, J.C., Wisner, B.: Disability and disaster. In: Handbook of Hazards and Disaster Risk Reduction, pp. 413–423. Routledge, London (2012)

American Red Cross Annual Report (2011). https://www.redcross.org/about-us/news-and-events/publications.html

Andersen, V., Rasmussen, J.: Decision Support Systems for Emergency Management. Risø-M, No. 2724 (1988)

Bonastre, L., Granollers, T.: A set of heuristics for user experience evaluation in e-commerce websites. In: The Seventh International Conference on Advances in Computer-Human Interactions, ACHI 2014, pp. 27–34 (2014)

Bunker, D., Smith, S.: Disaster management and community warning systems: inter-organisational collaboration and ICT innovation. PACIS 2009 Proceedings, p. 36 (2009)

Carver, L., Turoff, M.: Human-computer interaction: the human and computer as a team in emergency management information systems. Commun. ACM 50(3), 33–38 (2007). https://doi.org/10.1145/1226736.1226761

Christensen, K.M., Blair, M.E., Holt, J.M.: The built environment, evacuations, and individuals with disabilities: a guiding framework for disaster policy and preparation. J. Disabil. Policy Stud. 17(4), 249–254 (2007). https://doi.org/10.1177/10442073070170040801

Comfort, L.: Integrating information technology into international crisis management and policy. J. Contingencies Crisis Manag. 1(1), 15–26 (1993). https://doi.org/10.1111/j.1468-5973.1993.tb00003.x

Criddle, E.J., Fox-Decent, E.: Human rights, emergencies, and the rule of law. Hum. Rights Q. 34(1), 39 (2010). https://doi.org/10.2307/41345471

Cutter, S.L., Emrich, C.T., Adams, B.J., Huyck, C.K., Eguchi, R.T.: New information technologies in emergency management. In: Waugh Jr. W.L., Tierney, K. (eds.) Emergency Management: Principles and Practice for Local Government, 2nd edn. ICMA Press, Washington, DC (2007)

Fleischer, J.: Time and crisis. Public Manag. Rev. 15(3), 313–329 (2013). https://doi.org/10.1080/14719037.2013.769852

Giannoumis, G.A., Stein, M.: Conceptualizing universal design for the information society through a universal human rights lens. Int. Hum. Rights Law Rev. 8(1), 38–66 (2019)

Jaeger, P.T., Shneiderman, B., Fleischmann, K.R., Preece, J., Qu, Y., Wu, P.F.: Community response grids: e-government, social networks, and effective emergency management. Telecommun. Policy **31**(10–11), 592–604 (2007a). https://doi.org/10.1016/j.telpol.2007.07.008

Hamit, F.: GIS/GPS system for tire and EMS dispatch: imaging enabled. Adv. Imaging **12**(1), 1–4 (1997)

Hu, Q., Kapucu, N.: Information communication technology utilization for effective emergency management networks. Public Manag. Rev. **18**(3), 323–348 (2016). https://doi.org/10.1080/14719037.2014.969762

Hughes, A.L., Palen, L.: The evolving role of the public information officer: an examination of social media in emergency management. J. Homel. Secur. Emerg. Manag. **9**(1) (2012). https://doi.org/10.1515/1547-7355.1976

International Covenant on Civil and Political Rights, art. 41, 999 U.N.T.S. 171, 16 December 1966. https://www.ohchr.org/en/professionalinterest/pages/ccpr.aspx. Accessed 20 Dec 2018

Jaeger, P.T., Fleischmann, K.R., Preece, J., Shneiderman, B., Wu, P.F., Qu, Y.: Community response grids: using information technology to help communities respond to bioterror emergencies. Biosecurity and bioterrorism: biodefense strategy, practice, and science, **5**(4), 335–346 (2007b)

Kailes, J.I., Enders, A.: Moving beyond "special needs". A function-based framework for emergency management and planning. J. Disabil. Policy Stud. **17**(4), 230–237 (2007)

Kett, M., Van Ommeren, M.: Disability, conflict, and emergencies. Lancet **374**(9704), 1801–1803 (2009). https://doi.org/10.1016/S0140-6736(09)62024-9

Kuruczleki, E., Pelle, A., Laczi, R., Fekete, B.: The readiness of the european union to embrace the fourth industrial revolution. Management **11**(4), 327–347 (2016)

Kutty, F.: Review of Human Rights in Crisis: The International System for Protecting Rights During States of Emergency. By J. Fitzpatrick. University of Pennsylvania Press 1994 (2007)

Li, S.H., Yang, L.S.: Urban emergency service and tire facility apparatus distribution. In: Proceedings of GIS AM/FM Asia 1997 & Geoinformatics 1997, Taipei, Taiwan, pp. 329–334 (1997)

Lid, I.M.: Accessibility as a statutory right. Nordic J. Hum. Rights **28**(1), 20–38 (2010)

Mace, R.: Universal Design, Barrier Free Environments for Everyone, Designers West (1985)

Malizia, A., Astorga-Paliza, F., Onorati, T., Díaz Pérez, M.P., Aedo Cuevas, I.: Emergency alerts for all: an ontology-based approach to improve accessibility in emergency alerting systems. In: Proceedings of the 5th International ISCRAM Conference, pp. 197–207 (2008)

Mendonça, D., Beroggi, G., Wallace, W.A.: Decision support for improvisation during emergency response operations. Int. J. Emerg. Manag. **1**(1), 30–38 (2001)

Chang, N.-B., Wei, Y.L., Tseng, C.C., Kao, C.-Y.J.: The design of a gis-based decision support system for chemical emergency preparedness and response in an urban environment. Comput. Environ. Urban Syst. **21**(1), 67–94 (1997)

Nick, G.A., et al.: Emergency preparedness for vulnerable populations: people with special health-care needs. Public Health Rep. **124**(2), 338–343 (2009)

Nielsen, J.: Enhancing the explanatory power of usability heuristics. In: Proceedings of the ACM CHI 1994 Conference, Boston, MA, 24–28 April 1994, pp. 152–158 (1994)

Oraá, J.: Human Rights in States of Emergency. In: International Law, vol. 1 (1992)

Palen, L., Liu, S.B.: Citizen communications in crisis: anticipating a future of ICT-supported public participation. In: Proceedings of the ACM 2007 Conference on Human Factors in Computing Systems (CHI 2007), San Jose, California, USA, pp. 727–736. ACM (2007)

Perng, S.Y., Büscher, M., Halvorsrud, R., Wood, L., Stiso, M., Ramirez, L., Al-Akkad, A.: Peripheral response: microblogging during the 22/7/2011 Norway attacks. In: Proceedings of the Information Systems for Crisis Response and Management Conference (ISCRAM 2012), Vancouver, BC (2012)

Rauschert, I., Agrawal, P., Sharma, R., Fuhrmann, S., Brewer, I., MacEachren, A.: Designing a human-centered, multimodal GIS interface to support emergency management. In: Proceedings of the 10th ACM International Symposium on Advances in Geographic Information Systems (GIS 2002), pp 119–124. ACM, New York (2002). https://doi.org/10.1145/585147. 585172

Rowland, J.L., White, G.W., Fox, M.H., Rooney, C.: Emergency response training practices for people with disabilities: analysis of some current practices and recommendations for future training programs. J. Disabil. Policy Stud. 17(4), 216–222 (2007). https://doi.org/10.1177/ 10442073070170040401

Schwab, K.: The Fourth Industrial Revolution. Crown Publishing Group, New York (2016)

Shneiderman, B., Preece, J.: 911. Gov: Community Response Grids. Science 315(5814), 944 (2007). https://doi.org/10.1126/science.1139088

Steinfeld, E., Maisel, J.: The goals of universal design. Center for Inclusive Design and Environmental Access (2012). http://udeworld.com/presentations/oslo/Steinfeld.Goals of UD-Oslo_Final_web.pdf

Sutton, J., et al.: Emergent uses of social media in the California wildfires. In: Proceedings of the 5th International ISCRAM Conference – Washington, DC, USA, May 2008 (2008)

The Center for Universal Design: The Principles of Universal Design (Version 2.0). NC State University, Raleigh (1997)

Tufekci, S.: An integrated emergency management decision support system for hurricane emergencies. Saf. Sci. 20, 39–48 (1995)

Shakespeare, T.: Disability Rights and Wrongs. Routledge, London; New York (2006)

United Nations (open for signatures in 1966) Treaty Collection Database: Status of Treaties, Ch. IV (ICCPR). https://goo.gl/stiQgp. Accessed 23 Dec 2018

Veil, S.R., Buehner, T., Palenchar, M.J.: A work-in-process literature review: incorporating social media in risk and crisis communication. J. Conting. Crisis Manag. 19(2), 110–122 (2011). https://doi.org/10.1111/j.1468-5973.2011.00639.x

Vogt, M., Hertweck, D., Hales, K.: Strategic ICT alignment in uncertain environments: an empirical study in emergency management organizations. In: Proceedings of the 44th Hawaii International Conference on System Sciences (HICSS), pp. 1–11 (2011)

Wachtendorf, T.: Improvising 9/11: organizational improvisation following the world trade center disaster. Ph.D. dissertation, Department of Sociology, University of Delaware (2004)

Xu, K., Baron, F., Schnettler, J., Lewis, D.: GIS application for Miami transportation system: hurricane emergency preparedness. In: Proceedings of the 1996 Conference on Natural Disaster Reduction, Miami, FL, pp. 107–108 (1996)

Youngblood, N.E., Youngblood, S.A.: User experience and accessibility: an analysis of county web portals. JUS. J. Usability Stud. 9(1), 25–41 (2013)

Zhu, Q., Stillman, M.J.: Design of an expert system for emergency response to a chemical spill. J. Chem. Inf. Comput. Sci. 35(6), 956–968 (1995)

Digital Volunteers in Disaster Response: Accessibility Challenges

Jaziar Radianti[1] and Terje Gjøsæter[2]

[1] University of Agder, Grimstad, Norway
jaziar.radianti@uia.no
[2] Oslo Metropolitan University, Oslo, Norway
tergjo@oslomet.no

Abstract. The emergence of the Digital Humanitarian Volunteer (DHV) movements when disaster strikes have drawn the attention of researchers and practitioners in the emergency management and humanitarian domain. While there are established players in this rapidly developing field, there are still unresolved challenges, including accessibility of their digital tools and platforms. The purposes of this paper are twofold. First, it describes the background, impact and future potential of the DHV movement, and discusses the importance of universal design for the digital tools and platforms used for crowdsourcing of crisis information. Second, this paper shows how lack of concern for universal design and accessibility can have significant negative impact on the practical use of these tools, not only for people with disabilities, but also for anyone and in particular the DHVs who may be affected by situational disabilities in the field in an emergency situation. The insights from the findings serve as feedback on how to improve digital humanitarian response by broadening the base of potential volunteers as well as making the related tools and platforms more reliably usable in the field.

Keywords: Crisis mapping · Digital humanitarian volunteer · Crowdsourcing · Accessibility

1 Introduction

Collective action from the grassroots has changed response operations in a disaster [1]. Alexander [2] has noticed the shift on humanitarian response post-Indian Ocean Tsunami disaster due to the mass ownership of receiving devices, quicker international response to disasters. The author believes that existing scientific and technical know-how are promising to solve the global disasters issues. Rapid ICT technologies development have changed the landscape of humanitarian response [3], and have empowered non-first responder players to take an active role in responding crisis virtually. One of the fast-developing phenomena with respect to the shifting in the response operations is the presence of the digital humanitarian volunteers. They represent voluntary and technical communities, non-governmental organizations, expert

© Springer Nature Switzerland AG 2019
M. Antona and C. Stephanidis (Eds.): HCII 2019, LNCS 11573, pp. 523–537, 2019.
https://doi.org/10.1007/978-3-030-23563-5_42

groups, universities, research institutions and the private sector. Well-integrated emergency planning maps is a key device in promoting interagency and cross-jurisdictional coordination of emergency response [4]. Meier [5] describes this ICT role entering the humanitarian response domain as the shifting of humanitarian space from a traditional unipolar system to a more multipolar world order, allowing new actors to join and participate in providing additional support for crisis response.

Many approaches have been used worldwide to engage the digital humanitarian volunteers in the collection or analysis of crowdsourced data. Microblogging, crowd-sourcing, citizen journalism, social-media sharing, and mapping are examples of popular DHVs' methods of activities. In this paper, we try to look at specific examples, i.e. mapping and crowdsourcing-types of DIIVs.

The aims of this paper are twofold. First, to provide a solid overview for the phenomenon of DHVs, including history, terminology, and current state of the art; and second, to discuss challenges with the phenomenon from a universal design point of view.

This paper is organized as follow: Sect. 2 gives an overview of the terminology, history and early development related to DHVs. Section 3 elaborates different aspects of current DHVs. Section 4 examines the phenomenon from the perspective of challenges concerning accessibility and universal design. Conclusions and future directions are revealed in Sect. 5.

2 Digital Volunteering

Prior to explaining further both the competing views on the role of DHVs and our standpoint on perceiving these issues (Sect. 3), clarification on some basic terms is necessary as there are many different names which actually refer to the same meaning. The following terms are identified from the literature: Volunteered Geographic Information-VGI [6], crisis mappers [5, 7], digital volunteers [8], digital humanitarian organizations [1, 5], humanitarian cyber-space [9], and participatory mappers [10, 11]. These DHV actors can be mapper professionals and non-mappers. It can be recognized by observing if the organizations provide the geospatial or geographic information where mapping skills are obligatory, or they mostly harvest information and provide service digitally such as Tweet-based information, other crowdsourcing activities where cartographic knowledge is not always necessary, although eventually they may visualize their findings in a map. The VGI, for example, is referred to a phenomenon, where people without the cartographic skill of knowledge can report their geographic position and even make thematic maps [12].

The map platforms or map products are the core of the application of the digital volunteerism which are then used to communicate disaster information. Digital humanitarian organizations are grassroots organizations that mobilize a large number of individuals that share a set of open tools, practices, and ethical standards to create collective intelligence for providing information as aid [1]. Many of these organizations join a network called DHN (Digital Humanitarian Network), although admittedly some organizations do not fully operate in a digital fashion as they combine working remotely and working in the field such as MapAction[1] or CODE[2]. Phillips [13], however, differentiate Digital humanitarian networks (DHNs) and Digital Activist Networks. The former network creates maps, assesses building damage, builds missing person lists, monitors, and aggregates big crisis data, while the latter focuses on advocacy that relief who share text, video, and images about situation "on the ground" or uses petitions, email campaign or even hacking activities.

Regardless the terms used in the literature, there are a set of specific properties all authors want to point out: it is about information collected by digital volunteers which can be substantial and complementary to the data collected by official sectors, and it is concerning mobilization of digitally connected citizens and volunteers regionally or worldwide in acquiring specific crisis information. Hurricane Katharina response in 2005 has frequently used as an example, where DHVs impacts was more visible than ever. The volunteers bypassed official agencies and established a spontaneous digital assistant such as Katharina People Finder [12]. In the rest of our paper, we use the term Digital Humanitarian Volunteers as defined by Crowley [1]. In the next sections, we start briefly with the initial practice of Digital volunteering, the development, and the crowdsourcing practices.

2.1 Initial Practices

Meier [14] and Crowley [1] have discussed the historical development of DHVs extensively. Apparently, initial use of digital volunteering in a disaster was often associated with the mapping. Meier [5] recounts, crisis mapping itself is not new and dated back to the year 1668 as Louis XIV of France commissioned three-dimensional scale models of eastern border towns allowing his personnel plan realistic maneuvers. In several greatest crises in the history, however, the crisis map represented "the view from above" or the view from "who hold the control".

There is not so much information, when exactly the crisis mapping from grassroots representing "the view from below" started. We found, for example, an article from Dymon and Winter [4], elaborating the emergency mapping in the grassroots in the exercise context, when GIS technology was still expensive, not fully established part of emergency management practices as the stakeholders still tried to familiarize themselves with the technology. Indeed, the initial participatory, grassroots-based mapping approach was non-digital, non-real time and was a combination between a crude site sketch on-site and mental maps (mental images that have spatial attributes) of the evacuees. The process of on-the-spot map production during a disaster was considered as a new type of cartographic effort, a so-called "crisis mapping". The initial notion of crisis mapping activities is to supply critical information about the spatial dynamics dimension of the disaster. These crisis map sketches were made to: (1) help emergency managers to understand and learn the geographical setting in order to control the disaster conditions, (2) help to inform the media of risks, (3) help agencies to document

---

[1] MapAction, http://www.mapaction.org.

[2] CODE, Connected Development, http://connecteddevelopment.org.

history of an incident for better record keeping, and (4) help agencies to reconstruct the incident in order to establish lessons learned [4].

Meier [14] points out that new technologies are key enablers, facilitating both organization and collective action to be more rapidly deployed and provide more scalability than ever before. Citizens are more involved in putting crisis information into the map, and the citizen-produced maps began to appear. This new way of mapping is often referred as Neogeography [15–18], which can be summarized as a geography for everyone made by everyone [17]. Baker [19], however, has started using the NeoGeography term in relation to a space that is created by virtual communities formed of individuals who are far away from each other physically. As mentioned by Goodchild and Glennon [20], the technologies allow average citizens to determine position accurately without professional expertise; anyone has the ability to make maps with a cartographic design that previously possessed only by trained cartographers.

The launch of the Google Earth and Google Maps (2004) and broader access to satellite imaginary has brought further this new cartography direction to unlimited possibilities of making use of the map. In the meantime, the Harvard Humanitarian Initiative (HHI) at Harvard University launched a Crisis Mapping and Early Warning programme to study the potential use of live mapping technologies in humanitarian response, in 2007 [14]. Furthermore, in 2009, the HHI initiated the International Network of Crisis Mappers, a global network of members who are actively interested in the application of real-time mapping for crisis situations. Partnerships between traditional humanitarian actors and new informal networks became an important topic for discussion in the International Conference of Crisis Mappers 2010.

Another phenomenon frequently cited as an important milestone for live crisis mapping was the emergence and use of the web-based platform Ushahidi Map 2008 for reporting human rights violations during the post-election unrest [21]. Witnesses submitted these reports via web-form, email, and SMS. Reports from the mainstream media were also mapped. This enabled the 'crowd' to participate as witnesses and collectively uncover violence across the country [14]. Initially published as a free, open source software, Ushahidi becomes a popular live crowdsourcing mapping platform for crisis. Meier [5], Meier [14] presents exemplary deployment of crisis maps in the following disasters: Haiti, Chile, Pakistan, Russia, Syria, Tunisia, Egypt, New Zealand, Sudan, Libya, and, Somalia. Up to now, Ushahidi claims to have more than 90,000 deployments. Beginning from 2015, while keeping Ushahidi as an open source software, it is also offered as an easy to deploy service platform with elevating fees, depending upon the customer's feature preferences. From this brief explanation, apparently, the initial DHVs have a strong link to the mapping practices. However, the next question, should one really need to contribute in the mapping to be called digital volunteers? To put it differently: Are DHVs really about bottom-up driven crisis mapping only? The next section will discuss this development further.

2.2 Further Development of DHVs' Practices

While the use of a map is still the most popular DHVs' practices, tool variations, and the methods how to do it is developing into different directions. For example, when Malaysian Airlines flight MH370 was missing in 2014, the Tomnod MH370 website was set up[3], dedicated for uploading satellite imagery which was then used for searching for the wreckage in crowdsourced mode. It attracted approximately 2.3 million people who participated in the search mission by scanning more than 24,000 square kilometers of areas suspected as the incident location [8]. Since then, in 2015, apparently, Tomnod has been active in several disasters, i.e. Capetown wildfire, Nepal and Chile Earthquakes. Digital volunteers in the Tomnod network work together to identify and tag important objects in satellite images, e.g. burned building, damaged roads, fires, tents or shelters, major destruction, and so on.

The use of social media in disasters is another direction of recent DHVs practices [22, 23]. The exploitation of Twitter for crisis management has been growing in the last decade using different methods and various applications, ranging from natural disasters such as earthquakes or floods to major political conflicts. Although the attitudes towards the use of Twitter in formal emergency management still vary among scholars and practitioners [24], especially about the credibility of the information [25], the need for browsing and visualizing social media [26], in real-time and application of artificial intelligence [27] increases. In Haiti earthquake 2010, New York's 2012 hurricane Sandy, and Oklahoma's 2013 tornado humanitarian organizations and the networks of volunteers established platforms for crowdsourced information including those from social media which were furthermore transformed into live-web-based crisis maps. For longer term crisis such as Syrian [28] or refugee crises [29], advanced sentiment analysis is a popular approach. In this case, many scholarly works focus more on the development of analysis method than presenting social media as crowdsourced data for direct use as crisis decision support. By and large, organizations joining the DH-Network are good examples to grasp, what kind of activities that have been offered as a part of digital volunteering activities.

To see further development of DHVs activities, the next sections describe the DHVs Landscape and Players, Crowdsourcing technique, Technologies, Decision maker needs and deployment techniques.

3   DHVs Today

There are numerous DHV players worldwide, but people can easily identify the big players or strong network ones. Humanitarian Practice Networks, Digital Humanitarian Networks, Crisis Mappers, Standby Task Force (SBTF), and MapActions are just a few examples of DHVs joining DH-Network. Established in 2010, SBTF has partnerships between traditional humanitarian actors and new informal networks. Recently SBTF has over 1800 members in 100 countries[4].

Professionally, the DHV actors can be technology professionals (web and mobile developers, data scientists, social media analysts, wifi-network/vsat experts, geo-coding), geographers (infrastructure mapping, satellite imagery analysis, GIS analysts, spatial analysts, ground mappers), seasoned humanitarian practitioners, journalists,

---

[3] http://www.tomnod.com/.

[4] http://www.standbytaskforce.org/.

professional crowdsourced translators, and scholar. The output from the digital volunteering works can be map visualization, technical reports, social media monitoring and analysis, social media strategy, training, community's urgent needs, situational information, real-time online updates, and some other variants intended for decision supports and improving situational awareness.

We try to capture the landscape of DHVs as depicted in the charts in Fig. 1. The charts present a simplification of current approaches to digital humanitarian volunteering in a crisis. In the left figure, the horizontal axis represents the two opposite ways of operation of information collection, i.e. operating on-site and operating remotely. The vertical axis denotes the types of information sources preferably collected by different humanitarian actors which are simplified as:

(1) *Humanitarian aid workers* as a part of formal organizations sent in humanitarian missions in a disaster site; it could be affiliated to United Nations (UN) and government agencies, or NGO workers that are a part of network of collaborative partnerships to assist with humanitarian relief operations e.g. IFRC (International Federation of Red Cross and Red Crescent Societies). In short, this group consists of in-situ volunteers;

(2) *Emergency decision makers* are local, national or international authorities who responds directly on-site, and shall monitor overall crisis development, or those who operate slightly remotely e.g. in a command and control room to provide decision supports. These two formal actors are represented in the below part of the chart to indicate their working tendencies to make use of authoritative information sources.

(3) *Community volunteers* can be networks of affected communities. As a group identified as well by Phillips [13], the people directly affected by the disaster seek help to locate missing people, medical treatment, identify shelters, aid, and inaccessible zone, share information about the crisis to enhance situational awareness and seek for ways to provide support to those affected.

(4) *Digital volunteers* are representing two types of crowd. The first type is local citizens or geographically affected crowd who provide raw information directly. The second type is a remote volunteer crowd that coordinate and manage information to support a humanitarian response [12, 30]. They can be crisis mappers, like the so-called "Voluntweeters" [31] who were acting remotely to help the process, verify and direct information during the crisis, creating a mesh of interconnected volunteers from all over the world, or other humanitarian initiative organized digitally. The dashed line in the middle dividing the chart depicts the approach to humanitarian mission that much more grounded approach in the left side and more ICT supported on the right side and tendencies to work based on remote information, such as making use of non-authoritative information with the assistance of ICT technologies.

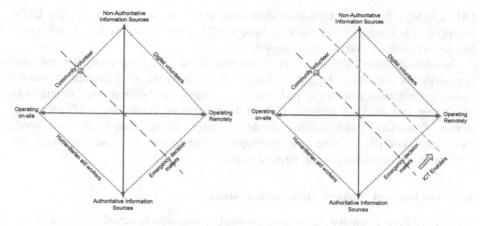

Fig. 1. Approaches to digital volunteering.

3.1 On Techniques and Technologies for DHVs

The most popular way of conducting digital volunteering is a crowdsourcing technique [32]. The term crowdsourcing describes a web-based business model that harnesses the creative solutions of a distributed network of individuals. Web2.0 technologies facilitate interactive information sharing, interoperability, and collaboration. Proponents of this approach assume that crowds of people can solve some problems faster than individuals or small groups, and rapidly generate data [33]. The term crowdsourcing itself used to be interpreted as:

> *"The act of taking a job traditionally performed by a designated agent (usually an employee) and outsourcing it to an undefined, generally large, group of people in the form of an open call."* In the mapping field, UNOOSA [34] defines crowdsource mapping as *"Reaching out to the unknown crowd for help in gathering geospatial information, visualizing that information on a map and gaining further insight by analyzing the data. Such a crowd would be supporting not only humanitarian and emergency crises but also all the phases of the disaster risk management cycle: prevention, preparedness, early warning, response, early recovery and reconstruction."*

As we see in the initial development, the smartphones, open-source software, GIS technologies and web-based platform are consistently used for conducting DHVs activities. Mobile and web-based applications, participatory maps & crowdsourced event data, aerial & satellite imagery, geospatial platforms are among the exploited technologies. The use of advanced visualization, live simulation, and computational & statistical models for early warning and rapid response are additional techniques can be applied in DHVs efforts. In addition, Meier [14] suggests GIS analysis, machine learning, pattern recognition and spatial econometrics as additional core skills to handle the big data stream originating from crowdsourcing activities. Ushahidi[5] and Kricket[6] are examples of open-source platforms for crowd-mapping. On information gathering,

---

[5] Ushahidi, https://www.ushahidi.com/.

[6] Kricket, http://kricket.co/.

DHVs believe that the information should be real time, and specifically for DHV mappers, it is important that the data have spatial component [14] for visualization, besides for analysis and decision support.

In addition, social networking platforms and social media especially Twitter gain popularity for collecting data. As information scientists we also attempt to extract meaning from mass volumes of real-time data exhaust. Furthermore, the combination of Twitter and machine learning [35], real-time remote coordination, Twitter data mining, data analysis and mapping, Social Network Analysis have further developed. However, most of the time, the geographical component is most important for humanitarian information in the form of maps.

3.2 On Decision Makers' Information Needs

There have been extensive discussions on what information is actually needed by the decision makers in the humanitarian mission. It has been highlighted in the literature that crowdsources information is not intended to replace the work already being carried out by established organizations and the private sector but rather as an additional support to the emergency management process and decision-making. It is also have been discussed elsewhere that information collection needs to be in line with the specific information needs of the end-user community who actually affected by the disasters and emergency response managers dealing with the problem. Specifically, these should be a clarity how that community was able to access and use the information provided by the volunteer and technical communities. Gralla, Goentzel and de Walle [36] and Kuusisto, Kuusisto and Yliniemi [37] has proposed a framework for information needs of decision makers in crisis.

3.3 On Deployment

In a more organized/coordinated digital volunteer deployment, there are several stages to enable the digital volunteering, as seen in the Fig. 2. We will use two big DHV organizations as an example to explain the volunteering operation, i.e. MapActions and StandBy Task Force (SBTF). Member recruitment is typically conducted voluntarily, e.g. by signing-up. Some organizations will provide their services upon request from formal humanitarian organizations. For example, MapAction has deployed two volunteers to map the Ecuador earthquake based on a request from the United Nations. Meanwhile, activation of SBTF can be based on the request of international agencies, local stakeholders operating in disaster setting or active support from SBTF to humanitarian organizations. However, a set of criteria has been outlined prior to fulfillment of SBTF activation request. These criteria are ranging from the capability of the organization to respond to disasters, its presence in the field, clear need for SBTF support, to clear plan of data collection, sharing and privacy, sustainability plan, monitoring and evaluation plan for SBTF support and understanding of potential risks for local people and the digital volunteers. Maximum two weeks is the typical duration of activation period, although an extension can be decided through a meeting between

SBTF and Activating Organizations on day 10^7. Note that SBTF also encourages its members to volunteer in deployments as individuals without any official engagement from SBTF.

On the Activation stage, the operation can be varied as well: onsite or remotely. MapAction will mobilize data collection and base map data gathering of the affected country, as well as field base deployment as a part of United Nations Assessment and Coordination (UNDAC) mission. In the SBTF operation, the members will get the notification when the activation starts, via email and the activation is announced in SBTF website (See for example the activation for Refugee Crisis[8]).

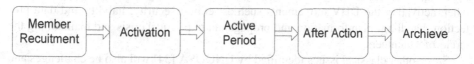

Fig. 2. Typical digital humanitarian deployment life cycle.

In the active period, the Map Action team will map the most crucial operational questions in the field such as movement of people affected by the disasters or the logistic capacity to reach them. During this time the situation maps will be circulated and updated, both as printed map or electronic distribution via websites. While in the SBTF deployments, the operation varies, but the volunteers gather data, put the collected information in a centralized workspace. The information collection does not only focus on those that can support map visualization, but also other sources including pictures in specific social media. Depending upon the crisis development, some updates might be sent in between the deployment, such as occurred in Refugee Crisis case.

At the end of the mission, MapAction will hand over the maps and collected data to UN organizations signify the completion of one digital volunteering mission. Archiving the maps of completed missions is a normal final sequence to preserve all results of each volunteering activities. In the case of SBTF, the collected data is handed over to the activating organization, and archiving public information in the website.

Finally, we return to the question if DHVs are really about bottom-up driven crisis mapping only. The explanation above has provided a clear answer that digital volunteering is not only about the mapping, but wider than that. However, map visualization is apparently the heart for digital humanitarian volunteer activities as it is the best way to communicate the location of all information being collected, gathered and analyzed during the disasters. To put it differently, spatial data is always an advantage to any crisis and any DHVs efforts that include spatial information will provide benefits to the affected people.

---

[7] SBTF, Our Activation Criteria, http://www.standbytaskforce.org/for-humanitarian-agencies/our-activation-criteria/.

[8] SBTF, Refugees in the Balkans. Activation Starts 16 September 2015. http://www.standbytaskforce.org/2015/09/13/refugees-in-the-balkans-activation-starts-16-september-2015/.

3.4 On DHVs' Role

The influential DHVs are well organized, have a huge number of members and strong connections to the formal humanitarian organizations such as Red Cross or UN agencies, and professional map producers and research institutions. However, we have also observed that the presence of strong network and players does not stop smaller players or newcomers (institutions and individuals, whether or not the initiators have experience with digital volunteering) to deploy their initiatives.

The majority of crisis maps are actually not launched by humanitarian organizations or digital volunteer networks; sometimes just ordinary individuals launching they own maps. Experienced organizations may have a better way of handling several issues when activating the DHV movements such as: ethics, security, liability and data protection; handling misinformation and propaganda in political crisis and maintaining the circle of volunteers.

Some other classical issues identified by UNOOSA [34] will also be challenges and unknown to newer players such as preparedness and prioritization regarding to the need for geospatial data to be readily available for support during any disaster event; data licensing, which ensured that satellite imagery was made available to the volunteer and technical communities; data scramble, or contribution to the definition and compilation of available geospatial data during a crisis; decision maker's needs; impact evaluation, and bringing together all those willing to volunteer their time and involving them in meaningful activities that contributed to the decision-making process.

4   Accessibility of Tools and Platforms for DHVs

From earlier sections, we learn that DHVs require a lot of considerations to make a great contribution on improving crisis communication and information sharing. We also see that DHVs rely on the ICT technologies both for digital volunteers (information collectors) and target groups (humanitarian workers and general public). In this case, the perspectives of accessibility and universal design are often missing in the DHVs' technology requirements or needs. As in most domains, the introduction of digital tools and ICT provide both opportunities and challenges, in particular for people with disabilities. A well-developed universally designed platform can provide access to information and means of communication, work and other productive activities that has made life for people with disabilities much easier. On the other hand, ICT can also bring potential barriers, if a system design process is not taking accessibility into account. In many digital systems, even those targeting the general public, lack of awareness of accessibility and even basic usability can be severely limiting the potential user base.

When the digital volunteers are primarily working through digital platforms, the usability and accessibility of these platforms are very important. However, the awareness of the importance of these factors among experts in ICT for emergency management is relatively low [38, 39].

Many of the online platforms in use for crowdsourcing of crisis information and crisis mapping among DHVs are Web-based. Web page accessibility can be evaluated based on conformance to the Web Accessibility Content Guidelines (WCAG) 2.0[9]. The guidelines are based on 4 principles for accessibility:

1. Perceivable - Information and user interface components must be presentable to users in ways they can perceive.
2. Operable - User interface components and navigation must be operable.
3. Understandable - Information and the operation of user interface must be understandable.
4. Robust - Content must be robust enough that it can be interpreted reliably by a wide variety of user agents, including assistive technologies.

There are several success criteria for each of these principles. that can be used to assess the accessibility of the digital platforms. Some success criteria can be automatically tested and others require human evaluation. In the previous study, automatic evaluation of a selection of typical web-based tools and platforms for crowdsourcing of information for DHVs have revealed that the tools and platforms tested are not accessible [39]. The main page as well as volunteer signup or information submission pages were tested as these typically are the first potential *barriers* (issues potentially blocking users from accessing some or all of the information or functionality of the web site) for new volunteers. Automatic testing showed that none of the tested tools[10] were fully compliant with WCAG 2.0.Common issues that were detected included:

- *Missing labels* that could affect screen reader users who rely on screen readers to understand the meaning and intention of the web elements (e.g. buttons).
- *Resizing issues* that affect users with visual impairments, making it more difficult to adapt the web site to their needs.
- *Lack of instructions or help* that are essential for users to learn how to use the web sites and get help when needed. They can also help users to prevent errors and understand error messages. Without appropriate instructions or help function, it is often difficult for users to understand what they should do and how to interact with certain functions.
- *Compatibility issues* - Lacking the robustness to ensure compatibility with current and future user agents, including assistive technologies.

To confirm the claims, we have examined three additional DHV-oriented websites for testing, all of them related to the ongoing refugee crisis. Two of the sites are map-based: The Refugee Project[11] providing information of refugees with origin and asylum country broken down by year, and The UN Refugee Agency (UNHCR)'s Refugees Operational Portal[12] providing maps and statistics regarding refugees. We have also tested one DHV-oriented site not related to mapping: Refugees on Rails[13] that aims «to

---

[9] https://www.w3.org/TR/UNDERSTANDING-WCAG20/conformance.html.

[10] Ushahidi Syria Tracker, Google Crisis Response Person finder demo, Quakewatch Prediction Center, Crisis Communication Wiki for professionals, and Emergency 2.0 Wiki.

[11] http://www.therefugeeproject.org.

[12] https://data2.unhcr.org/en/situations.

[13] http://refugeesonrails.org/en/.

help refugees to build and expand their qualification as software developers and provide them with valuable skills that will improve their chances on the labour market».

We find that both the Refugee project and the Refugees Operations Portal have many of the same issues as the previously tested sites, and in particularboth are affected by *missing labels* on various elements. In addition, the quite low contrast on the Refugee project site (see Fig. 3) can cause barriers for people with visual impairments as well as people attempting to use the site in poor light. Meanwhile, The Refugees Operations Portal has several cases of missing alternative text for images, so if the page is accessed using assistive technologies or blocking images for bandwidth reasons, the information carried by the images will be missed.

On the other hand, on the Refugees on Rails site, we detected only one significant barrier: a lack of explicit language identification code in the html source. Although this may seem like a non-issue, this can have consequences for people using assistive technologies. For example, screen readers are using this information to load the correct pronunciation rules.

Fig. 3. The refugee project world map showing a lack of contrast.

To summarize the findings, the overall picture is that the majority of the tested sites have significant barriers, meaning that people with disabilities may be discouraged from joining the voluntary efforts.

An exacerbating factor is that in disaster situations, people can be affected by situational disabilities that are making use of mobile digital equipment like smartphones more challenging than in normal day-to-day life, because of factors like stress, environmental factors like rain/heat/cold/wind, fear/panic, information overload, smoke, crowds, noise, etc. These factors can negatively impact the vision, hearing, cognitive abilities and not least the manual dexterity needed to manipulate a user interface on the small touch sensitive screen of a smartphone, and it may thus disrupt their situational awareness [38–40].

An additional challenge is that much of the work by DHVs relate to maps which by their nature are difficult to·make accessible since they are very visually oriented, and may in addition be difficult to navigate on the small screen of a smartphone [41].

5 Conclusion and Future Work

This paper has provided a thorough overview of the phenomenon of DHVs, and also highlighted some challenges when it comes to accessibility and usability. Lack of attention to these factors has led to tools and platforms with barriers that make them inaccessible to many potential users. In a disaster situation, active DHVs in the field may also be affected by situational disabilities that makes the accessibility and usability of their tools even more important.

These highlighted issues deserve a much bigger focus, as they can make a big difference in the DHVs ability to provide quality information in the field, and can potentially open the field to a much broader diversity of DHVs in the future. One should not underestimate the value of motivated people armed with well-designed digital tools, and we firmly believe the advantages from this supplement to traditional emergency response can be tremendous. However, to make the impact as powerful as possible, universal design should be a non-negotiable part of the design process of the DHVs' digital tools.

Future and ongoing work that will be highly relevant for the DHVs, includes studies on the cause and effects of situational disabilities in disaster situations, and how these situational disabilities can affect the ability to gain and communicate situational awareness. Finally, studies on best practices on accessible maps and alternative representations of geographical data are also in the pipeline.

References

1. Crowley, J.: Connecting grassroots and government for disaster response. Commons Lab, Wilson Center (2013)
2. Alexander, D.: Globalization of disaster: trends, problems and dilemmas. J. Int. Affairs **59**, 1–22 (2006)
3. Ferris, E.: Megatrends and the future of humanitarian action. Int. Rev. Red Cross **93**, 915–938 (2011)
4. Dymon, U.J., Winter, N.L.: Emergency mapping in grassroots America: a derailment evacuation case study. Geoforum **22**, 377–389 (1991)
5. Meier, P.: Crisis mapping in action: how open source software and global volunteer networks are changing the world, one map at a time. J. Map Geogr. Libr. **8**, 89–100 (2012)
6. Haworth, B.: Emergency management perspectives on volunteered geographic information: opportunities, challenges and change. Comput. Environ. Urban Syst. **57**, 189–198 (2016)
7. Ziemke, J.: Crisis mapping: the construction of a new interdisciplinary field? J. Map Geogr. Libr. **8**, 101–117 (2012)

8. Whittaker, J., McLennan, B., Handmer, J.: A review of informal volunteerism in emergencies and disasters: definition, opportunities and challenges. Int. J. Disaster Risk Reduct. **13**, 358–368 (2015)

9. Sandvik, K.B.: The humanitarian cyberspace: shrinking space or an expanding frontier? Third World Q. **37**, 17–32 (2016)

10. Plantin, J.C.: The politics of mapping platforms: participatory radiation mapping after the Fukushima Daiichi disaster. Media Cult. Soc. **37**, 904–921 (2015)

11. White, J.I., Palen, L.: Participatory mapping for disaster preparedness: the development & standardization of animal evacuation maps. In: Palen, L.A., Comes, T., Buscher, M., Hughes, A.L., Palen, L.A. (eds.) 12th International Conference on Information Systems for Crisis Response and Management, ISCRAM 2015, pp. 214–224. Information Systems for Crisis Response and Management, ISCRAM (2015)

12. Hung, K.-C., Kalantari, M., Rajabifard, A.: Methods for assessing the credibility of volunteered geographic information in flood response: a case study in Brisbane, Australia. Appl. Geogr. **68**, 37–47 (2016)

13. Phillips, J.: Exploring the citizen-driven response to crisis in cyberspace, risk and the need for resilience. In: 2015 IEEE Canada International Humanitarian Technology Conference (IHTC 2015), pp. 1–6 (2015)

14. Meier, P.: New information technologies and their impact on the humanitarian sector. Int. Rev. Red Cross **93**, 1239–1263 (2011)

15. Haklay, M., Singleton, A., Parker, C.: Web mapping 2.0: the neogeography of the GeoWeb. Geogr. Compass **2**, 2011–2039 (2008)

16. Hudson-Smith, A., Crooks, A., Gibin, M., Milton, R., Batty, M.: NeoGeography and Web 2.0: concepts, tools and applications. J. Location Based Serv. **3**, 118–145 (2009)

17. Rana, S., Joliveau, T.: NeoGeography: an extension of mainstream geography for everyone made by everyone? J. Location Based Serv. **3**, 75–81 (2009)

18. Turner, A.: Introduction to neogeography. O'Reilly, Sebastopol (2006)

19. Baker, N.: Weed-at talk at the library. In: Brook, J., Carlsson, C., Peters, N.J. (eds.) Reclaiming San Francisco: History, Politics, Culture, pp. 35–50. City Lights Publishers (1998)

20. Goodchild, M.F., Glennon, J.A.: Crowdsourcing geographic information for disaster response: a research frontier. Int. J. Digit. Earth **3**, 231–241 (2010)

21. Meier, P., Brodock, K.: Crisis Mapping Kenya's Election Violence: Comparing Mainstream News, Citizen Journalism and Ushahidi. Harvard Humanitarian Initiative, HHI. Harvard University, Boston (2008)

22. Middleton, S.E., Zielinski, A., Necmioğlu, Ö., Hammitzsch, M.: Spatio-temporal decision support system for natural crisis management with TweetComP1. In: Dargam, F., et al. (eds.) EWG-DSS 2013. LNBIP, vol. 184, pp. 11–21. Springer, Cham (2014). https://doi.org/10.1007/978-3-319-11364-7_2

23. Simon, T., Goldberg, A., Adini, B.: Socializing in emergencies—a review of the use of social media in emergency situations. Int. J. Inf. Manag. **35**, 609–619 (2015)

24. Cameron, M.A., Power, R., Robinson, B., Yin, J.: Emergency situation awareness from Twitter for crisis management. In: Proceedings of the 21st International Conference on World Wide Web, pp. 695–698. ACM (2012)

25. Mendoza, M., Poblete, B., Castillo, C.: Twitter under crisis: can we trust what we RT? In: Proceedings of the First Workshop on Social Media Analytics, Washington D.C., District of Columbia, pp. 71–79. ACM (2010)

26. Terpstra, T., de Vries, A., Stronkman, R., Paradies, G.: Towards a realtime Twitter analysis during crises for operational crisis management. Simon Fraser University (2012)

27. Imran, M., Castillo, C., Lucas, J., Meier, P., Vieweg, S.: AIDR: artificial intelligence for disaster response. In: Proceedings of the 23rd International Conference on World Wide Web, Seoul, Korea, pp. 159–162. ACM (2014)

28. Lynch, M., Freelon, D., Aday, S.: Syria's socially mediated civil war. United States Institute of Peace, vol. 91, pp. 1–35 (2014)

29. Coletto, M., et al.: Sentiment-enhanced Multidimensional Analysis of Online Social Networks: Perception of the Mediterranean Refugees Crisis. arXiv preprint arXiv:1605.01895 (2016)

30. Starbird, K.: Digital volunteerism during disaster: crowdsourcing information processing. In: Conference on Human Factors in Computing Systems, Vancouver, BC, Canada (2011)

31. Starbird, K., Palen, L.: Voluntweeters: self-organizing by digital volunteers in times of crisis. In: Proceedings of the SIGCHI Conference on Human Factors in Computing Systems, Vancouver, BC, Canada, pp. 1071–1080. ACM (2011)

32. Madry, S.: The emerging world of crowd sourcing, social media, citizen science, and remote support operations in disasters. In: Madry, S. (ed.) Space Systems for Disaster Warning, Response, and Recovery. SSD, pp. 117–121. Springer, New York (2015). https://doi.org/10.1007/978-1-4939-1513-2_9

33. Barbier, G., Zafarani, R., Gao, H., Fung, G., Liu, H.: Maximizing benefits from crowdsourced data. Comput. Math. Organ. Theory **18**, 257–279 (2012)

34. UNOOSA: Space-based information for crowdsource mapping Committee on the Peaceful Uses of Outer Space (2011)

35. Meier, P.: Next generation humanitarian computing. In: Proceedings of the 17th ACM Conference on Computer Supported Cooperative Work & Social Computing, Baltimore, Maryland, USA, p. 1573. ACM (2014)

36. Gralla, E., Goentzel, J., de Walle, B.: Understanding the information needs of field-based decision-makers in humanitarian response to sudden onset disasters. In: Proceedings of the 12th International Conference on Information Systems for Crisis Response and Management (ISCRAM), pp. 1–7 (2015)

37. Kuusisto, T., Kuusisto, R., Yliniemi, T.: Information needs of strategic level decision-makers in crisis situations. In: ECIW, pp. 187–194 (2005)

38. Gjøsæter, T., Radianti, J., Chen, W.: Universal design of ICT for emergency management. In: Antona, M., Stephanidis, C. (eds.) UAHCI 2018. LNCS, vol. 10907, pp. 63–74. Springer, Cham (2018). https://doi.org/10.1007/978-3-319-92049-8_5

39. Radianti, J., Gjøsæter, T., Chen, W.: Universal design of information sharing tools for disaster risk reduction. In: Murayama, Y., Velev, D., Zlateva, P. (eds.) ITDRR 2017. IAICT, vol. 516, pp. 81–95. Springer, Cham (2019). https://doi.org/10.1007/978-3-030-18293-9_8

40. Gjøsæter, T., Radianti, J.: Evaluating accessibility and usability of an experimental situational awareness room. In: Di Bucchianico, G. (ed.) AHFE 2018. AISC, vol. 776, pp. 216–228. Springer, Cham (2019). https://doi.org/10.1007/978-3-319-94622-1_21

41. Tunold, S., Gjøsæter, T., Chen, W., Radianti, J.: Perceivability of Map Information for Disaster Situations HCII 2019, Orlando, Florida, USA. Springer, Cham (2019, in press)

The Contribution of Social Networks to the Technological Experience of Elderly Users

Célia M. Q. Ramos[1]([✉]) [iD] and João M. F. Rodrigues[2] [iD]

[1] ESGHT, CIEO & CEFAGE, University of the Algarve,
8005-139 Faro, Portugal
omramos@ualg.pt
[2] LARSyS (ISR-Lisbon) & ISE, University of the Algarve,
8005-139 Faro, Portugal
jrodrig@ualg.pt

Abstract. Social networks have changed the way people and companies communicate. Nowadays, more and more elderly persons are using these platforms to communicate with friends and family, access news, entertainment and education. This study focuses on the elderly population and its use of social networks, and analyzes the contribution provided by the platforms to the users' technological experiences, and whether this interaction contributes to the quality of life of this population. A survey based on the experience economy theory was disseminated online through Facebook to gauge users' behavior. A Social Networks User Experience (SNUX) model was developed to study the elderly-user experience associated with the use of social networks, which was analyzed through structural equations modeling using SmartPLS 2.0. From the results obtained, it was concluded that social networks can contribute to an increased well-being of the older population, mainly from the technological experience associated with the use of these platforms, the environment of which contributes to entertainment and education of these users.

Keywords: Social networks · Elderly users · Adoption and use of technology · Technological experience · Well-being · Experience economy

1 Introduction

Society is becoming increasingly technology-enabled. Smart cities are becoming a reality. Through the use of sensors, it is possible to develop mobile applications for human–computer interaction, which contributes to the connection between humans and society, even for those who have physical limitations, such as those associated with old age. With the increase in life expectancy, the aging population is growing in developed countries. The use of technologies by elderly users can contribute to their personal satisfaction by allowing communication with others and providing access to educational, entertainment, cultural and distracting resources that contribute to the well-being of the population in this age group.

© Springer Nature Switzerland AG 2019
M. Antona and C. Stephanidis (Eds.): HCII 2019, LNCS 11573, pp. 538–555, 2019.
https://doi.org/10.1007/978-3-030-23563-5_43

The whole environment provided by information and communications technologies (ICT) has caused behavioral changes in users in general, and particularly in the elderly, who, despite having more difficulty learning and using technology [5, 45] than younger people, use this medium to communicate with family and friends as a way of keeping in touch with others while at the same time increasing their social inclusion and decreasing their loneliness [47].

The use of social networks can contribute to the diminishment of elderly isolation [19, 23], increase their cultural knowledge (which contributes to the increase of their personal valuation), provide entertainment as a way to pass the time, and at the same time contribute to an increase in their technological experience and the acquisition of digital skills. All of the above factors result in the enhancement of the quality of life and well-being of the elderly population.

The objective of the present study is to analyze the technological experience associated with the use of social networks by elderly users, taking into account the concepts of the experience economy of Pine and Gilmore [34]. The focus is based not from the point of view of the consumption of economic goods, but the consumption of information, to determine whether or not it contributes to the personal valuation of the elderly population. The study takes into account the dimensions associated with the economy of experience: evasion, education, entertainment and aesthetics.

This article is structured in four sections, in addition to the introduction and conclusions. The first section clarifies how social networks can contribute to the quality of life of elderly users and their well-being, followed by the concepts of the experience economy that are associated with technological experience. The second section presents the conceptual research model and the research hypotheses associated with the study. The third section defines the methodology used, which considers concepts of structural equation models. The fourth section analyzes and presents the results. In the final section, conclusions are drawn with implications, limitations, and future work.

2 Technological Experience for the Elderly

2.1 Contributions of Social Networks for Elderly Users

The SeniorNet website (www.seniornet.org), created in 1986 for adults over 55, receives more than 1 million visits per month. It was created with the mission to provide the elderly with education for, and access to, computer technologies, as a way to enhance their lives and enable them to share their knowledge and wisdom [42].

Internet access by this age group encourages the search for lifelong learning opportunities as the life expectancy of this population increases. For instance, in Portugal, according to the Pordata, the Aging Index was 125.8 in 2011 and 153.2 in 2017, which represents the number of older people per 100 young people [36]. As the older population seeks to stay active and find activities that improve their physical and emotional well-being, they have a tendency to adopt technology, especially the use of social networks. The perception of utility, security of use and frequency of internet use are the main motives that explain the adoption of social networks, as also reported in the work of Chakraborty, Vishik and Rao [8].

The potential of the internet to contribute to the well-being of the elderly population has led to the development of senior-friendly websites, created with more simplified designs to increase usability for older adults [2, 6, 12, 14].

The age at which a user is considered a senior or elderly has no consensus in the literature. Barnard, Bradley, Hodgson and Lloyd [3] find that "older adults" are an extremely diverse group—that "old age" may differ with context. For example, in the context of work, "old" refers to ages of 50 to 55, since this is the age range in which workers' capacities tend to decrease [26, 46]. Sinclair and Grieve [43] also point out that there is no consensus on what age one considers a person as an "older adult", but they say that the most referenced literature considers as elderly the persons whose age is greater than or equal to 55 years.

In older adults, women are more familiar with social networks and are more frequent users when compared to men [38, 44]. Gender, age and education are factors that have a significant impact on the use of social networks when compared to those of younger users.

The main reasons the elderly used social networks were [44]: (1) fun, which translates into feelings of pleasure [29]; (2) social inclusion, involving interacting with others, talking with family and socializing, thereby reducing feelings of loneliness, even for those with problems of physical mobility [14, 23, 30, 43]; and, (3) a sense of well-being, with a commensurate reduction in fear and anxiety, which is associated with the support and availability of care information for people who are physically or geographically isolated [28, 31–33].

In addition to the motives presented by Vošner et al. [44], social networks allow for the acquisition of new knowledge. Xiang and Gretzel [48] point out that these platforms serve as a means to learn about products, brands and services, since they include a large variety of information sources produced and shared by the users themselves, which increases credibility and trust in the shared content. In this context, the first hypothesis of this research is to validate whether the use of social networks is motivated by the personal valuation of the acquisition of knowledge and technological experience associated with their use.

Hypothesis #1(H_1): Increase personal valuation through the acquisition of knowledge acquired through social networks?

For some older adults, social networks are their first technological experience of a social nature [13]. When they understand that social networks are easy to use and useful, they are more likely to use them [5], which helps prevent a more pronounced mental decline [23].

2.2   Technological Experience in the Use of Social Networks by the Elderly

As already mentioned, in recent years, participation in online communities has increased, including among the elderly [7, 45]. The technological experience associated with communication with distant relatives and friends, as well as the access information and knowledge available through clicks, has contributed to a lessening of the obstacles and difficulties of older adults in using these communities.

In general, users participate in a technological experience [7, 45] when they have developed a habit of utilization [16] and trustworthiness in the technology [15]; however, the elderly can experience a sensation of discomfort when new technologies are introduced, with new scenarios and new situations that emerge in a context in which they lack knowledge. Reticent elderly users quickly adopted emerging social networks as a way to communicate [16], however, because of their ease of use and its usefulness [5].

To reduce the discomfort associated with adopting a new technology, innovative design methods were necessary, taking into account the age group of older adults and their physical limitations: decreased vision, hearing loss, decreased physical mobility, cognitive and social impairment [2], as well as the user experience concepts, defined by ISO 9241-210 [25], which includes indications associated with the emotions, beliefs, preferences, perceptions, physical and psychological responses, user behaviors and achievements that occur before, during, and after use.

Although the technological experience associated with social networks is not analyzed from the economics of experience point of view [34], i.e., from the business point of view, these authors considered that consumers seek to be involved and absorbed in the experience [27, 41], and from the marketing point of view, the experience is consumer focused. In this context, it is possible to consider, in light of information consumption by the elderly user, whether the achievement is the perception of the value of a product or service, or merely the ability to be involved and absorbed in a memorable experience that contributes to well-being.

In light of the foregoing, we analyze the factors that influence elderly users in the use of social networks for information consumption, and identify the aspects that contribute to creating a memorable experience that influences how the elderly communicate, use time and consume information, as well increase their personal valorization [1].

Aluri [1] applied the economics of experience theory to the use of the Pokémon GO app to investigate the factors that influence travelers to use the application and its influence on the experience of each individual. On the other hand, Radder and Han [37] examined the experience of visiting a museum through the theory of the "experience economy".

The experience economy by Pine and Gilmore [34] is a concept that unites the dimensions: educational, entertainment, aesthetics and evasion (see Fig. 1), on a scale from passive to active participation, and feelings of immersion to absorption.

In terms of the use of social networks by the elderly, and taking into account the concepts of the experience economy, from an information-consumption perspective (in different formats: text, photographs, videos), the dimension of entertainment can be considered in the viewing of videos and photos, communication with others, the sharing of comments, and more. The educational dimension is revealed by observing searches for information regarding products, services and news. The dimension of evasion can be observed while the user is viewing photos and videos, which can permit the viewer to overcome the physical barriers where he or she is while allowing for the creation of other creative, mental spaces. The dimension of aesthetics can be measured in terms of the environment provoked by the interface, i.e., whether it is pleasant, beautiful or intuitive.

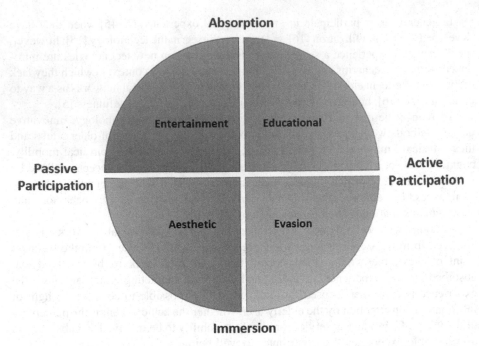

Fig. 1. The experience economy. Source: Adapted from Pine and Gilmore, 1999, p. 30.

In this context, we intend to analyze the contribution of the four dimensions of the experience economy to the user experience in association with social networks and elderly users. For this analysis, the following four research hypotheses are considered:

(1) Hypotheses #2 (H_2): Does the technological experience contribute to the educational of the elderly user of social networks?
(2) Hypotheses #3 (H_3): Does the technological experience contribute to the evasion of the elderly user of social networks?
(3) Hypotheses #4 (H_4): Does the technological experience contribute to the entertainment of the elderly user of social networks?
(4) Hypotheses #5 (H_5): Does the technological experience contribute to the aesthetics of the elderly user of social networks?

All four dimensions contribute to the overall experience [1]. The overall experience is associated with the motivation that prompted the use of social networks: entertainment, social inclusion, well-being and education. It also contributes to the personal valuation of the elderly user, as presented in H_1.

So far, and to our knowledge, no research has considered education as a reason why older adults use social networks, nor has there been an analysis of social networks on the basis of the concept of the experience economy from the point of view of information consumption or an educational perspective.

In this context, and in the present study, taking into account information consumption and consumer involvement, we intend to investigate the impact that the four dimensions of technological experience (according to the concepts of Pine and Gilmore [34]) has on older adults, as well as identify how the technological experience of using social networks contributes to an increase in the personal valuation and quality of life of elderly users.

3 Conceptual Research Model and Hypotheses

As the aim of the study is to analyze the contribution of social networks to the technological experience (user experience) of older adults, the concepts associated with the experience economy were considered.

3.1 Research Hypothesis

In order to investigate the objectives presented and the hypotheses formulated, a set of questions was considered to assess the experience associated with the use of social networks, taking into account the profile of the respondent.

A set of questions was considered for each dimension associated with the concept of the experience economy.

In the educational dimension, the use of social networks can contribute to: (1) becoming more cultured; (2) stimulating the curiosity to learn about new subjects; (3) learning new experiences; and, (4) increasing skills.

In the aesthetic dimension, the use of social networks can contribute to: (1) a harmonious experience; (2) a very attractive experience; (3) a carefully crafted interface design; and, (4) an appealing interface design.

In the entertainment dimension, the use of social networks can contribute to: (1) a captivating experience; (2) help passing the time; (3) enjoying the fun publications of friends; and (4) the discovery that friends' publications are interesting.

In the evasion dimension, the use of social networks can contribute to: (1) the feeling that the user is playing a different role when he or she uses social networks; (2) the feeling that the user lives in a different place; (3) complete distraction; and, (4) a way to help forget the daily routine. To investigate these hypotheses, a research model was developed to evaluate the elderly–user's experience with the use of the social networks, called the Social Networks User Experience Model (SNUX), as presented in Fig. 2.

3.2 Research Model

The SNUX model evaluates the dimensions of the technological experience associated with social networks (Social Networks Technological Experience (SNTE)), where the concepts of the experience economy were considered in the context of the consumption of information and its contribution to added personal value to the elderly user. Such was measured by the variable, add value to the user (AVU), which contributes to the well-being of the users that belong to these age groups.

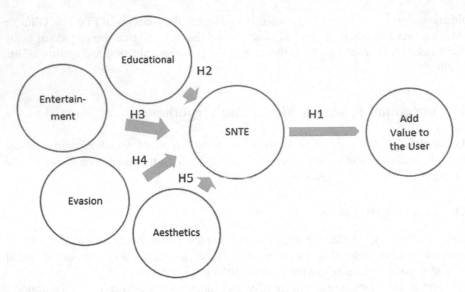

Fig. 2. Research model. Source: Author's elaboration.

4   Methodology

After the identification of the research hypotheses, the literature review and delimitation of the research problem, the proposed study methodology was based on the following steps: (1) construction of the survey; (2) data collection; (3) selection and codification of data; (4) selection of methods and techniques of data analysis; and, (5) analysis of the results.

Through the elaboration of the survey that was released in December 2017, of 60 survey responses, 58 valid responses were obtained.

The base sample was social network users over 55, using the probabilistic method of "convenience sampling", in which the sample is selected based on the availability and accessibility of the elements of the population. After data collection, responses were codified to make possible the analysis of the data by descriptive statistics to characterize the sample, and the structural equation model was applied to evaluate the SNUX research model. The interpretation of results was elaborated and the conclusions presented, along with the limitations and future work.

5   Results

5.1   Characterization of the Sample

Of the 58 respondents, the sample characterization is defined by a majority of women (55.2%) and possession of a university degree (32.8%). The respondents was divided in two groups: one group was aged greater than or equal to 60 (55.2%), and the second group was between 55 and 59 years of age (44.8%). In terms of occupation, the majority was retired (48.3%), followed by those employed on behalf of others (25.9%), self-employed (19.0%) and unemployed (6.9%). In terms of education level, the majority was graduated (32.8%), followed by those with a high–school diploma (29.3%) and PhD degree (17.2%). The majority of the respondents prefer to use computers (43.1%), followed by smartphones (36.2%) and tablets (19.0%).

Ninety-three percent of the elderly users interviewed had a Facebook profile, followed by 67.0% on YouTube, and a very close 64.0% on Google +. Snapchat and Twitter are the networks that had the lowest value of 7.0%, probably because they did not arouse the interest of this age group.

In response to the question "What kind of content do you share on social networks?", our analysis concluded that the most "often" shared content was images (36.2%), followed by texts (24.1%) and movies (13.8%). The answer "sometimes" was identical for all three of these types of information (39.7%).

In response to the question "What are the reasons why you use social networks?", among those who used social networks "many times", most respondents (55.2%) replied "to·communicate with their friends", followed by "find information" (39.7%) and "entertainment" (36.2%). What aroused less interest for this group was the "discovery of new trends".

In the group that "sometimes" used social networks, 39.7% responded that they use it to "find information", followed by 34.5% who used it for the "discovery of new trends". In this group, "entertainment" was the least valued use, with 24.1% of the answers.

5.2   Social Networks User Experience Model (SNUX) Model

Following the selection and codification of data, methods and techniques were identified to analyze the data. The structural equation model (SEM) was considered since it is the model indicated to overcome the need to measure multidimensional and not-directly-observable concepts, also called constructs or latent variables [4, 24]. According to the work of Gefen, Straub and Boudreau [18], the SEM "has become the rigueur in validating instruments and testing linkages between constructs".

The SEM considered is based on variance-based SEM or partial least squares path SEM [39], which permits the construction of the model in an exploratory phase, with a little portion of the sample that can be without normal distribution [9, 40].

Figure 3 shows the Social Networks User Experience Model, and Table 1 provides an overview of all variables, including the exogenous variables: education, entertainment, aesthetics and evasion; and endogenous variables: SNUX and AVU.

5.3   Model Estimation

After the definition of the variables and constructs of the SEM model, the next step is the model estimation to the structural equation modeling coefficients [35]. This study used SmartPLS 2.0 software, and the results are presented in Fig. 4.

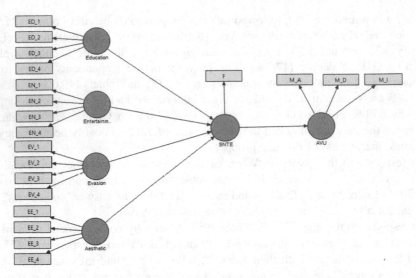

Fig. 3. The social networks user experience model (SNUX).

Table 1. List of the latent variables and their constructs.

Add Value to the User (AVU) - reflective	
M_A	Adquiring knowledge
M_I	Social Inclusion
M_D	Fun
Social Networks Technological Experience (SNTE) – reflective – single item	
F	Frequency of social networks use
Education dimension- reflective	
ED_1	Get more cultured
ED_2	Stimulates the curiosity to learn about new subjects
ED_3	Learning new experiences
ED_4	Increases the skills
Aesthetic dimension - reflective	
EE_1	A harmonious experience
EE_2	A very attractive experience
EE_3	The interface design is careful
EE_4	The interface design is appealing
Entertainment dimension – reflective	
EN_1	A captivating experience
EN_2	Help to pass the time
EN_3	My friend's publications are fun
EN_4	My friend's publications are interesting
Evasion dimension – reflective	
EV_1	The feeling that the user play a different role when he/she use social networks
EV_2	The feeling that the user live in a different place when use social networks
EV_3	The use of social networks distracts completely
EV_4	The social networks use help to forget the daily routine

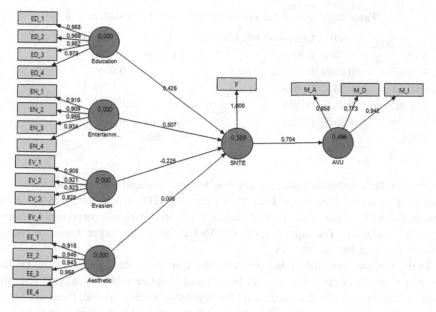

Fig. 4. Estimation of the SNUX model.

5.4 Evaluation of the Measurement Model

It is necessary to analyze the adjustment quality of the model through three steps: (1) evaluation of the measurement model to guarantee the convergent validity; (2) observation of internal consistency values; and (3) discrimination quality assessment [40].

The evaluation of the measurement model to guarantee the convergent validity is evaluated by observation of the average variance extracted (AVE) (Fornel and Larcker [17, 21]), where all the AVE values should be more than 0.5, which is a condition to guarantee that the model converges to a satisfactory result.

Observation of internal consistency values takes into consideration the values the Cronbach's alpha (CA) and composite reliability (CR), expressed by the rho of Dillon-Goldstein, which makes it possible to ascertain whether the sample is free of biases and whether, on the whole, it is reliable. The values of CA should be higher than 0.6, and values of 0.7 are considered adequate. Values of CR should be higher than 0.7, and values of 0.9 are considered satisfactory [11, 40].

Table 2 presents the estimated values of the adjustment quality of the SNUX model. By using the AVE values, it is possible to conclude that the SNUX model will converge to a satisfactory result, once all the values are higher than 0.5 [21].

Table 2 also permits analysis of the internal consistency of the model through the values of CA and CR. CA values higher than 0.6 and 0.7 and CR values higher than 0.7 and 0.9 indicate that the model is satisfactory.

In the third step, the discriminant validity assessment permits investigation of the independence between latent variables and other variables. This analysis can be done by observing cross loading, which should be the indicators with the highest factor

Table 2. Values of the adjustment quality of the SNUX model.

	AVE	Composite reliability (CR)	R square	Cronbachs alpha (CA)
AVU	0.7404	0.8947	0.4959	0.8239
Aesthetic	0.8849	0.9685	0	0.9566
Education	0.9420	0.9848	0	0.9795
Entertainment	0.8623	0.9616	0	0.9467
Evasion	0.8029	0.9421	0	0.9186
SNTE	1	1	0.5086	1

loadings in their respective latent variables, when compared with the observed variables [10], or by the criterion of Fornell and Larcker [21], which compares the square roots of the AVE values of each latent variable with the Pearson's correlations between the latent variables. The square roots of AVEs should be larger than correlations between those of latent variables.

In the first process, and taking into consideration the values presented in Table 3, the cross loadings of the observed variables in one latent variable is always higher than the cross loadings of the observed variables in another latent variable; these are not the variables that help to measure, which shows that the model has discriminant validity, in accordance with the work of Chin [10].

In the second process, and taking into consideration the values presented in Table 4, the variance identified in the AVE must exceed the variance that the overserved variables share with other latent variables of the model. In practice, discriminant validity exists when the squared root of the AVE of each construct is greater than the correlation values between the latent variables and the observed variables [35, 40]. Table 4 shows that the SNUX model has discriminant validity, as confirmed by the first process, once the values of the square roots of the AVE values, presented in the main diagonal, are higher than the correlation between the latent variables, in accordance with the work of Fornell and Larcker [21].

After guaranteeing discriminant validity in the evaluation of the measurement model (meaning that the adjustments in the measurement model have ended), the next step is the evaluation of the structural model.

5.5 Evaluation of the Structural Model

In the evaluation of the structural model, the first step is the evaluation of the Pearson's correlations coefficient (R^2) of the endogenous latent variables. In this study there are two endogenous latent variables: AVU with an $R^2 = 0.4959$ and SNTE with a $R^2 = 0.5086$, as presented in Table 2. These values are moderate and represent the portion of the variance of the endogenous variables that is explained by the structural model [9, 35].

Another aspect to analyze is the model's capacity to predict, which requires the calculation of the Stone-Geisser indicator (Q^2) and the Cohen indicator (f^2) associated to the effect size [9, 20].

Table 3. Values of cross loadings of the observed variables in the latent variables

	AVU	Aesthetic	Education	Entertainment	Evasion	SNTE
ED_1	0.6714	0.7853	**0.9553**	0.6471	0.6929	0.5805
ED_2	0.7286	0.7748	**0.9695**	0.6618	0.5819	0.6591
ED_3	0.6812	0.8027	**0.9821**	0.6756	0.6190	0.5911
ED_4	0.6854	0.8066	**0.9753**	0.6470	0.5777	0.6242
EE_1	0.7256	**0.9154**	0.7580	0.7180	0.6393	0.5652
EE_2	0.7154	**0.9457**	0.8015	0.7121	0.5622	0.6072
EE_3	0.7038	**0.9434**	0.7465	0.7209	0.6344	0.5452
EE_4	0.7050	**0.9579**	0.7592	0.7376	0.6788	0.5134
EN_1	0.7811	0.7771	0.6656	**0.9150**	0.6169	0.6609
EN_2	0.7734	0.6357	0.5385	**0.9088**	0.6085	0.5963
EN_3	0.7581	0.6975	0.6322	**0.9563**	0.7333	0.5206
EN_4	0.8086	0.7271	0.6743	**0.9335**	0.6885	0.5892
EV_1	0.5578	0.6433	0.6645	0.6549	**0.9081**	0.4162
EV_2	0.5210	0.5639	0.5532	0.5736	**0.9213**	0.3062
EV_3	0.7006	0.7004	0.5886	0.7315	**0.9231**	0.4265
EV_4	0.4761	0.4210	0.4269	0.5456	**0.8283**	0.2854
F	0.7042	0.5954	0.6339	0.6427	0.4111	**1.0000**
M_A	**0.8578**	0.7024	0.6950	0.6435	0.4771	0.6338
M_D	**0.7733**	0.5560	0.4813	0.7700	0.6316	0.4567
M_I	**0.9420**	0.6863	0.6430	0.7876	0.5797	0.6940

Table 4. Values of the correlations between the latent variables and the square roots of the AVE values (on the main diagonal)

	AVU	Aesthetic	Education	Entertainment	Evasion	SNTE
AVU	**0.8604**					
Aesthetic	0.7581	**0.9407**				
Education	0.7136	0.816	**0.9706**			
Entertainment	0.8423	0.7675	0.6778	**0.9286**		
Evasion	0.6405	0.666	0.6347	0.7099	**0.8960**	
SNTE	0.7042	0.5954	0.6339	0.6427	0.4111	**1.0000**

The Stone-Geisser indicator (Q^2) evaluates how close the model is to what was expected; the Cohen indicator (f^2) evaluates how useful each construct is for the model, as presented in Table 5. The Q^2 associated with the exogenous latent variables presents a value higher than zero, which means that both variables have predictive power, and the structural model has predictive relevance. The f^2 indicators associated with all the latent variables are higher than 0.35, which shows how useful each construct is for the model, which, in this case, all have a high impact on the structural model, as presented in Table 5.

Table 5. Indicators values of the predictive validity and effect size.

		Q^2	f^2
AVU	ENLV	114.0832	0.4795
Aesthetic	EXLV		0.7796
Education	EXLV		0.8691
Entertainment	EXLV		0.7314
Evasion	EXLV		0.6369
SNTE	ENLV	35.5643	0

Legend: *EXLV* – Exogeneous Latent
Variable; *ENLV* – Endogeneous Latent
Variable

The structural model analyses ends with the individual analysis of the coefficients of the respective model (path coefficients) [21, 35], where it is necessary to analyze the sign, the value and the statistical significance, which should be more than 1.96 (bi-lateral and with a 5% significance level).

Analysis of Fig. 5 shows that only the relationship between aesthetic and SNTE has a lower t-test value (lower than 1.96), which implies the acceptation of the H0 (null hypothesis), meaning that the structural coefficient is equal to zero, with a 5% significance level. This permits the conclusion that there is no empirical evidence to support the structural relationship between aesthetics and SNTE.

The model can also be analyzed by its direct, indirect and total effects. Table 6 presents the direct effect, which indicates, by the t-test value of 0.0476, that the H0 should be accepted and the direct coefficient should be zero. This means that the aesthetic dimension does not have an effect on the social network technological experience. Also, the evasion dimension decreases on the social networks once the sign is negative. However, as the SNTE increases the AVU, it can be concluded that the utilization of social networks contributes added value to the elderly user (personal value). The entertainment dimension is the concept associated with the experience economy [34], which contributes to an increase in the SNTE, as it has a higher value among the four dimensions of the experience economy.

In the SNUX model, there are no indirect effects; it is only necessary to analyze the total effects, presented in Table 7. In direct effects, the t-test value associated with the relationship between aesthetic and SNTE is lower than 1.96, which implies that the H0 should be accepted and the direct coefficient should be zero, such as in the relationship between aesthetic and AVU. The total effect between evasion and SNTE, and evasion and AVU contributes to a decrease in the SNTE and AVU.

With respect to the total effect of AVU, Table 7 shows that the SNTE has the most impact on the AVU (0.7042), followed by entertainment (0.3573) and education (0.3002). Taking into consideration the SNTE variable, the dimension that has the most impact is entertainment (0.5074), followed by education (0.4263).

There are studies that still consider the goodness of fit (GoF) to assess the quality of the overall reflective model as a whole, but Henseler and Sarstedt [22] have shown that it has no power to disrupt valid models or invalid models, so this indicator was not used.

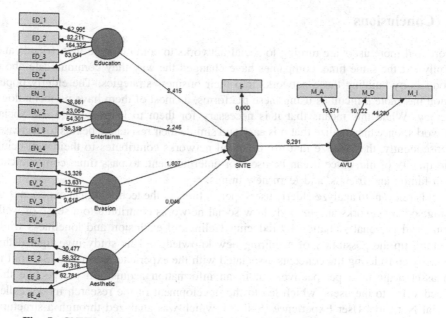

Fig. 5. SNUX model with the t-test values obtained by SmartPLS bootstrapping.

Table 6. Direct effects in the structural relationships between the latent variables

	Structural coefficient	Standard deviation (STDEV)	T-value	p-value
Aesthetic → SNTE	0.0079	0.1675	**0.0476**	0.9620
Education → SNTE	0.4263	0.1765	2.4154	0.0158
Entertainment → SNTE	0.5074	0.2261	2.2445	0.0248
Evasion → SNTE	**−0.2250**	0.1245	1.8073	0.0708
SNTE → AVU	0.7042	0.1119	6.2911	0.0000

Table 7. Total effects in the structural relationships between the latent variables

	Original sample (O)	Standard deviation (STDEV)	T statistics	P value
Aesthetic → AVU	0.0056	0.1957	**0.0287**	0.9771
Aesthetic → SNTE	0.0080	0.2736	**0.0291**	0.9768
Education → AVU	0.3002	0.1379	2.1767	0.0296
Education → SNTE	0.4263	0.1765	2.4154	0.0158
Entertainment → AVU	0.3573	0.1872	1.9091	0.0563
Entertainment → SNTE	0.5074	0.2518	2.0152	0.0439
Evasion → AVU	**−0.1585**	0.0958	1.6544	0.0981
Evasion → SNTE	**−0.2250**	0.1378	1.6332	0.1025
SNTE → AVU	0.7042	0.1119	6.2911	0

6 Conclusions

More and more users are turning to social networks to communicate with friends and family. At the same time, companies have changed the way they communicate about products by building these networks into their business strategies. The elderly population has more difficulties using these platforms, as most of them have no technology literacy. While this means that it is necessary for them to learn how to use social networks, once they realize that it is easy and simple their resistance begins to decrease. Consequently, they realize that use of social networks contributes to their well-being and quality of life, since it can be used for entertainment, to pass time, communicate with family and friends, and learn new things.

It is relevant to analyze elderly users' perceptions of the technological experience of using social networks and to study how social networks contribute to a sense of well-being and personal valuation by reducing feelings of exclusion and loneliness while opening up the possibility of acquiring new knowledge. This study investigated this issue by considering the concepts associated with the experience economy [34]; not in an asset acquisition perspective, but in an information acquisition perspective that added value to the users, which led to the development of the research model called Social Networks User Experience (SNUX), which was analyzed through a structural equations system.

The analysis of the research data allowed for the development of a model of the involvement of elderly users in social networks, which helps us understand the technological experience associated with the adoption and use of social networks, as well as develop strategies that should be considered for decreasing social isolation and increasing well-being and quality of life through communication, entertainment and the acquisition of knowledge.

This study's interviews revealed that Facebook is the social network where elderly users have the greatest presence (93.0%), followed by YouTube (67.0%) and Google + (64.0%). The content they prefer to share on social networks is images (36.2%), followed by texts (24.1%). Their stated reasons for using social networks are: "communicate with their friends" (55.2%), followed by "find information" (39.7%) and "entertainment" (36.2%), with the remainder choosing "discovery of new trends".

In terms of the technological experience associated with the concepts of the experience economy [34] and the use of social networks by elderly people, this research model found no empirical evidence to support a structural relationship between the aesthetic dimension and the SNTE. Another conclusion, achieved with the SNUX model, is that the evasion dimension of the social networks contributes to a decrease in the SNTE. On the other hand, the educational and entertainment dimensions contribute to the SNTE, and the effect of the SNTE in elderly people contributes to an increase in added value to these users, and consequently their well-being.

With the SNUX model, it is possible to conclude that with respect to the SNTE, entertainment and education contribute to added personal value for elderly users. Taking into consideration the research hypotheses, H_1 was proven to show that the use of social networks increases personal valuation through the acquisition of knowledge. H_2 and H_4 were proven to show that the technological experience contributes to the education and entertainment of the elderly user of social networks. However, H_3 and H_5 were rejected based on the responses of the interviewed, which showed that the technological experience does not contribute to the evasion and aesthetic dimension of the elderly user of social networks.

The overall experience is associated with the motivation to use social networks in terms of: entertainment, social inclusion, well-being and education, as well as the contributions made to the personal valuation of the elderly user. However, the SNUX model permits the conclusion that the entertainment and education dimensions have more impact on the technological experience associated with this age group. The results validate that the educational dimension is a main driver for the use of social networks.

The limitations with the present study include the fact that the members of the age group that use social networks have a secondary or higher education qualification level, which leaves out the elderly with less literacy and who do not use social networks, or even realize its potentialities. Also, the sample size of 58 responses is a small number to represent the entire elderly population, which is increasing in number.

In terms of future work, the analysis of the relationship between the reasons and the dimensions of the technological experience provided by social networks will be considered for different ages groups (for example, users aged from 60 to 90), and for different, less developed countries, where persons who are 70 to 80 years old are very common. These results will then be compared with the present work, in the context of the concepts of the experience economy.

Acknowledgments. This paper is financed by National Funds provided by FCT - Foundation for Science and Technology through project CIEO (UID/SOC/04020/2013) and project CEF-AGE (PEst-C/EGE/UI4007/2013). This work was supported by the Portuguese Foundation for Science and Technology (FCT), project LARSyS (UID/EEA/50009/2013) and CIAC.

References

1. Aluri, A.: Mobile augmented reality (MAR) game as a travel guide: insights from Pokémon Go. J. Hosp. Tourism Technol. **8**(1), 55–72 (2017)
2. Arch, A.: Web Accessibility for Older Users: A Literature Review. http://www.w3.org/TR/wai-age-literature/. Accessed 21 June 2018
3. Barnard, Y., Bradley, M.D., Hodgson, F., Lloyd, A.: Learning to use new technologies by older adults: perceived difficulties, experimentation behaviour and usability. Comput. Hum. Behav. **29**(4), 1715–1724 (2013)
4. Bollen, K.A.: Structural Equations with Latent Variables. Wiley, New York (1989)
5. Braun, M.T.: Obstacles to social networking website use among older adults. Comput. Hum. Behav. **29**(3), 673–680 (2013)
6. Bruder, C., Blessing, L., Wandke, H.: Adaptive training interfaces for less-experienced, elderly users of electronic devices. Behav. Inf. Technol. **33**(1), 4–15 (2014)
7. Castilla, D., et al.: Teaching digital literacy skills to the elderly using a social network with linear navigation: a case study in a rural area. Int. J. Hum. Comput. Stud. **118**, 24–37 (2018)

8. Chakraborty, R., Vishik, C., Rao, H.R.: Privacy preserving actions of older adults on social media: exploring the behavior of opting out of information sharing. Decis. Support Syst. **55** (4), 948–956 (2013)
9. Chin, W.W.: How to write up and report PLS analyses. In: Esposito Vinzi, V., Chin, W., Henseler, J., Wang, H. (eds.) Handbook of Partial Least Squares, pp. 655–690. Springer, Heidelberg (2010). https://doi.org/10.1007/978-3-540-32827-8_29
10. Chin, W.W.: The partial least squares approach for structural equation modeling. In: Marcoulides, G.A. (ed.) Modern Methods for Business Research, pp. 295–336. Lawrence Erlbaum Associates, London (1998)
11. Chin, W.W., Newsted, P.R.: Structural equation modeling analysis with small samples using partial least squares. In: Hoyle, R.H. (ed.) Statistical Strategies for Small Sample Research, pp. 307–341. Sage Publications, Thousand Oaks (1999)
12. Chung, J.E., Park, N., Wang, H., Fulk, J., McLaughlin, M.: Age differences in perceptions of online community participation among non-users: an extension of the technology acceptance model. Comput. Hum. Behav. **26**(6), 1674–1684 (2010)
13. Coelho, J., Duarte, C.: A literature survey on older adults' use of social network services and social applications. Comput. Hum. Behav. **58**, 187–205 (2016)
14. Coelho, J., Rito, F., Duarte, C.: "You, me & TV"—fighting social isolation of older adults with Facebook, TV and multimodality. Int. J. Hum. Comput. Stud. **98**, 38–50 (2017)
15. Dutton, W.H., Shepherd, A.: Trust in the Internet as an experience technology. Inf. Commun. Soc. **9**(4), 433–451 (2006)
16. Dutton, W.H., Reisdorf, B.C.: The internet through the ages. In: Nixon, P.G., Rawal, R., Funk, A. (eds.) Digital Media Usage Across the Life Course, pp. 16–28. Routledge, Abingdon (2016)
17. Fornell, C., Larcker, D.F.: Evaluating structural equation models with unobservable variables and measurement error. J. Mark. Res. **18**(1), 39–50 (1981)
18. Gefen, D., Straub, D., Boudreau, M.C.: Structural equation modeling and regression: guidelines for research practice. Commun. Assoc. Inf. Syst. **4**(1), 7 (2000)
19. Goswami, S., Köbler, F., Leimeister, J.M., Krcmar, H.: Using online social networking to enhance social connectedness and social support for the elderly. In: Proceedings of International Conference of, Information Systems (ICIS) (2010)
20. Hair, J.F., Hult, T.M., Ringle, C.M., Sarstedt, M.: A Primer on Partial Least Squares Structural Equation Modeling (PLS-SEM). Los SAGE, Angeles (2014)
21. Henseler, J., Ringle, C.M., Sinkovics, R.R.: The use of partial least squares path modeling in international marketing. Adv. Int. Mark. **20**, 277–319 (2009)
22. Henseler, J., Sarstedt, M.: Goodness-of-fit indices for partial least squares path modeling. Comput. Stat. **28**, 565–580 (2012)
23. Hill, R., Beynon-Davies, P., Williams, M.D.: Older people and internet engagement: acknowledging social moderators of internet adoption, access and use. Inf. Technol. People **21**(3), 244–266 (2008)
24. Hoyle, R.H.: The structural equation modeling approach: basic concepts and fundamental issues. In: Hoylw, R.H. (eds.) Structural Equation Modeling: Concepts, Issues, and Applications, pp. 1–15. Sage, Thousand Oaks (1995)
25. ISO 9241-210: Ergonomia da interação sistema humano - Parte 210: Projeto centrado no ser humano para sistemas interativos (anteriormente conhecido como 13407). International Organization for Standardization (ISO) (2009)
26. Kooij, D., deLange, A., Jansen, P., Dikkers, J.: Older workers' motivation to continue to work: five meanings of age, a conceptual review. J. Manag. Psychol. **23**(4), 364–394 (2008)
27. Moutinho, L.: Consumer behaviour in tourism. Eur. J. Mark. **21**(10), 5–44 (1987)

28. Nef, T., Ganea, R.L., Müri, R.M., Mosimann, U.P.: Social networking sites and older users–a systematic review. Int. Psychogeriatr. **25**(7), 1041–1053 (2013)
29. Nimrod, G.: Seniors' online communities: a quantitative content analysis. Gerontologist **50**, 382–392 (2010)
30. Nimrod, G.: The benefits of and constraints to participation in senior's online communities. Leisure Stud. **33**, 247–266 (2014)
31. Pfeil, U., Zaphiris, P.: Investigating social network patterns within an empathic online community for older people. Comput. Hum. Behav. **25**(5), 1139–1155 (2009)
32. Pfeil, U., Svangstu, K., Ang, C.S., Zaphiris, P.: Social roles in an online support community for older people. Int. J. Hum.-Comput. Interact. **27**(4), 323–347 (2011)
33. Pfeil, U., Zaphiris, P., Wilson, S.: Older adults' perceptions and experiences of online social support. Interact. Comput. **21**(3), 159–172 (2009)
34. Pine, J., Gilmore, J.: The Experience Economy. Harvard Business School Press, Boston (1999)
35. Pinto, P.: Modelos de equações estruturais com variáveis latentes, fundamentos da abordagem Partial Least Squares. Lisboa, Bnomics (2016)
36. PORDATA. População residente segundo os Censos: total e por grupo etário. https://www.pordata.pt/. Accessed 12 Dec 2018
37. Radder, L., Han, X.: An examination of the museum experience based on Pine and Gilmore's experience economy realms. J. Appl. Bus. Res. **31**(2), 455–470 (2015)
38. Ramos, C.M.Q., Mendonça, M.M., Rodrigues, J.M.F.: Analysis of adoption and habits of use of social networks in senior users. In: Proceedings of DSAI 2018, Thessalonika, Greece, June 2018 (DSAI 2018) (2018)
39. Reinartz, W., Haenlein, M., Henseler, J.: An empirical comparison of the efficacy of covariance-based and variance-based SEM. Int. J. Res. Mark. **26**(4), 332–344 (2009)
40. Ringle, C.M., Silva, D., Bido, D.D.S.: Modelagem de equações estruturais com utilização do SmartPLS. REMark **13**(2), 54 (2014)
41. Schmitt, B.: Experiential marketing. J. Mark. Manag. **15**(1–3), 53–67 (1999)
42. SeniorNet Education and Empowerment: SeniorNet facts sheet. http://www.seniornet.org/index.php?option.com. Accessed 12 Dec 2018
43. Sinclair, T.J., Grieve, R.: Facebook as a source of social connectedness in older adults. Comput. Hum. Behav. **66**, 363–369 (2017)
44. Vošner, H.B., Bobek, S., Kokol, P., Krečič, M.J.: Attitudes of active older Internet users towards online social networking. Comput. Hum. Behav. **55**, 230–241 (2016)
45. Vroman, K.G., Arthanat, S., Lysack, C.: "Who over 65 is online?" Older adults' dispositions toward information communication technology. Comput. Hum. Behav. **43**, 156–166 (2015)
46. Wagner, N., Hassanein, K., Head, M.: Computer use by older adults: a multi-disciplinary review. Comput. Hum. Behav. **26**(5), 870–882 (2010)
47. Yu, T.K., Lin, M.L., Liao, Y.K.: Understanding factors influencing information communication technology adoption behavior: the moderators of information literacy and digital skills. Comput. Hum. Behav. **71**, 196–208 (2017)
48. Xiang, Z., Gretzel, U.: Role of social media in online travel information search. Tour. Manag. **31**(2), 179–188 (2010)

Automatic Exercise Assistance for the Elderly Using Real-Time Adaptation to Performance and Affect

Ramin Tadayon[✉], Antonio Vega Ramirez, Swagata Das, Yusuke Kishishita, Masataka Yamamoto, and Yuichi Kurita

Hiroshima University, 1-4-1 Kagamiyama, Higashihiroshima, Hiroshima Prefecture 739-0046, Japan
rtadayon@hiroshima-u.ac.jp

Abstract. This work presents the design of a system and methodology for reducing risk of locomotive syndrome among the elderly through the delivery of real-time at-home exercise assistance via intensity modulation of a worn soft exoskeleton. An Adaptive Neural Network (ANN) is proposed for the prediction of locomotive risk based on squat exercise performance. A preliminary pilot evaluation was conducted to determine how well these two performance metrics relate by training the ANN to predict test scores among three standard tests for locomotive risk with features from joint tracking data. The promising initial results of this evaluation are presented with discussions for future implementation of affective classification and a combined adaptation strategy.

Keywords: Exercise assistance · Motion assessment · Artificial neural network · Locomotive syndrome · Soft exoskeleton · Real-time adaptation

1 Introduction

Locomotive Syndrome, a disability characterized by the degeneration of locomotive capability, affects a significant portion of the worldwide population and leads to reduced life expectancy, mobility and an individual's capability to independently complete activities of daily living (ADL) [21,31]. One of the primary contributors leading to locomotive syndrome progression is a lack of regular lower extremity exercise [20]. Exercises like the squat and lunge, when performed regularly and accompanied with proper diet and lifestyle changes, have been proven to reduce the risk for this disorder and to improve mobility and general well-being in the elderly [7,20,25]. Yet for many in this population, lack of a guided training environment can make it tedious to complete the exercises independently in the home [17].

To address these issues, the authors propose the development of an automated environment to provide adaptive, guided resistance training for completion of at-home squat exercises to improve quadricep function. In this case,

© Springer Nature Switzerland AG 2019
M. Antona and C. Stephanidis (Eds.): HCII 2019, LNCS 11573, pp. 556–574, 2019.
https://doi.org/10.1007/978-3-030-23563-5_44

the two-leg standing squat exercise was chosen as a target exercise task for such a system because it has proven to be highly beneficial in improving the lower extremity muscle strength necessary to improve locomotion [8, 10] and is a fairly prevalent exercise within this population. To provide adaptive resistance, a wearable, soft exoskeletal suit has been developed consisting of 10 pneumatic gel muscles (PGMs) distributed along the hip, knees and ankles which, when actuated, contract in a similar manner to human muscles to provide resistance during the standing (extension) phase of the squat. While the wearer is guided by a coach to complete squats in a virtual reality (VR) gym environment, these force-generating muscles are automatically controlled in real-time by adaptation software that uses an artificial neural network (ANN) to assess locomotive risk based on squat performance, and in future work, a second ANN that classifies emotion based on real-time physiological data.

It is argued that, by being observant of an individual's locomotive risk and emotional state in a manner similar to real coaches and physical trainers, this system can provide the appropriate level of resistance to build quadricep strength while completing routine squat exercise. To enable an automated system to perform this assessment in real-time, it is first necessary to examine the relationship between an individual's performance in the squat task and that individual's level of locomotive risk, which can then be utilized to assign the appropriate level of actuation to the PGMs along that individual's legs. This relationship is explored and evaluated in this work by preliminary training of a neural network using 13 healthy subjects in the age range of 20–35. Results of the training indicate the network's predictive accuracy for this age group in the three different tests of locomotive risk and lower extremity muscle function designed by the Japanese Orthopaedic Association (JOA) [15]. Future work is described that would validate these results by expanding the training set to include the target population of elderly users, and methods for classification of affective data, combined classification, and adaptation of PGMs based on this classification are presented.

2   Related Work

Several fields of research are encompassed by the proposed system for exercise assistance, including the application of ANNs to exercise performance evaluation and classification, the quantification and parameterization of squat performance based on real-time-measurable indicators, the quantification and assessment of risk for locomotive syndrome, and the classification of affective state in real-time. The relevant findings within these fields are summarized and their application toward the proposed system is discussed.

2.1   Neural Networks in Motion Performance

ANNs have proven invaluable and highly effective when applied toward the automated assessment and classification of exercise performance in several specific cases. Oniga and Suto successfully applied ANNs toward the recognition of a variety of daily activities as well as arm and body posture in the assistance of independence of sick and elderly individuals in their daily lives [23]. Cary, Postolache and Girão carried out pose and gesture recognition using ANNS and a Kinect camera to assist therapists in monitoring health information of patients for use in physiotherapy assessments [2]. Recent work has also emphasized the use of ANNs in predicting competitive performance in athletic tasks such as swimming [26].

These findings suggest that the application of these classifiers toward the assessment of locomotive risk based on squat performance may help to determine how well the two relate. In this case, the relationship can be examined by observing the accuracy attainable through iterative learning of a simple feedforward artificial neural network when the parameters of squat performance are provided as input and parameters of locomotive risk are provided as output. Hence, this preliminary evaluation is the focus of this study. Once a suitable means for real-time assessment by means of Artificial Intelligence has been determined, its application toward the control of assistive and rehabilitative robotics for exercise, particularly in the control of exercise intensity, is well-noted in recent review [22].

2.2   Parameters of Squat Performance

Recent work in squat performance classification and assessment is used as the basis for the selection of input features to the ANN. A primary restriction of the design of this system is that the system's method of tracking squat motion data should be as noninvasive and cost-effective as possible [18], which lends itself to the use of external joint tracking via the usage of a depth camera as in [2]. Hence, kinematic information that can be derived from real-time data on the joints of the lower extremity (hips, knees, trunk, spine, etc.) is of particular applicability. Based on the findings in Escamilla's observation of knee biomechanics within the squat exercise [10], the angular displacement (range of motion, from standing to squatting) of the knees and hips, in additional to the lateral motion (or shakiness) of the knees, are several useful indicators that can be externally tracked. Furthermore, an estimation of the Center of Mass (CoM) of a subject has been the focus of several studies observing balance during squat tasks and other exercises [12,14].

Finally, based on sit-to-stand testing as a common clinical evaluation technique [30], the time required to stand during a squat exercise and the number of squats completed within a time limit are also relevant features of performance. These indicators can be quantified only if a system can recognize the various phases of a squat task and, by extension, recognize a completed squat by a user, provided only with real-time joint tracking information on that user. Consequently, a simple method of differentiating between the phases of a squat is presented as a part of the design of this system.

2.3   Locomotive Risk Quantification

Classification of locomotive syndrome in individuals, particularly within the elderly population, is a recent accomplishment of the JOA, who has developed a 3-assessment method known as the Short Test Battery for Locomotive Syndrome (STBLS) [15] which can be completed independently by an individual to determine his or her locomotive risk. Each test provides a score relating to several different factors of individual performance and experience that relate directly to that individual's lower-extremity strength.

In the first test of the STBLS, known as the Stand-Up test, the subject must stand from the lowest height possible out of four seated heights of 40 cm, 30 cm, 20 cm, and 10 cm, using one leg or both legs, and to hold the standing position for 3 s or longer. This test typically begins with the easiest scenario of standing with both legs from a 40 cm height, then has the subject progressively lower the standing height, and if it is possible to stand from all heights using both legs, switch to using one leg with the same strategy. A score is assigned to the subject based on the lowest height from which he or she was able to stand and whether or not the subject was able to do so with both legs or one leg. A score of 1 through 4 denotes the ability to stand with both legs from a height of 40 cm, 30 cm, 20 cm, and 10 cm, respectively, while a score of 5–8 represents the ability to stand with one leg from the same respective range of heights. A score of 0 denotes the inability to stand and remain balanced for 3 or more seconds in any of these scenarios. The strength of several muscle groups in the lower extremity, especially the quadricep, and the subject's balance in the one-leg scenario, are all determined by the scoring of this assessment. It is perhaps the most closely related as an assessment task to the squat exercise.

In the second test, known as the Two-Step test, the subject begins by standing with both feet aligned, and must then take two strides forward, moving as far forward as possible with each stride while avoiding falling or losing balance. The distance from the subject's starting point to the ending point of the second stride is measured in centimeters, and then normalized by dividing by the subject's height in centimeters. This ratio represents the subject's score in this test. The score is invalidated if the user falls or loses balance during the strides. This assessment tests the gait stability, balance and lower extremity musculoskeletal strength of the subject.

The third test of the JOA STBLS is known as the Geriatric Locomotive Function Scale of 25 questions, or GLFS-25. It is a 25-question assessment that asks the subject a range of questions related to the level of pain or difficulty experienced in completing a variety of daily tasks, walking and other activities over the last month. Responses to each question are given as a value from 0 to 4 with 0 indicating the least pain, difficulty or discomfort and 4 indicating the greatest. These scores are totalled over all 25 responses to give a final score for the subject. In this case, a lower score indicates a healthier subject with a lower risk for locomotive syndrome. It provides a measure of the subject's mobility and motor ability as well as the effects of these factors on social participation.

To reach a final decision on a subject's clinical risk of locomotive syndrome, the JOA has also established clinical decision boundaries that relate the three scores above to a single locomotive "risk level" describing the subject [31]. Risk level 2, the highest level, is determined when the subjects scores 2 or less on the Stand-Up test, or less than 1.1 on the Two-Step test, or gets a GLFS-25 score of 16 or higher. Risk level 1 is attributed to individuals who meet none of the conditions of risk level 2, but score 4 or less on the Stand-Up test, less than 1.3 on the Two-Step test, or 7 or higher on the GLFS-25. Risk level 0 is assigned to healthy subjects who score higher than 4 on the Stand-Up test while also achieving a Two-Step score of 1.3 or higher and a GLFS-25 score lower than 7. Thus, using the STBLS, three levels of locomotive risk classification are possible for a subject.

2.4   Affective Classification

Although it has been made clear that adaptive coaching agents can greatly benefit from responsiveness to user performance [11,27], adaptation to affect is significantly less explored, as well as how affective adaptation should manifest itself from the perspective of an autonomous virtual trainer. Adjustment to affect is particularly important for the elderly population as recent studies have shown that it can be at least as effective as skill-based adaptation in exergames for this population [1,16].

Fortunately, recent work has supported the use of biometric signals (Heart Rate Variability or HRV, Galvanic Skin Response (GSR), and Skin Temperature (ST)) for real-time affect determination from the perspective of stress (negative affect) [29] and flow (positive affect) [3,6]. Given that real trainers respond and adapt training to emotional output [13,29], it is proposed in this work that an autonomous system should consider affective data in determining the amount of support to provide the user.

To facilitate this adaptation strategy, it is necessary to define how stress and flow are derived from the signals of HRV, GSR and ST, under the assumption that these two signals can be measured in real-time without interfering with exercise. Real-time stress detection has been achieved by Cho et al. in [4] from biosignals related to heart rate, skin response and skin temperature using a rapidly-trained kernel-based extreme learning machine (K-ELM) classifier across 5 distinct classes: baseline, mild stress, moderate stress, severe stress and recovery. One challenge with the approach is the invasiveness placed by the high quantity of finger-worn sensors on the ease of setup and exercise with the system. To alleviate this, the approach of Ciabattoni et al. [5] using a smartwatch may be considered. In this case, stress is reduced to a 2-class decision problem using features of Respiratory Rate (RR) frequencies and means, Galvanic Skin Response (GSR) mean and standard deviation, and Body Temperature (BT) mean and maximum. The features are used in the training of a 1-NN with Euclidean distance metric, and achieve a classification accuracy of almost 90% for stress classification indicating a potential tradeoff between sensor intrusiveness and classification accuracy.

The derivation of "flow", or a measure of the amount of focused engagement of the subject with the interface, is a far less validated approach. While there is plenty of evidence to support a strong effect of flow-state on positive learning experience, the methods of deriving it in real-time require further validation, particularly when such methods rely on non-intrusive sensing. Some studies have indicated a correlation between signals related to heart rate and skin response and flow-state [9]. Martinez et al.'s work [24] found that features extracted from heart rate and GSR signals such as average and minimum heart rate can serve as good indicators in the creation of an ANN designed to classify flow state in the standard 3-class set (boredom, flow, frustration). However, it is important to note that heart rate can increase naturally as a result of exercise. This presents the challenge of deriving a method to isolate the variation in mean and minimum heart rate induced as a result of emotional output from the component of this variation attributed to exercising.

3   Methodology

To address the challenges posed above, an at-home exercise environment has been developed to provide automated physical resistance and virtual guidance during the squat exercise. Components of this system include a wearable soft exoskeleton for the lower extremity for resistive strength training at the hip, knees and feet, a VR training gym with a virtual coach avatar to lead the subject in the completion of squats, a depth camera for real-time joint tracking, a feedforward ANN to evaluate the relationship between squat performance and locomotive risk, and a mechanism for real-time configuration of the pneumatic muscles on the soft exoskeleton based on the outputs of this neural network and the outputs of an emotional state classifier (to be implemented in future work) for real-time automatic squat exercise assistance. The design and configuration of each of these components are described as follows:

3.1   Squat Assistance Exoskeleton

A primary goal of at-home squat exercise is to build strength in the quadriceps and other muscle groups in the lower extremity, particularly in the elderly population wherein the musculoskeletal system can begin to degenerate. As such, a soft exoskeletal suit was designed as a wearable device for at-home strength training with a focus on the squat task. The suit, pictured in Fig. 1, consists of a series of soft pneumatic gel muscles introduced in previous work [28] attached at three key sections along the hips and legs to worn velcro bands. These PGMS are actuated by air flow through a network of tubes connecting to a single pressurized canister and a central controller, located on a worn belt, which are responsible for regulating air pressure to the valves and actuating the muscles. In this case, actuation refers to the contraction of a PGM, which causes it to shrink at both ends and, as a result, exert a pulling force at both ends to which it is attached. The same pumping mechanism can be used to expand the muscle, relaxing its pulling force at either end.

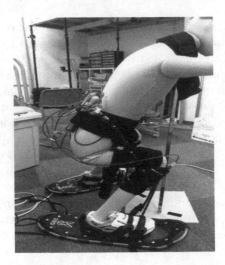

Fig. 1. First prototype of the soft exoskeleton for squat exercise. Pneumatic gel muscles are synchronized with a belt-worn controller which communicates wirelessly over Bluetooth with the training software.

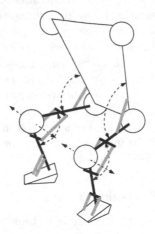

Fig. 2. Illustration of several squat performance parameters and the placement of PGMs across the lower extremity on the exoskeleton. Dotted lines represent the angles and motion values for knee angular displacement, hip angular displacement, and knee lateral motion. Locations of the 10 PGMs are shown in red. (Color figure online)

In total, 10 PGMs are distributed across the exoskeleton. Two are used to connect the upper thighs to the pelvis along the outside, while four are used to connect the thighs to the calves along the outside and inside, and four are used to connect the ankles to the outer and inner shin. The locations of these PGMS are illustrated in Fig. 2. Upon receiving a signal from the controlling software, the controller on the suit can actuate or relax the corresponding PGMs on the exoskeleton. When actuated while the subject is standing back up from a squatted position, these PGMS provide resistive force that increases the intensity of the task. For squatting down, this force becomes supportive, as it helps pull the subject into a squatted position. This supportive force, however, is unused in this system, as it is unnecessary and can even be dangerous if it moves the user beyond the appropriate range for a squat.

3.2   Virtual Reality Coach Software

To help guide the user through the completion of routine squat exercise in the home, a VR gym environment including a virtual coach avatar and player avatar have been developed in the Unity 3D platform. To interact with the environment, the user wears an Oculus VR headset and starts the application, at which point he or she is immediately placed in the VR gym space facing the virtual coach. A screenshot of this environment is provided in Fig. 3. The user's joints are tracked in this environment using the Intel RealSense D435 depth camera and NuiTrack joint tracking library. A Microsoft Kinect V2 camera was used in an earlier prototype, but was bulkier by comparison and tracked a large amount of unnecessary data for this application.

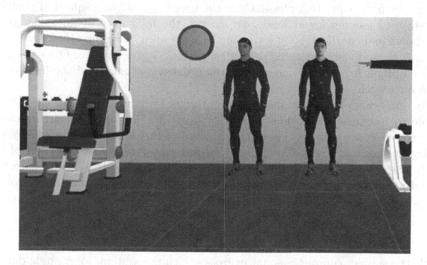

Fig. 3. Screenshot of the VR Gym environment, developed in the Unity 3D engine. The virtual coach avatar on the left leads the player through the squat task by demonstrating the next step of the squat, while the player avatar on the right (optionally hidden) reflects the player's movement as he or she follows the coach.

In the first few seconds of gameplay, the user is asked to stand upright without moving so that the system can calibrate the position of the joints for tracking. During this time, the light beside the virtual coach, pictured in Fig. 3, displays red. Once this light turns green, the coach begins guiding the subject by moving

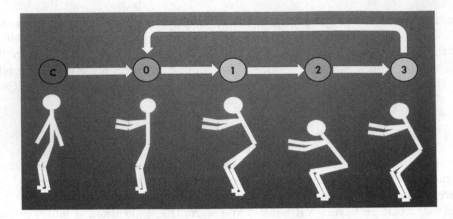

Fig. 4. Overview of the various phases of squat exercise as determined by the system. C represents the calibration phase while 0–3 represent the fully upright, squatting, fully squatted, and standing phases of the exercise, respectively.

into a squatted position. The user is then asked to follow along. As the user moves down into a squat, the light turns yellow to indicate that the system detects that the user is in the squatting phase. Once the user has reached a squatted position and stopped moving for at least one second, the light once again turns green and the virtual coach leads by returning to a standing position. Once again, the user follows and the light turns yellow as the user is standing up. Once the user is fully upright for at least one second, the process continues and the light turns green once again.

By observing the vertical movement of the user's spine base joint, the system is able to recognize the various phases of squat activity and the virtual coach leads the user accordingly. The entire process of the squat, as detected by the system, is shown in Fig. 4. An optional second avatar, shown to the right of the coach avatar in Fig. 3, copies the subject's motion rather than leading, giving the subject a second external point of reference to compare his or her motion to that of the virtual coach.

3.3   Joint Tracking for Squat Performance Data

While the subject completes squat exercises with the system, joint angle data is read from the depth camera via the Nuitrack library to provide joint angle and joint displacement data of the subject at each frame (approx. 60 frames/second). Based on the findings in Sect. 2.2, the following 9 values are selected as features for squat performance:

Input. Squat performance is represented as a vector of values $\{i_1, i_2, \ldots, i_9\}$, one per squat. The current configuration of this input vector is the following:

Input values i_1, i_2, i_3, and i_4 correspond to left knee angular displacement, right knee angular displacement, left hip angular displacement, and right hip angular displacement, respectively. These values are obtained by observing the corresponding joint angles at the fully upright and fully squatted positions of the subject on each squat, and finding the differences between each pair of joint angle values. Raw values fall within the range [0, 180] and are normalized to the range [0.00, 1.00].

Input values i_5 and i_6 correspond to the left and right knee lateral motion average. These values are obtained by observing the joint displacement values of the left and right knee along the frontal plane, sampled once per frame at roughly 60 frames per second and averaged over a single squat. Essentially, these values represent the shakiness of the knees or the load placed on the knees during exercise. Absolute value is used as we are interested in the magnitude of motion and not direction. The average of these values will lie in the range [0.00, 3.50] whose maximum represents the physical limitation of knee lateral motion expressed as a coordinate value within Unity coordinate space. This value is normalized to the range [0.00, 1.00].

Input value i_7 corresponds to the Center of Mass stability along the frontal plane. This value is roughly estimated by observing the vertical and horizontal differences between the left and right hip and knee joints to estimate a frontal plane CoM location, and measuring the displacement of that point over a single squat. A value of 1.0 represents perfect stability while 0.0 represents motion of 3.50 or higher as in the previous case with knee lateral motion.

Finally, input values i_8 and i_9 correspond to the time required, in seconds (including milliseconds) to move from fully squatted position to fully upright position, otherwise referred to in this work as "standing time" or "extension time", and the number of repetitions completed by the subject.

3.4   ANN for Locomotive Risk Based on Squat Performance

A feedforward ANN has been developed which accepts as input the squat performance features derived in Sect. 3.3 and attempts to predict the individual's scores on the three tests of the JOA STBLS described in Sect. 2.3. The structure of this neural network is shown in Fig. 5. The nine inputs for squat performance are supplemented with a bias node to create 10 total inputs and passed to the ANN with a single hidden layer of size 6 and an output layer of size 3. A leaky rectified linear unit (LReLU) activation function is utilized in this case to minimize the impact of vanishing gradient challenge on the learning process of the network:

$$f(x_i) = max(0, x_i) + 0.01 * min(0, x_i) \tag{1}$$

Outputs of the network are a 3-tuple $O = \{O_1, O_2, O_3\}$ and represent the three scores obtained by completing the JOA STBLS as detailed in Sect. 2.3.

3.5   Real-Time Adaptation

To configure the pneumatic muscles based on the output of the ANN, the range of possible configurations and support levels must first be defined formally. For

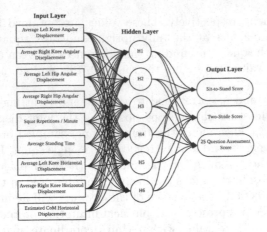

Fig. 5. Feedforward ANN used to predict locomotive risk as three scores (sit-to-stand, two-stride, 25-question assessment) representing the three tests of the JOA STBLS. Bias nodes, included in the input and hidden layer, are not shown in the diagram. Connections represent the weighted, activated output of the previous layer as it is passed as input into the next layer.

all pneumatic muscles present in the soft exoskeleton, the supportive capability of these muscles is represented as their strength of actuation discretized in the integer classification range $[0 \ldots n]$, where n is the number of distinct actuation levels derived by dividing the maximum force of actuation by a value representing the minimum change in force required for the average human to perceive a difference in the force felt in the lower extremity:

$$n = F_{max}/\delta F_{pmin} \tag{2}$$

Hence, a range of actuation forces $S = [s_0 \ldots s_n]$ can be defined wherein $S_i = i * \delta F_{pmin}$ in order to complete the discretization of pneumatic support into mappable values.

Squat Performance. Next, it is necessary to define the dimensional reduction function $f(O) = f(O_1, O_2, O_3)$ which outputs a singe value in the range $[0 \ldots m]$, where m is the number of distinct classifications of performance.

Fortunately, this dimensional reduction strategy is already derived by the JOA using the clinical decision boundaries in [31]:

$$f(O_1, O_2, O_3) = \begin{cases} 0 & O_1 > 4 \text{ and } O_2 >= 1.3 \text{ and } O_3 < 7 \\ 1 & O_1 <= 4 \text{ or } O_2 < 1.3 \text{ or } O_3 >= 7 \\ 2 & O_1 <= 2 \text{ or } O_2 < 1.1 \text{ or } O_3 >= 16 \end{cases} \tag{3}$$

The output of this function represents the corresponding stage of locomotive syndrome and the value ranges resulting in each output represent the clinical decision limits of these tests as proposed by the JOA. Hence, the outputs of the neural network can be reduced to a single value in the integer range $[0 \ldots 2]$ in which case $m = 3$.

At this point, it is possible to then map $f(O)$ to S as follows:

$$g(f(O), S) = \begin{cases} s_0 & f(O) = 0 \\ s_{n/2} & f(O) = 1 \\ s_n & f(O) = 2 \end{cases} \tag{4}$$

where $n = F_{max}/\delta F_{nmin}$.

For even values of n, we simply discard the highest force value s_n from S to revert to an even-valued n. In the case where $n = 1$, the single value in S is mapped to all outputs $f(O)$. For $n = 2$, the smaller value is assigned to scores of 0 and 1 while the larger value is mapped to the risk score of 2.

This provides a coarse mapping wherein 3 evenly distributed actuation values are selected from the range of distinct values available to adapt the system solely to the three-class "risk factor" of an individual as predicted by the ANN. However, this mapping alone does not take advantage of the dynamic nature of squat performance indicators, which can provide more detailed information about an individual's current skill level than the JOA assessment can alone. Perhaps one of the most powerful indicators of *dynamic* performance, or the performance of a specific squat attempt relative to an individual's standard, is the input value i_7 representing the lateral stability of the individual's CoM during a squat attempt.

Using the CoM value, one can derive an offset value z that is used to determine a more accurate value for $g(f(O), S)$, and can allow the output of this function to enable more dynamic adaptation between squat attempts. Assuming a Gaussian distribution of CoM values, z can be determined using the following function:

$$z = \begin{cases} 0 & n < 5 \\ i_7 * floor((n - 3)/2) & n >= 5 \end{cases} \tag{5}$$

This then allows us to define $g(f(O), S)$ as follows:

$$g(f(O), S) = \begin{cases} s_{0+z} & f(O) = 0 \\ s_{n/2+z} & f(O) = 1 \\ s_n & f(O) = 2 \end{cases} \tag{6}$$

Alternatively, we can represent the output values $\{O_1, O_2, \ldots, O_t\}$ (where t represents the number of squats performed) as 3D points within a centroid clustering space, where the value n from above represents the number of classes. To configure the classification, the first n values of output are obtained from the user after $t = n$ squat attempts and form the set Y. During this time, the system defaults to the maximum support level (or minimum resistance level) of s_n to ensure no risk to safety. The values of Y are then sorted as the first n centroids of the clustering space based on their distance from the origin $(0, 0, 0)$. Should there be less than n distinct output values among this set, then the number of classes are reduced to match the number of distinct values among this set.

From this point forward, the Nearest Centroid Classification [19] method is used to assign a support configuration to each future squat attempt. If the values $\{O_1, O_2, \ldots, O_t\}$ are highly similar, the dynamic adjustment strategy proposed earlier can be utilized to offset the classification and remap within the full range of S. Then, we can represent the final actuation value associated with an output O of the ANN as follows:

$$g(O, S) = s_{c+z} \tag{7}$$

$$c = argmin_{i \in Y} ||O_t - O_i|| \tag{8}$$

where z is defined as above.

Affect. This work proposes the classification of stress and flow-state using ANNs in real-time. Based on the work presented in Sect. 2.4, stress can be classified in the range {low stress, moderate stress, high stress} using the input features {HRV mean/std, GSR mean/std, deriv. GSR mean/std, BT min/max/avg.}, and flow-state can be classified in the range {boredom, flow, frustration} using the input features {HR mean/std/min, GSR mean/std}.

We then represent affective state as a two-tuple $AS = STR, FS$, where STR is stress level of integer index in the range $[0 \ldots 2]$ and FS is flow state of integer index in the range $[-1 \ldots 1]$. If the output of AS is represented as a point in two-dimensional space, a secondary goal of the coaching system's support then becomes to reduce the distance of this output from the optimal value $(0, 0)$, which represents the lowest output of stress and highest output of engagement/flow.

We can define a function that reduces AS to a single dimensional output reflecting the above goal, assuming that stress can be detected as a result of both boredom and frustration:

$$f(AS) = STR * FS \tag{9}$$

This output is given the range $[-2 \ldots 2]$ and reflects both the degree of adjustment necessary and the direction of that adjustment. Now we can define the output actuation value $s_t = g(f(AS))$ given the previous actuation value s_p as follows:

$$g(f(AS)) = s_i \tag{10}$$

$$i = clamp[0 \ldots n](p + \frac{f(AS)}{2} * n) \tag{11}$$

Combined Adaptation. It is of particular interest in this work to combine the adaptation based on performance and that based on affect into a single system so that the system can respond to both indicators. This combined approach has not been explored in previous work, which focuses on one type of dynamic difficulty adjustment or the other. Given two channels of individual response to a virtual exercise task (performance and affect), and a measure for each channel of the degree of deviation in the player's response from the optimal value for that individual, it is proposed that a system can provide the most effective dynamic difficulty adjustment (DDA) by adapting to the channel with the higher degree of deviation.

This implies that, as an adaptation strategy, given the values $g(f(O), S) = s_a$ and $g(f(AS), S) = s_b$, the system can choose between the two using the following strategy:

$$g(f(O), f(AS), S) = \begin{cases} g(f(O), S) & f(O) > \|f(AS)\| \\ g(f(AS), S) & f(O) < \|f(AS)\| \\ s_c(c = \frac{a+b}{2}) & f(O) = \|f(AS)\| \end{cases} \qquad (12)$$

4   Evaluation: Neural Network Training

To determine whether the ANN described above is capable of converging to a sufficiently reliable model for predicting locomotive risk based on squat performance, a preliminary pilot study was conducted wherein 13 members of the research team acted as subjects to train the ANN. The goal of this training was to determine to what error values the network converges in its prediction of each of the three locomotive assessment scores for the subject group as an approximate preliminary measure of the strength of the input feature set in relating to these assessments. In simpler terms, the potential relationship between squat performance and locomotive risk was explored through observation of post-training prediction results of the ANN.

4.1   Procedure

Thirteen healthy subjects (11 male and 2 female), in the age range 20–35 with no motor impairment, participated in this preliminary study. Each subject began by completing each of the three assessments of the JOA STBLS. The subject's three scores are written to a log file, and serve as real values against which the results of the ANN's prediction for that subject can be compared to determine error and perform gradient descent. After completing all three assessments, each subject then wears the Oculus VR headset and is asked to perform one minute of squat exercise in the virtual gym under the guidance of the virtual coach.

The exoskeleton is unused in this preliminary study as it is used to determine baseline performance for each subject. Once one minute of squat exercise is complete, the features of squat performance are collected and stored in a log file for each subject. This data is then passed as input into the ANN and used to generate three predicted score outputs. The error between these predictions and the subject's actual STBLS score is used to adjust the weights on the network. This process is repeated 100,000 times with the subject's performance data from the single exercise session.

4.2   Results and Discussion

Results of the ANN's predictive capability after training on the data of the 13 subjects is shown in Figs. 6, 7 and 8. For the GLFS-25, the ANN performed at an average error of 1.268 across all 13 subjects as shown in Fig. 6. This result is expected, as the GLFS-25 assesses factors such as pain, activity completion and mobility over a one-month period that are potentially not well-represented by execution of the squat task. In some cases, as in Subject 11's case, this error is high enough to misclassify the subject's locomotive risk level, which should be considered in future reconfiguration and pruning of the network. For the Standing test, the ANN converged to an error rate of 0.387. The remaining error may reflect the fact that the standing test is a measure of balance as much as strength, and therefore includes a one-leg standing task, which is not included in the standard squat task.

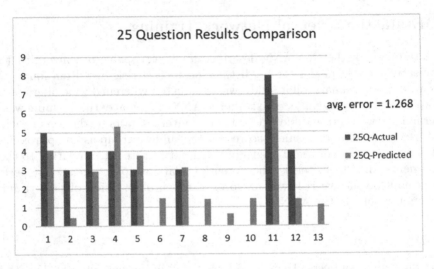

Fig. 6. Results of the ANN's prediction of 25Q score compared to actual score on 13 subjects after training.

Finally, the 2-Stride score prediction was significantly more accurate than the other two scores, converging to an average error of 0.063 over all subjects. While this may seem unusual since the 2-Stride test is a walking task and measure of gait stability, it can be noted that many of the features selected as inputs into the NN, such as angular displacement of the knees and hip and CoM displacement, are also useful as indicators of gait posture and can therefore relate well to this measure. These initial results were verified by testing the ANN on a 14th subject whose data had not been used to train it. The subject completed the same procedure as the training subjects, and the results of the ANN's output for that subject are given in Fig. 9. Error rates of 0.268, 1.131 and 0.010 were obtained for the system's prediction of the subject's Standing test score, GLFS-25 score and Two-Step test score, respectively.

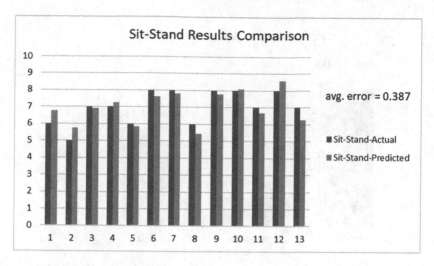

Fig. 7. Results of the ANN's prediction of Sit-Stand score compared to actual score on 13 subjects after training.

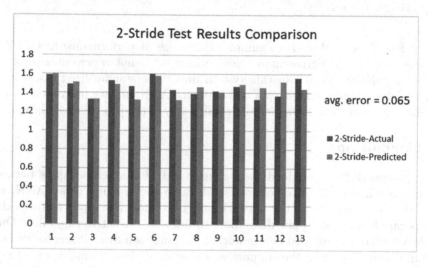

Fig. 8. Results of the ANN's prediction of Two-Stride score compared to actual score on 13 subjects after training.

4.3   Limitations and Future Work

Several limitations were present in this initial study. The first is in the variability of scoring among subjects. Since all subjects were young, healthy subjects, they all scored fairly highly on all three assessments, with very little variablity among their assessment scores. This resulted in a poor training dataset, so the scores needed to be normalized and rescaled based on the best and worst performance in each test among the group. This scaling is reflected in the vertical

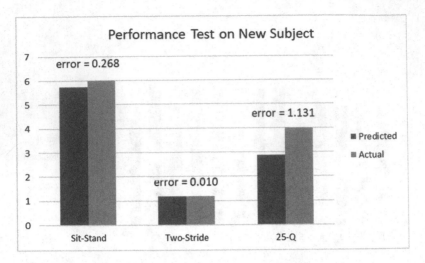

Fig. 9. Results of the ANN's prediction of Sit-Stand, Two-Stride, and 25-Q score compared to actual score (after training) on a single new subject with which the ANN was not previously trained.

axis of Figs. 6 through 8. For example, the range of performance in GLFS-25 was rescaled to 0–9. Furthermore, this sample set is not representative of the elderly population, for whom this system and these standardized assessments were designed. Hence, future work will train the ANN on data from users in this population to determine whether the results may extend to this population.

5   Conclusions

Having successfully determined the stronger and weaker relationships between squat performance and locomotive risk through the training of an ANN, the next step is to reconfigure the network and introduce new input features that incorporate balance and pain/discomfort to improve predictive capability. Once this is achieved, as proposed in this work, a dynamic adjustment mechanism can be implemented wherein the supportive strength of the pneumatic muscles on the wearable suit are adjusted on different levels based on the system's assessment of the user's risk. To evaluate this system, biometric data such as heart rate and galvanic skin response will be recorded to form an estimate of the affective response (negative-positive) of these subjects during exercise and compared with self-reported responses to post-exercise surveys to determine whether the implementation of adaptive support increases the quality of a subject's exercise experience during the squat task. The targeted population for this study will be elderly subjects, particularly those with locomotive difficulty, with the ultimate goal of improving physical and emotional response to at-home squat exercise.

Acknowledgments. The authors would like to thank the New Energy and Industrial Technology Development Organization (NEDO, Japan) for their support.

References

1. Catellier, J.R.A., Yang, Z.J.: The role of affect in the decision to exercise: does being happy lead to a more active lifestyle? Psychol. Sport Exerc. **14**(2), 275–282 (2013). http://www.sciencedirect.com/science/article/pii/S1469029212001343
2. Cary, F., Postolache, O., Girão, P.S.: Kinect based system and artificial neural networks classifiers for physiotherapy assessment. In: 2014 IEEE International Symposium on Medical Measurements and Applications (MeMeA), pp. 1–6, June 2014
3. Chen, J.: Flow in games (and everything else). Commun. ACM **50**(4), 31–34 (2007). https://doi.org/10.1145/1232743.1232769
4. Cho, D., et al.: Detection of stress levels from biosignals measured in virtual reality environments using a kernel-based extreme learning machine. Sensors (Basel, Switzerland) **17**(10), 2435 (2017)
5. Ciabattoni, L., Ferracuti, F., Longhi, S., Pepa, L., Romeo, L., Verdini, F.: Real-time mental stress detection based on smartwatch. In: 2017 IEEE International Conference on Consumer Electronics (ICCE), pp. 110–111, January 2017
6. Csikszentmihalyi, M.: Creativity: flow and the psychology of discovery and invention. In: Creativity: Flow and the Psychology of Discovery and Invention. Harper-Collins Publishers, New York, NY, US (1997)
7. DiBrezzo, R., Shadden, B.B., Raybon, B.H., Powers, M.: Exercise intervention designed to improve strength and dynamic balance among community-dwelling older adults. J. Aging Phys. Act. **13**(2), 198–209 (2005)
8. Domire, Z.J., Challis, J.H.: The influence of squat depth on maximal vertical jump performance. J. Sports Sci. **25**(2), 193–200 (2007)
9. Drachen, A., Nacke, L.E., Yannakakis, G., Pedersen, A.L.: Correlation between heart rate, electrodermal activity and player experience in first-person shooter games. In: Proceedings of the 5th ACM SIGGRAPH Symposium on Video Games, Sandbox 2010, pp. 49–54. ACM, New York, NY, USA (2010). http://doi.acm.org/10.1145/1836135.1836143
10. Escamilla, R.F.: Knee biomechanics of the dynamic squat exercise. Med. Sci. Sports Exerc. **33**(1), 127–141 (2001)
11. Gerling, K.M., Miller, M., Mandryk, R.L., Birk, M.V., Smeddinck, J.D.: Effects of balancing for physical abilities on player performance, experience and self-esteem in exergames. In: Proceedings of the 32nd Annual ACM Conference on Human Factors in Computing Systems, CHI 2014, pp. 2201–2210. ACM, New York, NY, USA (2014). http://doi.acm.org/10.1145/2556288.2556963
12. Hahn, M.E., Chou, L.S.: Age-related reduction in sagittal plane center of mass motion during obstacle crossing. J. Biomech. **37**(6), 837–844 (2004)
13. Hardy, C.J., Rejeski, W.J.: Not what, but how one feels: the measurement of affect during exercise. J. Sport Exerc. Psychol. **11**(3), 304–317 (1989). https://doi.org/10.1123/jsep.11.3.304
14. Hernández, A., Silder, A., Heiderscheit, B.C., Thelen, D.G.: Effect of age on center of mass motion during human walking. Gait Posture **30**(2), 217–222 (2009). http://www.sciencedirect.com/science/article/pii/S0966636209001386
15. Ishibashi, H.: Locomotive syndrome in Japan. Osteopor. Sarcopenia **4**(3), 86–94 (2018). http://www.sciencedirect.com/science/article/pii/S2405525518300608
16. Kaplan, O., Yamamoto, G., Taketomi, T., Plopski, A., Sandor, C., Kato, H.: Exergame experience of young and old individuals under different difficulty adjustment methods. Computers **7**(4), 59 (2018). https://www.mdpi.com/2073-431X/7/4/59

17. Lewthwaite, R., Wulf, G.: Optimizing motivation and attention for motor performance and learning. Curr. Opin. Psychol. **16**, 38–42 (2017). http://www.sciencedirect.com/science/article/pii/S2352250X1630152X

18. Li, S., Pathirana, P.N.: Cloud-based non-invasive tele-rehabilitation exercise monitoring. In: 2014 IEEE Conference on Biomedical Engineering and Sciences (IECBES), pp. 385–390, December 2014

19. Manning, C.D., Raghavan, P., Schutze, H.: Vector space classification. In: Introduction to Information Retrieval. Cambridge University Press, Cambridge (2008). http://ebooks.cambridge.org/ref/id/CBO9780511809071

20. Nakamura, K.: The concept and treatment of locomotive syndrome: its acceptance and spread in Japan. J. Orthop. Sci. **16**(5), 489 (2011). https://doi.org/10.1007/s00776-011-0108-5

21. Nakamura, K., Ogata, T.: Locomotive syndrome: definition and management. Clin. Rev. Bone Miner. Metab. **14**(2), 56–67 (2016). https://www.ncbi.nlm.nih.gov/pmc/articles/PMC4906066/

22. Novak, D., Riener, R.: Control strategies and artificial intelligence in rehabilitation robotics. Ai Mag. **36**(4), 23 (2015)

23. Oniga, S., Suto, J.: Activity recognition in adaptive assistive systems using artificial neural networks. Elektronika ir Elektrotechnika **22**(1), 68–72 (2016). http://eejournal.ktu.lt/index.php/elt/article/view/14112

24. Perez Martínez, H., Garbarino, M., Yannakakis, G.N.: Generic physiological features as predictors of player experience. In: D'Mello, S., Graesser, A., Schuller, B., Martin, J.-C. (eds.) ACII 2011. LNCS, vol. 6974, pp. 267–276. Springer, Heidelberg (2011). https://doi.org/10.1007/978-3-642-24600-5_30

25. Rousseau, A.C., Begon, M., Bessette, R.C.: Squatphy: Assessing squats with low-cost technologies. In: 2017 International Conference on Virtual Rehabilitation (ICVR), pp. 1–2, June 2017

26. Silva, A.J., et al.: The use of neural network technology to model swimming performance. J. Sports Sci. Med. **6**(1), 117–125 (2007)

27. Tadayon, R., et al.: Interactive motor learning with the autonomous training assistant: a case study. In: Kurosu, M. (ed.) HCI 2015. LNCS, vol. 9170, pp. 495–506. Springer, Cham (2015). https://doi.org/10.1007/978-3-319-20916-6_46

28. Thakur, C., Ogawa, K., Tsuj, T., Kurita, Y.: Unplugged powered suit with pneumatic gel muscles. In: Hasegawa, S., Konyo, M., Kyung, K.-U., Nojima, T., Kajimoto, H. (eds.) AsiaHaptics 2016. LNEE, vol. 432, pp. 247–251. Springer, Singapore (2018). https://doi.org/10.1007/978-981-10-4157-0_42

29. Williams, D.M., Dunsiger, S., Ciccolo, J.T., Lewis, B.A., Albrecht, A.E., Marcus, B.H.: Acute Affective Response to a Moderate-intensity Exercise Stimulus Predicts Physical Activity Participation 6 and 12 Months Later. Psychol. Sport Exerc. **9**(3), 231–245 (2008)

30. Yamada, S., Aoyagi, Y., Yamamoto, K., Ishikawa, M.: Quantitative Evaluation of Gait Disturbance on an Instrumented Timed Up-and-go Test. Aging Dis. **10**(1), 23 (2019). http://www.aginganddisease.org/EN/10.14336/AD.2018.0426

31. Yoshimura, N., et al.: Association between new indices in the locomotive syndrome risk test and decline in mobility: third survey of the ROAD study. J. Orthop. Sci. **20**(5), 896–905 (2015). https://doi.org/10.1007/s00776-015-0741-5

EEG Systems for Educational Neuroscience

Angeliki Tsiara[✉], Tassos Anastasios Mikropoulos,
and Panagiota Chalki

University of Ioannina, Ioannina, Greece
atsiara@cc.uoi.gr, amikrop@uoi.gr, pahalki@gmail.com

Abstract. Numerous studies suggest that digital technology has an important role to play in physical and mental functioning and generally in the quality of life of elderly people. Many digital serious games have been developed to enhance cognitive functions. These games incorporate a multitude of multi-media elements that are perceived as sensory stimuli. To implement an effective digital environment, all sensory representations have to be investigated in order to be compatible with the visual, acoustic and tactile perception of the user. An effective way to examine those stimuli is to study the users' brain functioning and especially the electric activity by using electroencephalographic recording systems. In recent years, there has been a growth of low-cost EEG systems. These are used in various fields such as educational research, serious games, mental and physical health, entertainment, etc. This study investigates whether a wireless low-cost EEG system (EPOC EMOTIV) can deliver qualitative results compared to a research system (G.tec) while recording the EEG data at an event-related brain potentials setup regarding the differentiation of the semantic content of two image categories. Our results show that, in terms of signal quality, the Emotiv system lags G.tec, however, based on the answers of the participants in the questionnaire, Emotiv excels in terms of ease of use. It can be used in continuous EEG recordings in game environments and could be useful for applications such as games in ageing.

Keywords: Digital environment · Electroencephalographic recording systems · Event-related brain potential · Ageing

1 Introduction

1.1 ICT and Learning

The advent of ICT is a key factor that allows significant improvements in the area of research and education. Simulations, virtual environments, serious games and other innovative technological means can facilitate and enhance people's skills in everyday activities. The ICT-supported learning process mainly involves visual perception, decision-making, rational thinking and executive discrimination tasks that require skills which are not only exclusively connected to formal education, but also concern everyday life skills.

In many countries, an effort is being made to provide ICT courses for older people. As Dickinson & Gregorargue [1], just providing ICT skills does not automatically

© Springer Nature Switzerland AG 2019
M. Antona and C. Stephanidis (Eds.): HCII 2019, LNCS 11573, pp. 575–586, 2019.
https://doi.org/10.1007/978-3-030-23563-5_45

increase the quality of life and skills of the elderly. Learning plays a key role in ageing societies as it can help older people to address many challenges such as re-skilling and up-skilling in the knowledge-based society. As the majority of ICT tools are not user-friendly for older people, it is difficult for them to use this technology either for learning purposes, or as a part of everyday activities. The basic problem is often the user interface, which is rarely designed for older people. A tool that makes the user feel frustrated cannot be a motivating environment for learning. Considering the needs of older learners and the cognitive abilities that are connected with skills, such as working memory, reasoning, and speed of information processing is essential in developing ICT-supported learning for the elderly.

In recent years, many digital tools have been developed in order to educate about disaster prevention and protection [2]. As it is commonly accepted, education, targeting disaster preparedness, should be extended throughout the society in order to foster a more resilient population. Shaw & Kobayashi present a serious game that provides an engaging virtual environment for training on disaster communication and decision-making processes [3]. Based on the concept of problem-based learning, players develop communication and group decision-making skills. Also, Mitsuharaet al. propose a web-based system for designing game-based evacuation [4] The players can learn about disaster prevention measures by viewing the materials and real-world scenery and making appropriate decisions during a virtual evacuation. In [5] authors compared students' views on three different web-based learning tools, an educational game, a dynamic simulation and a digital concept map. These three learning tools were used for educational purposes aiming at natural disaster readiness. The results showed that students remained highly-engaged while using all three learning objects. These examples suggest that educational material that is embedded in digital environments for disaster risk reduction should be designed and implemented in order to capitalize on the interactivity and the high sense of presence to promote the development of skills, such as critical thinking and problem-solving that users would carry in a real world situation. However, the basic principle for implementing an effective digital learning environment is to design sensory representations to be compatible with visual, acoustic and tactile perception of the user [6].

1.2 Educational Neuroscience

Neuroscience aims in the interpretation of human behavior through the study of the brain functions and more specifically, in the understanding of how neurons collaborate in order to produce a given behavior and how the environment affects them [7]. In this field, learning is defined as a process of creating neural connections in response to external stimuli and education as the process of creating and controlling these stimuli [8, 9]. Educational neuroscience is based on the general idea that anything that influences learning has its foundation in the human brain, therefore, the understanding of the human brain functioning could affect educational practices, create improved teaching methods and provide a conceptual framework for an in-depth understanding of how the human brain creates cognitive schemata based on the input of sensory stimuli.

Electroencephalography (EEG) is a popular method in recording the electrical brain activity, as it is non-invasive and provides an excellent temporal resolution [10, 11]. Event-Related Potentials (ERP) are the electrical responses of the cortex to a sensory, cognitive or emotional event . ERP components, as parts of the EEG signal, enable the recording and analysis of neural responses with high temporal resolution to specific visual events, which appear in a digital environment. The P300 and N200 signals are the two most prominent ERP components for decision-making tasks, as the one under study. The P300 wave is an event-related potential component with a parieto-central scalp distribution elicited in the processes of stimulus evaluation and decision-making [12]. Authors in [13] argue that the P300 latency is an indicator of the duration of the stimulus categorization. The N200 wave is considered to reflect processes involved mainly in the detection of novelty or mismatch [14].

The research objective of the present study was to compare two EEG systems, namely the G.tec EEG system with passive electrodes and the EMOTIV EPOCmobile EEG system. A visual decision-making task in constituting an earthquake survival kit from images of useful items and non-useful items was used for the experimental procedure. For the comparison of the two EEG systems, we used an online questionnaire and the measurements of the EEG systems. As the timeline of constituent processing steps in visual decision-making is below the earliest response time for a motor response (around 700 ms), the study hinges on measuring the electrical brain activity related to visual processing in terms of visual awareness and semantic recognition during the identification of ten different images depicting useful items (UI) and non-useful items (NUI) that they have to bring with them in their earthquake survival kit, in case of an earthquake.

2 Method

2.1 Research Questions

The research goal of this study was to record the electrical brain activity of male and female volunteers in order to compare the neurophysiological measurements received by using Emotiv's and Gtec's EEG systems. To reach this goal, the following research questions were set separately for both the EEG measurements:

- Which brain regions are associated with semantic recognition of the visual stimuli under study?
- How much time is needed for the semantic recognition of the visual stimuli under study?

2.2 Participants

To compare the two EEG systems, we recorded the brain activity of seven male (28 to 31 years old) and twenty female (20 to 21 years old) volunteers. All participants had normal vision, were right–handed native Greek speakers, without certain diagnosed learning difficulties or mental disease. None of the participants received any medication or substances that affected the operation of the nervous system and they had not consumed quantities of caffeine or alcohol in the last 24 h before the experiment. The alpha rhythm of all the participants was checked and found to be normal (8–12 Hz, 10 Hz peak).

2.3 Stimuli

Ten different images depicting useful items (UI) and non-useful items (NUI) in case of an earthquake, were presented to each subject (Fig. 1). The five useful items were a cereal bar, a flashlight, a pocketknife, a bottle of water and a whistle, whereas the non-useful items were an ice cream, a hamburger, a laptop, a bottle of milk and a tool kit. Each participant was comfortably seated at eye level and 100 cm away from α 17" TFT monitor, passively observing the displayed images. In the beginning of the experiment, each participant had a few minutes to adapt to the specific conditions, to relax and reduce the movements of their eyes. Before the EEG recording the researchers gave a briefing to each participant about earthquakes and the relevant precaution measures. The participant was familiarized with the useful and non-useful objects.

Fig. 1. Useful items (up) and non-useful items (bottom)

2.4 Procedure

The EEG data were recorded during the observation of images, incorporated in an educational digital environment and depicting useful and non-useful objects with which they should provide a survival bag in the event of an earthquake. The experimental procedure completed in one session that included the presentation of 300 images of which 15% depicted a useful item and 85% a non-useful object (oddball experimental paradigm).

The visual stimuli, five different images of each of the two categories, appeared randomly at the center of the screen for 2000 ms. Between the sequential appearance of two stimuli of any category, a blue cross with a display duration of 1000 ms appeared at the center of the screen (Fig. 2). The participants were instructed to observe the images displayed on the screen and respond mentally only to the stimuli that depicted a useful object. For each participant the experimental procedure lasted 30–40 min.

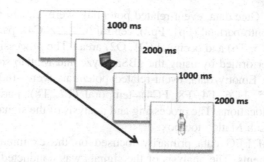

Fig. 2. Schematic view of the experimental procedure

The experimental procedure was performed twice for each participant, one for each EEG system. In order not to affect the results from the participants' previous experience, the participants were divided into two groups where one was initially recorded with g.tec and then with Emotiv and vice versa. For every participant, there was at least a period of two weeks between the two recordings.

2.5 EEG Systems

G.tec. EEG was recorded by using a g.tec 36-channel amplifier with 256 Hz sampling rate. The digital EEG data acquisition system had a 1–100 Hz band pass filter. EEG activity was monitored from 19 Ag/AgCl electrodes using an electrode cap with a standard 10–20 International Electrode Placement System layout. Raw EEG data was recorded from Fp1, Fp2, F7, F3, Fz, F4, F8, T3, C3, Cz, C4, T4, T5, P3, Pz, P4, T6, O1 and O2 electrode positions. All leads were referenced to linked ear lobe and a ground electrode was applied to the forehead. Horizontal and vertical eye movements were recorded simultaneously using four electrodes round the eyes. The electrodes impedance was kept below 5 KΩ.

Emotiv. EEG was recorded using the 16 electrodes of the Emotiv system, according to the standard 10–20 International Electrode Placement System layout: AF3, F7, F3, FC5, T7, P7, O1, O2, P8, T8, FC6, F4, F8, FC4, M1and M2, with the maximum sampling rate of 128 Hz. M1 sensor was used as ground and M2 contributed to lowering the external electrical interference. The electrodes impedance was kept below 5 KΩ.

2.6 Offline Analysis

For the process of the offline analysis of the signal, after removing eye movement and other artefacts by inspection, individual subject EEG data were filtered with a 30 Hz low-pass digital filter and divided into epochs. Each epoch began 100 ms prior to stimulus onset and continued for 600 ms thereafter. Single trials were then averaged per UI and NUI and subject using a 100 ms pre-stimulus baseline. Finally, the grand mean for each group of objects across all subjects was calculated.

Concerning the Gtec data, event-related potentials were studied within frontal (F7, F3, Fz, F4, F8), fronto-parietal (Fp1, Fp2), central (C3, Cz, C4), parietal (P3, Pz, P4), temporal (T3, T4, T5, T6) and occipital (O1, O2) areas. The processing and analysis of the signals were performed by using the gBSanalyze and Matlab software packages.

Concerning the Emotiv data, event-related potentials were studied within frontal (AF3, F7, F3, FC5, FC6, F4, F8, FC4), temporal (T7, T8), parietal (P7, P8) and occipital (O1, O2) locations. The processing and analysis of the signals were performed by using the EEGLab Matlab toolbox.

The analysis of EEG data primarily focused on the examination of P300 and N200 ERP components. The analysis of the signals was conducted separately for the male and the female participants to examine possible gender differences in brain activity. After the exclusion of the artifacts that were included in the data due to eye movements or other factors, the raw EEG data were filtered applying a 30 Hz low pass filter and divided into epochs. Each epoch started 100 ms before the stimulus onset and extended for 600 ms after the stimulus onset. The single average and the grand average from the individual ERP waveforms for the two stimuli categories were calculated for all participants, using a 100 ms baseline before the stimulus appearance. The grand average for each item category was calculated taking into account the signals from all participants, but separately for men and women. Based on data collected from the visual inspection of ERP waveforms of grand averages and previous studies [13, 15, 16], we defined the P300 as the greatest positive point of the ERP waveform of total averages in the interval between 250 ms–600 ms (Fig. 3).

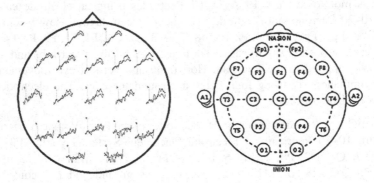

Fig. 3. Layout of electrode array, based on gtec's measurements from male participants, utilized for grand-average ERPs inspection for each stimulus category (green signals: UI (target), red signals: NUI (non-target)) (Color figure online)

2.7 Questionnaire

After the recordings, the participants were asked to fill in a questionnaire regarding the evaluation of both systems in terms of usability, comfort and speed of placement. Apart from some personal data i.e. gender, age etc. the participants had to answer the following questions in a five-level scale:

1. How do you evaluate the installation speed of the g.tec? (1 = Very fast/Fast/ Moderate/Slow/Very slow)
2. How do you evaluate the installation speed of the emotiv? (Very fast/Fast/ Moderate/Slow/Very slow)
3. How do you evaluate the comfort of the g.tec installation? (Very comfortable/ Comfortable/Moderate/Inconvenient/Annoying)
4. How do you evaluate the comfort of the emotiv installation? (Very comfortable/ Comfortable/Moderate/Inconvenient/Annoying)
5. How do you evaluate the difficulty of installing the g.tec system to the head? (Very Easy/Easy/Moderate/Difficult/Very Difficult)
6. How do you evaluate the difficulty of installing the emotiv system to the head? (Very Easy/Easy/Moderate/Difficult/Very Difficult).

3   Results

3.1   G.tec Data

EEG data from male and female participants gave almost the same results for P300 and N200 ERP components.

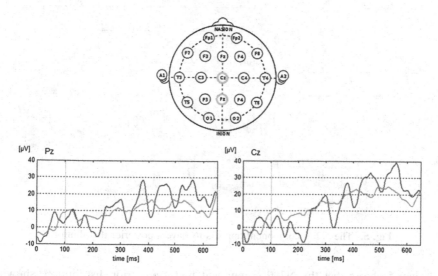

Fig. 4. Grand-average ERP waveforms at electrode locations Pz and Cz for each stimulus category. Time 0 corresponds to the presentation of the stimulus (green signals: UI (useful items), red signals: NUI (non-useful items) (Color figure online)

The stimuli depicting useful items (UI) appear to elicit a negative N200 component between 200 ms and 250 ms after the stimulus, and a positive P300 component between 500 ms and 600 ms. The stimuli depicting non-useful items (NUI) did not elicited any of the above-mentioned components. The P300 component appeared to have a greater amplitude in the parietal-central region of the skull and in particular at the Cz and Pz electrode locations. The latency of P300 component for both Cz and Pz appeared prolonged with the average latency for useful objects being approximately 550 ms (Fig. 4). The amplitude of the P300 component in the Cz and Pz locations was approximately 40 μV and 30 μV respectively.

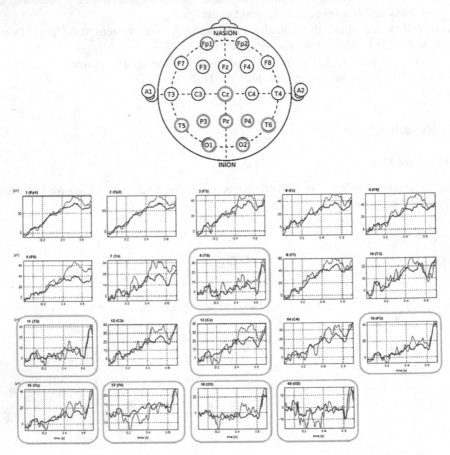

Fig. 5. The N200 component appears to have a posterior distribution

Figure 5 shows that the N200 component had a posterior distribution, showing greater amplitude in the parietal, temporal and occipital regions of the skull (i.e., in Pz, P3, P4, T5, T6, O1, O2 and Cz).

3.2   EMOTIV Data

Based on the literature, the P300 component is mainly located at Cz and Pz electrode locations. However, according to [18] it also can be found in other places, i.e. P7 and

O1. In our results, the P300 component was not detected in any electrode location, neither for UI nor for NUI signals. Moreover, we did not manage to have a good signal quality in most of the locations for almost all participants, especially for electrode locations F4 and P7, which were excluded.

3.3 Questionnaire Data

The absolute frequencies of the answers were calculated for each question separately. Regarding the evaluation of the installation speed of the two EEG systems, the participants' answers are presented in Fig. 6.

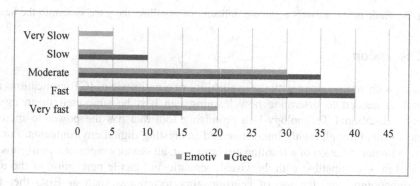

Fig. 6. Participants' answers regarding the installation speed of g.tec and emotiv system

Regarding the installation comfort of the two EEG systems, the participants' answers are presented in Fig. 7.

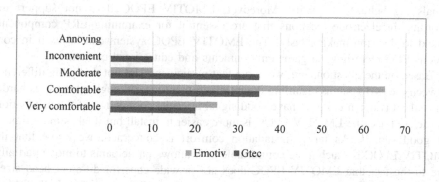

Fig. 7. Participants' answers regarding the comfort of the g.tec and emotiv installation

Regarding the difficulty of adjustment of the g.tec and EMOTIV system to the participants' head, the participants' answers are presented in Fig. 8.

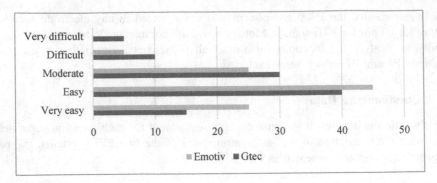

Fig. 8. Participants' answers regarding difficulty of installing the g.tec system to the head

4   Discussion

Learning is changing, along with the availability of a multitude of ICT applications and research is needed to determine how learning can best be supported in an ageing society. Educational Technology is a promising tool that has the power to increase people's skills through highly interactive and interesting digital environments. For the effective implementation of a learning environment, all sensory representations must be designed to be compatible with the visual, acoustic and tactile perception of the user. Brain functioning and the use of neuroimaging techniques such as EEG therefore seems to give tools beyond those of social sciences research methodology.

The present study examined two EEG systems in terms of signal quality and usability. Based on the EEG data, we argue that the G.tec system can be used to study cognitive processes in educational environments either in a continuous EEG recording or in ERP experimental set up. For EMOTIV EPOC system we argue that the recorded signals lag behind in quality. Moreover, EMOTIV EPOC does not support the recording in electrode locations that are essential for examining ERP components related to decision-making tasks. The EMOTIV EPOC system can be used in continuous EEG recordings in game environments and educational content.

Based on the questionnaire, we argue that participants did not find a big difference between the two systems regarding the installation speed. Although, gtec cap is harder to install, it is much easier to have good signal after the completion of the installation. On the contrary, the EMOTIV EPOC is more easier to install but it takes more time to get good signals. As far as installation comfort is concerned, we argue that the EMOTIV EPOC is much more comfortable as it allows participants to move partially. Due to the fact that EMOTIV EPOC does not provide caps for different head sizes, makes it sometimes hard to adjust all semi-rigid "fingers" to touch the skull. Table 1 presents comparative data for the two EEG systems.

Table 1. Comparison of Gtec and Emotiv EEG systems

	G.tec	Emotiv Epoc
Sampling frequency	Up to 512 Hz	128 Hz
Purchase cost	~20.000 €	~800 €
User comfort	No discomfort over time, but the participant must be immobilized during the recording process as the system is wired	Comfort is satisfactory for up to 40 min. The semi-rigid "fingers" push the skull causing discomfort after that time. The user has the possibility of partial movement as EPOC is wireless (Bluetooth 4.0)
Adaptability to different head sizes	Caps in different head sizes are provided	Difficult to adjust to large head sizes
Variation of electrode positions on the skull	19 electrode positions according to International System 10–20	14 electrode positions according to International System 10–20
Stability of electrical connection between the skull and electrodes	Good stability in electrical connection	Stability agitates over time as it depends on the battery duration of the device (about 2 h)
Timing capability between EEG system, computer and external events	Full timing capability	Satisfactory timing capability
Usability of systems for researchers	Handy system although it is necessary to individually wet all the electrodes	Handy system, there is a difficulty in placing the device and receive quality signals
Signal quality	Very good	Satisfactory, is not recommended for medical EEG signal research
Software for signal analysis	G.tec software for signal analysis	No software provided. Researcher should link it to third-party software such as Simulink and Matlab

Acknowledgments. The research is implemented through the Operational Program "Human Resources Development, Education and Lifelong Learning" and is co-financed by the European Union (European Social Fund) and Greek national funds.

References

1. Dickinson, A., Gregor, P.: Computer use has no demonstrated impact on the well-being of older adults. Int. J. Hum Comput Stud. **64**, 744–753 (2006)
2. Shaw, R., Kobayashi, M.: Role of schools in creating earthquake-safer environment. http://www.preventionweb.net/files/5342_SesiRoleSchoolsEQSafety.pdf. Accessed 09 June 2017
3. Haferkamp, N., Kraemer, N.C., Linehan, C., Schembri, M.: Training disaster communication by means of serious games in virtual environments. Entertain. Comput. **2**(2), 81–88 (2011)

4. Mitsuhara, H., et al.: Web-based System for Designing Game-based Evacuation Drills. Proc. Comput. Sci. **72**, 277–284 (2015)

5. Natsis, A., Hormova, H., Mikropoulos, T.A.: Students' views on different learning objects. In: Chova, L.G., Martínez, A.L., Torres, I.C. (eds.), INTED 2014 Proceedings, 8th International Technology, Education and Development Conference, pp. 2363–2372. IATED Academy Valencia (2014)

6. Alkire, S.A: Conceptual Framework for Human Security. http://www3.qeh.ox.ac.uk/pdf/crisewps/workingpaper2.pdf Accessed 16 June 2018

7. Kandel, E.R., Schwartz, J.H., Jessell, T.M.: Essentials of Neural Science and Behavior, 2nd edn. McGraw-Hill Education-Europe, Maidenhead (2011)

8. Ferrari, V., Bradley, M.M., Codispoti, M., Lang, P.J.: Detecting novelty and significance. J. Cogn. Neurosci. **22**(2), 404–411 (2010)

9. Koizoumi, H.: Brain-Science Based Cohort Studies. Educ. Philos. Theor. **43**(1), 48–55 (2011)

10. Fisch, B.J.: Fisch and Spehlmann's EEG Primer: Basic Principles of Digital and Analog EEG, 3rd edn. Elsevier, Amsterdam (1999)

11. Savoy, R.L.: History and future directions of human brain mapping and functional neuroimaging. Acta Psychol. **107**, 9–42 (2001)

12. Duncan-Johnson, C.C., Donchin, E.: On quantifying surprise: the variation of event-related potentials with subjective probability. Psychophysiology **14**, 456–467 (1977)

13. Gray, H.M., Ambady, N., Lowenthal, W.T., Deldin, P.: P300 as an index of attention to self-relevant stimuli. J. Exp. Soc. Psychol. **40**, 216–224 (2004)

14. Folstein, J.R., Van Petten, C.: Influence of cognitive control and mismatch on the N2 component of the ERP: A review. Psychophysiology **45**(1), 152–170 (2008)

15. Azizian, A., Watson, T.D., Parvaz, M.A., Squires, N.K.: Time Course of Processes Underlying Picture and Word Evaluation: An Event-Related Potential Approach. Brain Topogr. **18**(3), 213–222 (2006)

16. Olofsson, J.K., Nordin, S., Sequeira, H., Polich, J.: Affective picture processing: An integrative review of ERP findings. Biol. Psychol. **77**(3), 247–265 (2008)

A Soft Exoskeleton Jacket with Pneumatic Gel Muscles for Human Motion Interaction

Antonio Vega Ramirez[✉] and Yuichi Kurita

Graduate School of Engineering, Hiroshima University, Hiroshima, Japan
{tonovr, ykurita}@hiroshima-u.ac.jp

Abstract. This work proposes to use an assistive device to augment the perception of human motion force applied from one subject to another through an avatar. This experiment presents an avatar augmentation that can provide the feeling of motion with two degrees of freedom at the elbow and shoulder acquired with algorithmic detection of input angles. To generate the operation of the appropriate motion, the interface is comprised of a depth sensor, an ESP32 embedded board, electric valves, and pneumatic gel muscles. An evaluation is presented which confirms the performance of the suit by measuring the latency of the system. The experimental results demonstrate that the developed suit can convey the motion of one user to another with a delay of 670 ms.

Keywords: Pneumatic gel muscle · Human augmentation · Tele existence · IoT

1 Introduction

Elderly society is generally afforded very little free time [1], and in most cases it may be difficult to go to a hospital when an injury occurs. Furthermore, hospitals generally include a limited number of therapists relative to the demand for rehabilitative therapy, which necessitates the implementation of alternative means of supervision and training.

Due to technological advances in modern science, this challenge is being addressed through externally worn assistive force output devices or "exoskeletons". These devices are capable of providing external support to augment human motion, and in many cases, to improve performance in a variety of tasks. Scientists have explored the application of this technology to other fields of study beyond force-based human augmentation; for example, exoskeletons have also been applied in the areas of rehabilitation and tele-operation. Typically, these exoskeletal assistive devices have consisted of bulky, rigid and heavy electric machinery; recently, however, "soft" lightweight alternatives to these interfaces have been developed. This new generation of exoskeletons, called "soft exoskeletons", utilize lighter, more flexible materials to generate externally-delivered force along the body. Unlike their counterparts, soft exoskeletons lack an external rigid frame, which serves as both an advantage and a disadvantage [2]. The primary drawback of this design is that these wearable devices have difficulty in transferring power from any area of the body to the ground. In

M. Antona and C. Stephanidis (Eds.): HCII 2019, LNCS 11573, pp. 587–603, 2019.
https://doi.org/10.1007/978-3-030-23563-5_46

addition, motors and sensors are more difficult to mount. Furthermore, torque and force generated by actuators enter the user's body. Imagine for example an elderly individual who uses an exoskeleton to go the store and back, in which case the exoskeleton must assist with balance, usage of stairs, entering and exiting a vehicle (currently no product on the market has this capability although technology is rapidly overcoming this limitation). A soft exoskeleton can provide the additional force to supplement a user's muscles [3], but most if not all of that energy will take a toll on that individual's body. The elderly and disabled populations already have weakened bones and muscles, so adding more strain to their bodies may be counterproductive.

With the relative advantages provided by soft exoskeletons, it is possible to provide a device to a patient to assist with the performance of rehabilitative exercises at home. One challenge with this approach is that without the physical presence of the therapist, the force-guidance protocol must be as intuitive as possible to avoid misguidance or injury. To address this, the proposed system is an integration of a soft exoskeleton suit with a natural user interface including as Kinect sensor, and is designed to prove an interaction that is direct and consistent with our 'natural' behavior, without requiring additional sensor attachments on the learner. The advantage of Natural User Interfaces (NUIs) is that the user interaction feels fun, easy and natural because the user can utilize a broader range of basic skills compared to more traditional graphical user interface interaction. In this case a camera-based motion tracking method can be used by the therapist to supervise a patient's posture during exercise, process and analyze this information, and activate the appropriate valves of Pneumatic Gel Muscles (PGMs) attached in specific locations along the exoskeleton to generate corrective gestures and guide this patient.

In addition, Internet of Things (IoT) boards, such as Raspberry Pi or ESP32, can be used as embedded systems to add their many capabilities in the provision of services such as tele-rehabilitation [4, 5]. This suggests that if the output user (the user receiving the assistance) wears a soft exoskeleton, he or she can receive tele-rehabilitative coaching to correct posture during exercises, receive direct and active routine guidance by physiotherapists [6], and participate in direct communication with the hospital from within the user's home. In addition, if a set of output users assigned to a single therapist each wear different jackets, the therapist can potentially provide rehabilitative sessions to several of these users at the same time, assuming that these patients are assigned the same routine. This process is entitled "mimic motion".

Several works have developed soft exoskeletons and telerehabilitation systems with ingenious strategies to improve the performance of these procedures. For example, several works have documented the use of Kinect and the IoT to acquire remote patient data and perform therapy studies without the need for the subject to personally attend the hospital [7, 8], and in assistive suits to support elbow motion with soft materials [9, 10] to provide a light alternative to traditional hard exoskeletons.

One of the objectives of these technological areas is the development of innovative systems and/or applications that offer services that were not available until recently.

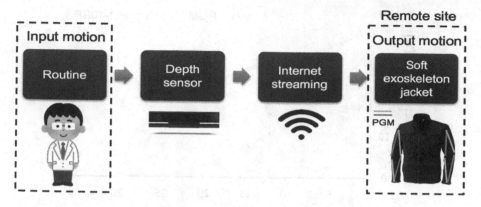

Fig. 1. Framework of the presented system.

For example, the process of introducing new technologies that offer new applications for human augmentation as wearable devices has been explored from the design stage to the prototyping of terminal equipment. Undoubtedly the field of modern human augmentation devices is one of the most dynamic and fastest growing segments in the areas of wearable assistive devices [11]. This paper proposes the use of new technologies for human augmentation as a tele-coaching system generating gestures in a VR environment. It consists of a jacket that equips PGMs developed by Daiya Industry Co. Ltd. [12] in various joints along the upper body, along with their corresponding valves, a microcomputer and an Oculus VR system. An individual wearing the proposed soft exoskeleton jacket can feel the motions of another individual providing guidance from a remote location. Figure 1 depicts the framework of the purposed system.

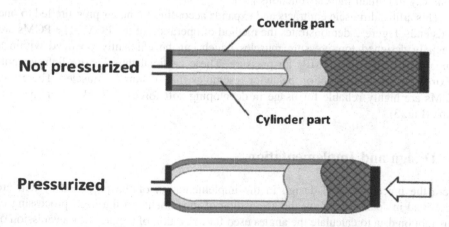

Fig. 2. Pneumatic gel muscle actuation [15].

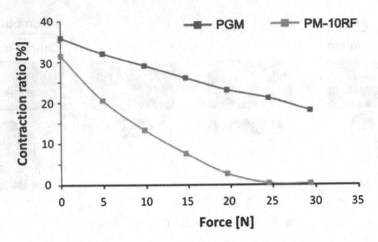

Fig. 3. Comparison of results between PGM and PM-10RF [15].

2 Pneumatic Gel Muscles

In this system, McKibben type artificial muscles (PGMs) [14] were used. The McKibben artificial muscle is the most popular PGM. Its behavior resembles biological muscles. The McKibben artificial muscle is comprised of an inner tube surrounded by braided mesh. The inner tube is made of a stretchable rubber tube and the braided mesh provides protection and controls contraction of the artificial muscle. The pantograph structure of the braided mesh allows the muscle to extend and contract easily. The contraction of the actuator depends on the air pressure and volume and the non-extensibility of the braided mesh shortens and produces tension if the endpoint is attached to a load. Such actuators are favorable for motion assistance because of their similarity to human muscle functions [15].

This artificial muscle contracts and expands according to the air pressure fed to one of its ends. Figure 2 demonstrates the method of operation of the PGM. The PGMs are specially designed low-pressure muscles which can be efficiently operated within a range of 0.1 MPa to 0.3 MPa air pressure. These artificial muscles are light-weight, flexible, and their contraction behavior resembles that of human muscles. Therefore, PGMs are highly reliable for usage in developing soft force-feedback providing systems (Fig. 3).

3 Design and Implementation

Here the procedures and design in the implementation of the proposed system are described in detail, including the acquisition of movements of the user, processing of this motion data to calculate the angles used for activation of the jacket, transmission of activation data, and the usage of this data by the microcontroller to activate the appropriate PGMs. The challenges and limitations encountered during development and the steps taken to overcome these issues are also explained. Figure 4 presents a block diagram of the design implemented for the system.

Fig. 4. Block diagram of the system.

3.1 Acquisition of the User Data

The sensing environment relies upon a Kinect V2 sensor, a depth sensor used to acquire the information of the input subject (the individual providing the motion guidance). The acquired information is processed in the PC to estimate the position of the input user joints based on the Kinect SDK algorithm. This information is then displayed in a Unity 3D environment to provide a visual experience to both the input user and output user. The output user can then view the motion of the input user in the first-person perspective by wearing an Oculus helmet that displays a VR environment as shown in Fig. 5.

Angle Calculation. The primary objective at this stage is to calculate the angles between each limb and its adjacent body parts. The PGM will then be activated to reproduce the angles of the user's shoulder. The "angle of the shoulder" refers to the angle that the shoulder joint creates between the upper arm and the torso. Figure 6 depicts the angle that to the system must track emulate gestures involving shoulder motion. The desired values can be derived by calculating the input user's shoulder angle against the vertical line of the torso and the elbow angle about the orientation of the upper arm. Therefore, before calculating these angles, the vectors corresponding to each of these axes of orientation must be created. Subtracting two vectors produces a third vector representing the orientation between them. The axes of orientation can be created in this way by subtracting two pairs of skeleton joints. For the shoulder joint, a vertical axis that points up along the orientation of the user's torso is necessary. It is calculated by subtracting the right hip from the right shoulder and left shoulder. Since those joints are approximately vertically aligned on the body, the vector between them captures how the user is leaning away from vertical. Once these orientation axes are formed, the calculation of the angles of the limbs can be completed. The logic of the function that was applied is demonstrated in Fig. 7.

3.2 Receptor

The receptor is deployed within the exoskeletal jacket. Here, one socket connection is created to generate the bridge between the client and the server. A socket connection provides a near real-time communication, also known as synchronous [16]. Synchronous communication is also necessary for live performance applications that need real-time interaction between elements. A network socket connection is a continuous

Fig. 5. Avatar displayed in the VR environment developed to emulate the tele-existence of the input user.

connection between two programs across a network, each consisting of an IP address, the numeric address of a machine on the network, and a port number, a number used to interact between elements. One example of a simple socket connection is shown in Fig. 8. The project tests were carried out with one embedded system. The ESP32 card, with Wi-Fi capabilities, was programmed as a server whose task is to receive the information of the client and activate the outputs that are connected to the valves to actuate the PGM located in the jacket. This card is stable because the routine is

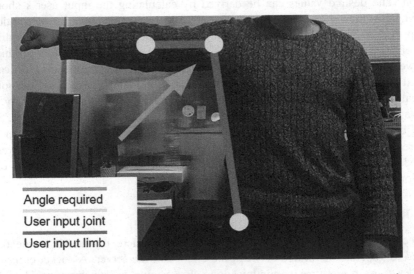

Angle required
User input joint
User input limb

Fig. 6. Joints and limb required to calculate the angle which generates the gestures in the output user.

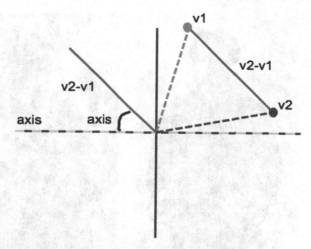

Fig. 7. Demonstration of conversion of the two vectors representing the ends of a limb into a single vector representing that limb in relation to a given orientation.

dedicated to generating the server and receiving the information that is continuously sent, and it is not necessary to halt this routine with any delay when the port is receiving information from the client to control the valves. The primary issue in this case is that in order to change the routine of the card, it is necessary to reprogram from the PC. In addition, because this card is smaller, the power source was also embedded within it to connect it directly with the board.

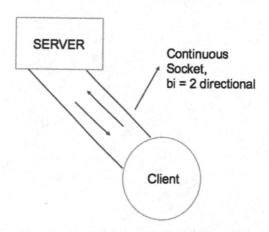

Fig. 8. Socket connection overview.

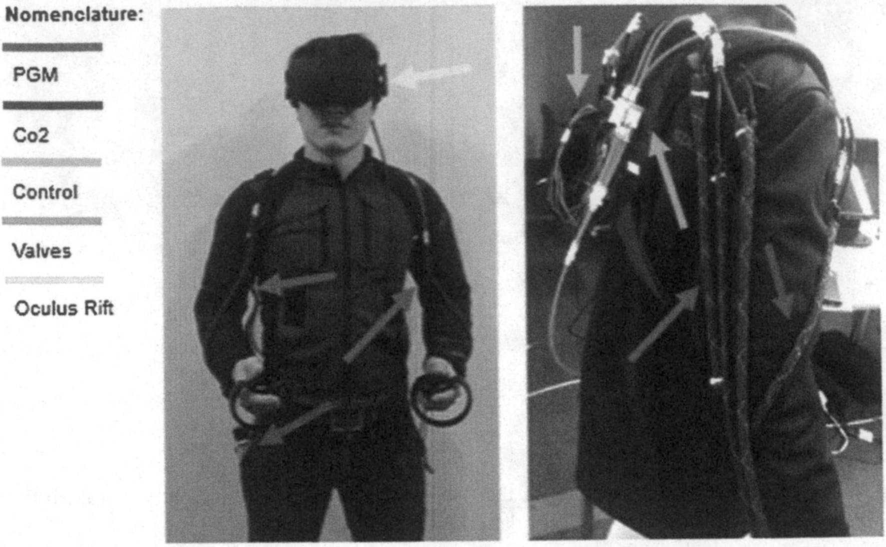

Nomenclature:

PGM

Co2

Control

Valves

Oculus Rift

Fig. 9. Elements that contain the soft exoskeleton jacket.

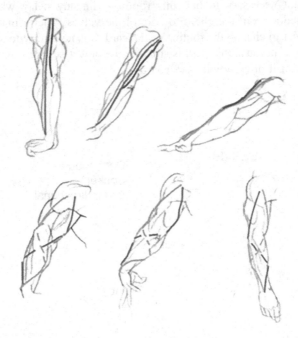

Fig. 10. PGM's configuration viewed from different perspectives.

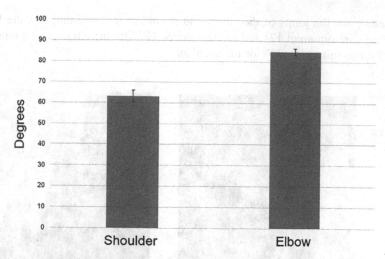

Fig. 11. Maximum angle of joints after PGM actuation.

3.3   Soft Exoskeleton Jacket

The proposed soft exoskeleton consists of a soft lightweight jacket, pneumatic gel muscles located in the arms, the receptor as previously described, a set of connective tubes, a CO_2 tank, and a power bank to energize the system. Figure 9 provides an overview of the elements within the developed suit. In robotics terminology, it can be stated that this jacket has two degrees of freedom (i.e., it can only move along two axes). The jacket holds 3 PGMs of 42 cm and one of 27 cm dedicated to the shoulder's extension motion. In this approach, the PGMs are placed in a straight alignment. For the elbow's flexion motion, unlike the previous configuration, 2 PGM's of 42 cm were twisted at the height of the user's elbow to provide more torque and a better feeling to the user. Figure 10 indicates the configuration followed for the PGMs in the elbow and shoulder locations.

The system, once all of the previous steps are completed, can actuate the PGMs in the elbow and shoulder joints so that the motion of elbow flexion is supported, and the extension of the shoulder is supported. This provides 2 degrees of freedom. Min. angle elbow

One significant consideration in this design strategy is to maximize the mobility of the user while wearing the exoskeleton to avoid issues with executing these motions.

4   Evaluation

A preliminary set of experiments was conducted to confirm the delay that exists in the system and to determine the degree of force felt by the user through the force-feedback system. Four male adults and one female adult participated in this evaluation. None of the participants in the studies had any previous knowledge of the system. Before evaluation, the task was to measure the minimum and maximum angle detection of the

Kinect sensor for the joints of the elbow and shoulder. It was found that the Kinect could detect a minimum of 17° and a maximum of 177° for the elbow, and a minimum of 5° and a maximum of 178° for the shoulder.

PGM

Transducer

Fig. 12. Shoulder extension measurement.

After determining the effective range of the sensor, the PGMs of the suit were activated in both the shoulder and elbow configurations to calculate the maximum angles that could be achieved using the suit. Figure 11 displays the maximum. angle that can be reached in both configurations of the presented system after PGM actuation.

PGM

Transducer

Fig. 13. Elbow flexion measurement.

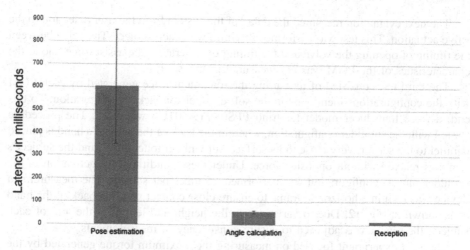

Fig. 14. Latency presented in the system.

The horizontal axis of Fig. 11 shows the conditions and the vertical axis the angles, in degrees. The evaluation found that the presented system can reach 62° for the shoulder extension configuration and 84° approximately for the elbow flexion configuration. This includes the points adjusted by the PGMs. After knowing these maximum angles, various threshold values were formed to emulate gestures of the input user in the output user. Two threshold intervals were derived to emulate pulling up of the elbow of the output user; the first is between 25° and 30° and the second is over 30°. For the second mimic, a threshold of 25° was set to activate the valves to emulate pulling up of the shoulder of the output user by the motion of the input user.

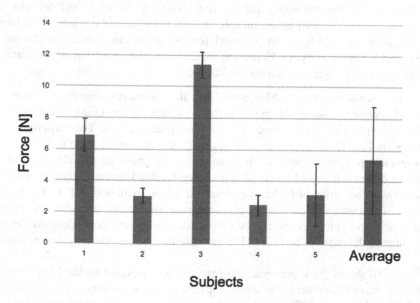

Fig. 15. Force-feedback experiment results for shoulder.

The next evaluation measured the delay of the system from the user detection to the valve actuation. This test was performed 30 times with each subject. The delay between the timing of opening the valve and the timing of generating the assist force due to the characteristics of the PGM was not measured in this study.

The third test consisted of measuring the maximum force generated by each PGM with the configurations mentioned in the soft exoskeleton jacket configuration. To this end, a force transducer model Leptrino PFS080YA501U6 was used. The procedure started with the shoulder configuration, wherein the back of the subject's hand is placed parallel to the sensor, very close to its surface but without touching it, and the subject's arm was relaxed without opposing force. Under these conditions the activation of the shoulder muscle configuration was performed 15 times per subject. The measurement device was held in a horizontal frame to ensure close contact with the back of the hand as is shown in Fig. 12. Due to variations in the height and length of the arm of each subject, the sensor was adjusted to ensure consistency of measurements.

The next experiment focused on measuring the maximum torque generated by the elbow configuration. Each subject was asked to place his or her arm horizontally to the trunk as is shown in Fig. 13. Once this requirement was met, the PGM configuration was actuated, and the palm of the hand was aligned horizontally with the transducer to achieve the measurement.

5 Results

The results from the previous evaluations are as follows:

Latency. Figure 14 shows the latency presented by the system when the calculations for pose estimation, angle derivation and reception of data were executed. The horizontal axis shows the conditions and the vertical axis is timing in milliseconds. It is possible to reach a delay of approximately 602 ms to estimate the joints of the current user, approximately 55.6 ms are required for the angle calculations to control the valves, and approximately 13.43 ms are required for the reception of the data and transmission of the signal to control the valves.

Maximum Assistive Force Measured for the Gestures. Figure 15 shows the graphical representation of the physical force exerted by the PGM actuation by the shoulder configuration in the shoulder extension measurement. This experiment was illustrated in Fig. 12. The vertical axis represents the maximum force in Newtons and the respective perceived forces of the 5 subjects. We observed that different subjects perceived different levels of forces with the force-feedback system. The physical force induced by the actuation of PGMs was observed as approximately 5.4 N. As discussed in previous research, muscle activity plays a significant role in distinguishing different weights or forces (in this context). Perceived force is greatly dependent on the operating range of the muscles involved [17], and also depends on the length of each user's arm.

Figure 16 shows the representation of the torque generated by the PGM actuation by the elbow configuration in the shoulder extension measurement, an experiment that was illustrated in Fig. 13. The vertical axis represents the maximum force in Newtons,

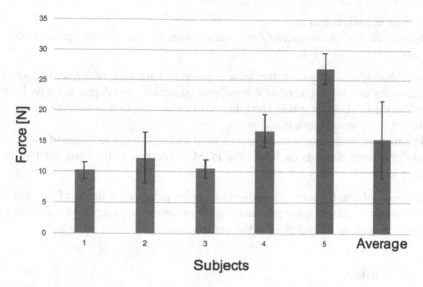

Fig. 16. Force-feedback experiment results for elbow.

and the horizontal axis represents the respective force perceived by the 5 users. As observed in the previous experiment (Fig. 15), in this experiment the subjects perceived different levels of forces with the force-feedback system. However, the physical force induced by the actuation of PGMs for this configuration was differed from the previous by approximately 15.3 N. Muscle torque depends on musculoskeletal geometry [18], and the anatomy of a muscle has a pronounced effect on its force capacity, range of motion, and shortening velocity. Various examples exist demonstrating the characteristics leading to the relative change in force in this experiment compared with the previous one:

Fig. 17. Mimic gesture displayed in a VR environment.

- Change in muscle length
- Rate of change in contractile force varies with the ratio of change in sarcomere length.

In addition, movement, or the muscle-controlled rotation of adjacent body segments, means that the capacity of a muscle to contribute also depends on its location relative to the joint that it spans. The rotatory force exerted by a muscle about this joint is referred to as muscle torque.

The moment arm usually changes as a joint rotates though its range of motion; the amount of change depends on where the PGM is attached to the body relative to the skeleton joint.

Restrictions of the System. The solenoid valves presented a delay of 3/6 mS [19]. Furthermore, the nature of the pneumatic gel muscles presented a hardware delay that cannot be reduce as a restriction of the current actuators.

6 Application

As was mentioned in the introduction of this work, most of the soft exoskeletons that were developed include sensors attached on the user's body and don't use completely soft actuators, such as the pneumatic gel muscles that were used in the developed work. Solutions that were developed for telerehabilitation with the Kinect sensor tend to focus on obtaining data during remote therapy sessions without force feedback interaction. Taking into account these previous considerations, the current application that was developed in this project takes advantage of natural user interface systems, such as the Kinect sensor, to directly project the motion of an input user to an output user using a soft exoskeleton jacket through mimic motion.

The system can be used in rehabilitation or telerehabilitation to provide motion force feedback during remote physiotherapy exercise for elbow and shoulder rehabilitation after upper extremity injuries such as fractures, dislocations and tears. In addition, due to the soft and lightweight nature of pneumatic gel muscles capable of providing resistive force, this implementation may be useful in low-intensity resistive training for elderly individuals with cardiac problems as an example [20]. Figure 17 shows an example of mimic motion for elbow flexion wherein the input user generates a gesture in the VR environment to pull up the hand of the output user. Once the threshold angle is reached, the control signal will reach the receptor in order to actuate the valves so that the output user can feel the gesture.

7 Discussion

In the evaluation, an effective transmission of motion toward an output user wearing an exoskeletal jacket was observed. However, there are several factors which can be improved to yield more reliable indicators.

- The first such improvement is a reduction in the delay of the system by migrating to another SDK or acquisition method.
- Secondly, in addition to the Wi-Fi via LAN connection, a WAN connection can be implemented. This will allow the output user to experience the motion of an input user from a more remote location.
- Third, as the Kinect sensor can track the joints of the legs, it is also possible to actuate PGMs with the angles of the ankles or legs.
- Fourth, while PGMs can provide sufficient force-feedback to generate the gestures, the length of the user's arm can directly impact that user's perception of the force due to the static size of the jacket prototype. To remedy this, various sizes of the jacket can be implemented.
- Finally, a future evaluation can determine how the current system could play a physiological role in the current patient by studying how the perception of force guidance from the jacket could improve a user's performance in rehabilitative exercise.

8 Conclusion

In this study, a soft exoskeleton jacket was developed with pneumatic artificial muscles for human motion interaction by using low-pressure-driven artificial muscles.

Future work includes the improvement of the algorithm to reduce the delay in the pose estimation process, and the reduction in the size of the control step. We plan to design a rehabilitative exergame with the developed system.

The experimental results allow us to conclude that:

- The pneumatic gel muscles are a good alternative for hard actuators, regarding price, flexibility and naturalness in the adaptation of the human motion.
- The system presents robustness in the acquisition of the motion of an input user to project to an output user.
- The system has flexibility in tracking various motions of an input user, due to the flexibility of the PGM, such that new gestures can be implemented which were previously difficult to deploy.
- The implementation of this system represents a strong relationship between cost and benefit.

Acknowledgment. This work is based on results obtained from a project of Development of Core Technologies for Next-Generation AI and Robotics, commissioned by the New Energy and Industrial Technology Development Organization (NEDO), Japan.

References

1. United Nations World Population Ageing [Highlights], https://bit.ly/2zHW0hc. Accessed 02 Jan 2019
2. Tsagarakis, N.G., Caldwell, D.G.: Development and control of a 'soft-actuated' exoskeleton for use in physiotherapy and training. Auton. Robots **15**, 21–33 (2003)
3. Thakur, C., Ogawa, K., Tsuji, T., Kurita, Y.: Unplugged powered suit with pneumatic gel muscles. Augmented Human a23 (2017)
4. Ackerman, M.J., Filart, R., Burgess, L.P., Lee, I., Poropatich, R.K.: Developing next-generation telehealth tools and technologies: patients, systems, and data perspectives. Telemed J. E-Health **16**(1), 93–95 (2010). https://doi.org/10.1089/tmj.2009.0153
5. Rogante, M., Grigioni, M., Cordella, D., Giacomozzi, C.: Ten years of telerehabilitation: a literature overview of technologies and clinical applications. NeuroRehabilitation **27**(4), 287–304 (2010). https://doi.org/10.3233/NRE-2010-0612
6. Zampolini, M., et al.: Tele-rehabilitation: present and future. Ann. Ist. Super. Sanita. **44**(2), 125–134 (2008)
7. Antón, D., Berges, I., Bermúdez, J., Goñi, A., Illarramendi, A.: A telerehabilitation system for the selection, evaluation and remote management of therapies. Sensors **18**(5), 1459 (2018). https://doi.org/10.3390/s18051459
8. Antón, D., Goñi, A., Illarramendi, A., Torres-Unda, J.J., Seco, J.: KiReS: a kinect-based telerehabilitation system. In: IEEE 15th International Conference on e-Health Networking, Applications and Services (Healthcom 2013) (2013). https://doi.org/10.1109/healthcom.2013.6720717
9. O'Neill, C.T., Phipps, N.S., Cappello, L., Paganoni, S., Walsh, C.J.: A soft wearable robot for the shoulder: design, characterization, and preliminary testing. In: International Conference on Rehabilitation Robotics (ICORR), London, pp. 1672–1678 (2017). https://doi.org/10.1109/icorr.2017.8009488
10. Xiloyannis, M., Cappello, L., Binh, K.D., Antuvan, C.W., Masia, L.: Preliminary design and control of a soft exosuit for assisting elbow movements and hand grasping in activities of daily living. J. Rehabil. Assistive Technol. Eng. **4**, 205566831668031 (2017). https://doi.org/10.1177/2055668316680315
11. Xiloyannis, M., Chiaradia, D., Frisoli, A., Masia, L.: Characterization of a soft exosuit for assistance of the elbow joint. In: International Symposium on Wearable Robotics (2017)
12. Ogawa, K., Thakur, C., Ikeda, T., Tsuji, T., Kurita, Y.: Development of a pneumatic artificial muscle driven by low pressure and its application to the unplugged powered suit. Adv. Robot. **31**(21), 1135–1143 (2017)
13. Tondu, B.: Modelling of the McKibben artificial muscle: a review. J. Intell. Mater. Syst. Struct. **23**(3), 225–253 (2012)
14. Fujimoto, S., Ono, T., Kazumasa, O., Lei, Z.Q.: Modeling of pneumatic rubber artificial muscle and control system design of antagonism driven system. Transact. Japan Soc. Mech. Eng. **73**(730), 1777–1785 (2007)
15. Ogawa, K., Thakur, C., Ikeda, T., Tsuji, T., Kurita, Y.: Development of a pneumatic artificial muscle driven by low pressure and its application to the unplugged powered suit. Adv. Rob. **31**(21), 1135–1143 (2017)
16. Shiffman, D.: Learning Processing, 1st edn, p. 358. Morgan Kaufmann, Burlington (2008). ISBN: 9780080920061
17. Jones, L.A.: Perceptual constancy and the perceived magnitude of muscle forces. Exp. Brain Res. **151**(2), 197–203 (2003)

18. Kandcl, E.R., Schwarts, J.H., Jessell, T.M., Siegelbaum, S.A., Hudspeth, A.J.: Principles Of Neural Science, 5st ed., pp. 780–783. Mc Graw Hill
19. Koganei Company: https://official.koganei.co.jp/downloader/catalog/G010_ALL/1. Accessed 05 Jan 2019
20. Frontera, W.R.: Exercise in Rehabilitation Medicine, 2st ed. Human Kinectics, pp. 123–124. ISBN: 0-7360-5541-X

Author Index

Printed in the United States
By Bookmasters